T0390139

Springer Praxis Books

Astronomy and Planetary Sciences

Series Editors

Martin A. Barstow, Department of Physics & Astronomy, University of Leicester, Leicester, Baden-Württemberg, UK

Ian Robson, UK Astronomy Technology Centre, Royal Observatory, Edinburgh, UK

Steven N. Shore, Dipartimento di Fisica "Enrico Fermi", Università di Pisa, PISA, Pisa, Italy

Derek Ward-Thompson, Jeremiah Horrocks Institute of Maths, Physics and Astronomy, UCLAN, Preston, UK

Textbooks and monographs published in this series from 2013 onward are written for advanced undergraduate students in astronomy and the planetary sciences and advanced amateur astronomers. The editors insist on good readability and encourage new approaches to teaching astronomy and planetary sciences. Books published before 2013 serve a spectrum of readership: some are at advanced amateur to advanced undergraduate level. Others are targeted at PhD students and researchers. Topics covered in the series include • Astronomical telescopes and instrumentation • Astronomical techniques, software and data • Astrophysics, Astrochemistry, Astrobiology • Solar system science (excluding the Earth sciences proper) and exoplanets • Stellar physics and black hole astrophysics • Galactic astronomy • Extragalactic astronomy and cosmology. The books are well illustrated with line diagrams and photographs throughout, with targeted use of colour for scientific interpretation and understanding. Many feature worked examples or problems and solutions.

A. E. L. Davis · J. V. Field · T. J. Mahoney
Editors

Reading the Mind of God

Johannes Kepler and the Reform of Astronomy

Royal
Astronomical
Society

Springer

Editors
A. E. L. Davis (Deceased)
London, UK

T. J. Mahoney
Research Department
Instituto de Astrofisica de Canarias
La Laguna, Spain

J. V. Field
History of Art
Birkbeck, University of London
London, UK

Springer Praxis Books
ISSN 2366-0082   ISSN 2366-0090 (electronic)
Astronomy and Planetary Sciences
ISBN 978-94-024-2248-1   ISBN 978-94-024-2250-4 (eBook)
https://doi.org/10.1007/978-94-024-2250-4

Jointly published with Royal Astronomical Society

The Royal Astronomical Society Series. A series on Astronomy & Astrophysics, Geophysics, Solar and Solar-terrestrial Physics, and Planetary Sciences.
Cover figure: Background: Detail of Kepler's Mars orbit drawing.

This Springer imprint is published by the registered company Springer Nature B.V.
The registered company address is: Van Godewijckstraat 30, 3311 GX Dordrecht, The Netherlands

If disposing of this product, please recycle the paper.

To
Arthur Beer
(1900–1980)

Preface

This book grew out of a special session on the life and work of Johannes Kepler at the General Assembly of the International Astronomical Union held in Rio de Janeiro (Brazil) in 2009. The session, organized by T. J. Mahoney, was designed to mark the four-hundredth anniversary of the publication of the work that contained Kepler's first two laws of planetary motion: his *New Astronomy* (*Astronomia nova*), published in Heidelberg in 1609. Since the work is notoriously technical, professional astronomers seemed the ideal audience for such a commemoration. So indeed it proved.

Following this very successful session, discussion among the speakers and other participants led to the setting up of the Johannes Kepler Working Group (JKWG) of the IAU. It was decided that one of the projects to be undertaken by this Working Group would be to publish a multi-authored conspectus of research on Kepler that was hitherto available only in specialized learned journals and was thus not likely to be read by astronomers or by undergraduate and postgraduate students of astronomy, or by students taking a general course in the history of science. We feel that this book will fill an important gap in Kepler-related literature.

We are grateful to the International Astronomical Union for hosting the Johannes Kepler Working Group and its successor, the Johannes Kepler Project Group, and to the Royal Astronomical Society (London) for its support of our work, not least in suggesting the volume might form part of Springer's Praxis series.

Most of all, we are, of course, grateful to the authors of the individual chapters, whose contributions in several cases extended beyond the chapters for which they are named as directly responsible. We are also grateful for the encouragement of a number of scholars whom circumstances prevented from contributing to the present volume more directly.

The volume is dedicated to the late Arthur Beer (b. Görlitz, Bohemia, 28 June 1900, d. Cambridge, UK, 21 October 1980), founder and for many years editor of *Vistas in Astronomy*. Volume 18 of *Vistas*, edited jointly by Arthur Beer and his son Peter and published in 1975, printed selected papers from the many conferences that had been held in 1971 in celebration of the four-hundredth anniversary of Kepler's birth. Being drawn into the editorial process, largely as a translator, gave one of us

(JVF) an interest in Kepler and an acquaintance with the current literature that later proved very useful. Moreover, Arthur Beer was the first editor to print papers by two of the editors of the present volume (AELD and JVF), when we were at the very beginning of what turned out to be long commitments to Kepler scholarship. There is thus a personal edge to the dedication of the present volume. The wider scholarly reason is that the Beers' publication of the *Vistas* volume marks the beginning of a new phase in Kepler studies. Our dedication thus expresses gratitude not only from the present editors but also from the historical community as a whole.

London, UK   J. V. Field
La Laguna, Spain   T. J. Mahoney
October 2023

A Note on A. E. L. Davis

Dr. Davis, a dedicated and assiduous scholar of *Astronomia nova*, was diagnosed with terminal cancer in April 2020, about a fortnight after the start of the 'lockdown' imposed on London in response to the COVID-19 pandemic. The following obituary was printed in the IAU Division C newsletter.

A. E. L. Davis
7 December 1928 (UK)–23 November 2020 (Canberra, Australia)

Dr. Davis was educated at Badminton School, Bristol (UK), and at St. Anne's College, Oxford. After a period teaching mathematics, in a school and in numeracy classes organized by local education authorities, Davis's interest in the history of the subject led to registration as a Research Student at Imperial College, London, under the supervision of Prof. A. R. Hall.

The subject of Davis's doctoral thesis, and indeed all the research on Kepler that followed from it, was 'A mathematical elucidation of the bases of Kepler's laws'. The thesis was accepted by the University of London in 1981. The external examiner was Prof. D. T. Whiteside (Cambridge). Working with two distinguished Newton scholars in this way gave Dr. Davis a lifelong determination to distinguish sharply between the approaches adopted by Kepler and by Newton.

Dr. Davis, who was also an active member of the British Society for the History of Mathematics and of the Fawcett Society, was the author of articles about various aspects of Kepler's astronomy in *Centaurus*, *Archive for History of Exact Sciences*, *Journal for the History of Astronomy*, and elsewhere.

Dr. Davis served as a co-vice chair of the International Astronomical Union's Johannes Kepler Working Group set up in 2009. It was largely thanks to the efforts of Dr. Davis that the text of the modern edition of Kepler's Complete Works published under the auspices of the Bavarian Academy of Sciences (Beck, Munich, 1938–2010) is now available online.

Dr. Davis will be much missed as an active member of the history of science community and specifically as one who had the mathematical skills required to follow up historical insights concerning astronomy.

J. V. Field

Introduction: Kepler's Place in the History of Science

Johannes Kepler (1571–1630) played a major part in the development of the mathematical science of astronomy and the 'mixed' (that is partly mathematical) science of optics. Moreover, similar historical significance can be seen in his work as a whole, including what might be called his 'philosophy of science'; that is, his outlook on problems and his approach to seeking solutions. One example is his commitment to examining the agreement between theory and observations and, if necessary, jettisoning the former—most famously in his rejection of candidate orbits in *Astronomia nova* (Heidelberg, 1609). This attitude marks Kepler as making a fundamental contribution to the emergence of the idea of observational error and a recognizably 'modern' approach to the treatment of data in the sciences. Another aspect of this is Kepler's openness about his work, for instance in optics. When Galileo Galilei (1564–1642) greatly improved the optical performance of a spyglass of which he had been given a description, and then started to make important discoveries when he used the instrument to observe the heavens, he wrote a book about his work, *Messenger from the stars* (*Sidereus Nuncius*, Venice, 1610). The book, which proved to be a best-seller and made Galileo famous, merely mentions his expertise in optics and concentrated on describing the astronomical discoveries. This is of course understandable, but Galileo never went on to discuss the optics of his telescopes. He tells us little more than that they used a concave and a convex lens. This too is understandable since he was making and distributing telescopes. However, the attitude harks back to the practices of members of trade guilds rather than pointing forward to emerging science. It was Kepler who wrote about the optics of the telescope, in his *Dioptrice* (Frankfurt, 1611), and in the process of thinking things through invented a new design for a telescope using two convex lenses, a design now known simply as 'the astronomical telescope' (see Chap. 7 of this volume).

This is to say that Kepler was one of the leading figures in what is now generally known as the Scientific Revolution of the seventeenth century, from which many branches of knowledge, in particular astronomy and subjects related to it, emerged in a recognizably modern form.[1]

---

[1] For an explanation of the origin of the term 'Scientific Revolution', see Hall (1993).

Kepler was a convinced Copernican, apparently from his days as a student at the University of Tübingen in the 1580s, and his astronomical work did much to establish that the planetary system was indeed centred on the Sun. He also made significant contributions to other branches of the mathematical sciences. In particular, his work marks a turning point in the study of optics. His analysis of the mode of vision, in his *Optical part of astronomy* (*Astronomiae pars optica*, Frankfurt, 1604) proved the eye saw by receiving light rather than by emitting eyebeams, thereby establishing a new relationship between what would now be called physiological and physical optics. This 'intromission' theory of vision had been put forward by Ibn al-Haytham (Hassan al-Hassan Ibn al-Haytham, *c*. 965–*c*. 1040; in Latin Alhacen or Alhazen), but was not widely accepted at the time. It was, however, adopted by René Descartes (1596–1650) in his *Dioptrics* (Leiden, 1637) and thereafter seems to have been regarded as obviously correct.

The present volume is intended to provide a general introduction to Kepler's work as a whole and to give an account of the place he is now seen to occupy in the history of science.

Astronomy

As has been mentioned in our preface, the idea of a collection of introductory essays on Kepler and his works arose from a meeting organized as part of the General Assembly of the International Astronomical Union in Rio de Janeiro in 2009. The meeting was organized to mark the four-hundredth anniversary of the publication of Kepler's *New Astronomy* (*Astronomia nova*, Heidelberg, 1609), the work which contained his first two laws of planetary motion: that a line joining the planet to the Sun sweeps out equal areas in equal times as the planet moves round its orbit, and that the shape of the path taken by the planet is an ellipse with the Sun in one focus. The *New Astronomy* deals only with the path of the planet Mars, covering work that Kepler referred to as 'my battle with Mars' though, when he later has cause to refer to it, he more soberly calls the book 'my Commentaries on Mars'. These words, which give an unconventional title for a book on planetary motion, appear in the long full title of the work.[2] They emphasize that Kepler saw Mars itself—that is, the observations of the planet—as determining the structure of the work.[3] In the course of investigating the motion of Mars, Kepler had of course also needed to find the orbit of the Earth, and he later showed that elliptical orbits and the area law would also account for the observed motions of the other planets. Historians of science see Kepler's work as marking the beginning of modern astronomy, because it abandons the apparatus of circles and uniform motions that had been current since

---

[2] On the title see Gregory, 'The translation of the title of Kepler's *Astronomia nova*' (this volume, Chap. 5).

[3] See Davis, 'Kepler's discovery of the planetary Orbit: the Goldilocks solution' (this volume, Chap. 4).

ancient times. Kepler's third law—that the squares of the periods of planets vary as the cubes of the mean radii of their orbits—was first published ten years after the first two, in his *Harmony of the World* (*Harmonice mundi*, Linz, 1619).[4]

In his *Mathematical Principles of Natural Philosophy* (*Philosophiae naturalis principia mathematica*, London, 1687), Isaac Newton (1642–1727) was to show—using his laws of local motion (which were markedly different from Kepler's)—that Kepler's third law can be deduced from the first two. And further (again together with the new understanding of physics developed since Kepler's time), that Kepler's first two laws prove that the planetary system is governed by an attractive force pulling the planets toward the Sun, and that this force obeys an inverse square law of distance. Kepler's third law appears only briefly in Newton's work, being used to show that the four satellites of Jupiter (discovered by Galileo in 1610) move under a similar force attracting them to Jupiter. (The orbits of the satellites, as measured, were too close to circles for Newton to check that they followed Kepler's first two laws.) Numerical results obtained by experimentation showed that the force that pulls the Moon to the Earth, which also obeys an inverse square law of distance, is the same as the force that pulls heavy objects toward the Earth, a force long known as 'gravity'.[5]

Newton's explicit reliance on his work clearly establishes that Kepler is an important figure not only in the history of astronomy but in the history of science as a whole. However, Kepler's place in that story is more subtle and more significant than this, and the scope of the present book, which is conceived as an introduction to Kepler's work as a whole, is consequently much wider than that of the meeting held for astronomers from which the project of writing the book arose.

For instance, in the *New Astronomy* it is not only the results that are radically new, so too is Kepler's approach to finding the motion of Mars, which is shaped by his Copernicanism, his belief that his task is to find the true path of the planet in space rather than merely constructing a model for predicting further positions of the planet in the sky, and his belief that in using the observations made by Tycho Brahe (1546–1601) he is privileged to be working with planetary positions that are considerably more accurate than those available to any of his astronomical predecessors. Moreover, unlike those predecessors, Kepler deliberately starts with no assumptions about the orbit beyond those imposed by his Copernicanism. For example, one assumption is that since the orbit of the Earth, which we see as the motion of the Sun along the ecliptic, is obviously a closed plane curve, all planetary orbits must also be closed plane curves. Further, at what he recognizes as a crucial point in his argument Kepler rejects a possible model orbit because it is not in adequately good agreement with Tycho's observations.[6] That is, Kepler—or Tycho, or maybe the two of them together—had invented the hitherto unknown concept of observational error,

[4] See Field, 'Kepler's cosmology' (this volume, Chap. 2).

[5] On Newton's use of Kepler's laws see Field, 'The long life of the *Rudolphine Tables*' (this volume, Chap. 9).

[6] Kepler (1609), ch. XIX, pp. 113–14; Kepler (1990), pp. 177–78; Kepler (2015), p. 211. See Mahoney, 'Measuring the heavens: Tycho Brahe's reform of observational astronomy' (this volume, Chap. 3), Davis, 'Discovery of the planetary orbit: the Goldilocks solution' (Chap. 4), and Field, 'The long life of the *Rudolphine Tables*' (this volume, Chap. 9).

and Kepler is convinced of the robustness of the concept to the extent that he allows his argument to turn upon it.

Reading his other works shows that this strong-mindedness is characteristic of the way Kepler thinks: if he believes in something he does really believe in it. This is of course most obvious in his adherence to heliocentric astronomy but it is also seen in more conventional matters. For instance, in this period, everyone believes, in some sense, that God is a Geometer, and that He created the Universe in number, measure, and weight (*Wisdom*, 11:20), but Kepler, who is deeply religious, goes so far as looking to Euclid's *Elements* (*c.* 300 BC) to explain the number of the planets and the overall structure of the planetary system. It helps, of course, that the theory he proposes is in good numerical agreement with the observations (to express it in today's terms: the theory agrees with observation to about 10 per cent).[7]

Optics

Another area in which Kepler thought that mathematics could be relied upon to give him correct answers (that is, ones that corresponded with the way things were in the physical world) was optics. As it turned out, he was right. In his first book on optics, *The optical part of astronomy* (*Astronomiae pars optica*, Frankfurt, 1604),[8] is written partly in response to the fairly recent publication of a reasonably good Latin translation of the work of Ibn al-Haytham, together with an edition of a treatise by Witelo (b. 1220, d. after 1278) that drew on Ibn al-Haytham.[9] In his treatise, Kepler discusses the formation of images in the *camera obscura* (a customary observing instrument for astronomers) and the working of the human eye. He proves that Ibn al-Haytham was right about the process of vision: the eye sees by the reception of light into the eye rather than by the emission of eyebeams from it. The choice between these possibilities, and various pick-and-mix combinations of them, had been a matter of dispute among physicians and natural philosophers for about two millennia, with most people believing in the emission of eyebeams and some scholars considering the question undecidable. But Kepler's solution—in which an inverted image is formed on the retina at the back of the eye—was accepted very rapidly, possibly because physicians recognized it as having no implications for medical practice.

Kepler's treatise of 1604 is now seen as marking the beginning of modern optics because it definitively established a division between the study of seeing (now called 'physiological optics') and the study of the properties of light ('physical optics'). One of the most influential natural philosophers of the seventeenth century, René

---

[7] See Methuen 'Kepler, religion and natural philosophy: A theological biography' (this volume, Chap. 1) and Field 'Kepler's cosmology' (this volume, Chap. 2).

[8] The full title of the work is 'Things left out by Witelo that are dealt with in the optical part of astronomy' (*Ad Vitellionem paralipomena quibis astronomiae pars optica traditur*).

[9] Risner (1572).

Descartes, adopts Kepler's model of the eye in his short treatise on optics of 1637, giving a full description of it as if it were entirely uncontroversial.[10] We know from his other writings that Descartes was a heliocentrist and, after the condemnation of Galileo in 1633, was anxious not to offend the religious authorities. That may explain his omitting to mention Kepler by name when passing on his description of the eye. In any case, Kepler's own treatise seems to have been widely read.

Galileo's telescopic discoveries, published in May 1610,[11] turned Kepler's attention back to optics, leading to his invention of a telescope which used two convex lenses and proved to have many advantages over the instruments used by Galileo (which employed a combination of a convex and a concave lens).[12]

Other Contributions

Kepler's astronomical and optical work obviously marks him out as a major figure in the development of science. There are also a number of less well-known parts of his work or his thinking that show the originality and reforming tendency of Kepler's turn of mind. For instance, he conceived the idea of a non-convex polyhedron; that is, one whose faces are not all completely visible on the outside of the solid. And he thought out a way of finding volumes of solids by slicing them up and reassembling the slices into a different solid whose volume could be calculated relatively easily. This technique is an extension of ideas found in works by Archimedes (*c*. 287–212 BC) and is now recognized as a step toward the invention of the integral and differential calculus.[13] Both these innovative pieces of mathematics involve the use of a power of visualizing structures in three dimensions that we can also see at play in Kepler's ability to grasp the complex motion of Mars as seen from the Earth, where both planets are moving at variable speeds along paths that lie in different planes. This power of visualization is also apparent in Kepler's thinking about the problem of stacking equal spheres (now known as 'close packing') that arises when he considers possible origins of the six-cornered shape of snowflakes, but ends up explaining the honeycomb instead.[14] The concept of 'close packing' was later used extensively by crystallographers (incautious historians occasionally hail Kepler, who did not believe in atoms, as having invented crystallography, an area of physics that depends entirely on considering atoms).

Kepler's works contain many such relatively minor discoveries and inventions. There are also omissions that in context may look surprising. For instance, despite giving the matter some thought, and carrying out experiments, Kepler never arrived at a satisfactory mathematical formulation of the law of refraction of light, though (as we

---

[10] Descartes (1637).

[11] Galileo (1610). See Field, 'Kepler and Galileo' (this volume, Chap. 8).

[12] On Kepler's optical works see Donahue, 'Kepler's contributions to optics' (this volume, Chap. 7).

[13] See Knobloch 'Kepler's contributions to mathematics' (this volume, Chap. 11).

[14] Kepler (1611); Kepler (1941), pp. 259–80; Kepler (1966).

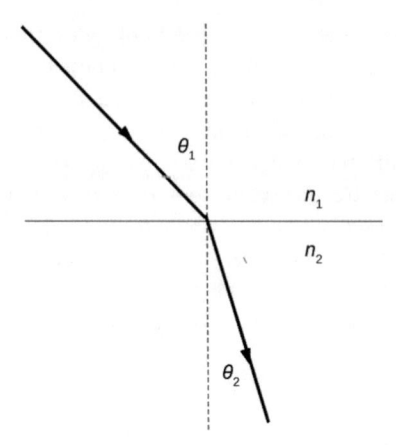

Fig. 1 The law of refraction ('Snel's law'). The law of refraction states that if we have two media of different optical densities, such as air and water, or empty space and air, then the sine of the angle of incidence, shown as θ_1 in the diagram, is equal to a constant times the sine of the angle of refraction, shown as θ_2 in the diagram. The medium through which the incident ray passes is denoted by the constant n_1 and that of the refracted ray by the constant n_2. Snel's law is then given by the relation $\sin \theta_1 = n_1/n_2 \sin \theta_2$

now know) the law is rather simple: that for transmission between two specific media the sine of the angle of refraction is equal to a constant times the sine of the angle of incidence (see Fig. 1). The law of refraction was probably discovered by Ibn Sahl (*c.* 940–1000 AD), but his work did not find its way into the optical tradition of Western Europe. His law is now usually named after Willebrord Snel (1581–1626), though it seems to have been known earlier to one of Kepler's correspondents, Thomas Harriot (1560–1621). Like many of his contemporaries, though not Kepler, Harriot seems to have been inclined to regard such information as a private possession and in this case he did not choose to share it. The law of refraction was significant for astronomy because it was sometimes important to make corrections for the effects of atmospheric refraction; for instance, when objects were observed relatively low in the sky. Tycho Brahe's tables of corrections for refraction are printed in Kepler's *Rudolphine Tables* (1627), though Kepler seems to have been more than a little doubtful about their usefulness.[15]

With hindsight, it again looks like a blind spot that, in mathematics proper, Kepler apparently did not recognize the power of algebraic methods. Like many of his contemporaries, he regarded algebra as essentially an extension of arithmetic. He thus saw it as yielding merely numerical answers when what the problem required was exact geometrical constructions. This goes with his conviction that the matter of the Universe was continuous, meaning that there were no limits to how finely it could be divided (that is, no indivisible 'atoms'). This conviction was a conventional Aristotelian notion, one of many that Kepler never saw any good reason to reject.

---

[15] See Mahoney, 'Measuring the heavens: Tycho Brahe's reform of observational astronomy' (this volume, Chap. 3) and Field, 'The long life of the *Rudolphine Tables*' (this volume, Chap. 9).

Kepler is also conservative in his attitude to music theory, particularly in his attitude to composers' use of dissonance.[16] This was a more or less defensible position at the time, and we do not know what music Kepler actually heard, but it turned out that dissonance was there to stay and theory eventually caught up.

Another example of Kepler's conservatism is his belief in astrology. Experience and the evidence that planets had no intrinsic light but shone only by reflecting sunlight made Kepler change his mind not only about details but also about some of the basic tenets. For example, in his *Harmony of the World* (1619) he describes his own birth chart and comments.

> Yet in this my stars were not Mercury as morning star in the angle of the seventh house, in quartile with Mars, but they were Copernicus, they were Tycho Brahe, without whose books of observations everything which [has] now been brought by me into the brightest daylight would lie buried in darkness; not Saturn the overlord of Mercury, but Rudolf and Matthias, each a Caesar Augustus, my overlords; ... .[17]

And he goes on to dismiss arguments connected with signs of the zodiac. But he continued to believe that heavenly bodies could exert force, action at a distance, that, for instance, enabled the Moon to affect the sea.[18] As it turned out, a modified version of Kepler's traditional notion of force turned out to play a part in the Newtonian picture of the world. But Newton had given a specific mathematical description of force; Kepler's mathematics of force went only as far as considering the geometrical relationships between the rays coming to the Earth from planets. It is not wise to be too eager to construct Kepler (or indeed any person from the past) as a through and through 'modern' figure. He was certainly thoroughly rational, but some of his ideas do not stand the test of time.

Reform

Nevertheless, in almost all Kepler's works there are ideas that prove to be fruitful in significant ways for future generations of natural philosophers. Kepler's discoveries, particularly in astronomy and optics, are so important that unless we are willing to fall back on the essentially empty explanation of 'genius' or saying that Kepler was 'ahead of his time'—a phrase that seems absurd to any historian, since people obviously live in their own time—we need to look at what made the discoveries possible at that time and to Kepler himself. There is, of course, an element of simple chance, as is implied in the passage just quoted from Kepler's comments on his horoscope. But essentially we need to try to place the discoveries in a proper historical context. We need to ask wider questions about what was known or accepted as true that led Kepler in this or that particular direction. As we shall see, some of the discoveries have deep roots. And many of them seem to have been made possible by something

---

[16] Kepler (1619), book 3; see Field (1988), esp. ch. 6.

[17] Kepler (1619), book 3; see Field (1988), esp. ch. 6.

[18] See Rabin, 'Kepler's reform of astrology' (this volume, Chap. 6).

in Kepler's character that made him continue to follow a line of thought until he either got a satisfactory answer, such as the orbit of Mars, or could at least explain why he had failed to do so, as in the case of the shape of the snowflake or the problem of inscribing a regular heptagon in a circle.[19] This is to say that Kepler followed through on his chosen method of approach to problems. And that tended to pay off since his strong preference was for mathematical methods. And he presumably recognized he was good at getting answers when a problem was cast in mathematical form.

Sometimes, of course, twenty-first century judgements of what constitutes a satisfactory answer are not the same as Kepler's. A spectacular example of this is that he sees the 'musical' ratios he found among planetary motions—that is small-number ratios among the greatest and least speeds of neighboring or individual planets—as indicative of a plan and thus as explaining the elliptical orbits, not simply as being ellipses (as Newton was to do) but as the specific ellipses deduced from Tycho's observations.[20] In the parlance of modern celestial mechanics such ratios are referred to as 'orbital resonances'. They seem to be fairly common, but are the subject of various competing explanations.

The Contents of This Volume

The editors of the present volume have not tried to impose a unified view of Kepler and his work on the authors who wrote the chapters. The writers are established scholars working on the areas concerned and as such are entitled to their individual opinions. Nor have we attempted to include a discussion of every work Kepler wrote. There are too many of them for that. What we have done is to try to put the chapters into an order that follows a pattern of Kepler's own priorities while also taking account of chronological order.

Both priority and chronology place religion first. Kepler lived through a period in which religious convictions and the divisions between Catholics and Protestants, and, within Protestantism, between Lutherans and Calvinists, led to religious and social persecution, social unrest and, in 1618, to the outbreak of the brutal and hugely destructive Thirty Years War. Many of the external circumstances of Kepler's life were determined by his commitment to Lutheranism. He was deeply religious, and his faith played a very important part in his thought on all subjects. Thus the material of Chap. 1 in this volume, 'Kepler, Religion, and Natural Philosophy: A Theological Biography', by Charlotte Methuen, is relevant to everything that follows. Chapter 2, 'Kepler's Cosmology', by J. V. Field, reflects the influence of Kepler's religion: he proposes a heliocentric cosmogony, describing how the structure of the Universe shows the nature of its Creator. This was the subject of Kepler's first published work

---

[19] See Rabin, 'Kepler's reform of astrology' (this volume, Chap. 6).

[20] See Field, 'Kepler's cosmology' (this volume, Chap. 2).

(1596), and it was in search of better observations that would better confirm his theory that Kepler turned to Tycho Brahe, whose observations provided the basis for calculating the more exact orbits that appear in the later version of Kepler's cosmogonic model (1619).

In Chap. 3, 'Measuring the Heavens: Tycho Brahe's Reform of Observational Astronomy', by T. J. Mahoney, we turn to the work of Tycho Brahe, including the design of the instruments he used to make very accurate observations in his purpose-built observatory on the island on Hven in Copenhagen Sound. Chapter 4, 'Kepler's Discovery of the Planetary Orbit: The Goldilocks Solution', by A. E. L. Davis, describes the mathematical route by which Tycho's observations led Kepler to his first two laws of planetary motion (published in 1609). Chapter 5, 'The Translation the title of Kepler's *Astronomia nova*', by Andrew Gregory, sheds new light on what Kepler meant by the long full title of his *New Astronomy*, specifically the Greek word that appears in it. The meaning ascribed to the title has played an important part in some scholars' understanding of the method Kepler employs in the work itself.

In Kepler's time, the duties of astronomers almost always included the practice of astrology. For example, universities which taught prospective physicians employed professors of mathematics (that is the four mathematical sciences: arithmetic, geometry, music, and astronomy) to teach medical students elementary (geocentric) astronomy as a preparation for using astrology in what was considered to be the correct manner in their medical practice. Galileo was employed by the University of Padua, from 1592 to 1610, for exactly this purpose. Kepler, who was regularly asked for predictions of one kind or another, also compiled astrological calendars that predicted the weather. Consequently, Chap. 6 in this volume, 'Kepler and the Reform of Astrology', by Sheila Rabin, examines Kepler's astrological writings and his increasingly unconventional ideas on astrology.

We then turn to optics. Chapter 7, 'Kepler's Work on Optics', by W. H. Donahue, considers Kepler's optical treatises of 1604 and 1611. The former includes his work on the human eye and the latter his new design for a telescope with two convex lenses. As we mentioned earlier, these works, particularly the second, raise the question of relations between Kepler and Galileo. Their interactions are examined in Chap. 8, 'Kepler and Galileo', by J. V. Field, which also compares their reactions to astronomical events such as the New Star of 1604 and the controversy over the comets of 1618–19.

We then move forward in time to consider Kepler's legacy. Chapter 9, 'The Long Life of the *Rudolphine Tables*', by J. V. Field, discusses the book by which Kepler was to be best known in the century following his death, the astronomical tables that he based on Tycho's observations and his own laws of planetary motion. The tables, published in 1627, remained reliable for much longer than any earlier tables had done, thus lending credibility to Kepler's laws and hence to Copernicanism. This volume of tables was almost certainly the direct or indirect route by which Kepler's laws came to the attention of Newton, whose use of them is discussed briefly at the end of the chapter. Chapter 10, 'Johannes Kepler, the Kepler spacecraft and transits' by Jay Pasachoff, deals with transits; that is, events in which one of the inner planets, Mercury or Venus, is seen to move across the face of the Sun. Observations of transits, which were made possible by improved telescopes and by the accurate predictions of ecliptic latitudes given by the *Rudolphine Tables*, provided considerably more accurate estimates of the distances of these planets and hence of the overall size of the Solar System.

Chapter 11, 'Kepler's Contributions to Mathematics', by Eberhard Knobloch, describes some of the interesting mathematical results Kepler obtained when explicitly working on mathematics as such or in the course of his other researches, for instance his work on a new method of finding volumes that he developed from the work of Archimedes that we mentioned earlier. Chapter 12, 'Kepler's Dream and Lunar Astronomical Phenomena', by Jarosław Włodarcyk, discusses Kepler's account of astronomical phenomena as they would be observed from the surface of the Moon and explains the circumstances in which the story was written and eventually published after Kepler's death. Chapter 13, 'On Translating Kepler', by W. H. Donahue, gives a translator's view of Kepler's writings and in particular of the various challenges encountered in producing his English version of Kepler's *Astronomia nova*.[21]

There are two appendices: a detailed chronology of Kepler's life and a glossary of technical terms. These are followed by a detailed index.

A Note on Calendar Reform

In the church calendar, which was also used for many civil purposes, Easter Day is the first Sunday after the first Full Moon after the Spring equinox. So finding the date of Easter requires prediction of the date of the equinox. By the fifteenth century, it was clear that prediction of the date of the Spring equinox required a reform of the calendar. The calendar then in use was the Julian calendar—named after Julius Caesar (100–44 BC) who, having taken advice from the foremost experts of the day, had introduced a new calendar in 45 BC to replace the chaos of local calendars in use in the lands under Roman rule.

Calendar reform became a major driver of the study of astronomy in the Renaissance, but the solution to the problem was unspectacular. It turned out that the Julian calendar was merely showing its age. Caesar had made the year a little too long, and after so many years the Spring equinox had moved from 21 March, the date given by Claudius Ptolemy (*fl.* 128–141 AD) in the Almagest, to about 31 March. The solution, imposed on the Catholic Church by Pope Gregory XIII in 1582, was to lose ten days and to decree that henceforth leap years (in which an extra day is added to February) would occur every four years but with the proviso that when the year in question marked the end of a century, that is if its number was a whole number of hundreds, it would not be a leap year unless the number of centuries was divisible by four. So 1600 was a leap year, as everyone had been expecting it to be, but the next such century year was 2000.

---

[21] Kepler (1609, 2015).

Caesar's reform seems to have been adopted slightly patchily, and the same was true of Gregory's, particularly in lands ruled by Protestants. However, astronomers, even if Protestant, were generally in favour of having a more stable equinox, and the reform was accordingly widely adopted in technical contexts. Kepler regularly uses Gregorian dates and points out when he is not doing so.

There is, however, a further complication in interpreting dates, this time concerning the numbering of years. The Julian year started on 1 January. However, some states, particularly in Italy, started the New Year not on 1 January but on Lady Day, 25 March.

London, UK   J. V. Field
La Laguna, Spain   T. J. Mahoney

References

Descartes, R. (1637). *Discours de la méthode pour bien conduire la raison … plus la Dioptrique, les Météores et la Géometrie* ….Leiden: Jan Maire.

Field, J. V. (1988). *Kepler's Geometrical Cosmology*. London and Chicago: Athlone Press and University of Chicago Press.

Galilei, G. (1610). *Sidereus nuncius.*Venice: Thomas Baglioni.

Hall, A. R. (1993), Retrospection on the scientific revolution. In J. V. Field & F. A. J. L. James (Eds.), *Renaissance and Revolution: Humanists, Craftsmen and Natural Philosophers in Early Modern Europe* (Reprinted 1997, pp. 239–249). Cambridge: Cambridge University Press.

Kepler, J. (1609). *Astronomia nova aitiologêtos seu physica coelestis tradita commentariis de motibus stellae Martis* …. Heidelberg: E. Vogelin.

Kepler, J. (1611). *De nive sexangula*. Prague: Godefried Tampach.

Kepler, J. (1619). *Harmonice mundi libri V*. Linz: Joannes Plank.

Kepler, J. (1941). *Johannes Kepler Gesammelte Werke. Band IV: Kleinere Schriften 1602–1611/ Dioptrice* (M. Caspar and F. Hammer (Eds.)). Munich: C. H. Beck.

Kepler, J. (1966). *The Six-Cornered Snowflake* (several reprints). Edited and translated from the Latin by C. Hardie, with essays by L. L. Whyte and B. F. J. Mason. Oxford: Oxford University Press.

Kepler, J. (1990). In M. Caspar & K. Kepler (Eds.), *Johannes Kepler Gesammelte Werke. Band III: Astronomia nova*. Munich: C. H. Beck.

Kepler, J. (2015) *Johannes Kepler. Astronomia nova* (Rev. Edn.). Translated from the Latin by W. H. Donahue. Santa Fe: Green Lion Press.

Risner, F. (1572) *Opticae thesaurus: Alhazeni Arabis Libri Septem … Item Vitellionis Thuringopoloni Libri X*. Basel: Episcopii.

Contents

Contributors

A. E. L. Davis University College London, London, UK

W. H. Donahue St. John's College, Santa Fe, NM, USA

J. V. Field Birkbeck, University of London, London, UK

Andrew Gregory University College London, London, UK

Eberhard Knobloch Berlin Academy of Sciences and Humanities, Berlin, Germany

T. J. Mahoney Instituto de Astrofísica de Canarias, La Laguna, Santa Cruz de Tenerife, Spain

Charlotte Methuen University of Glasgow, Glasgow, UK

Jay M. Pasachoff Williams College, Williamstown, MA, USA; Caltech, Pasadena, CA, USA

Sheila J. Rabin Emerita, Saint Peter's University, Jersey City, NJ, USA

Jarosław Włodarczyk Institute for the History of Science, Polish Academy of Sciences, Warsaw, Poland

Chapter 1
Kepler, Religion and Natural Philosophy: A Theological Biography

Charlotte Methuen

Surprising as it may seem to a twenty-first century reader, religion permeated Johannes Kepler's life. Caught up in the religious struggles between Catholics and Protestants which succeeded the Reformation and dominated the late sixteenth and early seventeenth centuries, Kepler's education, career and biography were moulded by the religious and theological interests of his day. Religion shaped his scientific ideas as well. Kepler was convinced that by studying the natural world, he could gain a better understanding of God. Indeed, in his astronomy and his natural philosophy, Kepler saw himself as interpreting not only God's world but God himself.

1.1 Childhood and Education (1571–1594)

According to his own horoscope,[1] Kepler was born in the early afternoon of 27 December 1571 (in the old Julian calendar[2]) in the South German city of Weil-der-Stadt. From the beginning his confessional identity was complex. His mother, Katharina (1546–1622), was almost certainly Lutheran: she was the daughter of Melchior Guldenmann (1520–1601), an innkeeper in nearby Eltingen, which lay in the Lutheran duchy of Württemberg which surrounded Weil-der-Stadt. Kepler's father, Heinrich (1547–c. 1586), had grown up in Weil-der-Stadt, a free imperial city

---

[1] Hübner (1975), p. 3.

[2] According to the Gregorian calendar, introduced in Catholic areas in 1582, Kepler's birthday fell on 5 January 1572 (Boockmann and di Liscia, 2009, p. 13, n. 3). Kepler cast horoscopes for nearly all the members of his immediate family (his horoscope for his sister, if it ever existed, seems not to have survived) so that—unusually for a family of this status at this date—their dates of birth are known quite accurately. See Kepler (2009), pp. 5–48.

C. Methuen (✉)
University of Glasgow, Glasgow, Scotland
e-mail: charlotte.methuen@glasgow.ac.uk

© Springer Nature B.V. 2024
A. E. L. Davis et al. (eds.), *Reading the Mind of God*, Springer Praxis Books,
https://doi.org/10.1007/978-94-024-2250-4_1

that was not under the jurisdiction of the Duke of Württemberg. At the reformation its city council had chosen to remain Catholic, so Heinrich was probably originally Catholic, although he seems later to have become Lutheran. Certainly, Heinrich's brother Sebald became a Jesuit.[3] The infant Johannes was almost certainly baptized by a Catholic priest in a Catholic service: he later attested in his correspondence with the Jesuit Paul Guldin (1577–1643) that his parents had carried him to a Catholic church and sprinkled him with holy water, and hinted that he had also been anointed with oil or chrism.[4] The Jesuits had encouraged Kepler to convert: in his letter Kepler argued that since he had been baptized a Catholic and had never left the true church, conversion was unnecessary.[5] For the adult Kepler, the true church was not unambiguously to be associated with any of the confessions he knew: it was neither Lutheran, Catholic nor Reformed, but transcended their differences. This conviction would shape his theological beliefs and lead to his excommunication by the Lutheran church in Württemberg.

Johannes was the eldest of six children born to Heinrich and Katharina Kepler. Four survived to adulthood: Johannes himself, Heinrich jun. (1573–1615), their younger sister Margarete (b. 1584) and Christoph (1587–1633).[6] When Johannes was nearly three, his father became a mercenary in the Spanish Netherlands, and Katharina soon followed him, leaving Johannes in the care of his grandparents, who were not kind to him. On his parents' return the family moved to Leonberg, about 15 km (9 statute miles) from Weil-der-Stadt, which fell under the jurisdiction of the Duchy of Württemberg. His father soon rejoined the army in the Netherlands. Returning in 1580, he sold the Leonberg house to pay the family's debts and took over the inn at Ellmendingen near Pforzheim. By 1583 the Kepler family was back in Leonberg, but in 1588 Heinrich sen. once again left for the wars, this time to join the imperial navy near Naples. He never returned, probably dying on his return journey. Kepler later recorded that his parents had had a difficult relationship: his father treated his mother badly and there were constant quarrels.[7]

In Leonberg, Kepler entered the German school, part of the educational provision put in place in Württemberg as a result of the Reformation.[8] Here he was taught to read and write; lessons (assuming they followed the instructions laid down in Württemberg's 1559 'Great Church Order') would have focused on religious instruction, including the study of scripture, especially the books of Psalms, Proverbs, Ecclesiasticus (or Jesus Sirach) and the New Testament, and the catechism, probably that of Martin Luther (1483–1546). The disruption to the family's life meant that

---

[3] Boockmann and di Liscia (2009), p. 10, n. 4.

[4] Kepler to Paul Guldin, 24 February 1628, in Kepler (1959), letter 1072, lines 39–45. For this correspondence, see also Schuppener (1997).

[5] Cf. Hübner (1975), pp. 2–3.

[6] Caspar (1959), p. 35, including a summary of Kepler's account of his brother Heinrich's life (Kepler, 2009, pp. 17–18). Kepler records the names of two other younger brothers, Sebald (birthdate unknown), and Bernhard (born 23 June 1589), neither of whom survived (Kepler, 2009, p. 33).

[7] Caspar (1959), p. 34.

[8] The school system in Württemberg is outlined in Methuen (1994).

Johannes's schooling was interrupted, but he was nonetheless admitted to Leonberg's Latin school in 1578. On 17 May 1583 he took the examination for the Duke of Württemberg's scholarship system, which educated the sons of poor Württemberg families to serve as loyal Lutheran teachers, pastors and civil servants in the Duchy, and was successful. On 16 October 1584, aged thirteen, Kepler entered one of Württemberg's prestigious monastery schools,[9] the lower school, or grammar school, in the former monastery of Adelberg; two years later, on 26 November 1586, he transferred to the upper school in the former Cistercian monastery of Maulbronn, where he completed the requirements for his Baccalaureate in September 1588 and undertook a further year's schooling. On 3 September 1589 Kepler took up his place in Tübingen's *Stift*, the official residence for the beneficiaries of the Duke's scholarship whilst they studied at the University of Tübingen, and on 5 October he matriculated at the University.[10]

Academically, Kepler's schooling at Adelberg and Maulbronn focused on the learning of Latin and Greek and study of classical texts, but religion and religious observance were also taken very seriously. Württemberg's Church Order instructed that early morning prayers should take place at 4.00 a.m. in summer and 5.00 a.m. in winter, followed by morning prayers at 8.00 a.m. or 9.00 a.m.; the day finished with evening prayers at around 4.00 p.m. There was a daily period of *lectio theologica*, or theological reading, which included the theological and grammatical analysis of scriptural texts and the preaching of sermons.[11] The religious aspects of Kepler's schooling served to intensify his already deep interest in questions of faith. A pious child, he sought to be disciplined in his prayers and worried about the purity of his life and the state of his soul. As a ten-year-old boy he resolved that 'Jacob and Rebecca'[12] should serve as the model for his marriage, should he later marry.[13] Aged twelve, Kepler was shaken by a series of sermons by the new deacon in Leonberg who sought to demonstrate the error of Calvinist teachings by appealing to Paul's Epistle to the Romans. The young Johannes was devastated to discover that the Church was divided, but he also resolved to study the biblical texts at home, only to discover that the opinions that he had heard condemned seemed to him to be in fact the more convincing.[14] Soon after he arrived at Adelberg, the question of predestination began to interest him: he wrote to the University of Tübingen asking for a copy of a disputation on free will, probably an exploration of Luther's *De servo arbitrio* ('On the unfree will');[15] later as a student at Tübingen he found himself asking whether God's mercy might not allow those who did not know Christ to be

---

[9] These boarding schools were located in the monasteries which before the Reformation had been home to monastic communities.

[10] See previous note.

[11] Methuen (1998), pp. 46–47.

[12] He must have meant either Isaac and Rebecca (Genesis 24) or Jacob and Rachel (Genesis 29–30); the former is more likely since Rebecca was a popular role model for early-modern women.

[13] Kepler (1990b), p. 315; Hübner (1975), p. 3.

[14] Kepler (1990b), p. 315; Hübner (1975), p. 3.

[15] Kepler (1990b), p. 315; Hübner (1975), p. 4.

saved.[16] In Adelberg, Kepler also began to consider the theology of the Eucharist, which had proved highly controversial amongst Protestants during the Reformation, leading to deep divisions between the Lutheran and the Reformed churches. The young theologians—the *praeceptors*—who taught and preached to Kepler and the other boys at Adelberg emphasized the Lutheran doctrine of the real presence—that the bread and wine received at the Eucharist were truly and physically the body and blood of Christ—and rejected both the Zwinglian understanding that the bread and wine symbolized the presence of Christ and the Calvinist position that Christ was present in the bread and wine spiritually but not physically. Here too Kepler found himself inclined to favour precisely the Calvinist views that were being condemned from the pulpit.[17] He was admitted to communion, but increasingly found receiving the bread and wine to be a test of conscience. As an adult he came to hold a position which sought to mediate between that of the Lutherans and that of the Calvinists,[18] but these questions had become important to him long before he began his formal study of theology.

As a student at Tübingen, Kepler had first to complete his *Magister Artium* (Master of Arts), which he did on 11 August 1591. Thereafter he undertook nearly three years of theological studies before leaving Tübingen on 1 April 1594 to take up a post as mathematics teacher at the Lutheran school in Graz. (The length of the theological studies undertaken by a student in the Duke's scholarship depended on when a suitable post became available to them.) It was probably during his time in the Arts Faculty that Kepler first encountered astronomy and the study of celestial mathematics. His mathematics teacher, Michael Maestlin (1550–1631), himself a product of the Duke's scholarship system, was an important influence on Kepler and became a life-long friend. Maestlin was not only a talented mathematician and astronomer who introduced Kepler to the Copernican hypothesis—that the Universe was heliocentric (centred on the Sun) and not geocentric (centred on the earth)[19]— but was also interested in the theological implications of what he observed and calculated.[20] The Arts curriculum included the study of rhetoric and dialectic (or logic), moral philosophy (ethics) and natural philosophy (or physics), as well as continuing engagement with Latin and Greek, classical texts, and with the Old and New Testaments.[21] Kepler clearly profited from his studies of natural philosophy, which considered, amongst other questions, the true nature of the Cosmos. He must also have engaged productively with dialectics, which introduced him to considerations about the nature of proofs. His studies continued to be placed in a firmly

---

[16] Hübner (1975), p. 5.

[17] Hübner (1975), pp. 4–5.

[18] Hübner (1975), p. 5. He described his position in a poem, *De omnipraesentia Christi* ('On the ubiquity of Christ'; Kepler, 1990a, 1990b, p. 7) and a catechetical text written primarily, but not only, for his children, *Unterricht vom h. sacrament* ('Teaching on the Holy Sacrament'; Kepler, 1990b, pp. 11–18).

[19] Cf. Methuen (1996), p. 230.

[20] Methuen (1996), p. 230.

[21] Methuen (1998), pp. 47–50; cf. also Hofmann (1982).

theological context, not least because life in the *Stift* followed a disciplined pattern similar to that which Kepler had experienced at school, with compulsory prayers morning and evening, and additional teaching on scripture and theology provided by the officers of the *Stift*.[22] Here Kepler heard lectures based on Jakob Heerbrand's *Compendium theologiae* (Tübingen, 1573), probably including an exposition of the theology of creation which was to prove a strong factor in Kepler's astrological and astronomical work.[23] Once he had completed his *Magister Artium*, Kepler's theological studies were shaped by Tübingen's professors of theology Jakob Heerbrand (1521–1600); Matthias Hafenreffer (1561–1619), who also became Kepler's close friend and correspondent; Stephan Gerlach (1546–1612), the Superintendent of the *Stift*; and Johannes Georg Sigwart (1554–1618), second Superintendent of the *Stift*. The official curriculum focused on Scripture and the central theological themes of the day, although Kepler recorded that he found the commentaries of Ägidius Hunnius (1550–1603) more useful when studying the New Testament than the lectures he was supposed to attend.[24]

The teaching at Tübingen whilst Kepler was there was fundamentally Aristotelian, although this was a version of Aristotelianism designed to foster the principles of Lutheranism, and the curriculum had a profoundly theological slant. This had some influence on the teaching of astronomy. Maestlin drew on theological arguments to support the conclusions he drew from his own observations—for instance, that comets were further away from the Earth than the Moon was—over the assumptions of Aristotelian physics—that comets and other unpredictable changes in the heavens could only happen below the Moon. For Maestlin, scripture included the call to study the natural world and the heavens directly in order better to understand how God had created them. Direct observation of the created world, leading to a better understanding of God's divine plan, overrode the hypotheses put forward by Aristotle (384–322 BC).[25] Hafenreffer was also interested by the interactions between theology and mathematics: in his commentary on Ezekiel published in 1613 (for which Maestlin supplied mathematical and geometrical insights which informed a theoretical reconstruction of the Jerusalem temple), he described God as a geometer and as architect of the world.[26] However, Hafenreffer was also cautious about the predictability of the order of the Universe, maintaining that God might choose at any time to intervene in its running.[27] Despite their friendship, Hafenreffer and Kepler disagreed fundamentally about Christology.

Kepler used the three years of his theological studies also as an opportunity to deepen his understanding of mathematics and astronomy, taking advantage of the

---

[22] Methuen (1998), pp. 50–51.

[23] For Heerbrand's theology, see Methuen (1998), pp. 132–152.

[24] Hübner (1975), p. 8.

[25] For Maestlin's mathematical teaching see Methuen (1996). For the theological basis of his teaching, see Methuen (1998), pp. 129–130 and 153–158.

[26] Hafenreffer (1613); Mehl (2010), pp. 207–212.

[27] Methuen (1998), pp. 150–152.

university library to read widely. He recorded reading the *Exotericarum exerci-
tationum liber de subtilitate* ('Exoteric Exercises on Subtlety') of Julius Caesar
Scaliger (1484–1558), which laid out an empirical approach to natural philosophy
based upon experience and observation.[28] Kepler may also have taken the opportunity to study works by earlier teachers at Tübingen, such as the *Libellus geographicus*
(Tübingen, 1562) and *De usu partium coeli oratio* (Tübingen, 1563) published by
Samuel Eisenmenger or Siderocrates (1534–1585), a predecessor of Maestlin as
Professor of Mathematics and Astronomy at Tübingen from 1557 to 1567, whose
writings drew heavily on Plato (428/427 or 424/423–348/347 BC) in their emphasis
on the need to study the heavens and the natural world.[29] He almost certainly encountered the writings of Marsilio Ficino (1433–1599), whose edition of Plato's works
in the 1532 Basel edition seems to have been in the University Library at Tübingen,
and Giovanni Pico della Mirandola (1463–1494), who applied the insights of neo-
Platonism in his critique of astrology. It was probably through his reading, therefore,
rather than through the teaching he received, that Kepler encountered the Platonism
which became so fundamental to his own interpretation of the heavens.

1.2   Graz (1594–1600) and the *Mysterium Cosmographicum*

In 1594, Kepler moved to Graz to take up the post of district mathematician and
teacher of mathematics at the Protestant (Lutheran) school. On accepting the appointment, he announced his intention of later returning to Tübingen in order to continue
his theological studies and seek a position as a pastor,[30] although he later conceded
that his theological qualms would have made him unsuitable for such a post.[31] He
had in fact left Tübingen for good.

Kepler's duties in Graz included teaching not only mathematics (a subject in
which, in his second year there, he had no students), but also rhetoric and the poetry of
Virgil, and he was also responsible for producing astrological prognostications.[32] In
addition, he found the time to work on his first book, the *Mysterium cosmographicum*,
an unambiguously Copernican work which was published in Tübingen 1596 with the
explicit support of the University authorities.[33] In it Kepler presented an astounding
discovery: the distances between the orbs of neighbouring planets were given by
the ratios between the radii of the circumspheres and inspheres of the five regular
polyhedra, so that the orbs could be seen as nested between successive (imaginary)

---

[28] Kepler (1963), p. 15, line 8. Cf. Field (1988), p. 34; Grafton (1992), p. 563.

[29] See Methuen (2010), pp. 193–195; Mehl (2008), pp. 361–365.

[30] Kepler to the Theology Faculty in Tübingen, 28 February 1594 (Kepler, 1945, letter 8); cf. Hübner
(1975), p. 9.

[31] Kepler (1949), letter 132, lines 532–535; cf. Hübner (1975), p. 9.

[32] Hübner (1975), p. 9; Caspar (1959), pp. 54–60.

[33] Rosen (1975), pp. 324–326.

polyhedra.[34] He heralded this as an a priori geometrical truth,[35] the revelation of a cosmological structure given to the Universe by God:

> what else remains except to say with Plato, 'God is a geometer' and in this structure of moving stars he has inscribed solids within spheres, and spheres within solids, until no further solid was left which was not robed outside and inside with moving spheres.[36]

Moreover, he argued, the congruence between the planetary distances and Platonic ratios offered evidence that Copernicus' heliocentric hypothesis must be a physical representation of the truth:

> For what could be said or imagined which would be more remarkable, or more convincing, than that what Copernicus established by observation, ex *phaenomēnois*, from the effects, *a posteriori*, by a lucky rather than a confident guess, like a blind man depending on a stick as he walks, and believed to be the case, all that, I say, is discovered to have been quite correctly established by reasoning derived *a priori*, from the causes, from the idea of the Creation.[37]

For Kepler, such a discovery could only come from God: this was divine revelation and he was charged with transmitting it, and in doing so to contributing to the better knowledge of God and the reconciliation of the church.[38] He would become, he determined, not a priest of the book of scripture, but a 'priest of the book of nature'.[39]

The claims made by Kepler in the *Mysterium cosmographicum* indicate not only the underlying theological intentions of his astronomical investigations, but also the way in which he was already beginning to elide the distinction between astronomy and physics through, as Rhonda Martens puts it 'the unconventional mixing of three disciplines: celestial physics, mechanics, and mathematics'.[40] The title of Kepler's work identified it as cosmographical, a term commonly used to refer to geographical works considering the physical nature of the world. However, Kepler's treatment expanded the term to consider the nature of the Universe, deliberations which traditionally belonged to natural philosophy, but he also brought the mathematical

---

[34] Kepler (1596), chapter II, pp. 18–25; Kepler (1938), pp. 23–28; Kepler (1963), pp. 44–49; Kepler (1981), pp. 93–101; cf. Field (1988), pp. 35–72. See also Field's discussion of Kepler's cosmology in Chapter 2 of this volume. Kepler has an unconventional definition of a planetary orb as being the shell contained between two concentric spherical surfaces, the inner surface being that of the sphere that will just fit inside the path of the planet, the outer one being the sphere that will just fit round it. From Chap. 15 of the *Mysterium cosmographicum* onwards all spheres are centred on the Sun.

[35] That is, a truth showing the true nature or cause of things: see Barker and Goldstein (2001), pp. 91–93. Cf. Martens (2000), pp. 50–51.

[36] Kepler (1596), p. 22; Kepler (1938), p. 26; Kepler (1963), p. 47; Kepler (1981), p. 97. For Kepler's arguments see Jardine (2008).

[37] Kepler (1596), pp. 22–23; Kepler (1938), p. 26; Kepler (1963), p. 47; Kepler (1981), pp. 97, 99.

[38] Rothman (2011), p. 120.

[39] See, for instance, Kepler to Herwart von Hohenburg, 26 March 1598; Kepler (1945), letter 91, lines 182–184: 'Indeed I am of the opinion that since we astronomers are priests of God, the most high, in regard to the book of nature, it is fitting not to receive praise for our talent, but above all to look upon the glory of the Creator' (*Ego vero sic censeo, cum Astronomj, sacerdotes dej altissimj ex parte librj Naturae simus: decere non ingenij laudem, sed Creatoris praecipue gloriam spectare*).

[40] Martens (2000), pp. 102, 184 n. 8.

approaches of astronomy to bear on these. Kepler's aim, as Kenneth Howell has observed, was 'to represent the various parts of the world system in relation to one another'.[41] However, his aim in doing so was theological. Kepler's greeting to the reader announces that his book reveals, 'the nature of the universe, God's motive and plan for creating it'.[42] Howell has also said that Kepler wanted to 'plumb the depths of the divine will in the cosmos by using the appropriate tools of interpretation'.[43] Kepler believed that he was revealing a divine mystery, almost a sacrament: in the Vulgate, the Latin word *sacramentum* translates the Greek word *mysterion*: a deeper, spiritual reality expressed through a physical truth.[44] He was not claiming to 'save the appearances'—that is, to offer a mathematical explanation for the planetary motions observed by astronomers without making any claims for their relations to the underlying physical reality—but rather, on the basis of mathematical calculations based on astronomical observations, to have revealed a fundamental (indeed, divine) truth about the structure of the Universe.[45] This melding of mathematics, astronomy and natural philosophy was a significant step towards considering the natural world as susceptible to mathematical explanation. It was an approach which would guide all Kepler's astronomical and scientific work.

Kepler's life was also taking shape in other ways. In spring 1597, he married Barbara, the 23-year-old daughter, already twice widowed, of a local mill-owner, Jobst Müller. Barbara's five-year-old daughter, Regina, also joined their household. Difficult years followed. On 2 February 1598, Johannes and Barbara's first son, Heinrich, was born, only to die aged two months. Their daughter, Susanna, born in June 1599, lived for just 35 days.[46] Alongside these tragedies, another development was brewing which would prove disastrous for Kepler: the Austrian Counter-Reformation.[47] The Lutheran church in Graz had already been under pressure from the Counter-Reformation when he arrived, and it was hard to attract good pastors to its congregations: Kepler frequently complained about the low standard of the preaching there. In 1598 the situation deteriorated when Ferdinand, Duke of the Steiermark (Styria) vowed to return his territories to Catholicism. In September 1598 Lutheran teachers and preachers were ordered to leave not only Graz but Austria. Kepler and his colleagues made their way to Hungary, from where, after a month, Kepler was allowed to return to Graz. Lutheran services and preaching became increasingly difficult to arrange, even in private. The following year, Kepler was fined for having his daughter Susanna baptized according to a Lutheran rite and forced to

---

[41] Howell (2001), pp. 110–113 (quote pp. 111–112).

[42] Kepler (1596), <unnumbered fol. A1>ᵛ; Kepler (1938), p. 4; Kepler (1963), p. 14; Kepler (1981), p. 49.

[43] Howell (2001), p. 113.

[44] Howell (2001), p. 114; Barker and Goldstein (2001), p. 99.

[45] This claim caused the Tübingen theologians, and particularly Matthias Hafenreffer, some concern: Rosen (1975), pp. 326–330.

[46] Caspar (1959), p. 71–77.

[47] For the Reformation and Counter-Reformation in the Austrian lands, see von Schlachta (2015), pp. 68–91; compare also Thaler (2020), p. 97, especially Chaps. 1 and 2.

pay the fine before he was allowed to bury her.[48] He resolved to visit Tycho Brahe (1546–1601), now the Imperial Mathematician in Prague, with whom he had been corresponding since 1597. After a difficult series of negotiations (Tycho wanted to ensure that Kepler supported him in a conflict with his predecessor, Nicolaus Ursus [Nicolai Reymers Baer, 1551–1600][49]), Tycho agreed to offer his support to secure Kepler's position in Graz. Back in Graz, Kepler refused the demand of the Catholic 'Reformation commission' that he should convert to Catholicism, and asked to be relieved of his duties. On 4 September 1600, Kepler left Graz, a refugee from the Catholic Counter-Reformation. Tycho Brahe encouraged him to come to Prague.

1.3 Prague (1600–1612) and the *Astronomia nova*

Kepler's first months in Prague continued to be difficult. He and his wife were both ill; Kepler had no post; and Brahe, upon whom he was entirely dependent, was reluctant to give Kepler access to his observations. However, on 24 October 1601, Tycho Brahe died, and Kepler was appointed his successor. Kepler saw God's hand in this: 'If God is concerned with astronomy, which piety desires to believe, then I hope that I shall achieve something in this domain, for I see how God let me be bound with Tycho through an unalterable fate and did not let me be separated from him by the most oppressive hardships.'[50] His time in Prague would prove the most productive period of his life in terms of his astronomical work, with the composition of works including his *Apologia* for Tycho against Ursus (written 1600–01, it remained unpublished);[51] his exploration of optics and astronomy, *Astronomiae pars optica* (published 1604);[52] his study of the 'New Star' of 1604, *De stella nova* (1606);[53] his *Astronomia nova*, or *New Astronomy* (1609);[54] his defence of Galileo's discovery of the moons of Jupiter with the help of a telescope, *Dissertatio cum nuncio sidereo* (*Conversation with a Starry Messenger*; 1610);[55] a further study of optics, the *Dioptrice* (1611);[56] and a study of snowflakes, recognizing them as hexagonal, *De nive sexangula* (also 1611).[57]

The most significant of Kepler's publications whilst he was in Prague was his *Astronomia nova* of 1609, which has been described as 'the first modern astronomy

---

[48] Caspar (1959), p. 97.

[49] The negotiations are detailed in Jardine (1984), pp. 9–28.

[50] Kepler to Maestlin, 10/20 December 1601; Kepler (1949), letter 200: 102–7; cited according to Caspar (1959).

[51] See Jardine (1984) for an English translation and commentary.

[52] Kepler (1604).

[53] Kepler (1606a).

[54] Kepler (1609).

[55] Kepler (1610).

[56] Kepler (1611a).

[57] Kepler (1611b).

book'.[58] In it, Kepler recounted his discovery of what have come to be known as his first two laws: that the orbit of a planet is an ellipse with the Sun at one of the two foci, and that a line joining a planet and the Sun sweeps out equal areas during equal intervals of time.[59] The second law led Kepler to the first; he derived both laws from his work on Tycho's observations of Mars and did not at this stage generalize his findings to all planets. Working with Tycho's results, which he believed to be highly accurate, Kepler discovered a discrepancy of eight minutes of arc between where his calculations suggested Mars should be and where Tycho's observations placed the planet. Kepler concluded that the theory underlying his calculations must be wrong, and that he should seek a theory that would explain them more accurately, in order better to understand the Universe as God had created it:

> Since the divine benevolence has vouchsafed us Tycho Brahe, a most diligent observer, from whose observations the 8′ error in this Ptolemaic computation is shown, it is fitting that we with thankful mind both acknowledge and honour this blessing from God. And therefore we should search out the true form of the heavenly motions.[60]

Kepler's calculations took several years, and led him down several blind alleys.[61] His recognition that Mars moved in an ellipse with the Sun at one focus[62] arose from his conviction that the planet's motion focused on the actual Sun, and did not, as Copernicus (following Ptolemy) had suggested, relate to the so-called mean Sun.[63] This emphasis on the importance of the Sun as the focus of Mars' motion had physical reasons,[64] but it also had a theological underpinning: Kepler held that the structure of the Universe was Trinitarian, identifying the Sun with God the Father, the fixed stars with Jesus Christ, the Son, and 'the heavens or the intermediate ether' with the Holy Spirit.[65] As the Son and the Spirit proceeded from the Father, Kepler was convinced that the motive power of the Universe emanated from the Sun; although he was unclear as to what this power was, he thought 'it will be some natural—or

---

[58] Caspar (1959), p. 139.

[59] As Gingerich (2011, p. 50) points out, Kepler 'didn't point them out as laws, and never spoke of "laws of nature".' On Kepler's derivation of the first two laws, see Davis, 'The discovery of the planetary orbit: the Goldilocks solution' in this volume (Chap. 4) and Gregory, 'The Translation of the title …' in this volume (Chap. 5).

[60] Kepler (1609), p. 113 (http://diglib.hab.de/drucke/n-29-2f-helmst-2/start.htm, accessed 30.8.13); Kepler (1990a), p. 178; Kepler (2015), p. 28 (translation amended by CM).

[61] Caspar (1959), pp. 126–134, and Martens (2000), pp. 68–98, both outline the process which let Kepler to this conclusion.

[62] Kepler (1609), p. 285; Kepler (1990a), 366; Kepler (2015), p. 576.

[63] Martens (2000), pp. 71–76; Caspar (1959), p. 129.

[64] Kepler (1609), pp. 3, 9–10, 169–170; Kepler (1990a), pp. 63, 70, 237–239; Kepler (2015), pp. 118, 128, 378–379; cf. Martens (2000), pp. 73, 81; Gingerich (2011), p. 47.

[65] Kepler to Maestlin, 3 October 1595, Kepler (1945), p. 35. Cf. Howell (2001), p. 128; Kozhamthadam (1994), pp. 15–22.

better, corporeal—faculty and not a planetary mind',[66] such as light or magnetism, the solution which at this stage he preferred.[67]

Significantly, although originally Kepler had used his Trinitarian analogy to argue that the Universe was spherical, and he continued to argue for the importance of the sphere, as the most perfect solid, in understanding the Universe,[68] his commitment to exploring the underlying realities revealed by Tycho's observations showed that not all heavenly motion was circular. Once again, Kepler had shown that a geometrical form (albeit not the circle he had expected) underlay the structure of the Universe. For Kepler, this was not just a question of mathematical fit, but revealed the Universe's true, physical—and God-given—reality. Maestlin was sceptical about the validity of this approach, but Kepler asserted:

> I shall accept only that which cannot be doubted as truly real, and therefore physical, keeping in mind the nature not of the elements, but of the heavens. If I wholly reject the perfect eccentrics and epicycles, I do so because they are purely geometrical assumptions that do not correspond to any body in the sky.[69]

Kepler claimed that he had transformed 'the whole of astronomy from fictitious circles to natural causes'.[70] Through the approach taken in *Astronomia nova*, and particularly his development of the concept of the orbit,[71] and his combination of insights from a range of disciplines, including not only natural philosophy and mathematics but also optics,[72] Kepler had changed the nature of astronomy. Indeed, Giora Hon argues that with the *Astronomia nova*, 'astronomy is now celestial physics.'[73]

In the introduction to the *Astronomia nova*, Kepler included a discussion of the objections to the motion of the Earth, and thus to the Copernican hypothesis. Three objections were drawn from physics: these related to the motion of heavy bodies, the assumed consequences of the 'swiftness of the Earth's motion', and the immensity of the heavens.[74] Two were drawn from theology: from Scripture and from 'the authority of the saints' (or the pious), by which Kepler meant particularly the Church

---

[66] Kepler (1609), p. 269; Kepler (1990a), pp. 348; Kepler (2015), p. 548.

[67] Kepler (1609), p. 273; Kepler (1990a), pp. 352; Kepler (2015), p. 554. Kepler argued that the Earth must possess a soul and explored the possibility that comets and planets might be moved by the interaction between their own souls or spirits and that of the sun: see Field (1984), pp. 197–198, 223–224; Boner (2012). Indeed, the concept of the world soul became the subject of a major disagreement between Kepler and Robert Fludd: see Schmidt-Biggemann (2008); Boner (2006). The soul and the mind were closely related for Kepler and foundational to his epistemology: see Escobar (2008).

[68] Hübner (1975), pp. 186–192; Howell (2001), pp. 127–129.

[69] Kepler's notes on Maestlin to Kepler, 21 September 1616, Kepler (1955), letter 744, note to line 29 (p. 188); cf, Danilov and Smorodinskii (1975), p. 703.

[70] Aiton (1975), p. 57.; cf. Aiton (1978), p. 175n8.

[71] See Goldstein and Hon (2005). Maestlin had already developed the concept of the orbit when discussing comets: Methuen (1998), pp. 173–176.

[72] Hon (2014).

[73] Hon (2014), 157.

[74] Kepler (1609), 'Introductio' (with unnumbered pages); Kepler (1990a), pp. 24–28; Kepler (2015), pp. 54–59.

Fathers. Here Kepler pointed out that piety did not necessarily go hand in hand with knowledge of natural philosophy: 'Lactantius [Lucius Caecilius Firmianus, *c.* 250–*c.* 325] is holy [*sanctus*], who denied that the world is round, Augustine of Hippo [354–430] is holy, who, though admitting the roundness, denied the antipodes, and the Inquisition nowadays is holy, which, though allowing the Earth's unimportance, denies its motion.' For Kepler, however, 'with all respect to the Doctors of the Church, the truth about the world—that it is round, the existence of its antipodes, that it is insignificantly small and hurries through the stars—demonstrated by philosophy, is holier still.'[75] More important, however, was his discussion of Scripture. Having been persuaded by Maestlin and Hafenreffer to omit any discussion of the compatibility of Scripture and the Copernican hypothesis in the *Mysterium cosmographicum*,[76] Kepler offered a detailed and careful consideration of the structure of Scripture and its role with respect to astronomy.[77] He considered the proper interpretation of Joshua 10:12–14 and its claim that 'the Sun stood still', arguing that the verse represented what Joshua observed, rather than the physical reality.[78] Kepler pointed out that Psalm 104 was a hymn in praise of the Creator, in the form of a commentary on the creation account in Genesis 1, and should be read accordingly; it was not intended to give accurate knowledge of natural philosophy.[79] Indeed, Kepler argued that this was true of Scripture generally. Because its central message is theological and moral, Scripture uses every-day language and images comprehensible to 'the simplest mind', in order that it may 'speak to the ignorant as well as to the learned'.[80]Consequently, because 'to absolutely all men, the Sun appears to move and not the Earth', Scripture was formulated in these terms: deeper knowledge was needed in order to recognise the reality of what was being observed:

> It is therefore impossible for a previously uninformed reason to imagine anything but that the Earth, along with the arch of heaven set over it, is like a great house, immobile, in which the Sun, so small in stature, travels from one side to the other like a bird flying in the air.[81]

Only those learned in astronomy could grasp this truth. Kepler had some strongly-worded advice to those who are untutored (*idioti*) in the questions of astronomy:

> To whoever is too stupid to understand astronomical science, or too weak to believe Copernicus without affecting his faith, I would advise him that, having dismissed astronomical studies and having damned whatever philosophical opinions he pleases, he mind his own business and betake himself home to scratch in his own dirt patch, abandoning this wandering

---

[75] Kepler (1609), 'Introductio'; Kepler (1990a), pp. 33–34; Kepler (2015), p. 66; cf. Methuen (2013), pp. 770–771.

[76] Rosen (1975), pp. 327, 330; Rothman (2011), pp. 116–121.

[77] Kepler (1609), 'Introductio'; Kepler (1990a), pp. 28–33; Kepler (2015), pp. 59–65.

[78] Kepler (1609), 'Introductio'; Kepler (1990a), p. 30; Kepler (2015), p. 61; cf, Rosen (1975), pp. 331–333.

[79] Kepler (1609), 'Introductio'; Kepler (1990a), pp. 31–33; Kepler (2015), pp. 63–65. Cf. Howell (2001), pp. 119–125; Hübner (1975), 168–170.

[80] Kepler (1609), 'Introductio'; Kepler (1990a), pp. 31–32; Kepler (2015), p. 63; cf. Methuen (2008), pp. 85–90.

[81] Kepler (1609), 'Introductio'; Kepler (1990a), p. 30; Kepler (2015), p. 62.

about the world. He should raise his eyes (his only means of vision) to this visible heaven and with his whole heart burst forth in giving thanks and praising God the Creator. He can be sure that he worships God no less than the astronomer, to whom God has granted the more penetrating vision of the mind's eye, and an ability and desire to celebrate his God above those things he has discovered.[82]

Only the astronomer, Kepler implied, could know the full truth about God as it was revealed in the heavens.

Alongside his theological work, Kepler also used his time in Prague to deepen his engagement with confessional theological questions. Although according to an Edict of 1602 only the Catholic and Utraquist churches were officially recognized in Bohemia, in reality other confessions, including the Bohemian Brethren, Calvinists and Lutherans, were tolerated, and Kepler's circle included friends from all these groups.[83] The three children born to Kepler and Barbara in Prague—Susanna, born 9 July 1602; Friedrich, born 3 December 1604; and Ludwig, born 21 December 1607—were all baptized by Utraquist clergy.[84] Kepler engaged particularly with Calvinism, and especially with questions of predestination and divine providence;[85] he further explored questions of Christology;[86] he received some encouragement to think about becoming a Catholic; and from 1609—as his confessional situation in Prague became more precarious—he engaged in correspondence with the authorities in Württemberg about the possibility of employment at the University of Tübingen.[87] This quickly became an examination of his confessional stance, and in particular of his position in relation to the *Formula of Concord*, the Lutheran confession of faith which all Württemberg's pastors, school teachers and university postholders were required to sign. The case was referred to the Consistory, whose members concluded that Kepler was in reality a 'sly Calvinist', and confessionally suspect. On 25 April 1611, the Duke of Württemberg endorsed a document put forward by the consistory to this effect. Kepler would not be eligible for a post in Tübingen, or indeed anywhere in Württemberg, and the possibility of succeeding Maestlin at Tübingen was closed to him.

The decision was one aspect of what would prove another disastrous year for Kepler. On 19 February, his six-year-old son, Friedrich, died of smallpox. Political trouble culminated in the abdication of Emperor Rudolf on 23 May 1611, to be succeeded by his brother Matthias. Kepler's position in Prague was not secure, and on 11 June 1611 he accepted a post as District Mathematician in Linz, hoping that his wife would be happier closer to home. However, Barbara had contracted a form of typhus and on 3 July she died. Her funeral was conducted by a Lutheran theologian,

---

[82] Kepler (1609), 'Introductio'; Kepler (1990a), p. 33; Kepler (2015), pp. 65–66: *'consilium pro idiotis'*.

[83] Hübner (1975), pp. 15–19.

[84] Caspar (1959), p. 174.

[85] Hübner (1975), pp. 20–22.

[86] Hübner (1975), pp. 24–28.

[87] Hübner (1975), pp. 22–24; Caspar (1959), pp. 204–205.

Matthias Hoe.[88] Kepler remained in Prague until after the death of the former Emperor Rudolf on 20 January 1612. In March, Emperor Matthias confirmed him in his post as imperial mathematician, giving him permission to live in Linz. In April Kepler and his children left Prague for Upper Austria.

1.4   Linz (1612–1626) and the *Harmonice mundi*

Kepler's situation in Linz was something of a sinecure: his post had been created to allow him to continue his astronomical work, and his duties at the local school were light. Kepler's priority was to complete the *Rudolphine Tables*, which he finally did in 1624; they were published at Ulm in 1627 after protracted negotiations with Tycho's heirs, the imperial treasury and the printers. Considerably more reliable than earlier versions of such tables (the margin of error was less than 10 arcsec), and incorporating logarithmic tables to aid the necessary calculations, the *Tables* were based on Kepler's heliocentric system, using elliptical planetary orbits, and made it possible to calculate planetary positions at any point in the past or future.[89] Whilst in Linz Kepler also wrote, amongst other works, the *Harmonices mundi libri V* (1619)[90] and the *Epitome astronomiae Copernicanae* (published in three parts: 1618 [books I–III],[91] 1620 [book IV],[92] and 1621 [books V–VII[93]]). In 1620, the *Epitome astronomiae Copernicanae* was placed on the Catholic church's index of prohibited books, on account of its defence of the Copernican hypothesis, although this did not prevent Kepler's appointment as Imperial Mathematician. However, it was also in Linz that Kepler's differences with the Lutheran Church in Württemberg on the question of Eucharistic theology came to a head.

Daniel Hitzler (1575–1635) had been pastor of the Lutheran congregation in Linz since June 1611. Like Kepler, Hitzler had studied theology at Tübingen; he was an engaged and well-informed pastor, who wrote theological works, revised the liturgy, and edited a hymnbook.[94] He was also in close contact with the Württemberg church authorities. Kepler, as was his custom, requested admission to communion and explained his reservations relating to the Lutheran doctrine of the Eucharist. Hitzler refused to admit him to communion, and the Stuttgart Consistory, to which Kepler appealed, confirmed this decision.[95] With his family, Kepler continued to attend the services at which Hitzler presided,[96] and he seems to have continued to

---

[88] Caspar (1959), pp. 204–207.

[89] Kususkawa (1999).

[90] Kepler (1619).

[91] Kepler (1618).

[92] Kepler (1620).

[93] Kepler (1621).

[94] Hübner (1975), p. 31.

[95] Hübner (1975), pp. 32–33; Caspar (1959), pp. 213–214.

[96] Hübner (1975), p. 37.

receive communion from other preachers, including chaplains appointed by local noblemen with Calvinist sympathies and the Lutheran pastor Garthius in Prague.[97] A pamphlet 'Teaching about the Holy Sacrament', published anonymously in Prague in 1617, indicates the care he took to teach these questions in his household; it also shows that his objection related to the Lutheran doctrine of ubiquity—the teaching that Christ could be omnipresent—rather than the Lutheran teaching that the bread and wine were truly the body and blood of Christ. In the bread and wine, Kepler taught, the body and blood of Christ were indeed received, and those who received became members of the spiritual body of Christ. The Eucharist offered comfort and strengthened faith, rather than representing the forgiveness of sins.[98] Kepler's pamphlet was received critically by the authorities in Württemberg, amongst them his friend and former teacher Matthias Hafenreffer. A long correspondence ensued.[99] In 1623, Kepler published a confession of faith in which he explained that he did not regard himself as belonging to any of the 'three great factions'—Catholic, Lutheran or Reformed[100]—and outlined the theological problems which he saw arising from the Lutheran 'innovations'; however, he also confirmed that he could accept the *Formula of Concord* in as far as it agreed with the Augsburg Confession (1530).[101] This changed nothing: Kepler had been excommunicated by his own church in Württemberg.

In October 1613, after a protracted process of considering possible candidates, Kepler married again. His new wife, Susanna Reuttinger, was the 24-year-old orphaned daughter of a cabinet-maker. Her step-daughter Regina, now 21, who had herself married aged 16 in 1608, considered Susanna too young to take responsibility for Kepler's other children,[102] but Kepler was sure that he had been 'guided by divine decree … so that I should learn to despise high society, riches, relations, … and to aspire calmly to other simple traits.'[103] He brought his two elder children to Linz. Together, Susanna and Johannes had a further six children, the first three of whom died young: Margareta Regina (7 January 1615–8 September 1617); Katharina (31 July 1617–9 February 1618); Sebald (28 January 1619–15 June 1623). Cordula (b. 22 January 1621), Fridmar (24 January 1623–1634?) and Hildebert (6 April 1625–18 October 1635) outlived their father, although neither boy survived to adulthood.[104]

Other family events also took up a good deal of Kepler's time in this period. In August 1615, Kepler's mother, Katharina, was accused of witchcraft. Protracted

---

[97] Hübner (1975), pp. 33–37, 44–45.

[98] 'Unterricht vom H. Sacrament'; Kepler (1990b), pp. 9–18; cf. Hübner (1975), pp. 37–45.

[99] Hübner (1975), pp. 45–68.

[100] *Glaubensbekenntnis*, Kepler (1990b), pp. 19–38; here 27; cf. Hübner (1975), pp. 68–78. Compare also Lanzinner (2003), pp. 531–532.

[101] *Glaubensbekenntnis*, Kepler (1990b), p. 25; cf. Hübner (1975), p. 70.

[102] Regina Ehem (née Lorenz) to Kepler, 3 September 1612; Kepler (1955), letter 635, line 16; Caspar (1959), p. 223.

[103] Kepler to anon., 23 October 1613; Kepler (1955), letter 669, lines 273–276: Caspar (1959), p. 222.

[104] Caspar (1959), p. 223.

processes of accusation and interrogation ensued, during which Kepler's mother fled to Linz, where she spent much of 1617; in October 1617 she returned to her home with her daughter Margareta in Leonberg to face trial, and in July 1620 was arrested and imprisoned. She was held until September 1621, when she was led into the torture chamber and threatened with torture if she did not confess. Katharina Kepler maintained her innocence. Kepler travelled to Stuttgart to defend his mother and eventually the Duke of Württemberg pronounced her innocent, freeing her on 4 October 1621. Kepler's mother died on 13 April 1622, not having dared to return home to Leonberg.[105] In October 1617, in the midst of his efforts for his mother, Kepler was informed of the death of his step-daughter; his daughter Susanna, 15 years old, was sent to look after Regina's young children.

By now, Linz had been drawn into the religious and political conflicts which came to be known as the Thirty Years War. The Calvinist Frederick of the Palatinate (1596–1632, reigned 1610–1623), who had laid claim to the throne in Bohemia, was vigorously opposed by the Habsburgs. Emperor Matthias (1557–1619, reigned 1612–1619) died on 20 March 1619, and was succeeded by his cousin, the uncompromisingly Catholic Ferdinand, who was supported by Duke Maximilian of Bavaria. In July 1620, Maximilian's troops entered Linz, introducing the Counter-Reformation. Over the next few years, measures were introduced into Linz as they had been into Graz; the Lutheran school was closed down; Lutheran public worship banned—and (ironically) Daniel Hitzler was arrested on suspicion of being a Calvinist.[106] Kepler's efforts on his mother's behalf had meant that he and his family were absent from Linz when the Bavarians arrived, and he considered carefully whether to return. He had a good deal of sympathy for Frederick of the Palatinate, the 'Winter King', and son-in-law of James VI/I of Scotland and England, who on 8 November 1620 had suffered a decisive defeat at the battle of the White Mountain and had been outlawed. Kepler had been offered a post in England and was unsure whether or not to accept the invitation. However, in autumn 1621, he returned to Linz, and on 30 December 1621 the Emperor Ferdinand confirmed his appointment as court mathematician.[107] For the time being—and despite the Counter-Reformation measures being imposed in Upper Austria—his position was secure.

Despite all the turmoil, these were productive years for Kepler. Early in his time in Linz, he completed a treatise on the birth-date of Christ, *De vero anno natali Christi* ('On the True Year of Christ's birth', Linz, 1614),[108] a question which he had originally raised in *De Iesu Christi vero anno natalitio* ('On the true natal year of Jesus Christ'), written at the same time as his work on the New Star of 1604, *De stella nova* (Frankfurt, 1606), a star which he thought was probably of the same kind as that which had been seen over Bethlehem. His calculations had attracted some controversy, and he now defended his thesis, arguing on the basis of the gospels, the accounts of the

---

[105] Caspar (1959), pp. 240–258. For a recent discussion of the case see Rublack (2015); still useful is Sutter (1979). An account of the trial can be found in Kepler (1990b), pp. 63–100.

[106] Caspar (1959), pp. 251–252, 257–258; Hübner (1975), pp. 83, 87–88.

[107] Caspar (1959), pp. 257–258.

[108] Kepler (1614).

Jewish historian Flavius Josephus (*c.* 37–*c.* 100) and Roman historians, and his own calculation of astronomical events, in particular a lunar eclipse, that Jesus had been born in AD 4 (by the Julian calendar).[109] He worked on the *Rudolphine Tables* and the *Epitome of Copernican Astronomy*, his most substantial work, which offered a detailed introduction to his astronomical method. In the aftermath of the death of his daughter Katharina in February 1618, he returned to an idea which had been at the back of his mind since 1599, before he left Graz: the contemplation of the harmony of the world. The 'short book' he had originally planned now grew to a folio of over 320 pages,[110] incorporating insights from geometry, music, astrology and astronomy.[111] Kepler proposed that the 'fundamental melody of the planets' could be attained from the angular speed of each planet in its movement around the Sun, which, measured in seconds of arc gave the pitch of each tone.[112] He also elaborated a theory of archetypes which he saw as the means by which the human mind recognized the geometrical structures through which the mind of God had shaped the world:

> Geometry, which before the origin of things was coeternal with the divine mind and is God himself (for what could there be in God which would not be God himself?), supplied God with patterns for the creation of the world, and passed over to Man along with the image of God; and was not in fact taken in through the eyes.[113]

The conception was deeply Platonic, influenced by Kepler's reading of Ptolemy's *Harmonics* in 1607 and Proclus' commentary on the first book of Euclid's *Elements*.[114]

As Owen Gingerich suggests, Kepler's hope was to find a connection between the periods of the planets and their distance from the sun (now known as his third law). He was seeking the underlying principles by which 'the Creator had set all of the orbital eccentricities and the maximum and minimum speed of each planet in a tightly logical, harmonic fashion'. His resulting theory proposed a connection between the orbital periods of the planets and their distance from the Sun which, as Gingerich further observes, in Kepler's view revealed the underlying principles by which 'the Creator had set all of the orbital eccentricities and the maximum and minimum speed of each planet in a tightly logical, harmonic fashion'.[115] Similarly, Patrick J. Boner finds that Kepler was in search of a geometric explanation for all that he had observed, and he believed not only that he had at last found it, but that he had done so in a way which made it possible to 'locate heavenly and earthly occurrences within the same explanatory terrain' so that 'faraway phenomena such as comets and

---

[109] Caspar (1959), pp. 227–228.

[110] Field (1988), p. 98.

[111] For a useful summary, see Stephenson (1998), pp. 4–8, 118–127.

[112] Proust (2009) p. 362; cf. Stephenson (1998), pp. 125–127.

[113] Kepler, J. (1619), p. 119; Kepler (1940), p. 223; Kepler (1997), p. 304. For the role of archetypes in Kepler's thought, see Martens (2000); Samsonow (1986), esp. pp. 86–89.

[114] Caspar (1959), p. 266; Stephenson (1998). p. 5.

[115] Gingerich (2002), p. 235; here 235; cf. Stephenson's (1998, pp. 128–241) discussion of the *Harmonice mundi*, Book V. See also Field (1988) and Field, 'Kepler's cosmology' in this volume (Chap. 2).

new stars could thus be seen as essentially comparable to more accessible curiosities on the Earth'.[116]

At the beginning of Book IV of the *Epitome of Copernican Astronomy*, published in 1620, Kepler returned to his comparison of the Trinity and the sphere, now drawing on the language of archetype to define the 'principal parts of the world' and with them the foundation of celestial physics:

> In the sphere, which is the image of God the Creator and the Archetype of the world ... there are three regions, symbols of the three persons of the Holy Trinity—the centre, a symbol of the Father; the surface, of the Son; and the intermediate space, of the Holy Spirit.[117]

Kepler also applied his new insights to a revised edition of the *Mysterium cosmographicum*, published in 1621, into which he introduced the language of harmonies and archetypes which had not been present in the first edition. In the first edition he had argued that 'it was with a definite intention that the straight and the curved were chosen by God to delineate the divinity of the Creator in the universe'.[118] In the revised edition, in his notes to Chapter 2, he commented on the need to adjust his language to that of archetypes, exhorted that 'in describing motions, which take place along lines, let us not despise lines and surfaces which are the only origin of Harmonic proportions,' and affirmed that 'the five regular geometrical solids ... are, so to speak, the archetypes.'[119] Affirming his insight that the 'creator ... is a mind',[120] Kepler explained: 'The reason the Mathematicals are the cause of natural things ... is that God the Creator had the Mathematicals with him as archetypes from eternity.'[121] The human mind was also created so as to recognize these geometrical truths.[122]

By the time Kepler published the *Harmonice mundi* and the second edition of the *Mysterium cosmographicum* Linz was embroiled in the Bohemian war. The prefaces to these works recorded Kepler's deep dismay at this disruption to the harmony intended by God for the world. Kepler dedicated the *Harmonice mundi* to James VI/I of Scotland and England, who, he affirmed, had 'removed in the happiest way the hereditary discord between two extremely hostile nations.'[123] Linz was soon to witness severe disruption of the harmony for which Kepler longed. From 1624 Counter-Reformation measures began to be applied in Linz, and on 10 October 1625, the implementation of measures to expel 'preachers and un-Catholic schoolmasters'

---

[116] Boner (2006), p. 32.

[117] Kepler (1620), pars I.i, p. 438; Kepler (1991), p. 258; Kepler (1995), pp. 13–14; cf. Gingerich (2011), p. 45.

[118] Kepler (1596), p. 20; Kepler (1938), p. 24; Kepler (1963), p. 45: Kepler (1981), p. 95.

[119] Kepler (1621), p. 27; Kepler (1963), p. 50; Kepler (1981), p. 149. Cf. Gingerich (2002), p. 233; Field (1988), p. 8.

[120] Kepler (1596), p. 36; Kepler (1621), p. 37; Kepler (1938), pp. 37–38; Kepler (1963), p. 61; Kepler (1981), p. 123.

[121] Kepler (1621), p. 38; Kepler (1963), p. 63; Kepler (1981), p. 125.

[122] Field (1984), p. 221; Kozhamthadam (1994), pp. 18, 24–26.

[123] Kepler (1619), 'Dedicatio' (with unnumbered pages); Kepler (1940), p. 10; Kepler (1997), p. 4. Cf. Methuen (2008), p. 80.

was announced; Protestant preaching was to cease and Protestant schools to close; 'heretical' books were to be surrendered.[124] As Imperial Mathematician, Kepler was initially exempt from many of these measures, and he and his printer, Plank, who also held to the Augsburg Confession, were allowed to continue the printing of the *Rudolphine Tables*.[125] However, he and his family attended Catholic services, and in April 1625 his son, Hildebert was given a Catholic baptism.[126] At the turn of the year, Linz's 'Reformation Commission' confiscated Kepler's books.[127] He was accused of teaching Lutheran children in secret, to which he replied that, since he had been excommunicated by his own church, even if he were to offer such classes, no-one would attend them. He was, of course, teaching his own children.[128] Then, in June 1626, the city was besieged by rebels against the Bavarian regime. Kepler and his family survived unscathed, although recently printed copies of the *Rudolphine Tables* were destroyed when Plank's house and the printing press were burned. Kepler had had enough; as soon as the siege was broken in August, he petitioned the Emperor for permission to leave for Ulm. The Emperor assented, and on 20 November 1626, Kepler and his family took a boat up the Danube to Passau.[129]

1.5 Final years (1626–1630)

Having settled his family in Regensburg, Kepler travelled on to Ulm, where he supervised another printing of the *Rudolphine Tables*,[130] and then to Frankfurt and Strasbourg, where he sought to sell them. In November 1627, he returned to Regensburg, and from there went to Prague. The question of the future was becoming urgent. On the way to Frankfurt, he had entered into negotiations with Landgrave Philip of Hesse, but in Prague he was urged to retain his post as Imperial Mathematician and return to Prague. One condition was attached: that he should become a Catholic. The Jesuit Paul Guldin, himself a convert from Protestantism, urged him to do so. Kepler responded that a conversion was out of the question: he was already catholic in the truest sense of the word, and there were aspects of the doctrine of the church of Rome, and particularly its teaching on Eucharistic sacrifice, which he could not accept.[131] Kepler was granted the considerable sum of 4000 Gulden in recognition of his work on the *Rudolphine Tables*; he was confirmed in his position of Imperial Mathematician; but it was clear that he would not be returning to Prague.

---

[124] Caspar (1959), p. 316.

[125] Caspar (1959), pp. 316–317; Hübner (1975), p. 87.

[126] Hübner (1975), p. 88.

[127] Caspar (1959), 317; Hübner (1975), p. 87–88.

[128] Caspar (1959), 316–317; Hübner (1975), p. 88.

[129] Caspar (1959), pp. 318–320.

[130] Kepler (1627).

[131] Kepler to Paul Guldin, 24 February 1628, in Kepler (1959), letter 1072, lines 80–97, and cf. 114–124: Hübner (1975), pp. 94–96; cf. Caspar (1959), pp. 334–338; Schuppener (1997), p. 241.

Unexpectedly, another opportunity presented itself. Kepler's work on astrological prognostications and horoscopes had attracted the attention of General Albrecht von Wallenstein (Albrecht Václav Eusebiusz Valdštejna, 1583–1634), who had emerged as a key player in the Imperial campaign, and who had in return been granted the Duchy of Sagan in Silesia (now known by its original name of Żagań, Western Poland). Wallenstein appeared to have little interest in Kepler's confessional stance, and Silesia, although in the Empire, was still relatively tolerant in questions of religion. Kepler resolved to accept Wallenstein's offer. Having wound up his affairs in Linz, he and his family made their way to Sagan, where they arrived in July 1628. However, as a South German, he found it hard to settle in North Germany, commenting 'I barely understand the dialect, while I myself am considered a barbarian.'[132] Wallenstein also proved reluctant to pay Kepler the money which had been granted him by the Emperor. To make matters worse, soon after his arrival, confessional conflicts erupted in Sagan, as Wallenstein sought to ensure that his own lands were loyal to the Catholic church. Kepler was exempt from the measures imposed on the townspeople, but once again he found his options restricted because of his religious views. Moreover, until he had set up a printing press, he could not work, and Wallenstein's Counter-Reformation measures made it difficult to attract a good printer. Eventually Kepler was able to print the *Ephemerides* for the years 1621 to 1639. Kepler also negotiated the marriage of his eldest surviving daughter, Susanna, to his former assistant Jakob Bartsch (*c.* 1600–1633) who had been promised an appointment as professor of mathematics in Strasbourg; the wedding took place in Strasbourg on 12 March 1630. Kepler and his wife Susanna were unable to be present: Strasbourg was too far away, and Susanna was pregnant: their daughter, Anna Maria, was born on 18 April. Kepler was still trying to arrange his financial affairs, his problems compounded by Wallenstein's fall from favour in August 1630. He needed in any case to visit Linz to attempt to obtain payment of the interest of bonds that he held there, and he resolved to travel via Regensburg, where the Imperial Diet was meeting. On 2 November 1630, he arrived in Regensburg. Two days later he was seriously ill. He was attended by physicians and by the local pastors, but on 15 November he died. He was buried in the Protestant cemetery of St Peter, either on 17 or 18 November. A memorial service was held for him by Imperial order, at which the Protestant pastor and author of the Regensburg Chronicle, later Superintendent, Sigmund Christoph Donauer (1593–1655), gave the funeral oration. To the last, Kepler did not accord easily with confessional boundaries.

1.6   Conclusion

Kepler's biography was profoundly shaped by the religious struggles through which he lived. Always resistant to confessional categorization, he was nonetheless recognizably and consistently a Protestant, holding to the Augsburg Confession and

---

[132] Caspar (1959), p. 346.

resisting the increasingly acrimonious divisions between Lutheran and Reformed. His refusal to compromise on questions of theology led to his exclusion from the Eucharist by his own church, as well as to his inability to take up potentially lucrative positions of employment. However, the recognition of his undoubted talents by a succession of Emperors and others in authority served to exempt him from some—albeit no means all—of the requirements and restrictions placed on other Protestants in the Hapsburg heartlands. Kepler is a fascinating example of an articulate, theologically educated man who refused to bow to the requirements of the confessional age in which he lived.

In addition, Kepler's theology and his experiences also shaped his intellectual endeavours, driving him to seek in the heavens the order and unity which he found so wanting in the world. From early in his career, he was convinced that his cosmological discoveries could help to bring unity to a divided church. His conviction that God had created the human mind in such a way that it could discover and recognize this order impelled him to the discovery, not only of the three laws of planetary motion which now bear his name, but of important aspects in optical theory and mathematical method. Without the confessional conflicts in which he lived, perhaps Kepler would have become a pastor, and his trail-blazing astronomical work would never have been undertaken.

References

Aiton, E. J. (1975). Johannes Kepler and the astronomy without hypotheses. *Japanese Studies in the History of Science, 14*, 49–71.

Aiton, E. J. (1978). Kepler's path to the construction and rejection of his first oval orbit for Mars. *Annals of Science, 35*, 173–190.

Barker, P., & Goldstein, B. R. (2001). Theological foundations of Kepler's astronomy. *Osiris, 16*(2001), 88–113.

Boner, P. J. (2006). Kepler's living cosmology: Bridging the celestial and terrestrial realms. *Centaurus, 48*, 32–39.

Boner, P. J. (2012). Beached Whales and Priests of God: Kepler and the cometary spirit of 1607. *Early Science and Medicine, 17*, 589–603.

Boockmann, F., & di Liscia, D. A. (Eds.). (2009). *Johannes Kepler Gesammelte Werke. Band XXI.II.II: Manuscripta astrologica / manuscrita pneumatica*. Munich: C. H. Beck.

Caspar, M. (1959). *Kepler*. Translated from the German and edited by C. D. Hellman. London and New York: Abelard-Schuman.

Danilov, Y. A., & Smorodinskii, Y. A. (1975). Kepler and modern physics. *Vistas in Astronomy, 18*, 699–707.

Eisenmenger, S. (1562). *Libellus geographicus*. Tübingen.

Eisenmenger, S. (1563). *De usu partium coeli oratio*. Tübingen.

Escobar, J. M. (2008). Kepler's theory of the soul: A study on epistemology. *Studies in History and Philosophy of Science Part A, 39*, 15–41.

Field, J. V. (1984). A Lutheran astrologer: Johannes Kepler. *Archive for History of Exact Sciences, 31*, 189–272.

Field, J. V. (1988). *Kepler's geometrical cosmology*. London: The Athlone Press.

Gingerich, O. (2002). Kepler then and now. *Perspectives on Science, 10*, 228–240.

Gingerich, O. (2011). Kepler's Trinitarian cosmology. *Theology and Science, 9*, 45–51.

Goldstein, B. R., & Hon, G. (2005). Kepler's move from *orbs* to *orbits*: Documenting a revolutionary scientific concept. *Perspectives on Science, 13*, 74–111.

Grafton, A. (1992). Kepler as a reader. *Journal of the History of Ideas, 53*, 561–572.

Hafenreffer, M. (1613). *Templum Ezechielis sive in IX. postrema prophetae capita: Commentarius*, Tübingen.

Heerbrand, J. (1573) *Compendium theologiae*. Tübingen.

Hofmann, N. (1982). Die Artistenfakultät an der Universität Tübingen. *Contubernium, 28*, 122–124.

Hon, G. (2014). 'Kepler's revolutionary astronomy: Theological unity as a comprehensive view of the world. In T. Demeter, K. Murphy, & C. Zittel (Eds.), *Conflicting values of inquiry: Ideologies of epistemology in early modern Europe* (pp. 155–175). Leiden: Brill.

Howell, K. (2001). *God's two books: Copernican cosmology and biblical interpretation in early modern science*. Notre Dame: University of Notre Dame Press.

Hübner, J. (1975). *Die Theologie Johannes Keplers zwischen Orthodoxie und Naturwissenschaft*. Tübingen: Mohr Siebeck.

Jardine, N. (2008). Kepler, God, and the virtues of the Copernican Hypothesis. In M. A. Granada & E. Mehl (Eds.), *Nouveau Ciel, Nouvelle terre: La révolution copernicienne dans l'Allemagne de la Réforme 1530–1630* (pp. 269–277). Paris: Belles Lettres.

Jardine, N. (1984). *The birth of history and philosophy of science: Kepler's a defence of Tycho against Ursus with essays on its provenance and significance*. Cambridge: Cambridge University Press.

Kepler, J. (1596). *Mysterium cosmographicum*. Tübingen: G. Gruppenbach. http://diglib.hab.de/drucke/40-astron/start.htm. Accessed Aug 30, 2013.

Kepler, J. (1604). *Astronomiae pars optica*. Frankfurt.

Kepler, J. (1606a). *De stella nova*. Prague.

Kepler, J. (1606b). *De Iesu Christi vero anno natalitio*. Frankfurt.

Kepler, J. (1609). *Astronomia nova aitiologêtos seu physica coelestis*. Heidelberg: E. Vogelin.

Kepler, J. (1610). *Dissertatio cum Nuncio Sidereo*. Prague.

Kepler, J. (1611a). *De nive sexangula*. Frankfurt.

Kepler, J. (1611b). *Dioptrice*. Augsburg.

Kepler, J. (1614). *De vero anno natali Christi*. Linz.

Kepler, J. (1618). *Epitomia astronomiae Copernicanae* lib. I–III, Linz.

Kepler, J. (1619). *Harmonice mundi*. Linz: Johannes Planck. http://diglib.hab.de/drucke/150-quod-2f-1/start.htm. Accessed Aug 30, 2013.

Kepler, J. (1620). *Epitomia astronomiae Copernicanae* lib. IV, Linz. http://daten.digitale-sammlungen.de/~db/bsb00008958/images/. Accessed Aug 30, 2013.

Kepler, J. (1621). *Epitomia astronomiae Copernicanae* lib. V–VII. Linz.

Kepler, J. (1627). *Tabulae Rudolphinae*. Ulm.

Kepler, J. (1938). In M. Caspar (Ed.), 1938, *Johannes Kepler Gesammelte Werke. Band I: Mysterium cosmographicum/De stella nova*. Munich: C. H. Beck.

Kepler, J. (1940). *Johannes Kepler Gesammelte Werke. Band VI: Harmonice mundi*. M. Caspar. Munich: C. H. Beck.

Kepler, J. (1945). *Johannes Kepler Gesammelte Werke. Band XIII: Briefe 1590–1599*. M. Caspar. Munich: C. H. Beck.

Kepler, J. (1949). *Johannes Kepler Gesammelte Werke. Band XIV: Briefe 1599–1603*. M. Caspar. Munich: C. H. Beck.

Kepler, J. (1955). *Johannes Kepler Gesammelte Werke. Band XVII: Briefe 1612–1620*. M. Caspar. Munich: C. H. Beck.

Kepler, J. (1959). *Johannes Kepler Gesammelte Werke. Band XVIII: Briefe 1620–1630*. W. von Dyck and M. Caspar. Munich: C. H. Beck.

Kepler, J. (1963). *Johannes Kepler Gesammelte Werke. Band VIII:Mysterium cosmographicum (editio altera cum notis)/De cometis, Hyperaspides*. F. Hammer. Munich: C. H. Beck.

Kepler, J. (1981) *The secret of the universe: Mysterium cosmographicum*. Translated from the Latin by A. M. Duncan with introduction and commentary by E. J. Aiton. Norfolk CT: Abaris Books.

Kepler, J. (1990a). *Johannes Kepler Gesammelte Werke. Band III: Astronomia nova*, M. Caspar and Kepler-Kommission. Munich: C. H. Beck.

Kepler, J. (1990b). *Johannes Kepler Gesammelte Werke. Band XII: Theologica / Hexenprozess Itus-Übersetzung, Gedichte*. V. Bialas. Munich: C. H. Beck.

Kepler, J. (1991). *Johannes Kepler Gesammelte Werke. Band VII: Epitome astronomiae Copernicanae*. M. Caspar and Kepler-Kommission. Munich: C. H. Beck.

Kepler, J. (1995). *Epitome of Copernican astronomy & Harmonies of the world*. Translated from the Latin by C. G. Wallis. New York: Amherst.

Kepler, J. (1997). *Memoirs of the American Philosophical Society, Vol. 209: Harmony of the world*. Translated from the Latin with and Introduction and Notes by E. J. Aiton, A. M. Duncan and J. V. Field. Philadelphia: American Philosophical Society.

Kepler, J. (2009). *Johannes Kepler Gesammelte Werke. Band XXI.II.II: Manuscripta astrologica/manuscrita pneumatica*, F. Boockmann and D. A. di Liscia. Munich: C. H. Beck.

Kepler, J. (2015). *New astronomy: New revised edition*. Translated from the Latin by W. H. Donahue. Santa Fe: Green Lion Press.

Kozhamthadam, J. (1994). *The discovery of Kepler's laws: The interaction of science, philosophy, and religion*. Notre Dame: University of Notre Dame Press.

Kusukawa, S. (1999). Kepler's Rudolphine Tables. http://www.hps.cam.ac.uk/starry/keplertables. html. Accessed Aug 29, 2013.

Lanzinner, M. (2003). Johannes Kepler: A man without confession in the age of confessionalization? *Central European History, 36*, 531–545.

Martens, R. (2000). *Kepler's philosophy and the new astronomy*. Princeton: Princeton University Press.

Mehl, E. (2008). Héliocentrisme et eschatologie: Citations et usages du Sol justitiae (Mal 4, 2). In M. A. Granada & E. Mehl (Eds.), *Nouveau Ciel, Nouvelle terre: La révolution copernicienne dans l'Allemagne de la Réforme 1530–1630* (pp. 355–380). Paris: Belles Lettres.

Mehl, E. (2010). La science capitale: Johann Valentin Andreae et les mathématiques. In K. von Greyerz, T. Kaufmann, K. Siebenhüner, & R. Zaugg (Eds.), *Religion und Naturwissenschaften im 16. und 17. Jahrhundert* (pp. 198–216). Gütersloh: Gütersloher Verlagshaus.

Methuen, C. (1994). Securing the reformation through education: The Duke's scholarship system of sixteenth-century Wurttemberg. *Sixteenth Century Journal, 25*, 841–851.

Methuen, C. (1996). 'Maestlin's teaching of copernicus: The evidence of his university textbook. *Isis, 87*, 230–247.

Methuen, C. (1998). *Kepler's Tübingen: Stimulus to a theological mathematics*. Aldershot: Ashgate.

Methuen, C. (2008). *Science and theology in the reformation: Studies in theological interpretation and astronomical observation in sixteenth-century Germany*. London: T and T Clark.

Methuen, C. (2010). To delineate the divinity of the Creator: The search for Platonism in late sixteenth-century Tübingen. In K. von Greyerz, T. Kaufmann, K. Siebenhüner, & R. Zaugg (Eds.), *Religion und Naturwissenschaften im 16. und 17. Jahrhundert* (pp. 186–197). Gütersloh: Gütersloher Verlagshaus.

Methuen, C. (2013). Kepler, Johannes. In W. Otten, & K. Pollmann (Eds.), *The Oxford guide to the historical reception of Augustine*. Oxford: Oxford University Press.

Proust, D. (2009). The harmony of the spheres from Pythagoras to Voyager. *Proceedings of the International Astronomical Union, 5*, 358–367.

Rosen, E. (1975). Kepler and the Lutheran attitude towards Copernicanism in the context of the struggle between science and religion. *Vistas in Astronomy, 18*, 317–338.

Rothman, A. (2011). From cosmos to confession: Kepler and the connection between astronomical and religious truth. In P. J. Boner (Ed.), *Change and continuity in early modern cosmology* (pp. 115–133). Dordrecht: Springer.

Rublack, U. (2015). *The astronomer and the witch: Johannes Kepler's fight for his mother*. Oxford: Oxford University Press.

von Samsonow, E. (1986). *Die Erzeugung des Sichtbaren: Die philosophische Begründung naturwissenschaftlicher Wahrheit bei Johannes Kepler*. Munich: W. Fink.

Scaliger, J. C. (1557). *Exotericarum exercitationum liber de subtilitate*. Lyon.

Schmidt-Biggemann, W. (2008). Der Streit um Kosmologie und Harmonie zwischen Robert Fludd und Johannes Kepler. In M. Zywietz (Ed.), *Buxtehude jenseits der Orgel* (pp. 119–150). Graz: Akademische Druck.

Schuppener, G. (1997). Kepler's relation to the Jesuits: A study of his correspondence with Paul Guldin. In *NTM Zeitschrift für Geschichte der Wissenschaften, Technik und Medizin N.S.* (Vol. 5, pp. 236–244).

Stephenson, B. (1998). *The music of the heavens: Kepler's harmonic astronomy*. Princeton: Princeton University Press.

Sutter, B. (1979). *Der Hexenprozeß gegen Katharina Kepler*. Weil der Stadt: Kepler-Gesellschaft.

Thaler, P. (2020). *Protestant resistance in counterreformation Austria*. Abingdon.

von Schlachta, A. (2015). The Austrian Lands. In H. Louthan, & G. Murdock (Eds.), *A companion to the reformation in Central Europe*. Leiden: Brill.

Chapter 2
Kepler's Cosmology

J. V. Field

To astronomers of Kepler's time, the Cosmos consisted of the planetary system and beyond it a pattern of 'fixed' stars, so called because they did not appear to move relative to one another. It was against the unchanging pattern of the fixed stars that the changing positions of planets were measured. In conventional geocentric astronomy, the fixed stars were attached to a sphere which, like everything else (apart from the Earth), was subject to the diurnal rotation of the Universe as a whole. Copernicanism changed that. But not all of it. The planetary system still seemed special and since the fixed stars showed no change in position with respect to one another in the course of the Earth's annual motion about the Sun—that is, they did not show annual parallax—Copernicans deduced that the stars must be very distant indeed. Geocentrists tended to regard this as physically ridiculous. Worse still, in geocentric astronomy the sphere of the fixed stars was considered to be finite, since Aristotle (384–322 BC) had given a proof that an infinite body could not move. However, the argument based on his proof collapsed if the stars were at rest, as they were in the Copernican system. Copernicus himself had noted this but merely commented that he left it to natural philosophers to decide whether the Universe was finite or infinite.[1]

In his book about the New Star of 1604, *On the new star in the foot of Serpentarius* (*De stella nova in pede Serpentarii*, Prague, 1606), Kepler attempted to estimate the distances of the fixed stars, trying out the hypothesis that the stars were scattered evenly in space and then the hypothesis that they were all of equal intrinsic brightness so that their apparent brightness indicated distance. But he came to no definite conclusion.[2] This uncertainty about the distribution of the fixed stars did

---

[1] Copernicus (1543), bk 1, ch. 8, esp. f6r, l. 2 ff. See also Field (1988), p. 18 and van Helden (1985).

[2] Kepler (1606); reprinted in Kepler (1938). For a fuller discussion of Kepler's argument see Field (1988), esp. ch. 2.

J. V. Field (✉)
Birkbeck, University of London, London, UK
e-mail: jv.field@hart.bbk.ac.uk

© Springer Nature B.V. 2024   25
A. E. L. Davis et al. (eds.), *Reading the Mind of God*, Springer Praxis Books,
https://doi.org/10.1007/978-94-024-2250-4_2

not, however, affect the status of the planetary system (to Copernicans, the Solar System). Thus for Kepler the planetary system is special. And it is the structure of the Solar System that he sets out to describe in his two works on cosmology: the *Mystery of the Cosmos* (*Mysterium cosmographicum*, 1596) and the *Harmony of the World* (*Harmonice mundi*, 1619).[3] Both works were very important to Kepler.

In the *Mysterium cosmographicum* Kepler puts forward the theory that the number of the planets (six) and the spacing between their paths are derived from the ratios between the radii of the inspheres and circumspheres of the five regular polyhedra: the regular tetrahedron (four triangular faces), the cube (six square faces), the regular octahedron (eight triangular faces), the regular dodecahedron (twelve pentagonal faces) and the regular icosahedron (twenty triangular faces). Alone among his works, the *Mysterium cosmographicum* was reprinted in Kepler's lifetime, and for the second edition, in 1621, Kepler let the main text stand, bringing it up to date by adding notes. These notes contain references to almost everything he had written in the interim and justify the comment that he made in his new dedicatory letter that

> almost every one of the astronomical works I have written since that time [1596] could be referred to some particular chapter of this little book, and be seen to contain either an illustration or a completion of what it contains; … .[4]

He adds that the ideas he put forward in 1596 were confirmed by the trustworthy observations of Tycho Brahe (1546–1601).

The *Harmonice mundi* takes account of what Kepler had deduced from Tycho's observations in constructing orbits of the planets.[5] It incorporates a slightly modified version of the theory put forward in the *Mysterium cosmographicum* and extends it to account for not only the number and spacing of the planetary orbs but also their thickness; that is, the eccentricities of the orbital ellipses. Here the agreement with observation is so good that it has left today's astronomers with the task of explaining how the 'musical' ratios Kepler found can be accounted for in terms of modern celestial mechanics (in which such ratios are called 'resonances').

Given the good agreement with values deduced from observation, it is not surprising that Kepler thought highly of both these works on cosmology. And it seems that some of his contemporaries agreed with him. For the twenty-first century reader, there are two obvious difficulties. First, Kepler is trying to explain things that today are not regarded as particularly important or standing in need of explanation: details such as the number of the planets and the exact shapes of their elliptical orbits. These things are now seen as accidental properties of our local planetary system, rather than as characteristics of the Universe as a whole. Second, Kepler draws his theories directly from a belief that mathematical truths are fundamental and physical truths can be deduced from them. In Kepler's time, everyone would have agreed that God is a Geometer, or that, in the words of the Book of Wisdom, God created the world in Measure, Number and Weight.[6] Kepler went further than

---

[3] Kepler (1596); reprinted in Kepler (1938); Kepler (1619), reprinted in Kepler (1940).

[4] Kepler (1621), first paragraph of Dedicatory Letter; Kepler (1963), p. 9, l. 25 ff.

[5] Kepler (1609); reprinted Kepler (1990). See Chap. 4 of this volume.

[6] *Wisdom of Solomon* 11:20.

this. In both his cosmological works he looked to Euclid's *Elements* to explain the structure of the Universe. In the first book of the *Harmonice mundi* he says not that God is a geometer, but simply that God is Geometry since Geometry is co-eternal with Him.[7] One could say that Kepler believed what everyone believed, the difference being that he really believed it and built cosmological theories upon it. In fact, he went a little further: these are not simply cosmologies in the modern sense, they represent cosmogonies. That is, they set out to describe the model according to which God created the Universe.

2.1 Mysterium cosmographicum

Kepler studied at the University of Tübingen, where he proposed to take a degree in theology. This involved studying mathematics—that is, the four mathematical sciences: arithmetic, geometry, astronomy and music.[8] Every student thus learned something of standard geocentric astronomy. Kepler's astronomy teacher, Michael Maestlin (1550–1631), also taught his more able students about the work of Nicolaus Copernicus (1473–1543), *On the revolutions of the heavenly orbs* (*De revolutionibus orbium coelestium*, 1543).[9] It seems Maestlin also pointed out (at least to Kepler) that the unsigned preface, in the original edition of 1543 and reprinted in subsequent editions, was not written by Copernicus but by the editor, Andreas Osiander (1498–1552), who saw the book through the press. This was significant because the preface suggests that the book is merely about mathematics; that is, about finding positions of planets against the fixed stars, and not proposed as representing physical reality. Copernicus certainly believed otherwise, and so did Kepler.[10]

Kepler seems to have adopted Copernicus' theory at once. To get an idea why he did so, we need to look back at the traditional geocentric model of the planetary system, whose classic description was given by Claudius Ptolemy (*fl.* AD 129–141) in a work that, through its Arabic translation, later slightly modified and transliterated into Latin, became known as the *Almagest*.

2.1.1 The Traditional Geocentric Cosmos

The Ptolemaic system had the Earth at its centre. Working outwards, the order of the orbs of the 'wandering stars' was Moon, Mercury, Venus, Sun, Mars, Jupiter, Saturn. The planets below the Sun were called the 'inferior planets', those above it, the 'superior planets'. Above the system of planets was the sphere of the fixed stars.

---

[7] Kepler (1619), book 1, prop. XLV, marginal note p. 38; Kepler (1940), p. 55; Kepler (1997), p. 74.

[8] See Chap. 1 of this volume. See also later in this chapter.

[9] Copernicus (1543); reprinted Basel, 1566.

[10] See Field (2009a).

The whole system was moved from the outside through the motion of further spheres that, like the sphere of the fixed stars, are (properly speaking) spherical shells. The planetary orbs each have an internal structure of rolling spheres and spherical shells to account for the rather complicated motions known from observations. Details of the construction of orbs were given in textbooks such as that by Georg Peurbach (1423–1461), which was first printed in the 1490s and regularly reprinted well into the seventeenth century.[11] Mathematically, these models reduced to using combinations of circles to produce the motion of each planet.[12] The outermost spheres also imposed a diurnal (24 h) rotation on the sphere of the fixed stars and all the planetary orbs.

In this system the order of the orbs, working outwards from the Earth, is that of increasing period. The fact that the Moon was the closest object was apparent from the fact that it was seen to occult other heavenly bodies. Since observations extended back about eight centuries before Ptolemy, there had been plenty of time to observe this phenomenon. The ordering of the planets about the Earth was rational, but involved an additional cosmological assumption. In this system, there is no independent way of finding the distance of any orb from the Earth. Distances are deduced from the assumption that each orb is in contact with its neighbours—a reasonable assumption if one believes the orbs are solid and the system is turned from the outside.

2.1.2 The Copernican System

Copernicus' diagram of his own heliocentric model of the Universe, both in his manuscript of *De revolutionibus* and in the printed book, looks very like a diagram of Ptolemy's system, with the difference that the Earth and Sun have exchanged places (Fig. 2.1). The Earth is now a planet with an orb, and the Moon is the only object that moves round it. Everything else moves round the Sun.

For Copernicus, the Sun is at rest, but mathematical details given later in the work show that the centre of the planetary system is the centre of the orb of the Earth, which does not exactly coincide with the Sun. Thus we have a heliostatic planetary system rather than a properly heliocentric one. However, Copernicus explicitly places the Sun at the centre of the Universe. Closer inspection shows that Copernicus has devised a heliostatic planetary system by transferring the Ptolemaic motion of the Sun to the Earth. It was Kepler who was to take the further step of referring all planetary motion to the Sun, giving a heliocentric planetary system. This step is important because it means that planetary motion is referred to a body instead of to an empty geometrical point. Kepler makes the change to heliocentrism in Chap. 15 of the *Mysterium cosmographicum*.

Copernicus' diagram of the system as a whole, showing what seem to be evenly-spaced orbs, is in the introductory first book of his work. It becomes apparent later on that Copernicus realizes that his system allows one to use observations to calculate

---

[11] See Aiton (1987).

[12] For a detailed account see, for example, Neugebauer (1975); Ptolemy (1984).

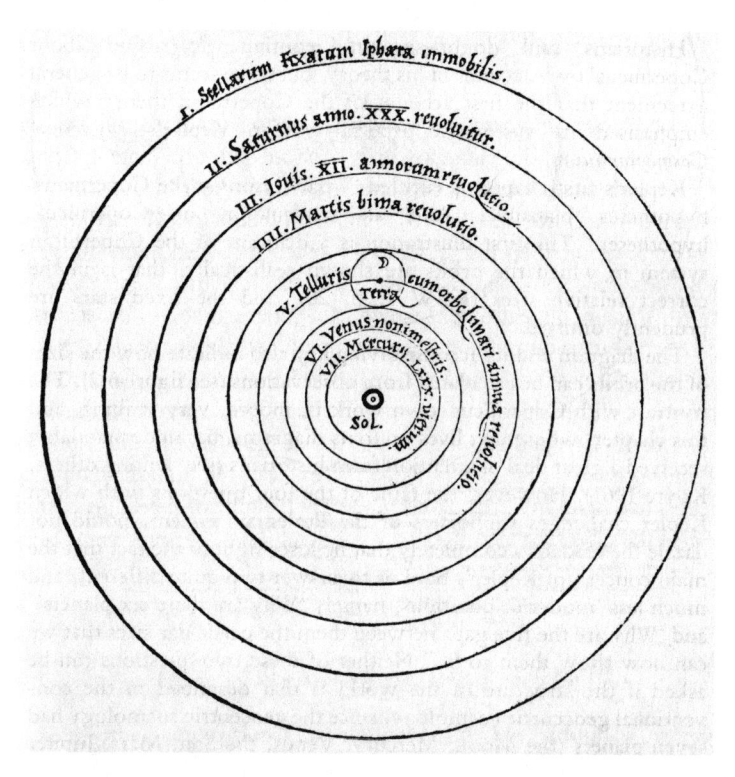

Fig. 2.1 The Copernican planetary system (Reproduced from N. Copernicus, *De revolutionibus orbium coelestium*, Nuremberg, 1543, fol. 9 *verso*)

the distance of each planet from the centre of the system, in terms of the size of what Copernicus calls 'the great orb' of the Earth; that is, the orb that defines its annual motion round the Sun. The reasoning for the 'superior' planets (those whose orbs lie outside the orb of the Earth) is that the small loops in their observed motion against the fixed stars are images of the annual motion of the Earth, so their angular size, which can be measured, shows how large the orb of the Earth would look from the distance of the planet concerned. This then gives the distance of the planet. A slightly different argument is required for the 'inferior' (inner) planets, but again their distances can be found.

The planets proved to be widely spaced, far more widely than would be required by orbs constructed merely to accommodate the combinations of rolling spheres required to construct their motion. This does not seem to have greatly troubled astronomers of the following generation, who were mainly concerned with using Copernicus' models of planetary motions to try to make predictions of the observed positions of planets.[13] The first diagram of the Copernican planetary system showing the correct relative sizes of the orbs is provided by Kepler in *Mysterium cosmographicum*.

---

[13] Westman (1975, 2011).

2.1.3 Explaining the Number and Spacing of the Orbs

The format of *Mysterium cosmographicum* is quarto. Kepler supplies a diagram to show the relative sizes of the planetary orbs as a fold-out folio plate in Chap. 14 (see Fig. 2.2). However, the book is much better known for the fold-out plate in Chap. 2, showing the planetary orbs and the five regular polyhedra that Kepler imagines lying between them to define their number and the spaces between them (Fig. 2.3).

From the diagram for Chap. 14 (Fig. 2.2) it is clear that Kepler's definition of a planetary orb is unconventional. Instead of a spherical shell that encloses all the spheres used in constructing the motion of the planet, Kepler envisages a shell that merely encloses the path traversed by the planet. So once the orbs are truly heliocentric we have a shell bounded by two spherical surfaces centred on the Sun, whose radii are the perihelion and aphelion distances of the planet in question. This definition of an orb is unambiguous in Kepler's text, but the draughtsman responsible for the decorative plate in Chap. 2 has made the mistake of including a small circle within one of the orbs (see Fig. 2.3).

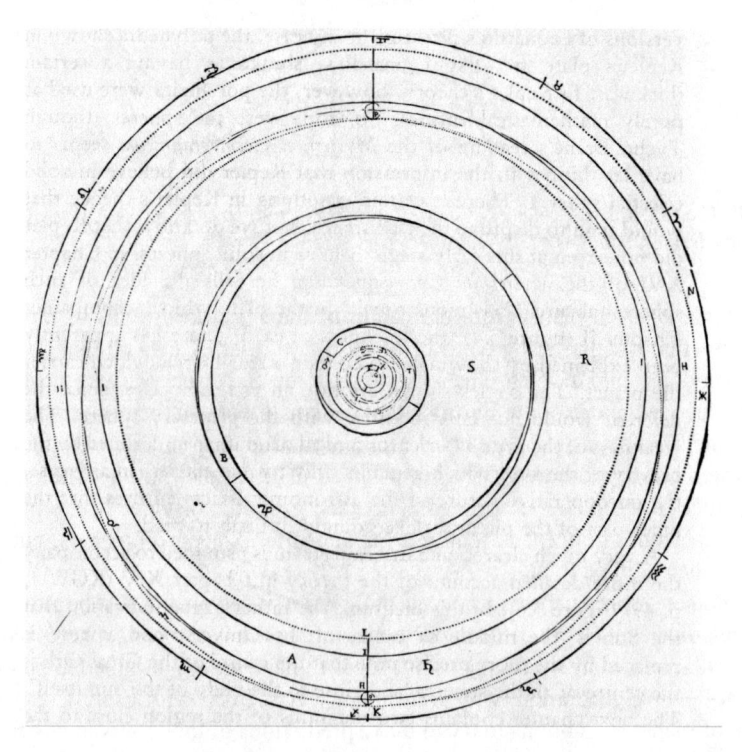

Fig. 2.2 Copernican planetary orbs (Reproduced from J. Kepler, *Mysterium cosmographicum*, Tübingen, 1596, fold-out plate in Chap. 14. See Sect. 1.6 below)

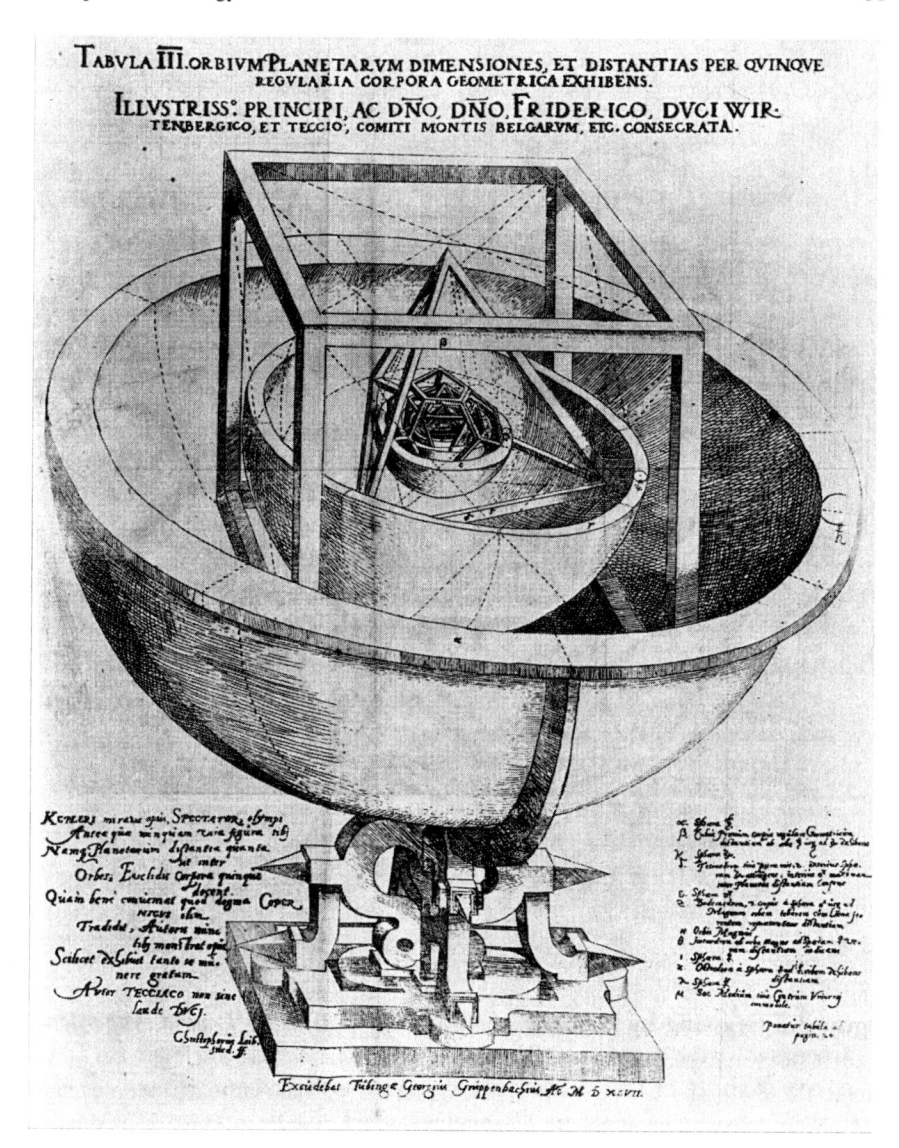

Fig. 2.3 Planetary orbs and the five regular polyhedra (J. Kepler, *Mysterium cosmographicum*, Tübingen, 1596, fold-out plate, dated 1597, in Chap. 2)

The diagram in Chap. 2 is also misleading in another respect: the orbs and the polyhedra separating them look decidedly material. When Kepler sent a copy of the work to Tycho Brahe, Tycho's letter of thanks says that he rejects planetary orbs:

> I consider that all reality of orbs, whatever they are taken to be, should be eliminated from the heavens.[14]

He goes on to say that comets have no orbs. Beside these lines Kepler has written in the margin 'This is allowable (*licet*) for me and for my book'.[15] Kepler has clearly taken 'reality' to refer to material orbs—whereas Tycho may have meant to be more sweeping in his condemnation. In any case, in Chap. 16 of *Mysterium cosmographicum* Kepler refers to the suggestion that solid orbs exist in the heavens as 'absurd and monstrous'[16] And a little later he is straightforwardly down-to-earth about them, dismissing solid orbs with

> As for this Earth, which we, with Copernicus, take to be in motion, what bars, what chains fix it into its adamantine celestial orb?[17]

answering himself that the planet is entirely surrounded by air.

Nevertheless, Kepler does adopt orbs in the sense in which we see them defined in the figure in Chap. 14 (Fig. 2.2): the orb is the region that contains the path of the planet. In *Harmonice mundi* he gives a striking image for this

> ... and thus it comes about gradually by the linking and accumulation of a great many revolutions that a kind of concave sphere is displayed, having the same center as the Sun, just as by a great many circles of silken thread, linked with each other and wound together, the dwelling of a silkworm is made.[18]

Maybe Kepler had watched a silkworm spinning its cocoon. As the thread is very fine the process takes a considerable time and at first the body of the cocoon remains transparent, allowing one to watch the motion of the animal inside it.

However, conceived, the orbs in the Copernican planetary system differ from those in the Ptolemaic one by seeming to have wide spaces between them (thereby raising questions about why or how they move) and the number of planets is different. In the Ptolemaic system, seven bodies moved round the central Earth, in the Copernican system there are six planets, the Moon having become a subsidiary body that moves round the Earth and travels with it. So a defender of geocentric astronomy might ask a Copernican to explain both the number and the spacing of the orbs in the new system.

Kepler's answer is that there are six planets because there are exactly five regular polyhedra; that is, polyhedra whose faces are all regular polygons of the same kind: the tetrahedron (whose faces are four triangles), the cube (six squares), the octahedron

---

[14] Tycho Brahe to Kepler, 1 April 1598, Kepler (1945), letter 92, lines 61–63, pp. 198–99.

[15] Kepler (1945), p. 201.

[16] Kepler (1596), ch. 16, p. 55; Kepler (1938), p. 56, lines 17–19.

[17] Kepler (1596), ch. 16, p. 55; Kepler (1938), p. 56, lines 28–31.

[18] Kepler (1619), book 5, ch. 9, IV. Axiom, p. 216; Kepler (1940), p. 332; Kepler (1997), p. 453.

(eight triangles), the dodecahedron (twelve pentagons) and the regular icosahedron (twenty triangles). The fact that there are exactly five solids with these properties is proved by Euclid in the final proposition of *Elements* book 13. The five regular solids are described by Plato (427/423–348 BC) in his dialogue *Timaeus*, where he associates them with the 'elements': earth (cube), water (icosahedron), air (octahedron), fire (tetrahedron) and aether, the material of the heavens (dodecahedron). The reference to them in the *Timaeus* led to the bodies sometimes being known as the 'Platonic solids'.

Each of the five solids can be inscribed in a sphere; that is, a sphere can be drawn to pass through all the vertices of the solid; and a sphere can be inscribed within each solid to touch each of the faces of the solid at the centre of the face. For each solid, the ratio of the diameters of these two spheres is fixed. Thus the five solids give five ratios. And, as can be seen in Fig. 2.3, these are the ratios Kepler finds between the diameters of successive planetary orbs; that is, between the inner diameter of one orb and the outer diameter of the one immediately inside it. In fact the solids give rise to only three different ratios, so Kepler needs to give arguments for the placing of particular solids. But once this is done he has, to his mind, given an explanation for the number of the planets and the sizes of the spaces between their orbs.

Disconcertingly (to a twenty-first century reader) the agreement with values deduced from observation is rather good, as can be seen in Table 2.1.[19] Until the last decades of the twentieth century, most cosmologists would have been delighted to find their theories fitted the observations as well as this. Like his successors, Kepler hoped that better observations would improve the agreement, a hope that, with some help from Maestlin, took him to Prague to work with Tycho Brahe, one of the foremost observational astronomers of the day.

The good agreement with values derived from observation, which is the aspect of the theory that Kepler was at pains to set out when he first told Maestlin about the theory,[20] and which it turned out was not much changed by using Tycho's observations, explains why Kepler continued to believe his theory was basically sound. The mathematical form of Kepler's theory springs from his conviction that God expressed His own mathematical nature in the Universe He created. There is an indissoluble fusion between the Biblical story of Creation from nothing and the story in Plato's *Timaeus*, where the world is made out of pre-existing matter using mathematical ideas as models. Kepler has given us the earliest truly mathematical cosmology, but its philosophical roots lie in ancient texts. The immediate rationale is set out in Kepler's Preface to *Mysterium cosmographicum*, where he explains how he came to formulate his theory.

---

[19] For a more detailed discussion see Field (1982); see also Field (1988).

[20] Kepler to Maestlin, 4 September 1595; Kepler (1945), letter 22, pp. 32–33.

Table 2.1 Agreement of observed and theoretical values for dimensions of planetary orbs. Kepler works outwards from the aphelion of the Earth and then inwards from its perihelion. The thicknesses of the orbs are taken from observed values and are not given by the theory. The 'observed' values are from Copernicus, but the orbs have been made truly heliocentric

Planet		Observed distance	Theoretical No Moon	Theoretical With Moon	Error % obs no Moon	Error % obs with Moon
Saturn	aph	9.987	10.599	11.304	+ 6	+ 13
	peri	8.342	8.852	9.441		
Jupiter	aph	5.492	5.111	5.431	− 7	− 1
	peri	4.999	4.652	4.951		
Mars	aph	1.649	1.551	1.658	− 6	0
	peri	1.383	1.311	1.398		
Earth	aph	1.042	1.042	1.102	0, by defn	0, by defn
	peri	0.958	0.958	0.958		
Venus	aph	0.741	0.761	0.714	+ 3	− 4
	peri	0.696	0.715	0.671		
Mercury	aph	0.489	0.506	0.474	+ 4	− 3
	peri	0.233	0.233	0.279		

2.1.4 Rationale and Discovery

Kepler tells us what set off the train of thought that led to his theory, and the date on which this happened. He was teaching astronomy and on 19 July 1595 was drawing a diagram to show his students that successive 'great conjunctions', that is conjunctions of Jupiter and Saturn, take place in positions in the zodiac that are separated by almost 120°. So if one draws lines from one to the next, three successive great conjunctions give an almost exact equilateral triangle in the circle of the zodiac. If we consider not the exact positions of the conjunctions but merely the zodiac signs in which they occur then we have a perfect triangle. Signs lying at the vertices of an equilateral triangles were said to form a 'triplicity' and the four triplicities were named after the four terrestrial elements. Since the angles between the conjunctions were not exactly 120°, the sets of conjunctions eventually progressed to another triplicity. The conjunctions that started on the new triplicity were considered particularly significant. One such was the great conjunction of 1604 which took place at the beginning of the 'Fiery Trigon' (the triplicity of the 'fiery' signs Aries, Leo and Sagittarius) and close, in time and position, to the New Star of that year (now known as 'Kepler's supernova').[21] Accordingly, Kepler's book about the New Star—*On the new star in the foot of Serpentarius* (*De stella nova in pede Serpentarii*, 1606)[22]—includes a diagram showing the movement of great conjunctions (Fig. 2.4), which is easier

---

[21] On the New Star, see also Field, 'Kepler and Galileo' in this volume (Chap. 8).

[22] Kepler (1606); Kepler (1938), pp. 146–356.

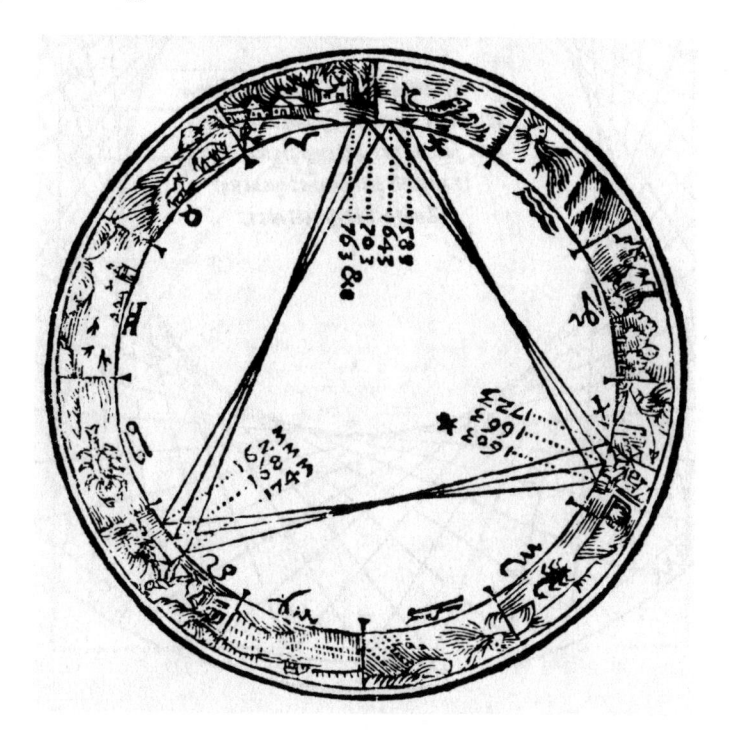

Fig. 2.4 The changing positions of great conjunctions in the zodiac (Reproduced from J. Kepler, *De stella nova*, Prague, 1606, ch. 6, p. 25)

to read than the one in *Mysterium cosmographicum*, though the latter shows more clearly what it was that caught Kepler's attention: that the quasi-triangles define a smaller circle within the circle representing the zodiac (Fig. 2.5).

The ratio of the radii of the outer and inner circles is determinate. Geometrical problems of inscription and circumscription of circles and polygons are discussed by Euclid in book 4 of the *Elements*, so Kepler was on well-trodden ground in proceeding to investigate the ratios that could be generated by the successive inscription of polygons and circles. There was, however, a serious objection: the process of inscription could be continued indefinitely. Thus Kepler could give no reason 'why there should be six moving orbs rather than twenty or a hundred'.[23] Since there were only six planets, he needed to choose five figures and could not decide what criteria to employ. Then he thought again:

> Why should there be plane figures between solid [i.e. three-dimensional] orbs? Solid bodies would be more appropriate. Here, Reader, is the discovery (*inventum*) and the matter of all of this little work. For if anyone who has some slight experience in Geometry is told about this in so many words, it will at once make him think of the five regular bodies with the ratios between their circumscribed and inscribed spheres: he will immediately see before his eyes the scholium to Euclid's proposition 18 of book 13.[24]

---

[23] Kepler (1596), Preface, p. 8; Kepler (1938), p. 12.

[24] Kepler (1596), Preface, p. 8; Kepler (1938), p. 13.

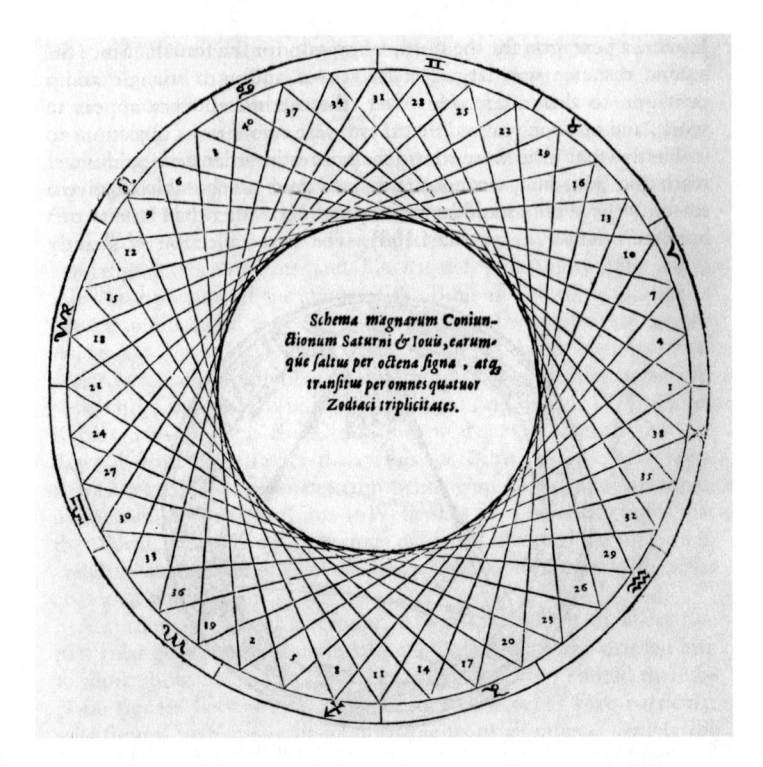

Fig. 2.5 Great conjunctions. The text in the central circle says 'Diagram of the great conjunctions of Saturn and Jupiter, and their leaps through eight signs, and transitions through all four triplicities of the zodiac' (Reproduced from J. Kepler, *Mysterium cosmographicum*, Tübingen, 1596, Preface, p. 8)

The scholium in question considers fitting regular polygons together round a vertex to prove that there are five, and only five, regular solids.

Although the details are different, it is clear that Kepler, like Plato in *Timaeus*, believes that the properties of mathematical entities can be seen as determining properties of the physical world built into it by the Creator. In *Mysterium cosmographicum* Kepler seems to take it for granted that his readers will recognize that there is ancient Greek authority for this approach. In *Harmonice mundi*, published about twenty years later, he goes back to first principles and gives a much fuller mathematical justification of his model.

2.1.5 Reasons for Adopting the Copernican System

As we have seen, by his own account, in 1595 Kepler was looking for an explanation for two specific characteristics of the Copernican model of the planetary system. There is, however, nothing in *Mysterium cosmographicum,* or in any of Kepler's surviving correspondence, to suggest that his success in finding such an explanation played any part in persuading him that the Universe, and the planetary system, were heliocentric. *Mysterium cosmographicum* is obviously conceived as supporting Copernicus' theory, though at this time, when many astronomers and natural philosophers rejected the theory, Kepler apparently sees his support for it as defensive. In 1598, when writing to a friend at the ducal court in Munich, Herwart von Hohenburg (1553–1622), Kepler notes that many people are offended by the apparent absurdity of Copernicus' theory and adds that for him

> It is enough of an honour that while Copernicus officiates at the high altar I am able to guard the door of the temple with my discovery (*inventione*).[25]

Nevertheless, *Mysterium cosmographicum* is clearly also intended to show that Copernicus' theory should be taken seriously. To say the least, it leaves the reader in no doubt about the author's beliefs.[26] The first chapter, which has the rather circumspect title 'The reasoning that accords with the Copernican hypotheses. And an exposition of Copernicus' hypotheses',[27] in fact makes out a case for the Copernican system. The 'reasoning' includes deducing the sizes of the planetary orbs from observations. Kepler considers that the possibility of carrying out this calculation constitutes an advantage of the Copernican system over the Ptolemaic one. But his first appeal is to the superior explanatory power of the Copernican planetary system. He proposes four questions that can be answered in the Copernican system but not in the Ptolemaic one. The passage is too long to cite in full but the questions can be summarized as follows:

1. Why do the largest of the circles used to construct the motions of the Sun, Venus and Mercury all have the same period, namely one year? The Copernican answer: Apparent annual motion is due to the motion of the Earth.
2. Why do the five planets Saturn, Jupiter, Mars, Venus and Mercury sometimes have a retrograde motion whereas the luminaries (the Sun and the Moon) do not? The Copernican answer: The observed retrograde motions of the planets are due to the motion of the Earth, which is added to their own motion, whereas for the Sun we see only the motion of the Earth, and for the Moon we do not see the motion of the Earth at all.
3. Why do the sizes of the circles used to construct the motions have the relations they do? The Copernican answer: The relative sizes of the various circles depend on the sizes of the orbs and on the distances of the planets from the Earth.
4. Why are the superior planets, Saturn, Jupiter and Mars, sometimes seen at opposition to the Sun (that is 180° from it), whereas the inferior ones, Venus and Mercury, are not? The Copernican answer: Venus and Mercury lie between the Earth and the Sun.[28]

---

[25] Kepler to Herwart von Hohenburg, 26 March 1598; Kepler (1945), letter 91, p. 193, lines 191–92.

[26] As Galileo Galilei recognized immediately, see Field, 'Kepler and Galileo' in this volume (Chap. 8).

[27] Kepler (1596), p. 11; Kepler (1938), p. 14.

[28] Kepler (1596), ch. 1, pp. 15–16; Kepler (1938), p. 18, l. 5–p. 19, l. 8; English translation Kepler (1981), p. 81.

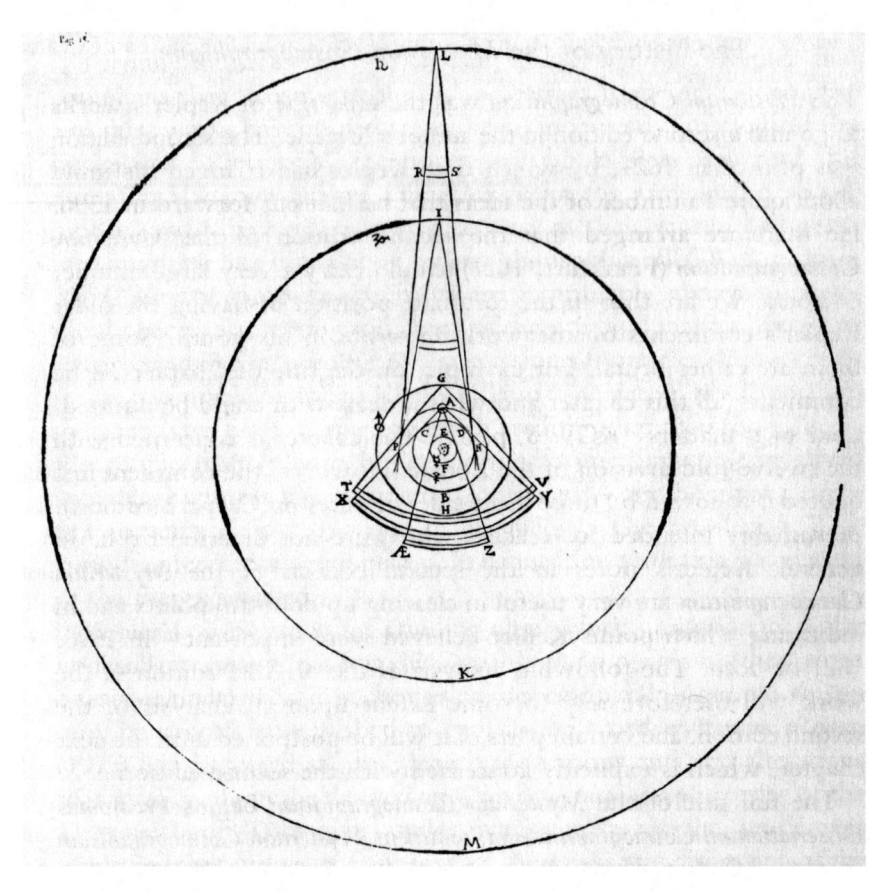

Fig. 2.6 The Copernican planetary system showing the angles subtended at each planet by the orb of the Earth (Reproduced from J. Kepler, *Mysterium cosmographicum*, Tübingen, 1596, Figure I, fold-out plate in ch. 1)

The questions are followed by the fold-out plates to which Kepler has referred the reader for some of the answers (see Fig. 2.6).

These four questions seem to be the best clues we have as to why Kepler adopted the Copernican system. To a twenty-first century reader their appeal to the superior explanatory power of heliocentrism may seem reassuringly 'modern', but it is by no means clear whether contemporary readers saw them as persuasive. The explanation of the relative sizes of the spaces between the planetary orbs by the five Platonic solids, which Kepler goes on to describe in his next chapter, and which is the main subject of the book, seems to have met with some approval from Kepler's contemporaries, since a publisher saw fit to bring out a second edition of the work in 1621, the only change made in this edition being, as we have already commented, the addition of notes.[29] In any case, the four questions make no further appearance. In Chaps. 3–8,

---

[29] Kepler (1621); reprinted in Kepler (1963).

Kepler gives arguments to show that his theory is mathematically coherent; that is, that there is a rationale to the placing of particular polyhedra between particular pairs of orbs. These chapters connect each planet with a regular polyhedron in rather the same way that in *Timaeus* Plato connected each solid with an 'element'.

The chapters accomplish Kepler's purpose of using his theory to explain the number of the planetary orbs and the distances between them. But in the first sentence of his Preface Kepler had announced that his theory would also explain the motions of the planets,[30] and it is from these later chapters in particular, the ones that as it were show the loose ends of the theory, that we can see Kepler's later work as developing—not only in cosmology but also in astronomy.

2.1.6 Astrology, Numbers, Music and Astronomy

The first subject Kepler tackles, in Chap. 9, is astrology. He uses the association of particular planets with particular polyhedra to provide an astrological character for each planet. The associations are Saturn-cube, Jupiter-tetrahedron, Mars-dodecahedron, Venus-icosahedron, Mercury-octahedron.[31] That is, each planet is seen as connected with the first polyhedron lying between its orb and that of the Earth. Astrology deals with the influence of heavenly bodies on the Earth, so some degree of geocentricity is inevitable. These associations with particular polyhedra are then seen as establishing affinities between the planets in their astrological effects. The first note that Kepler adds to this chapter in the 1621 edition of *Mysterium cosmographicum* says the chapter should be thought of as a digression.[32]

In Chap. 10 he again turns his attention to a subject on which he was to change his mind in the years that followed: the origins of the numbers that seem to play a special part in the structure of the Universe. In 1596, this chapter, entitled 'On the origin of the noble numbers', covers only a single page. In it, Kepler takes it as well known that certain numbers are of cosmological significance—that is they formed part of God's model in creating the Universe—and shows how these numbers are connected with the regular polyhedra, as the numbers of faces, sides or vertices of the solids. In the second edition (1621) this chapter has a note to say that geometry is prior to arithmetic, with a reference to *Harmonice mundi* (1619) for a more appropriate treatment.[33]

There follows a short chapter relating the regular polyhedra to the origins of the division of the zodiac into twelve signs.

---

[30] Kepler (1596), p. 6; Kepler (1938), p. 9; Kepler (1981), pp. 40–41.

[31] On the characters of the planets see Field (1984a) and Rabin in Chap. 6 of this volume.

[32] Kepler (1621), Chap. 9, unnumbered note, p. 35; Kepler (1963), p. 59; Kepler (1981), p. 119.

[33] Kepler (1621), Chap. 10, note 1 (referring to the title of the chapter), p. 36; Kepler (1963), p. 60; Kepler (1997), p. 121.

The next chapter, Chap. 12, on astrological aspects is rather more substantial. When heavenly bodies were 'at aspect' with one another—that is when their separation on the zodiac was a specified fraction of the complete circle, such as a third or a quarter—then the combined effect of the bodies was considered to be enhanced or diminished (depending on the angle of separation). It is easy to see how this belief would be sustained by observations of the effect of the Moon in causing tides, which varies noticeably with the phase of the Moon, that is with its angular distance from the Sun. Astrologers sometimes considered not the exact angular separation of the bodies but the separation of the signs in which they were located. Also, there was some disagreement over the relative power of various aspects. However, it was accepted that aspects were properties of the World (what in the twenty-first century might be called 'laws of nature'), so it is reasonable for Kepler to try to explain them by means of the regular solids, or even by the polygons that form their faces. Like many of his predecessors, he sees aspects as defined by drawing a diameter, a square, an equilateral triangle and an equilateral pentagon in a circle representing the zodiac. He also proposes to add two further aspects, defined by inscribing a star pentagon and a star octagon. However, before making this connection with regular polygons, he notes that the divisions aspects mark out on the zodiac are like those made in a string to form musical consonances (that is to give two parts whose notes form a consonant chord). This connection dates back to ancient times and is found in a number of Renaissance treatises on music. As Kepler points out in his notes on this passage in the 1621 edition of *Mysterium cosmographicum*, the idea points forward to the much fuller treatment of music in book 3 of *Harmonice mundi* of 1619.[34] In the 1621 edition of *Mysterium cosmographicum*, the notes added to this chapter are considerably longer than the chapter itself. As we shall see, this was a subject Kepler continued to think about, and the ideas he put forward in 1596 later seemed incoherent.

The remaining chapters of *Mysterium cosmographicum* are concerned with astronomy, sometimes directly in connection with the agreement between Kepler's theory and the sizes of the orbs deduced from observations, sometimes with discussion of technical details of the paths of particular planets. The final chapter, Chap. 23 'On the astronomical beginning and end of the Universe and the Platonic year', concerns Plato's 'Great Year', the time after which the planets return to their original configuration.

Three of the astronomical chapters are of particular interest in the present context: Chaps. 14, 15 and 20. Chapter 14, which looks at the agreement between Kepler's theory and the sizes of the orbs as found by Copernicus (illustrated by the fold-out plate shown in Fig. 2.2) makes only rudimentary use of the Copernican apparatus that is presented by Maestlin as a supplement to Kepler's calculations. Maestlin had adapted one of his long letters in reply to Kepler's technical queries to form an appendix describing the apparatus of circles that Copernicus used to construct the apparent motions of the planets. This was no doubt helpful to Kepler's contemporary readers, but (with hindsight) the fact that it appears merely as an appendix reinforces

---

[34] Kepler (1621), Chap. 12, notes 8 to 14, p.46; Kepler (1963), pp. 72–73; Kepler (1981), p. 141.

the impression that, while Kepler was interested in the paths themselves, he was much less interested in the apparatus used to construct them, perhaps even a little impatient with it. His preoccupation with the actual path was to be a significant component in his detailed work on the motion of Mars.

Moreover, Kepler is slightly inconsistent: the orb of the Earth is centred on the Sun whereas the other planetary orbs are centred on the centre of Copernicus' orb of the Earth.[35] In the following chapter, Chap. 15, Kepler repeats the comparison with all the orbs made heliocentric (see Table 2.1). However, the tables printed in both editions of the *Mysterium cosmographicum* include errors—that is, they show differences between values predicted by theory and those deduced from observations. The following chapter considers the Moon and the material of the orbs and planets; it is here that Kepler explicitly rejects solid orbs (see Sect. 2.1.2 above). Chapter 17 briefly discusses Mercury, which presented particular problems, because its closeness to the Sun made it difficult to observe. The next two chapters, 18 and 19, return to examining the discrepancies between Kepler's theory and values deduced from observations. In his cosmology as in his astronomy, Kepler was always much concerned with agreement between theory and observation.

Kepler's later astronomical work was to provide much more exact dimensions of the planetary orbs. It seems likely that the reason why Kepler left his tables unchanged in the second edition of *Mysterium cosmographicum* was that by 1621 he had decided the theory was only approximate, as he had by then explained in *Harmonice mundi* (1619).[36] In any case, Kepler's definition of a planetary orb, in Chaps. 14 and 15 of *Mysterium cosmographicum*, shows his concern with the actual path of the planet and his making the orbs truly heliocentric points forward to his later explicitly relating the paths of planets to the Sun. Both of these attitudes represent departures from the details of Copernicus' work and are crucial to the derivation of Kepler's first two laws of planetary motion, published in 1609.[37]

Thus Chaps. 14 and 15 show the beginnings of the path that would lead Kepler to his first two laws. Chapter 20 provides our first indication of the concern with relating periods to distances that was to lead Kepler to his third law, discovered in 1618 and published in *Harmonice mundi* the following year.

Thus Chap. 20 has significant ramifications in Kepler's further work on the overall structure of the planetary system. The title of the chapter promises an investigation of the proportions between the distances of the orbs and their motions. In 1620, Kepler notes that this is the question he addresses at length in book IV of his *Epitome of Copernican astronomy* (Linz, 1621) and later incorporated into Chap. 3 of book 5 of his *Harmonice mundi* (Linz, 1619). This remark serves as a reminder to would-be historians that the order of publication of books does not always reflect the order in which the authors carried out the work they describe. Moreover, as befits the first publication of a result, the fourth book of the *Epitome of Copernican astronomy* not only states the law but puts it in a physical context, proposing possible explanations

---

[35] Aiton (1977); Kepler (1997).

[36] For a more detailed discussion, see Field (1988), esp. Chap. 3.

[37] Kepler (1609); see Davis in Chap. 4 of this volume.

for it, in terms of the planets further from the Sun being more massive than those closer to it. Kepler suggests their weights could be those of various metals.[38] In contrast, what was intended to be the second publication of the third law, in the context of mathematical astronomy in the last book of *Harmonice mundi* (though by the time he wrote it he must have known it would appear first) presents the relationship simply as one mathematical rule among many (see below).

As we have seen, Chap. 12 of *Mysterium cosmographicum*, 'Division of the zodiac and Aspects', contains references to music, and the notes added in 1621 refer the reader to the third book of *Harmonice mundi*, whose subject is theory of music. Kepler's correspondence shows that his dissatisfaction with the content of Chap. 12 started almost immediately after his book was published. At this time, his correspondents happened to include Galileo Galilei (1564–1642), who wrote to thank Kepler for a copy of *Mysterium cosmographicum*, received through a friend of Kepler's who was travelling to Italy and seems to have been asked to give copies to people teaching mathematics in universities there.[39] Those with a taste for the might-have-been may like to wonder what would have happened if Kepler had realized he was in correspondence with the son of the famous music theorist Vincenzo Galilei (*c*. 1520–1591)—whose work was to play an important part in *Harmonice mundi* book 3.

In the years following the publication of *Mysterium cosmographicum*, Kepler's letters frequently mention music and its relationships with arithmetic, geometry and astronomy. For example a long letter to one of his regular correspondents, Herwart von Hohenburg, dated 6 August 1599, contains a discussion of musical harmonies, with examples written out in formal musical notation, together with some of the mathematics and mathematical explanations, involving polygons individually and in tessellations, that later appears in the first three books of *Harmonice mundi*.[40] Thus we do not have to rely only on Kepler's comments of 1621 to tell us that he was not satisfied with his treatment of aspects, polygons and musical harmonies in Chap. 12 of *Mysterium cosmographicum*. But Kepler's letters also make it clear that he believed the approach he had adopted was valid. His later cosmological work again employs mathematical properties of polygons to deduce the musical, astrological and astronomical properties of bodies in the physical Universe. The process of deduction is set out clearly and at some length in the five books of his *Harmony of the World* (*Harmonices mundi libri V*, Linz, 1619).[41]

---

[38] Kepler (1620), part 1, pp. 488–90; Kepler (1991), p. 284.

[39] See Chap. 8 of this volume.

[40] Kepler to Herwart von Hohenburg, 6 August 1599; Kepler (1949), letter 130, pp. 21–41; see esp. p, 27, lines 220 et seqq of letter.

[41] Kepler (1619, 1940); English translation Kepler (1997).

2.2 Harmonice mundi

His correspondence shows that from the first Kepler had it in mind to write another work on cosmology to develop the mathematical model he had put forward in *Mysterium cosmographicum*. But he could not set about the task immediately. The most obvious impediment was that he was being paid to do other things with his time, carrying out calculations as Tycho's assistant and, after Tycho's death in 1601, as his successor as Imperial Mathematician. In the latter capacity Kepler's principal task was to produce a set of astronomical tables of outstanding excellence, to confer immortality on the name of his employer the Holy Roman Emperor Rudolf II (1552–1612, reigned from 1576), as the *Alphonsine Tables* had done for the name of Alfonso X King of Castile and Leon (1221–1284, reigned from 1252). For these tables, which Kepler completed in 1624—they were eventually published as the *Rudolphine Tables* (Ulm, 1627)—Kepler used Tycho's observations of planetary positions to calculate the paths of the planets in space, their orbits, which provided the basis for the further development of his cosmological theory. In 1609 Kepler introduced the term 'orbit' (*orbita*) to astronomy, taking it from the Latin term for the orbit of the eye.[42]

The orbits derived from Tycho's observations were of unexampled accuracy, as (it turned out) were the tables Kepler calculated from them. The enduring reliability of these tables did much to establish the credibility of the laws of planetary motion that Kepler had deduced from Tycho's observations (laws he had used in the calculation of the tables) and gradually increased the credibility of Kepler's heliocentric modal of the planetary system.

2.2.1 The Structure of Harmonice mundi

Harmonice mundi is an extremely orderly work, set out in definitions and axioms (some of which are not quite axioms in today's sense of the word), followed by long series of theorems each depending on its predecessors. That is, the modal for its style is Euclid's *Elements*. The contents of the five books of Kepler's work look as if they were intended to demonstrate an underlying unity in the four mathematical sciences of the university curriculum, that is the four sciences of the quadrivium (arithmetic, geometry, music and astronomy), though Kepler's ordering of them, designed to show causal relationships, is slightly different from the conventional one. Like Plato, but in a more radical style, Kepler regards the fundamental truths, and the fundamental entities, as being those of geometry. So geometry forms the subject matter of the first two books, book 1 being concerned with the construction of regular polygons in a circle, book 2 with grouping regular polygons round a vertex to form either an infinite pattern in the plane (a tessellation) or a closed body in space (a polyhedron).

Considering these properties leads to two slightly different hierarchies of regular polygons. The first hierarchy, from book 1, is used in book 3 to provide a basis for the

---

[42] See Davis's description in Chap. 4 of this volume.

theory of music. The second, from book 2, is used in book 4 to explain the number and characteristics of astrological aspects. In book 5, in a condensed recap of the theory put forward in the *Mysterium cosmographicum*, the number of the planets and the spacing between planetary orbs (as before conceived as spherical shells containing the path of the planet) is established from the five regular polyhedra, which were constructed in book 2. The musical theory of book 3 is then used to establish a model of the planetary system in which all the notes of musical scales are found in the ratios between speeds of planets at extreme points of their orbits. Kepler's third law, the law relating periods of planets to their distances from the Sun, appears near the end of book 5.

2.2.2 Book 1: Constructing Regular Polygons in a Circle

In book 4 of the *Elements* Euclid shows how to construct regular polygons with circumscribed and inscribed circles. The propositions come in sets of four. For example, for the square we have: to inscribe a square in a circle, to circumscribe a square about a circle, then a circle in a square and a square about a circle. There is no discursive text. Euclid simply proceeds as far as the regular pentagon, adds an equilateral triangle that shares a vertex with the pentagon and thus derives vertices of a regular 15-gon. This concludes book 4. Earlier constructions have shown that bisection, that is drawing perpendicular bisectors of sides, can produce higher polygons, for instance a regular dodecagon from a hexagon, so Euclid's regular polygons do not simply stop at some particular number. But there are gaps in the sequence and later geometers tried to fill them by providing straightedge and compasses constructions for, say, a regular heptagon or enneagon. Only some of these constructions were explicitly proposed as approximate.[43] Euclid's constructions are of course exact. Kepler is in search of exact constructions, but he is aware of some of the approximate ones that were on offer, for instance the construction for the regular heptagon provided by Albrecht Dürer (1471–1528) in his *Treatise on Measurement* (1525).[44]

Kepler's original readers would have known Euclid. Kepler's purpose is to establish which polygons can be inscribed by 'geometrical means' (that is, using straightedge and compasses) the means employed by Euclid. Only inscribable polygons will play a part in the cosmological model, since only they can be truly known. This kind of 'knowledge' is the subject of a definition.[45] After dealing with the regular polygons considered by Euclid, Kepler gives a proof of the impossibility of knowing the sides of some other regular polygons. The statement of the proposition is

XLV. Proposition

---

[43] See annotations to book 1 in Kepler (1997).

[44] Dürer (1525). See Kepler (1619), bk 1, prop. 45, pp. 32–39 esp. p. 38; Kepler (1940), pp. 47–56, esp. p. 55; Kepler (1997), pp. 60–79 esp. p. 75 and note 243.

[45] Kepler (1619), bk 1, def. 7, p. 8; Kepler (1940), p. 21; Kepler (1997), pp. 18–19.

The Heptagon and all figures the number of whose sides are Primes (so-called), and their stars, and the complete classes [of figures] derived from them, have no Geometrical description independent of the circle: in the circle, although the quantity of the side is determinate, it is equally impossible to evaluate.[46]

This is the earliest example I know of proof of impossibility, and we should note immediately that in proposing it Kepler has a cosmological ulterior motive: just as he chose five solids to define the spacing of the six planetary orbs in *Mysterium cosmographicum*, in *Harmonice mundi* he is employing mathematical properties to select polygons that will play a part in determining the various forms of harmony to be found in the Universe.

The mathematical result Kepler is trying to prove is in fact true, and a modified form of his theorem was eventually proved by Carl Friedrich Gauss (1777–1855) in his *Disquisitiones arithmeticae* (1801). Euclid's 'geometrical means' will solve only problems that lead to linear or quadratic equations, whereas constructing the heptagon leads to a cubic. Kepler's geometrical investigation is too long and intricate to describe here. However, it is a matter of some historical interest that Kepler eventually, anticipating Gauss, turns to algebra. Algebra, which was not taught in university courses of mathematics, was recognized as having been invented by Islamic mathematicians and in Kepler's time was generally regarded as a higher form of arithmetic.[47] Kepler says he learned about algebra (*cossa*) from the court clock and instrument maker Jost Bürgi (1552–1632). He duly obtains what we would now regard as the correct answer. But he regards it as unsatisfactory because it is expressed as a number. The numerical value can be found to any assigned degree of accuracy, but not with absolute accuracy as when a magnitude is constructed geometrically.[48] To put it in today's terms: arithmetic deals with discrete quantities and Kepler is looking for continuous ones.

Kepler goes on to classify regular polygons according to the number of steps required to construct their side from the radius of the circle in which the figure is inscribed. This corresponds closely with Euclid's classification of magnitudes in *Elements* book 10, the book that had long been regarded as the most difficult part of Euclid's work and for which, in Renaissance editions, even that by the very learned humanist translator Federico Commandino (1509–1575), editorial explanations all have recourse to algebra. Except in the proof of impossibility just mentioned, Kepler uses only geometry, and gives many references to propositions in *Elements* book 10. The rank assigned to a polygon in his concluding classification determines the importance of the figure in the cosmological model.

As this cosmological model is also cosmogonic, that is it is seen as governing the process of Divine Creation, a theologian might be tempted to suggest Kepler is

---

[46] Kepler (1619), bk 1, p. 32; Kepler (1940), p. 47; Kepler (1997), p. 60.

[47] See al-Khwarizmi (2009).

[48] For a fuller account, see Field (1994). Unfortunately, near the end of this paper I claimed it was impossible for Kepler to find areas in an elliptical orbit. That is untrue; I misunderstood what I was told by Dr Davis. See Davis in Chap. 4 of this volume. A more general overview of Kepler's mathematics is given by Knobloch in Chap. 11 of this volume.

imposing restrictions on the power of the Creator. Kepler has thought of this and confronts it explicitly when he first rejects a regular polygon because its side cannot be constructed in the circle by 'geometrical means'. The polygon in question is the regular heptagon, but the defence of his criterion is general. It appears in a very long marginal note, part of which reads:

> these formal ratios of geometrical entities are nothing else but the Essence of God and geometrical entities, because whatever in God is eternal, that thing is one inseparable divine essence.[49]

We are being told not that God is a Geometer, but that God is Geometry.

Book 1 ends with a summary entitled 'Comparison of the Figures or divisions of the circle'. This sets out a hierarchy of regular polygons. First we have the diameter, second the hexagon (whose side is equal to the radius of the circle), third the square and the equilateral triangle, and so on down to the pentagon and decagon.[50]

2.2.3   Book 2: Fitting Regular Polygons Together Round a Vertex

Book 1 of *Harmonice mundi* is clearly addressed to professional mathematicians. In contrast, the mathematics of book 2 is very simple and the display of tessellation patterns looks like what can be found in Dürer's *Treatise on Measurement* (1525), which belongs to the tradition of practical works in the vernacular addressed to artisans.[51] It is highly likely that Kepler knew Dürer's book, but the method he employs in his investigation suggests that his starting point was Euclid.

In the last proposition of the *Elements*, proposition 18 of book 13, Euclid constructs the five regular solids. There follows a short scholium to prove that there are exactly five such solids. The proof considers grouping regular polygons round a vertex (bearing in mind that one needs at least three plane faces to enclose a solid angle): three triangles at a vertex give a tetrahedron, four an octahedron, five an icosahedron, six tessellate, three squares give a cube, four squares tessellate, three pentagons give a dodecahedron, four will not fit. Kepler recapitulates this proof in *Harmonice mundi* book 2, proposition 25 (see Fig. 2.7).[52] The method it uses is employed throughout the book.

---

[49] Kepler (1619), book 1, prop. 45, marginal note, and book 1, p. 38; Kepler (1940), p. 55; Kepler (1997), p. 74.

[50] Kepler (1619), book 1, p. 46; Kepler (1940), pp. 63–64; Kepler (1997), pp. 92–93.

[51] Dürer (1525), see note 45 above. The book appeared in Latin translation in 1532 (Dürer 1532, 1535). Dürer (1977). French translation (reliable, with very useful notes): Dürer (1995).

[52] Kepler (1619), book 2, prop. 25, pp. 57–60; Kepler (1940), pp. 78–82; Kepler (1997) pp. 112–16.

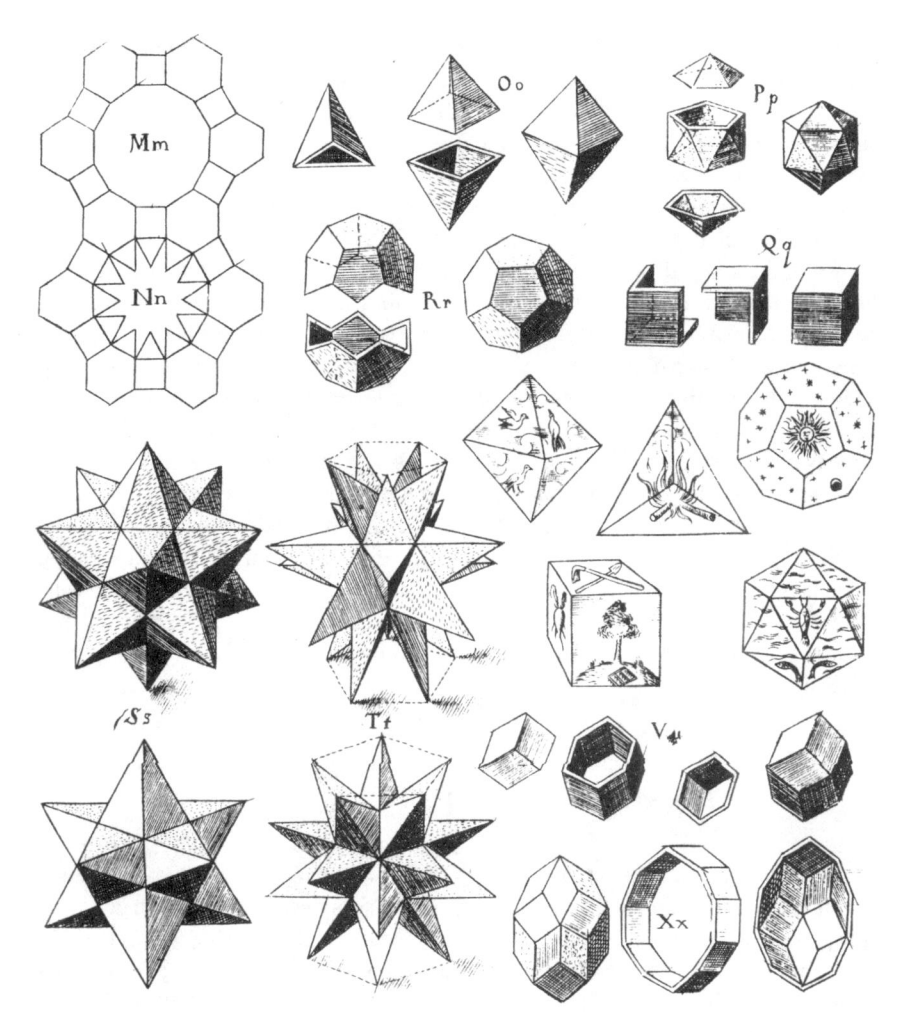

Fig. 2.7 Polyhedra and tessellations separate plate supplied to illustrate J. Kepler, *Harmonice mundi*, Linz, 1619, book 2)

The method looks plodding, since it is a method of exhaustion, proceeding by trying all cases, but it is effective, and it has at least two interesting consequences. First, though one cannot be sure what caused what, it gives a naturalness to Kepler's treating tessellations and polyhedra as the same kind of entity. He imagines fitting polygons together to form either, and seems to think of the closed finite pattern of the faces of the polyhedron as equivalent to the indefinitely extended pattern of a tessellation. This is of interest because at the time mathematicians show a marked reluctance to deal with entities that are not finite.[53] Second, once Kepler extends the

---

[53] On Kepler and points at infinity see Field (1986).

method to regular star polygons, it allows him to discover two new regular polyhedra, whose faces are regular star pentagons (see Fig. 2.7). Moreover, thanks to Kepler's impressive (and habitual) thoroughness, the tessellations he considers include the partial tessellations involving pentagons to which the name of Roger Penrose (b. 1931) has sometimes been attached.

In addition to tessellations, Kepler considers polyhedra whose faces are all regular polygons but of more than one kind. In particular he is interested in polyhedra whose vertices are all the same; that is, the pattern of faces round each vertex is the same (in today's terms, their vertices are uniform). These solids were known as the Archimedean solids because Pappus of Alexandria (*fl.* AD 300–350) says they were discovered by Archimedes (287–212 BC). But Pappus' description of the solids extends only to listing the number of faces of each kind, e.g. 'a tessarakaidecahedron with six square faces and eight triangular ones'. So on the technical level Kepler's work is a contribution to a process of rediscovery, with a large dash of visualization. Pappus also mentions that there are thirteen such solids, Kepler proves it. And somewhat disconcertingly the names he provides for the solids, which are still in use today, suggest the way he thought of them was different from the way in which he gives the proof (see Figs. 2.8 and 2.9).[54]

Like book 1, book 2 concludes with a classification of regular polygons, this time according to their capacity to contribute to 'congruences' (that is, tessellation or polyhedra).

2.2.4 Book 3: Music

As one of the four mathematical sciences taught in universities, music was seen as depending on arithmetic in the same way that astronomy was dependent on geometry. In this context the subject matter was mainly what would now be called theory of music, in which ratios of whole numbers expressed the ratios of the lengths of strings whose notes were separated by particular intervals. For instance, if the ratio of the lengths of the strings were 1:2 then the interval between the two notes was an octave. Great importance was given to whether some particular ratio defined an interval that was 'consonant', which in principle meant that the effect of the two notes being sounded together was pleasant. It was, however, accepted that the term 'consonance' was open to interpretation and when there was dispute over which numbers should be considered to contribute to consonances it was not always clear whether an argument from the judgement of the sense of hearing would necessarily prevail over one based purely on numbers. In addition to this, considerable difficulty was encountered in using the accepted consonances to construct scales, a problem given importance by the increasing use of pre-tuned instruments, such as the lute or viols or early forms of stringed keyboard instruments, to accompany singers or to play together. Some difficulties were eased by increasing the number of accepted consonances, but

---

[54] For more detail see Field (1997).

Fig. 2.8 Six Archimedean polyhedra. The solids shown are 8. cuboctahedron, 9. icosidodecahedron, 10. rhombicuboctahedron, 11. rhombicosidodecahedron, 12. snub cube, 13. snub dodecahedron (Reproduced from J. Kepler, *Harmonice mundi*, Linz, 1619, book 2, prop. 28, p. 63)

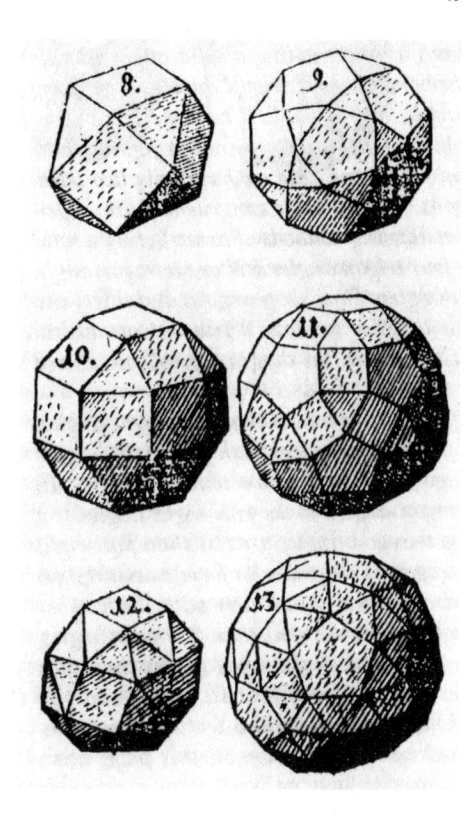

problems with scales remained. if, for example, one moved up three notes and then came down again by two, it was possible that the effect would not be the same as if one had simply moved up one note. That is, the scales were 'unstable'. Contrary to the hopes of learned humanists, recovering ancient texts on music, the principal ones being those by Ptolemy and Aristoxenus (*fl.* after 330–322 BC), did not help very much.[55]

The numbers chosen to generate the ratios corresponding to consonances, sometimes called 'sonorous numbers', were conventionally the Pythagorean Tetractys, that is the numbers 1 to 4, and from the later sixteenth century onwards the *senario* ('set of six') that is the numbers 1 to 6. Kepler regarded both of these sets of numbers as arbitrary and looked for a geometrical explanation.[56] He found such an explanation in the ratios of the arcs that are cut off from the circumference of the circle by inscribing regular polygons. Each consonance is then ranked according to the rank of the polygon in the hierarchy of 'inscribability' given at the end of book 1.

The basis of this theory is not novel. There was a history of presenting consonances in terms of this kind of division of the circle. It is for instance found in the standard

---

[55] English translations in Barker (1984, 1989).

[56] Further detail in Field (1984b).

Fig. 2.9 Seven
Archimedean polyhedra. The
solids shown are 1. truncated
cube, 2. truncated
tetrahedron, 3. truncated
dodecahedron, 4. truncated
icosahedron, 5. truncated
octahedron, 6. truncated
cuboctahedron, 7. truncated
icosidodecahedron
(Reproduced from J. Kepler,
Harmonice mundi, Linz,
1619, book 2, prop. 28, p. 64)

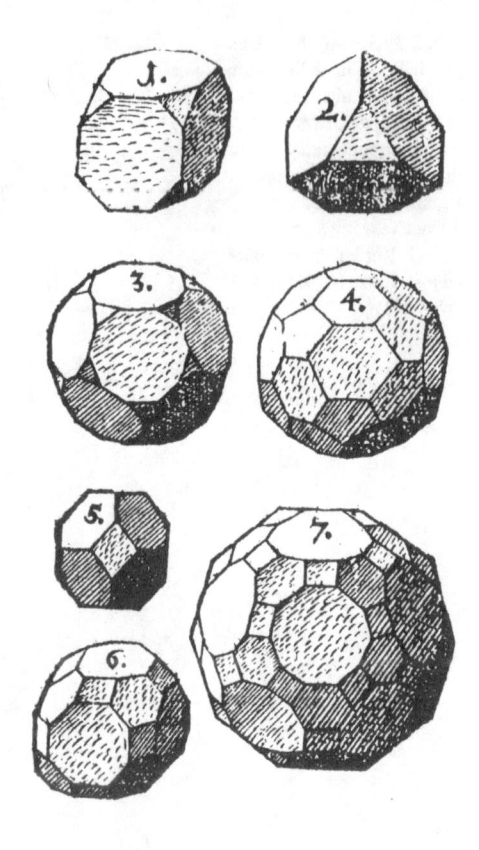

work on music theory *Rules of Harmony* (*Istitutioni harmoniche*, Venice, 1558)
written by Gioseffo Zarlino (1517–1590), a prominent theorist who became the
master of music (*maestro di capella*) at St Mark's Venice. (St Mark's was then the
Doge's private chapel so this was essentially a court appointment rather than an
ecclesiastical one.) In *Harmonice mundi* book 3 Kepler refers to Zarlino's treatise—
though he regards Zarlino's theory of consonances as being the same as Ptolemy's—
but it is not known at what stage he read the work. He could have been told about
it in the course of his university studies at Tübingen. In any case, diagrams like
Zarlino's, though less decorative, appear in Chap. 12 of *Mysterium cosmographicum*
and in some of Kepler's letters to Herwart von Hohenburg. However, none appears
in *Harmonice mundi* book 3.

It is difficult to find out what music Kepler may have heard in Prague, but Herwart,
being attached to the Bavarian court in Munich, was undoubtedly familiar with the
work of Orlandus Lassus (Orlande de Lassus, *c.*1532–1594), who came to Munich
as a choirboy and had made his career in the city, where some luxurious choir books
of his music are still preserved. Lassus became one of the most famous and highly
regarded musicians of Europe—indeed the uncertainty about his date of birth is partly
due to astrologers hailing him as the perfect musician and adjusting his date of birth

to make his horoscope fit their style of prediction. There is thus no mystery about the admiration for his work expressed by Herwart and apparently shared by Kepler, whose theory of music stresses the importance of the evidence of the senses.

In his younger days, Lassus experimented with dissonances and with quarter-tones (an idea derived from Aristoxenus) and some of his later works could also fairly be described as *avant garde* or cutting edge, but he also shows an understanding of the use of consonance and theorists tend to cite passages from his works as examples of how consonance and dissonance should be used.

Beyond the establishment of a geometrical basis for the consonances, Kepler's musical theory is largely conventional, though it is sufficiently fully worked out to have attracted a certain amount of attention from historians of music.[57] Kepler's interactions with the Ptolemy–Zarlino system seem to be partly connected with his reading of the *Dialogue on Music Ancient and Modern* (*Dialogo ... della musica antica e della moderna,* 1581, 1602) written by the important Florentine music theorist—also a lute player and a composer—Vincenzo Galilei,[58] who was a former pupil of Zarlino and later his opponent in a notably bitter controversy.[59] His son Michelangelo Galilei (1575–1631) became a professional musician of some standing in his day; Vincenzo's polemical tendencies seem to have been inherited by another son, Galileo Galilei, who surely must have read at least some of Kepler's *Harmonice mundi* on account of its repeated references to the work of his father (even if one of them, in the concluding paragraphs of book 3, Chap. 8, proposed a system for tuning the strings of a lute that was claimed to be better than Vincenzo's)? As fellow Copernicans, Kepler and Galileo exchanged personal letters and, largely on Kepler's side, publications that were effectively open letters. This went on for many years but the relations between them were never completely smooth except perhaps in regard to optics.[60]

What was unlucky about Kepler's music theory was his timing. In the early seventeenth century there was a divide between the various rival forms of consonance-based theory that had been current since Antiquity and, against them, the musical practice of leading composers and performers of music, who more and more favoured a style that employed dissonance in a much more radical manner. This division had been developing since the middle of the previous century. By Kepler's time it had reached the point that a distinction was made between the (conventional) 'first practice' (*prima prattica*) and the new 'second practice' (*seconda prattica*), terms apparently invented by one of the exponents of the latter, Claudio Monteverdi (1567–1643), in defending himself against an accusation that he had no theory at all. He promised to write a treatise on the theory behind his music, but seems never to have done so. One may see the *prima prattica* as having lost the contest in 1612 when Monteverdi became master of the music at St Mark's Venice, after which Venice gradually became the

[57] See for example Dickreiter (1973).

[58] Galilei (1581); reprinted 1602.

[59] See Walker (1978).

[60] See Field, 'Kepler and Galileo' (Chap. 8), and Donahue on Kepler's optics (Chap. 7) in this volume.

leading city in Europe for music. In music it turned out that Kepler was speaking for the past rather than the future. After constructing scales and discussing their characters, Kepler ends his book with some reflections on musical questions about melody and harmony.

However, there follows a long 'Political Digression about Three Means' that was almost certainly intended to appear earlier but the printer (wisely, in view of its length) put it at the end. The three means are the arithmetic, geometric and harmonic means and their relevance to political systems is based on *Six books on the republic* (*De republica libri sex*, Paris, 1586) by the French jurist and political philosopher Jean Bodin (1530–1596). (Bodin's work had originally been published in French as *Six livres de la république*, Paris, 1576, but it seems likely that Kepler read the Latin edition.) The title's apparent reference to Plato's dialogue *The Republic* is somewhat misleading, since Bodin is only incidentally concerned with Plato's central theme of justice.

2.2.5　Book 4: Astrology

The first three books of *Harmonice mundi* are concerned with the abstract mathematical entities, though with a nod to materiality in regard to music. The fourth book turns to the real world. In the twenty-first century it is difficult not to describe this fourth book as concerned with astrology. As can be seem in Fig. 2.10, the title page describes the content as 'On the harmonic configurations of the stellar rays on the Earth, and their effect on events in the sky and other natural phenomena'.[61] However, the introductory chapters show that Kepler is concerned with a wide range of natural phenomena. Indeed, the phrase translated 'the effects on events in the sky' (*effecta in ciendis Meteoris*) has strongly Aristotelian echoes and could perhaps also be translated as 'their effects in driving phenomena in the sublunary sphere'. Kepler is concerned not only with what we should now call terrestrial physics, including subterranean and atmospheric phenomena, but also with the behaviour of living things. That is, he proposes to identify harmonies in the world and to examine how they affect the processes of nature.

Apart from having a geometrical rather than a numerological basis, the music theory in *Harmonice mundi* book 3 is otherwise fairly conventional. In contrast, the astrology of *Harmonice mundi* book 4 contains much that is unconventional and original to Kepler, though readers who had been following his successive contributions to astrology might have seen some of this coming.[62] For example, as a Copernican Kepler was free to regard the constellations as accidental patterns made up of stars that are in reality distant from one another, and he had become steadily less inclined to ascribe any astrological significance to the signs of the zodiac. In *Harmonice*

---

[61] Kepler (1619), book 4, p. [104]; Kepler (1940), p. 206; Kepler (1997), p. 281.

[62] See Rabin in Chap. 6 of this volume and Field (1984a).

Fig. 2.10 Title page of J. Kepler, *Harmonice mundi*, Linz, 1619, book 4, p. 104, a translation of all the wording is given in Kepler (1997), p. 280 (Copyright Bavarian Academy of Sciences, Munich)

IO:KEPLERI
Harmonices Mundi
LIBER IV.

DE CONFIGVRATIONI-
BUS HARMONICIS RADIORUM
fideralium in Terra, earumq; effe-
ctu in ciendis Meteoris, alijfq;
Naturalibus.

Proclus Diadochus
Libro I. comment: in L Euclidis

*De Mathematicis ufu in Phyfiologia & Politica: qua potif.
fimùm partem illius Harmonicam de Radiatio-
nibus concernunt.*

Ad contemplationem Naturæ præcipua omnia fuppeditat , decla-
rans Rationum ordinem pulcherrimum,fecundùm quem fabricarum eft totum hocVniverfum;pro-
portionumq; Analogiam, quæ omnia mundana inter fe connectit, ut loquitur alicubiTimæus ,
quæq; amicitiam inter pugnantia, refponfum & mutuam affectionem inter longiffimè diffita, con-
ciliat. *Et poft pauca* Inde & angulationes commodas poffibile eft ratiocinando venati. *Rur-
fum*,Hoc opinor & Timæus fignificare voluit, dum paffim per voces mathematicas , tradit contemplationem de Natura totius univerfi, ortumq; elementorum, Numeris & Figuris depingit, faculta-
tesq; & affectiones illorum, etiamq; effectus,his *(figuris)* acceptos fert; angulorum acuta vel obtu-
fa,laterumq; afpera vel lævia , &c. caufas conftituens omni variarum mutationum.
Ad Politicam verò dictam doctrinam , qui negari poffit , illam plurima & mirabilia confer.
re; dum opportunitates rerum gerendarum dimetitur , ratiosq; circuitus totius univerfi &c.
Numerofq; Harmonicos, vitæ moderatores, aut incongruentiæ authores, & in
univerfum impetus aut remiffionis opitulatores
&c.

Cum S. C. M. Privilegio ad annos XV.

LINCII AVSTRIÆ,
Excudebat Johannes Plancus.
ANNO M.DC.XIX.

mundi book 4 astrology has become a matter only of Aspects—that is, of the angles made by rays reaching the Earth from the planets.

Aspects are defined at the beginning of Chap. 5, whose title is 'On the Causes of the Influential Configurations, and of their Degrees in Number and Order'.[63] There follows a definition

Definition 1

The word 'configuration' is used for the angle between two rays, each descending from its planet, the angle at which the rays meet here at the Earth (which is deemed to be a point); or, which comes to the same thing, it is used for the arc of the great circle drawn on the zodiac,

---

[63] Kepler (1619), book 4, p. 133; Kepler (1940), p. 239; Kepler (1997), p. 326.

the arc which is the measure of the said angle; or the arc which the two planets seem to mark out by the interposition of their bodies and, so to speak, cut off for us dwellers on Earth.[64]

This definition is conventional, but it leaves several loose ends, and Kepler duly follows it with about a page of amplifications. He begins with a discussion of the various names given to these configurations in earlier writings, noting that Ptolemy, in his *Tetrabiblos*, *Almagest* and *Harmonica*, calls them 'appearances',[65] and that the term 'aspects' is derived from Arabic translations. He notes that an equivalent term, with an implied reference to the face, is used in German. Here, at the risk of arousing Kepler's posthumous disgust, we shall generally use 'Aspects' as being less ambiguous than 'configurations'. He next briefly describes the standard astrological matter of angles between signs of the zodiac deemed to be 'looking at' one another before reiterating that he will be concerned with the angles rays of planets make at the Earth, taking the Earth as a point at the centre of the circle of the zodiac. Finally, Kepler points out that the angles between planetary rays are (effectively) the same for all points on or inside the Earth and for all creatures that live on it. There follows a definition of what is meant by saying a configuration is influential.

Next, we have two axioms relating these properties of configurations to the properties of polygons explored in the first two books: knowability (the capacity to be inscribed in a circle, considered in book 1) and congruence (the capacity to fit together, considered in book 2).

Axiom 1

The arc of the zodiac circle which is cut off by the side of a figure or of a star [*sc.* star polygon] which is congruent and knowable, measures the angle of an influential configuration.

Axiom 2

The angle of a figure or star [*sc.* star polygon] which is knowable and congruent is the gauge of the angle of an influential configuration.[66]

As the ensuing discussion shows, these are not axioms in today's mathematical sense, dating back to Archimedes, where an axiom is a statement whose truth will not be called in question in what follows. Kepler uses the word to mean an assumption or a statement to be accepted for the time being. As he certainly knew, this sense accords with its etymological connection to the verb 'to believe, consider or accept as true' (*axio*).[67] Plato, whose writings lie behind a great deal of Kepler's thinking, makes a sharp distinction between such changeable opinion and certain knowledge (*Republic*, book 6).

---

[64] Kepler (1619), book 4, p. 133; Kepler (1940), pp. 239–40; Kepler (1997), p. 326.

[65] Ptolemy (1940) bk 1, ch. 2 and 13; Ptolemy (1984), book 8, Chap. 4; Kepler (1619), book 3, Chap. 9. See Tester (1987).

[66] Kepler, (1619), book 4, p. 134; Kepler (1940), p. 241; Kepler (1997), p. 328.

[67] This is not exactly the same as what is found in a twenty-first century dictionary, though it refers to the same Greek root.

Fig. 2.11 The aspects of conjunction and opposition defined by the digon (diameter) (Reproduced from J. Kepler, *Harmonice mundi*, Linz, 1619, book 4, Chap. 5, prop. 10, p. 144)

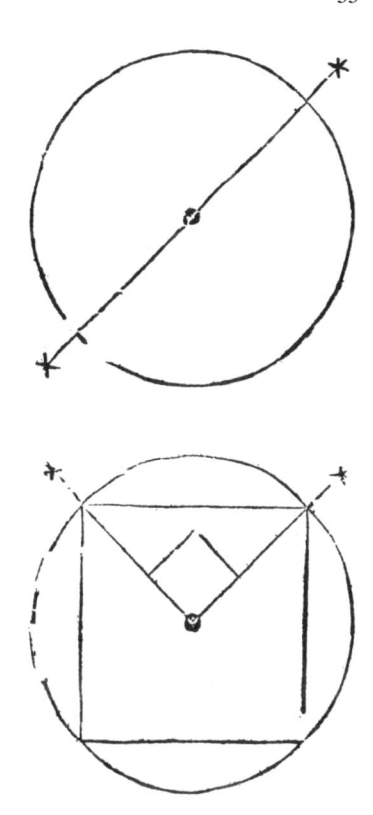

Fig. 2.12 The aspect of quadrature defined by the tetragon (square) (Reproduced from J. Kepler, *Harmonice mundi*, Linz, 1619, book 4, Chap. 5, prop. 11, p. 145)

There follows a series of propositions, with diagrams showing a circle representing the zodiac, and rays descending to the Earth at its centre from pairs of stars (see Figs. 2.11, 2.12, 2.13 and 2.14), the angle at which the rays meet being the angle of a congruent or knowable regular polygon while the arc of the circle between the stars corresponds to the side of such a regular polygon. The first two cases, propositions 10 and 11, show angles at the centre of zero, 180° and 90°, so in each case the angle at the centre and the side defined by the arc both correspond to the same regular polygon, the digon (conjunction and opposition, proposition 10) and the square (quadrature, proposition 11), see Figs. 2.11 and 2.12. For the remaining configurations the polygon at the centre and the one inscribed in the circle are different from one another. As Kepler had said immediately after stating the axioms, he had given two of them 'because there are two probable means by which souls and sublunary Natures can come to knowledge of the configurations which exist at a given time'.[68] Thus Kepler can use the hierarchies of polygons established in books 1 and 2 to construct a hierarchy of configurations, which is what he proceeds to do. It includes 'quintile' configurations associated with the pentagon, and the pentagonal star, aspects not found in standard astrology.

---

[68] Kepler (1619), book 4, p. 134; Kepler (1940), p. 241; Kepler (1997), p. 328.

Fig. 2.13 The aspect of trine defined by the regular trigon (equilateral triangle) (Reproduced from J. Kepler, *Harmonice mundi*, Linz, 1619, book 4, Chap. 5, prop. 12, p. 146)

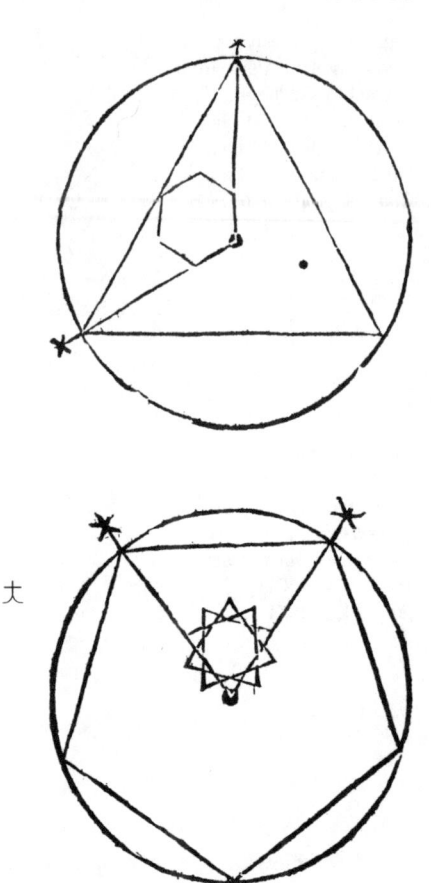

Fig. 2.14 The aspect of quintile defined by the regular pentagon (convex regular pentagon) (Reproduced from J. Kepler, *Harmonice mundi*, Linz, 1619, book 4, Chap. 5, prop. 12, p. 146)

Kepler's treatment of the subject gives us astrology in its most abstract form, far from specific predictions for individual human beings. However, there is a substantial final chapter, 'Epilogue on Sublunary Nature and on the Inferior Faculties of the Soul, Especially those on which Astrology Depends', that does deal with some more conventional matters such as the influence of aspects at the moment of birth on the course taken by the child's life.[69] There is a series of short discussions, in one of which Kepler makes use of information about himself, his family and people he knew as a way of showing how many factors other than astrology need to be taken into account in explaining a life. He cites a woman whose birth aspects were much like his own, as was her temperament, 'but by which she not only has no advantage in book learning (that is not surprising in a woman)' but also does not fit well into the society of her town. This is almost certainly a reference to his mother, Katharina Kepler (*née* Guldenmann, 1546-1622), who at the time was not yet free of a charge

---

[69] Kepler (1619), book 4, Chap. 7, pp. 157–76; Kepler (1940), pp. 264–86; Kepler (1997), pp. 358–85.

of witchcraft. Kepler contrasts this with his being born a man and, with help from relatives and financial support from local 'magistrates' receiving an education. He numbers the items that in his case affected the working out of the influence of the aspects.[70]

2.2.6   Book 5: Astronomy

The title page of *Harmonice mundi* book 5 says we shall be concerned with the motions of the planets (see Fig. 2.15):

ON THE MOST PERFECT HARMONY OF THE HEAVENLY MOTIONS and on the origin from the same of the Eccentricities, Semidiameters and Periodic Times. According to the precepts of the most thoroughly corrected astronomical teaching of the present day, and the hypotheses of Copernicus, but also those of Tycho Brahe, one or the other of which are today publicly accepted as true, superseding those of Ptolemy.[71]

In referring to 'thoroughly corrected astronomical teaching' Kepler presumably has in mind his own wholesale replacement of systems of circles by elliptical orbits like the one he found for Mars in *Astronomia nova* (Heidelberg, 1609).[72] This justifies the choice, already seen in *Mysterium cosmographicum*, of including in the planetary orb only the actual path of the planet and not the system of circles used to construct it. By defining the orb simply as enclosing the orbital ellipse, Kepler makes its thickness a measure of what would today be called the eccentricity of the planetary orbit concerned. It is also notable that the schema outlined on the title page of 1619 is very like the one Kepler set out in a letter to Herwart von Hohenburg in 1599 telling him about plans for a book called *De Harmonice Mundi*.[73] Kepler had been thinking about this material for a long time.

The text of *Harmonice mundi* book 5 starts with an explanation of the mathematical background to the theory that the five regular polyhedra determine the spacing of planetary orbs, the theory Kepler had first put forward in *Mysterium cosmographicum* in 1596. He sets out some properties of the five regular polyhedra and shows some relations between the solids, such as that a regular tetrahedron can be drawn to share some of the vertices of a cube. This is equivalent to saying that pieces can be cut off from a cube to leave a tetrahedron (see Fig. 2.16). This piece of mathematics is not found in *Mysterium cosmographicum*. It does, however, appear when the cosmological theory is presented in the fourth book of Kepler's *Epitome of Copernican Astronomy* (1620) which, as we have seen in connection with Kepler's third law, he expected to be published before *Harmonice mundi* (see comments on

---

[70] Kepler (1619), book 4, Chap. 7, p. 170; Kepler (1940), p. 279; Kepler (1997), p. 376.

[71] Kepler (1619), book 5, title page, p. [177]; Kepler (1940), p. 287; Kepler (1997), p. 387.

[72] See Davis (Chap. 4) in this volume.

[73] Kepler to Herwart von Hohenburg, 14 Dec 1599, Kepler (1949), letter 148, p. 100. See also Field (1988), pp. 142–43.

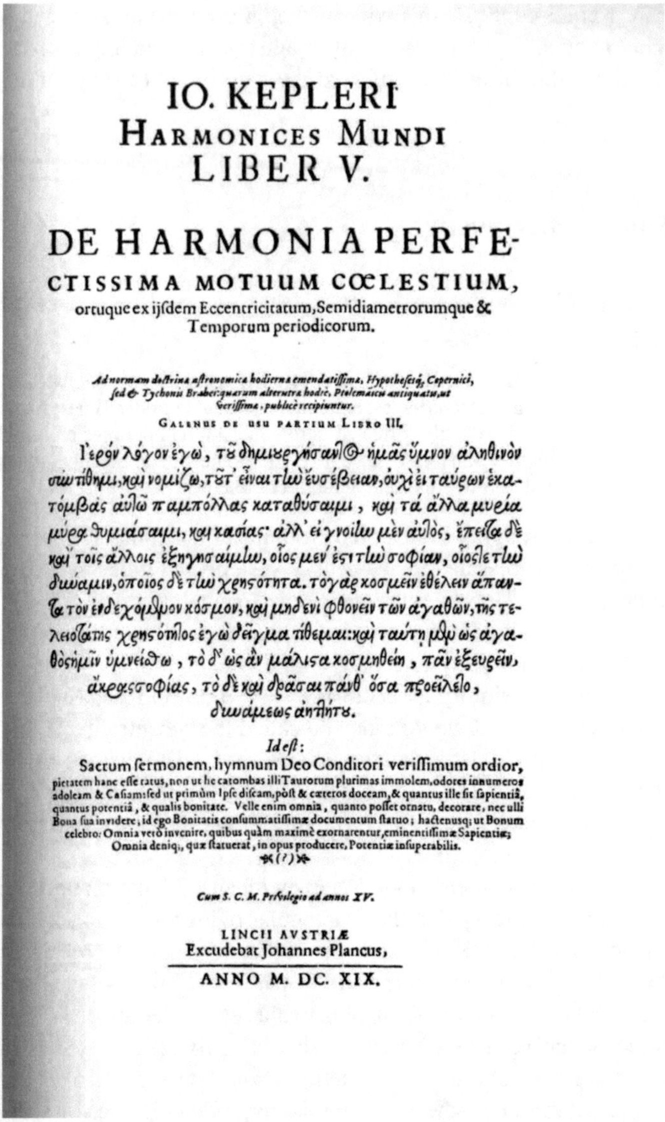

IO. KEPLERI
Harmonices Mundi
LIBER V.

DE HARMONIA PERFE-
CTISSIMA MOTUUM CŒLESTIUM,

ortuque ex ijſdem Eccentricitatum, Semidiametrorumque &
Temporum periodicorum.

Ad normam doctrina aſtronomica hodierna emendatiſſima, Hypotheſeis, Copernici,
ſed & Tychonis Braheiquarum alterutra hodiè, Ptolemaicis antiquatis, ut
veriſſimæ, publicè recipiuntur.

GALENUS DE USU PARTIUM LIBRO III.

Ἱερὸν λόγον ἐγώ, τῷ δημιεργήσανῷ ἡμᾶς ὕμνον ἀληθινὸν
σωτίθημι, καὶ νομίζω, τότ᾽ εἶναι τὴν εὐσέβειαν, οὐχὶ εἰ ταύρων ἑκα-
τόμβας αὐτῷ παμπόλλας καταθύσαιμι, καὶ τά ἄλλα μυρία
μύρα θυμιάσαιμι, καὶ κασίας· ἀλλ᾽ εἰ γνοίην μὲν αὐτὸς, ἔπειτα δὲ
καὶ τοῖς ἄλλοις ἐξηγησαίμην, οἷος μέν ἐςι τὴν σοφίαν, οἷός τε τὴν
δύναμιν, ὁποῖος δὲ τὴν χρηςότητα. τὸ γὰρ κοσμεῖν ἐθέλειν ἅπαν-
τα τὸν ἐνδεχόμενον κόσμον, καὶ μηδ᾽ ἐνὶ φθονεῖν τῶν ἀγαθῶν, τῆς τε-
λειοτάτης χρηςότητος ἐγὼ δεῖγμα τίθεμαι: καὶ ταύτῃ μὲν ὡς ἀγα-
θὸς ἡμῖν ὑμνείσθω, τὸ δ᾽ ὡς ἂν μάλιςα κοσμηθείη, πᾶν ἐξευρεῖν,
ἄκρας σοφίας, τὸ δὲ καὶ δρᾶσαι πᾶνθ᾽ ὅσα προείλετο,
δυνάμεως ἀνίκητα.

Id eſt:

Sacrum ſermonem, hymnum Deo Conditori veriſſimum ordior,
pietatem hanc eſſe ratus, non ut hecatombas illi Taurorum plurimas immolem, odores innumeros
adoleam & Caſiam: ſed ut primùm Ipſe diſcam, pòſt & cæteros doceam, & quantus ille ſit ſapientiâ,
quantus potentiâ, & qualis bonitate. Velle enim omnia, quanto poſſet ornatu, decorare, nec ulli
Bona ſua invidere, id ego Bonitatis conſummatiſſimæ documentum ſtatuo; hactenusq; ut Bonum
celebro: Omnia verò invenire, quibus quàm maximè exornarentur, eminentiſſimæ Sapientiæ;
Omnia deniq., quæ ſtatuerat, in opus producere, Potentiæ inſuperabilis.
⁂ (?) ⁂

Cum S. C. M. Privilegio ad annos XV.

LINCII AVSTRIÆ
Excudebat Johannes Plancus,

ANNO M. DC. XIX.

Fig. 2.15 Title page of J. Kepler, *Harmonice mundi* book 5, title page, p. 177 (Copyright Bavarian Academy of Sciences, Munich)

Kepler's annotations to *Mysterium cosmographicum* Chap. 20 in Sect. 2.1.6 above). Presumably, as Kepler was writing his *Epitome of Copernican Astronomy* he felt it inappropriate to refer his prospective readers to a book published so many years before, or perhaps his earlier introduction of the theory now seemed somewhat inadequate. In *Mysterium cosmographicum* the theory is introduced in a chapter concerned

Fig. 2.16 A regular tetrahedron drawn so that its four vertices coincide with four of the eight vertices of a cube (Reproduced from J. Kepler, *Harmonice mundi* book 5, Chap. 1, p. 181)

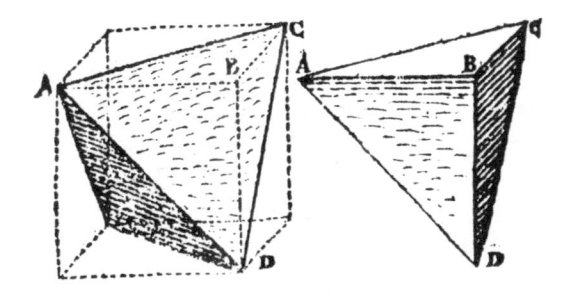

with astronomy, specifically with arguments in favour of the Copernican model of the planetary system, rather than with the geometry of the Platonic solids. The following chapter of *Harmonice mundi* book 5 is also about mathematics, describing relations between the five regular solids and the proportions that define harmonies.

Kepler describes his third chapter as providing a 'summary of astronomical theory, necessary for the study of the heavenly harmonies'.[74] There is a list of thirteen items, each marked with a marginal note. The first item describes the heliocentric planetary system.[75] The second provides a brief account of how the regular polyhedra can be fitted between the planetary orbs, together with a concise two-dimensional version of the elaborate fold-out engraved illustration of the theory in the *Mysterium cosmographicum* (see Figs. 2.17 and 2.3).[76] Item eight, with the marginal note 'VIII. What is the proportion of the periodic times to the distances from the Sun of any pair of planets?', gives Kepler's third law. But first he describes how he discovered it, starting by alluding to his having been in search of such a law for a long time—which any reader of *Mysterium cosmographicum* will know to be true—before sketching the final stages of discovery

> … and if you want the exact moment in time, it was conceived mentally on the 8th March in this year one thousand six hundred and eighteen, but submitted to calculation in an unlucky way, and therefore rejected as false, and finally returning on the 15th of May and adopting a new line of attack, stormed the darkness of my mind. So strong was the support from the combination of my labour of seventeen years on the observations of Brahe and the present study, which conspired together, that at first I believed I was dreaming, and assuming my conclusion among my basic premises.[77]

On the human level, this is a good story but on the historical level, in the context of *Harmonice mundi*, it tells us rather little about the discovery. Most obviously, it does not provide any kind of mathematical or physical justification for the proportion Kepler has found. We are offered only numerical confirmation. This aspect is stressed again as the account of the law continues

---

[74] Kepler (1619), book 5, Chap. 3, p. 184; Kepler (1940), p. 296; Kepler (1997), p. 403.

[75] Kepler (1619), book 5, Chap. 3, p. 186; Kepler (1940), p. 297; Kepler (1997), p. 405.

[76] Kepler (1619), book 5, Chap. 3, plate following p. 186; Kepler (1940), pp. 297–98; Kepler (1997), p. 405.

[77] Kepler (1619), book 5, Chap. 3, p. 189; Kepler (1940), pp. 301–302; Kepler (1997), p. 411.

Fig. 2.17 The planetary orbs with regular polyhedra fitting between them. Note the indication of the path given to the Sun in Tycho's system (Reproduced from J. Kepler, *Harmonice mundi*, 1619, book 5, Chap. 3, plate following p. 186)

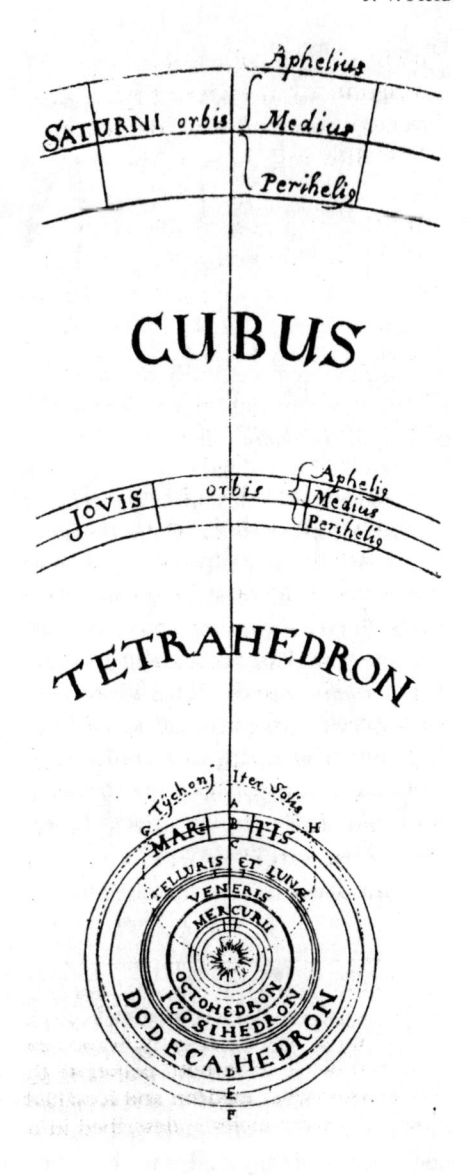

But it is absolutely certain and exact that *the proportion between the periodic times of any two planets is precisely the sesquialterate proportion of their mean distances, that is of the actual spheres*, though with this in mind, that *the arithmetic mean between the two diameters of the elliptical orbit is a little less than the longer diameter*.

The Euclidean language of this is perhaps a little awkward to understand. 'Sesquialterate' means 'one and a half', so Kepler is saying that the proportion between the times is the three halves power of that between the distances.

The account of the law in *Epitome of Copernican Astronomy*, book 4 shows that Kepler did try to find a rationale for his law. The matter is discussed in book 4, part 2 in a section headed 'On the causes of the proportion of the periodic times',[78] but Kepler presumably did not find it satisfactory, and it has been thoroughly neglected by historians. Their justification is that Kepler seems to have forgotten about it too. However, as with the elliptical orbits, even in the absence of a convincing rationale, that is a rationale capable of making mathematical predictions, Kepler was prepared to regard numerical agreement with observations as itself convincing. So it is no surprise that he tried out his new law on the orbits of the satellites of Jupiter. At this time, observations were not very reliable—Kepler prefers those of Simon Marius (Simon Mayr, 1570–1624) to those of Galileo—but the agreement seems to him to be adequately good.[79] Some 65 years later, Isaac Newton (1642–1727), using much better observations, was to apply Kepler's third law to check that Jupiter's satellites were moving under the influence of a force obeying an inverse square law, in their case attracting them to Jupiter.[80] The orbits were too nearly circular for him to use the first two laws.

At the end of *Harmonice mundi* book 5, following the word '*FINIS*' at the end of Chap. 10, 'Conjectural Epilogue on the Sun', Kepler or his printer added the date (as was usual at the time) and gave a little more detail:

> This work was completed on the 17/27th May in the year 1618, but Book V was revised (during the time when its printing was in progress) on the 9/19th February 1619.
>
> At Linz, the capital of Upper Austria.[81]

It was because Kepler was living in the same city as his printer that he was able to make last-minute changes. It seems likely that the revisions to book 5 were carried out after Kepler had realized that *Harmonice mundi* would be published before the fourth volume of his *Epitome of Copernican Astronomy*, but it is not possible to determine in what order the two accounts of the third law were written. In any case, Kepler makes very little use of the third law in *Harmonice mundi*.[82] His interest is in finding harmonies among the celestial motions. It is to these that he turns in Chap. 4, which has the rather long title 'In what features relating to the motions of the planets have the harmonic proportions been expressed by the Creator and how?'. The first sentence plunges straight in:

> When therefore the fantasy of retrogressions and stations has disappeared and the planets' proper motions, in their own true eccentric orbits, have been stripped to essentials, there still remain in the planets the following different features: 1. their distances from the Sun; 2. their periodic times; 3. their daily eccentric arcs; 4. the daily times expended on their arcs; 5. their angles at the Sun, or apparent daily arcs to observers, so to speak, on the Sun.[83]

---

[78] Kepler (1620), part 2, Sect. 4, pp. 530–34; Kepler (1991), pp. 306–308.

[79] Kepler (1620), part 2, Sect. 3, p. 449; Kepler (1991), p. 264.

[80] Newton (1687), book 3, prop. 1, p. 405.

[81] Kepler (1619), book 5, Chap. 10, p. 248; Kepler (1940), p. 368; Kepler (1997), p. 368.

[82] For details see Field (1982).

[83] Kepler (1619), book 5, Chap. 4, p. 192; Kepler (1940), p. 306; Kepler (1997), p. 417.

Apparentes diurni

Harmonia binorum		diurni. Prim Sec.	Harmonię singulorum propria Prim. Sec.	
Diver. Conc.				
a 1 b 1	♄ Aphelius	1.46.a.	Inter 1.48	est $\frac{4}{5}$ Tertia major.
	Perihelius	2.)5.b.	& 2.15,	
d 3 c. 2	♃ Aphelius	4.30. c.	Inter 4.35.	est $\frac{5}{6}$ Tertia mi- nor.
c 8 d 1	Perihelius	5. 30. d.	& 5.30.	
f 1 e 5	♂ Aphelius	26.14. e.	Inter 25.21.	est $\frac{2}{3}$ Diapente
e 5 f 2	Perihelius	3.81. f.	& 38.1.	
h 12 g 3	Tel.Aphelius	57.3.g	Inter 57. 28.	est $\frac{15}{16}$ Semitoni;
g 3 h 5	Perihelius	61.18. h	&61.18.	
k 5 i 8	♀ Aphelius	94.50. i.	Inter 94.50.	est $\frac{24}{25}$ Diesis
i 1 k 3	Perihelius	97.37. k.	& 98.47.	
m 4 l 5	☿ Aphelius	164.0.l.	Inter 164. 0.	est $\frac{5}{12}$ Diapason. cum tertia minore
	Perihelius	384.0. m.	& 394, 0;	

Fig. 2.18 Table of speeds of planets. The outer columns show ratios of extreme speeds of planets. The middle column shows these extreme speeds, that is daily motions in arc as seen from the Sun. Kepler works inwards through the system, starting with Saturn, indicating each planet by its symbol and giving both the minimum (aphelion) and maximum (perihelion) speeds. (Reproduced from J. Kepler, *Harmonice mundi*, 1619, book 5, Chap. 4, p. 195)

Kepler proceeds to try each possibility in turn, showing the results in tables. The fifth case is the one that yields a series of speeds whose ratios exhibit 'harmonies', that is ratios expressible in terms of small whole numbers that can be derived from the regular polygons given a high rank in the hierarchy of 'congruence' at the end of *Harmonice mundi* book 2. The table is shown in Fig. 2.18, in which the two outside columns show ratios, together with letters denoting the quantities from which they are formed. These letters appear against entries in the inner column, which displays apparent daily motions (as seen from the Sun) expressed in minutes and seconds of arc. The rows work downwards through the planets, the uppermost being Saturn. Each planet has two entries, one for its speed, one at aphelion (that is its minimum speed) and the other at perihelion (maximum speed), so each planet has two speeds. Thus the first ratio on the left is that of *a* to *d*, where *a* appears in the first row of the inner column as the daily motion of Saturn at aphelion and *d* appears in the next double row down, that is in the pair of entries for Jupiter, as the daily motion of Jupiter at perihelion. As can be seen, the various ratios in the outer columns are all expressible in terms of small numbers, which is to say they are what twenty-first century astronomers would, like Kepler, describe as 'musical'.

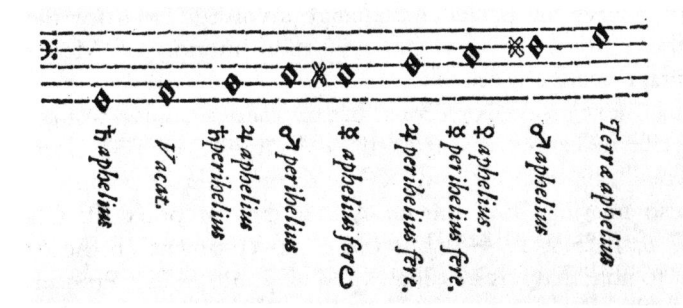

Fig. 2.19 Scale in *cantus durus*, starting from the aphelion speed of Saturn (Reproduced from J.Kepler, *Harmonice mundi* 1619, book 5, Chap. 5, p. 204)

These ratios may look too good to be true, but nevertheless they are true. Although the phenomenon is not well understood, which is to say there are competing theories, it is well known (in the twenty-first century) that systems of orbits tend to show such 'resonances'. They are, for instance, found among the satellites of Jupiter. The reason that Kepler is able to find the resonances in the orbits of the major planets is that the orbits he has deduced from Tycho's observations are very accurate—indeed, they are so accurate that in order to assess them one needs to make allowance for secular changes since the 1580s.[84]

Applying the music theory set out in *Harmonice mundi* book 3, Kepler then turns the ratios in his table into musical intervals and shows that these can be used to construct two scales, of the two types recognised at the time, *cantus durus* and *cantus mollis* (close to but not precisely the same as modern major and minor; see Figs. 2.19 and 2.20). In the following chapter, he shows the range of notes each planet, and the Moon, runs through as it moves round its orbit (see Fig. 2.21). The ranges indicate the eccentricity of the orbital ellipse, greatest for Mercury and very close to zero for Venus.

2.2.7 Cosmic Harmony

The musical ranges displayed in Kepler's figures are, of course, little more than a decorative illustration. But the existence of the musical ratios between extreme speeds of the planets has been used to explain why God chose to express His geometrical nature in constructing the planetary orbs to have the precise thicknesses that are observed. Kepler has filled in the blank left by the polyhedral model of the planetary system described in his *Mysterium cosmographicum*, where the spaces between the orbs were explained as showing the ratios between the radii of the inspheres and

---

[84] See Bialas (1971).

Fig. 2.20 Scale in *cantus mollis*, starting from the perihelion speed of Saturn (Reproduced from J. Kepler, *Harmonice mundi* 1619, book 5, Chap. 5, p. 204)

Fig. 2.21 Ranges of notes for planets and, at bottom right, the Moon as they move round their orbits (Reproduced from J. Kepler, *Harmonice mundi* 1619, book 5, Chap. 6, p. 207)

circumspheres of regular polyhedra, but no explanation was offered for the thicknesses of the orbs. It is clearly this victory of constructing a complete cosmological theory (in his terms a cosmogony) that Kepler has in mind when he writes in the final paragraph of the Introduction (*Proëmium*) to *Harmonice mundi* book 5

> … it is my pleasure to taunt mortal men with the candid acknowledgement that I am stealing the golden vessels of the Egyptians to build a tabernacle to my God from them, far, far away from the boundaries of Egypt. If you forgive me I shall rejoice; if you are enraged with me, I shall bear it. See, I cast the die and I write the book. Whether it is to be read by the people of the present or of the future makes no difference: let it await its reader for a hundred years, if God Himself has stood ready for six thousand years for one to study him.[85]

'I cast the die' is a reference to the well-known story of Julius Caesar (101–44 BC) saying 'the die is cast' (*alea iacta est*) on 10 January 49 BC when he led his army across the Rubicon, which constituted an act of rebellion against the authorities in Rome. The reference to the Egyptians is Biblical (the book of *Exodus*). Thus,

---

[85] Kepler (1619), book 5 *Proëmium*, pp. 178–79; Kepler (1940), p. 290; Kepler (1997), p. 391. See also Field (1988), p. 179 and footnote 1, p. 225.

before likening himself to Julius Caesar, Kepler is comparing himself to Moses leading the Israelites out of slavery in Egypt. I know no other example of Kepler employing this level of rhetoric. However, the reference to Egypt carries an echo of an intellectual debt: in looking to music in the heavens Kepler was building on the work of Ptolemy. Vague references to the music of the spheres, sometimes called '*musica mundana*', can be found in many authors. However, in his *Harmonica*, Ptolemy gives the only known substantial astronomical-cum-mathematical treatment of celestial music before Kepler's. Finding the available translations of Ptolemy's work unsatisfactory, Kepler went to considerable trouble to obtain a manuscript of the original Greek text.[86] Kepler had referred to the importance of Ptolemy's work earlier in his Introduction and seems to have believed, with a certain amount of justification, that Ptolemy's aims had been the same as his own: to give a model of phenomena in satisfactory numerical agreement with observations, the large differences (as Kepler saw them) being that the observations available to Ptolemy (in the second century AD) were very much less accurate than those available to Kepler himself and, when it came to cosmology, that Ptolemy's understanding of the Universe was deeply flawed by his not recognizing the existence of a Creator. Thus in *Harmonice mundi* Kepler saw himself as correcting and rewriting Ptolemy's *Harmonica* in the same way that, some years earlier, he had used Tycho's observations, and some new ideas about how nature worked, to rewrite the *Almagest* in the form of *Astronomia nova*.

2.3   In Conclusion

There can be no doubt that Kepler took his cosmological models seriously and subjected them to the same level of scrutiny as his astronomical work, in philosophical terms and in relation to agreement with observation. It was indeed his search for better observations, which he of course hoped would provide better confirmation of his theory, that Kepler first contacted Tycho Brahe, sending him a copy of the then newly-published *Mysterium cosmographicum*. Many years later, the agreement of the theory put forward in *Harmonice mundi* is in strikingly good agreement with observation. The mathematical results that appear in the last book have a firm mathematical foundation in the theoretical explorations of the first two books, which moreover contain quite a lot of original work. In another context, such work might have attracted attention from mathematicians and (eventually) from historians. In the event, most of *Harmonice mundi* was so thoroughly forgotten by mathematicians that in the early years of the nineteenth century, a mathematician at the École polytechnique in Paris, Louis Poinsot (1777–1859), rediscovered Kepler's two new regular polyhedra, whose faces are star pentagons, and discovered two further regular polyhedra, with convex faces but non-convex vertices (that is the vertex figures are star polygons). The four solids are now usually called the Kepler–Poinsot polyhedra.

---

[86] Klein (1971).

Nor, over the years, was the *Harmonice mundi* read for the third law—though that is usually the one result noticed by historians of science. In the generations of astronomers following Kepler, all three of his laws tended to be known through secondary sources, for example general textbooks of astronomy. One such was the *Astronomia Carolina* (London, 1661, 1664) of Thomas Streete (1621–1689). Streete mentions Kepler by name, but does not refer to the original contexts of the laws. That is, he treats them as live science. Streete seems to have been the source used by Isaac Newton.[87]

Kepler had a profoundly original mind that, together with his deep religious faith and his immense skill as a mathematician, is apparent in his cosmological writings as in his astronomical ones. Moreover, the cosmological writings and the astronomical ones show the same care in striving to achieve agreement with values derived from observation. This was by no means the norm in the cosmology of Kepler's time. Indeed, his are the first truly mathematical cosmological models in today's sense. His cosmological works were obviously exceedingly important to Kepler himself—not only in themselves but (as is made clear in the notes written for the second edition of *Mysterium cosmographicum*) in their influence on the development of his other work—and it is surprising that they have received so little serious attention from historians.[88]

References

Aiton, E. J. (1977). Johannes Kepler and the *Mysterium cosmographicum*. *Sudhoffs Archiv, 62*(2), 173–194.

Aiton, E. J. (1987). Peurbach's *Theoricae novae planetarum*. A translation with commentary. *Osiris (Second Series), 3*, 4–43.

Al-Khwarizmi (2009). *The beginnings of algebra*. Translated from the Arabic and edited with commentary Roshdi Rashed. Saqi.

Barker, A. (1984, 1989). *Greek musical writings*, 2 vols, *Vol. 1: The musician and his art, Vol. 2: Harmonic and acoustic theory*. Cambridge University Press.

Bialas, V. (1971). Die quantitative Beschreibung der Planetenbewegung von Johannes Kepler in seinem handschriftlichen Nachlaß. In *Kepler Festschrift 1971* (pp. 99–140). Regensburg.

Copernicus, N. (1543). *De revolutionibus orbium cœlestium*. Basel: Johannes Petreius. Nuremberg.

Dickreiter, M. (1973). *Der Musiktheoretiker Johannes Kepler*. Berne and München: Francke Verlag.

Dürer, A. (1525). *Underweysung der Messung mit dem Zirkel und Richtscheyt*. Hieronymus Andreae.

Dürer, A. (1532, 1535). *Institutionum geometricarum & Alberti Dureri pictoris et architecti*. Translated by Joachim Camerarius. Nuremberg.

Dürer, A. (1977). *The painter's manual: A manual of measurements for lines, areas, and solids by means of compasses and ruler assembled by Albrecht Dürer for the use of lovers of art with appropriate illustrations arranged to be Printed in the Year MDXXV*. Photostatic reproduction, translated and with commentary by Walter L. Strauss. New York: Abaris Books.

Dürer, A. (1995). *La Géométrie. Translated from the German and with an introduction by J*. Editions du Seuil.

---

[87] Whiteside (1970).

[88] There is a short historiographical survey in Field (2009a).

Field, J. V. (1982). Kepler's cosmological theories: Their agreement with observation. *Quarterly Journal of the Royal Astronomical Society, 23*, 556–568.

Field, J. V. (1984a). A Lutheran astrologer: Johannes Kepler. *Archive for History of Exact Sciences, 31*, 89–272.

Field, J. V. (1984b). Kepler's rejection of numerology. In B. W. Vickers (Ed.), *Occult and scientific mentalities in the renaissance* (pp. 273–296). Cambridge University Press.

Field, J. V. (1986). Two mathematical inventions in Kepler's *Ad Vitellionem paralipomena*. *Studies in History and Philosophy of Science, 17*(4), 449–468.

Field, J. V. (1988). *Kepler's geometrical cosmology*. London and Chicago: Athlone Press and Chicago University Press (reprinted London: Bloomsbury Press 2014).

Field, J. V. (1994). The relation between geometry and algebra: Cardano and Kepler on the regular heptagon. In E. Keßler (Ed.), *Girolamo Cardano: Philosoph, Naturforscher, Arzt* (pp. 219–242). Wiesbaden: Harrassowitz.

Field, J. V. (1997). Rediscovering the Archimedean polyhedra: Piero della Francesca, Luca Pacioli, Leonardo da Vinci, Albrecht Dürer, Daniele Barbaro, and Johannes Kepler. *Archive for History of Exact Sciences, 50*(3–4), 241–289.

Field, J. V. (2009a). The realism of Copernicus and Kepler. *Oriens-Occidens, 7*, 239–267.

Field, J. V. (2009b). Kepler's Harmony of the World. In R. L. Kremer, & J. Włodarczyk (Eds.), *Johannes Kepler: From Tübingen to Żagań. Studia Copernicana, 42*, 11–28.

Galilei, V. (1581). *Dialogo ... della musica antica e della moderna*. Florence: Giorgio Marescotti.

Kepler, J. (1596). *Mysterium cosmographicum*. Tübingen: G. Gruppenbach.

Kepler, J. (1606). *De stella nova*. Prague: Paul Sessius.

Kepler, J. (1609). *Astronomia nova aitiologêtos seu physica coelestis*. Heidelberg: E. Vogelin.

Kepler, J. (1619). *Harmonices mundi libri V*. Linz: Johannes Planck.

Kepler, J. (1620). *Epitomia astronomiae Copernicanae*, Book IV. Linz: Johannes Plank.

Kepler, J. (1621). *Mysterium cosmographicum* (2nd ed.). Frankfurt: Printed by Erasmus Kempfer on behalf of Godefried Tampach.

Kepler, J. (1938). *Johannes Kepler Gesammelte Werke. Band I: Mysterium cosmographicum/De stella nova*. M. Caspar. München: C. H. Beck.

Kepler, J. (1940). *Johannes Kepler Gesammelte Werke. Band VI: Harmonice mundi*. M. Caspar. München: C. H. Beck.

Kepler, J. (1945). *Johannes Kepler Gesammelte Werke. Band XIII: Briefe 1590–1599*. M. Caspar. München: C. H. Beck.

Kepler, J. (1949). *Johannes Kepler Gesammelte Werke. Band XIV: Briefe 1599–1603*. M. Caspar. München: C. H. Beck.

Kepler, J. (1963). *Johannes Kepler Gesammelte Werke. Band VIII: Mysterium cosmographicum (editio altera cum notis)/De cometis, Hyperaspides*. F. Hammer. München: C. H. Beck.

Kepler, J. (1981). *The secret of the universe: Mysterium cosmographicum*. Translated from the Latin by A. M. Duncan with introduction and commentary by E. J. Aiton. Norfolk, CT: Abaris Books.

Kepler, J. (1990). in *Johannes Kepler Gesammelte Werke. Band III: Astronomia nova*, M. Caspar and Kepler-Kommission.München: C. H. Beck.

Kepler, J. (1991). *Johannes Kepler Gesammelte Werke. Band VII: Epitome astronomiae Copernicanae*. M. Caspar and Kepler-Kommission. München: C. H. Beck.

Kepler, J. (1997). *Harmony of the world*. Translated from the Latin with and Introduction and Notes. In E. J. Aiton, A. M. Duncan, & J. V. Field (Eds.), *Memoirs of the American Philosophical Society* (Vol. 209). Philadelphia.

Klein, U. (1971). Keplers Bemühungen um die Harmonieschriften des Ptolemaios und Porphyrios. *Johannes Kepler Werk Und Leistung Linz, 1971*, 51–60.

Neugebauer, O. (1975). *A history of ancient mathematical astronomy (reprinted)*. Springer-Verlag.

Newton, I. (1687). *Philosophiae Naturalis Principia Mathematica*. London: John Streete (for the Royal Society).

Ptolemy, C. (1940). *Tetrabiblos* (reprinted). Edited and translated from the Greek by F. E. Robbins (Loeb Classical Library). London: Heinemann.

Ptolemy, C. (1984). *Almagest*. Translated from the Greek and Annotated by G. Toomer. London: Duckworth.

Tester, S. J. (1987). *A history of Western astrology*. Woodbridge: The Boydell Press.

van Helden, A. (1985). *Measuring the universe: Cosmic dimensions from Aristarchus to Halley*. The University of Chicago Press.

Walker, D. P. (1978). *Studies in musical theory in the late renaissance*. Warburg Institute (University of London) and Brill (Leiden).

Westman, R. S. (1975). The Melanchthon circle, Rheticus and the Wittenberg interpretation of the Copernican theory. *Isis, 66*, 163–193.

Westman, R. S. (2011). *The Copernican question: Prognostication, skepticism, and celestial order*. Berkeley, Los Angeles and London: University of California Press.

Whiteside, D. T. (1970). Before the *Principia*: The maturing of Newton's thoughts on dynamical astronomy, 1634 to 1684. *Journal for the History of Astronomy, 1*(1), 5–19.

Chapter 3
Measuring the Heavens: How Tycho Brahe Revolutionized Observational Astronomy

T. J. Mahoney

The observations made by Tycho (Tyge Ottesen) Brahe (1546–1601) were an order of magnitude more accurate than any that had gone before. It was Kepler's unwavering faith in the accuracy of Tycho's observations that led him to propose the ellipse as the correct description of planetary motion. After a summary of Tycho's early interest in astronomy we describe how he was given overlordship of the island of Hven, where he built the observatory of Uraniborg and embarked on a project that would transform observational astronomy. At Uraniborg and its later extension Stellaborg Tycho planned and executed a major programme of long-term astronomical observing and continuous improvement of his instruments.[1] This work laid the first solid observational foundations of high-precision positional astronomy. These observations provided Kepler with data of sufficient accuracy to enable him to develop his first two laws of planetary motion, which established the first mathematically convincing and observationally sound heliocentric theory (see Chap. 4 of this volume). We examine the accuracy of Tycho's instruments and observations, and end with Kepler's arrival at Prague.

---

[1] Brahe (1996).

T. J. Mahoney (✉)
Instituto de Astrofísica de Canarias, La Laguna, Santa Cruz de Tenerife, Spain
e-mail: tjm@iac.es

© Springer Nature B.V. 2024
A. E. L. Davis et al. (eds.), *Reading the Mind of God*, Springer Praxis Books,
https://doi.org/10.1007/978-94-024-2250-4_3

3.1   Tycho's Early Interest in Astronomy

Tycho was born on 14 December 1546 into an aristocratic family of high rank. He
was not raised by his parents (Otte Brahe and Beate Bille[2]) but 'fostered' by his uncle
Jørgen without seeking the consent of Otte and Beate. Jørgen and his wife Inger
Oxe proved to be caring and affectionate foster parents to their young ward, who
would receive all the advantages of an aristocratic upbringing. From the age of seven
until he was twelve years old Tycho studied at a residential cathedral school,[3] where
he learned to read, write and speak Latin. He then spent three years at university,
where he developed a keen interest in mathematics from his studies of the quadrivium
(arithmetic, geometry, astronomy—which included astrology—and music).

The educational reforms of Philipp Melanchthon (1497–1560) had become the
standard at Lutheran universities, where there would be at least one chair of mathematics. At Copenhagen, Tycho was introduced to the works of Johannes de Sacrobosco (c. 1195–c. 1256), whose *Tractatus de sphaera* (c. 1230) was the most
popular textbook on astronomy from the thirteenth to the sixteenth century,[4] and
Petrus Apianus (1495–1552), author of *Cosmographicus liber* (1524) and the lavish
Astronomicum Caesarium (1540).[5]

When Tycho left Copenhagen University at the end of 1561. Inger Oxe saw to it that
Tycho would continue his education by visiting universities rather than foreign courts.
Tycho was taken to Saxony in the care of a tutor (Anders Sørensen Vedel, 1542–
1616). There he would learn High German and go on a Lutheran pilgrimage. On 14
February 1562, Tycho left Copenhagen in the company of Vedel. They passed through
Wittenberg and stopped at Leipzig, where the university followed the principles of
Melanchthon. Tycho had developed a keen interest in astronomy and 'bought astronomical books secretly, and read them in secret'.[6] There was to be a certain amount of
friction with Vedel over Tycho's waywardness concerning his set curriculum, which
had no place for astronomy. Tycho acquired a small celestial globe, Albrecht Dürer's
(1471–1528) star maps, the *Alfonsine Tables* and the *Prutenic Tables*.[7] Even before
acquiring any instruments, he managed to convince himself, using only a taut string
to measure the motions of the planets, that both sets of tables were inaccurate in their
predictions of planetary positions.[8]

---

[2] It was the custom for aristocratic women in sixteenth century Denmark to retain their maiden
names after marriage.

[3] The identity of the Latin school is unknown. See Thoren (1990), p. 8.

[4] He later became known as John of Holywood. His country of birth is uncertain, there being some
evidence that he was born in the British Isles. See Thorndike (1949).

[5] His real name was Peter Bennewitz (or Bienewitz), which he Latinized to Petrus Apianus (*Bien*
is 'bee' in German; hence 'Apianus', from Latin *apis*, 'bee'). Later anglicized as 'Peter Apian'.

[6] Thoren (1990), p. 15.

[7] Anonymous (1483, 1492); Dürer (1515); Reinhold (1551); Dürer (1555).

[8] Thoren (1990), pp. 15–16.

Tycho acquired a book on astrology by Johannes Garcaeus the Younger (1530–1574)[9] and kept a notebook of horoscopes for important figures. He started the first of his observing logs in August 1563,[10] by which time he had acquired a pair of compasses (in the style of instrument **R** in Table 3.1) that enabled him to make more accurate measurements. The occasion that prompted Tycho to start an observing log was a conjunction of Jupiter and Saturn. He found that neither calculations based on the *Alfonsine Tables* nor those using the new *Prutenic Tables* gave an accurate prediction of so important an event and was aghast at such 'intolerable error' in the timing of the event (amounting to an entire month in the case of the *Alfonsine Tables*).[11]

Bartholäus Scultetus (Barthel Schulze, 1540–1614) showed Tycho the use of the cross-staff, and taught him geography, cartography and navigation. Tycho was quick to find systematic errors in the cross-staff and drew up a table of corrections for its use. Scultetus also introduced him to the use of transversals (see Sect. 3.3) for greater precision.

Tycho and Vedel left Leipzig on 17 May 1565 and reached Rostock on 25 May. Their stay was cut short on news reaching them of the death of Tycho's foster father, Jørgen Brahe. Tycho returned home and spent nearly a year with his family. His natural father insisted that Tycho abandon the life of a wandering scholar to take up duties more befitting a noble at court. But Tycho chose instead to set out for Wittenberg, where he arrived on 15 April 1566 to continue his education. His sojourn was interrupted five months later when an epidemic broke out, forcing the students to flee the town. Tycho left Wittenberg on 14 September and reached Rostock on the 24th of that month, where he matriculated at the University. While there, he observed a lunar eclipse on 28 October 1566.

The University of Rostock taught the Paracelsian doctrine, which was greatly to influence Tycho. According to Paracelsus (Philippus Aureolus Theophrastus Bombastus von Hohenheim, 1493–1541), God permeates all of Nature; the microcosm of the human body was an integral part of the macrocosm, and all that ailed the human frame could be cured by the studious application of medicines derived from the soil and plants. This blending of medicine with alchemy, astrology and astronomy thus produces an harmonious ensemble in the same way that single notes combine to produce musical chords. Tycho's course of study at Rostock was law, but he devoted most of his time there to medical alchemy.

Tycho had by now spent an entire decade as a university student; it was time for him to begin some sort of career. He was averse to following in the footsteps of his male Brahe relatives at court; luckily, there was an acceptable alternative in the form of canonries. After some string-pulling by Tycho's family, on 14 May 1568 royal letters of patent were issued granting Tycho the canonry of Roskilde Cathedral, the burial church of Danish monarchs, when it should next fall vacant.

---

[9] Probably Garcaeus (1556).

[10] Thoren (1990), p. 17.

[11] Brahe (1996), pp. 118–119; Brahe (1921), p. 107.

Table 3.1 Tycho's instruments on Hven (1576–1591)

Instrument	Location	Characteristics and use	Precision	Year built	Ref.[a]
(**A**) Smaller quadrant of gilt brass (*Quadrans minor orichalcicus instauratus*)		Solid (radius = 39 cm) with 44 nonius-type nested quadrants 5′ subdivisions Measured altitude	3′	1573	*Mech.* 11–15 [12–15]
(**B**) Medium sized azimuth quadrant of brass (*quadrans mediocris orichalcicus azimuthalis*)		Solid (radius = 58 cm) Measures altitude and azimuth	1′	1577	*Mech.* 16–20 [16–19] Thoren (1973b), 27
(**C**) Azimuth quadrant of brass (*quadans alius orichalicus etiam azimuthalis*)	Uraniborg (larger southern observatory)	Steel framework (radius = 58 cm) Portable Measures altitude and azimuth			*Mech.* 21–24, 149 [20–23, 144] Thoren (1973b), 39
(**D**) Astronomical sextant for measuring altitudes (*sextans astronomicus, prout altitudinibus inservit*)	Uraniborg (larger southern observatory)	Steel framework (radius = 155 cm) Measures altitude	1/4–1/3′	1584	*Mech.* 25–29, 149 [24–27, 144] Thoren (1973b), 39
(**E**) Mural, or Tychonian, quadrant (*quadrans muralis sive Tichonius*)	Uraniborg (SW ground-floor room)	Mural quadrant arc (radius = 194 cm) No alidade Measures dec	5″	1582	*Mech.* 30–35 [28–31]
(**F**) Revolving azimuthal quadrant (*quadrans volubilis azimuthalis*)	Stellaborg (NE basement)	Steel framework Radius = 155 cm Housed in a movable dome Measures altitude and azimuth	1/4′	1586	*Mech.* 36–39, 153 [32–35, 147] Thoren (1973b), 40
(**G**) Great steel quadrant (*quadrans magnus chalibeus, in quadrato etiam chalibeo comprehensus, unaque azimuthalis*)	Stellaborg (SW basement)	Steel framework held in a steel square Radius = 194 cm Housed in a dome Measures altitude and azimuth	10″		*Mech.* 40–43, 153 [36–39, 147] Thoren (1973b), 40

(continued)

Table 3.1 (continued)

Instrument	Location	Characteristics and use	Precision	Year built	Ref.[a]
(**H**) Great azimuth semicircle (*semicirculus magnus azimuthalis*)	Uraniborg (larger southern observatory)	Diameter = 233 cm Measures altitude and azimuth (made from end of diameter, not the centre)		1588	*Mech.* 44–47, 149 [41–43, 144]
(**I**) Parallactic or Ptolemaic ruler instrument (*instrumentum parallaticum sive regularum*)	Uraniborg (larger southern observatory)	3 rulers (vertical + 2 others of equal length = 155 cm with sliding connector	Approx. 3′		*Mech.* 48–52, 149 [44–47, 144] Thoren (1973b), 39
(**J**) Another parallactic or ruler instrument (*parallaticum aliud, sive regulae, &c.*)	Uraniborg (larger northern observatory)	Length of horizontal ruler = 330 cm Sliding ruler + alidade ruler Bears inscribed 6-figure sine table Measures altitude and azimuth		1583	*Mech.* 53–56, 149 [48–51, 144] Thoren (1973b), 35–37
(**K**) Zodiacal armillary instrument (*armillae zodiacales*)	Stellaborg (NW basement)	4 nested rings (fixed meridian, movable meridian, 2 zodiacal rings) Diameter = 117 cm Measures longitude and latitude	1–2′	1581	*Mech.* 57–61, 153 [52–55, 147] Thoren (1973b), 33, 35
(**L**) Equatorial armillary instrument (*armillae aequatoriae*)	Uraniborg (smaller northern observatory)	3 rings (fixed meridian), Diameter outer ring = 155 cm Measures dec. and RA		1584	*Mech.* 62–66, 150 [56–59, 144]

(continued)

Table 3.1 (continued)

Instrument	Location	Characteristics and use	Precision	Year built	Ref.[a]
(**M**) Another equatorial armillary instrument (*armillae aliae aequatoriae*)	Uraniborg (smaller southern observatory)	4 rings (fixed meridian, movable meridian, 2 equatorial) Diameter outer ring = 155 cm Measures dec. and RA		1584	*Mech.* 67–70, 149 [60–63, 144]
(**N**) Greatest equatorial armillary instrument (*armillae aequatoriae maximae, &c.*)	Stellaborg (S cellar)	Steel framework complete circle + 1 semicircle Diameter of circle = 272 cm. Measures dec	15″	1585	*Mech.* 71–74, 53, 154–155 [64–67, 147, 148] Thoren (1973b), 40
(**O**) Bipartite arc (*arcus bipartitus minoribus siderum distantis inserviens*)	Uraniborg (larger northern observatory	Measures small (< 30°) angular distance between stars	1/2–1′	1583	*Mech.* 75–78, 149 [68–71, 144] Thoren (1973b), 35
(**P**) Triangular astronomical sextant (*sextans astronomicus trigonicus, &c.*)	Stellaborg (SW basement)	Wooden framework Arm radius = 155 cm Measures angular distances	15″	1582	*Mech.* 79–82, 149, 153 [72–75, 144, 147]
(**Q**) Steel sextant for one observer (*sextans chalybeus pro distantiis &c.*)	Uraniborg (larger northern observatory)	Arm radii = 117 cm Measures angular distance	1′	1576	*Mech.* 83–86, 149 [76–79, 144] Thoren (1973b), 27
(**R**) Half-sextant (*alius instrumentum simile priori, pro distantiis*)	Uraniborg (larger northern observatory)	Radii of arms = 155 cm Measures angular distance			*Mech.* 87–90, 149 [80–83, 144]
(**S**) Sextant mounted (*instrumentum eisdem ut altitudinibus capiendis inserviat dispositio*)		Radii of arms = 155 cm Measures altitude			*Mech.* 91–95 [84–87]

(continued)

Table 3.1 (continued)

Instrument	Location	Characteristics and use	Precision	Year built	Ref.[a]
(**T**) The greatest quadrant (*quadrans maximus qualem olim prope augustam uindelicorum extstruximus*)	Augsburg (Paul Hainzel's observatory)	Wooden framework Radius 543 cm Measured altitude of Sun and planets	10″	1569	*Mech.* 96–99 [88–91]
(**U**) Greatest steel quadrant (*quadrans maximus chalybeus cuadrato inclusus*)	Stellaborg (SW basement)	Radius = 0.194 cm Measures altitude and azimuth	10″	1582	*Mech.* 100–102 [92–95]
(**V**) Bifurcated sextant (*sextans bifurcatus*)	Stellaborg (NE and NW corners?)	60° steel arc Alidade length = 155 cm Measures angular distances		1581	*Mech.* 103 [96] Thoren, 35
(**W**) Great semicircle (*semicirculus amplus pro maioribus distantis coelitus denotandis*)	Stellaborg (SE and SW corners)	Diameter = 233 cm Measures major angular distance			*Mech.* 103–104 [96–97] Thoren (1973b), 41
(**X**) Astronomical radius (*radius astronomicus*)	Portable	Radius = 117 cm	Low		*Mech.* 104–105 [97]
(**Y**) Astronomical rings (*annulus astronomicus*)	Portable	Three rings representing the celestial equator, meridian and declination	Low		*Mech.* 106–107 [98] Thoren (1973b), 41
(**Z**) Portable armilla (*armilla parotatilis*)	Portable	One (two) ring(s) Diameter = 117 cm To measure declination (declination and right ascension)	1′	1591	*Mech.* 107 [98–99]
(**AA**) Astrolabe (*astrolabium*)	Portable	Diameter of plates = 78 cm Never built	Low		*Mech.* 107–109 [99]
(**BB**) Great brass globe (*globus magnus orichalcicus*)	Uraniborg, Library (ground floor of S tower)	Diameter = 149 cm Stars mapped on globe	1′ on equator and zodiac	1580	*Mech.* 112–116, 146 [102–105, 142] Thoren (1973b), 30

[a] Unbracketed page numbers refer to the translation of *Mechanica* by Raeder et al. (Brahe, 1996). Bracketed page numbers refer to *TBOO* V. Other citations are denoted by author name, year and page number.

Tycho went to Basel during the summer of 1568, where he met the Dutch scholar and instrument maker Hugo Blotius (1533–1608). He then settled for a while at Augsburg, where he engaged in further instrument design and building. He built another half-sextant (**R** in Table 3.1), but by now Tycho's ambition was to achieve 1 arc minute accuracy, and he thought about building a much larger instrument. Paul Hainzel (1527–1581) pledged the funds for such a venture. Figure 3.1 shows the giant quadrant (**T** in Table 3.1) that was built on Hainzel's estate just outside Augburg. The lifetime of the instrument was very brief (observations were made from 1 April until 15 May 1569), after which the instrument was abandoned as too cumbersome to operate.

Tycho left Augsburg on receiving news of the poor state of his father's health and arrived in Denmark in November. His father died on 9 May 1571. The eventual settling of Otte's estate left Tycho with a small fraction of the whole, but sufficient to enable him to live in comfort. It was in 1572 that Tycho met his future wife, Kirsten

Fig. 3.1 The 'Greatest Quadrant' (*Quadrans maximus*) built on Paul Hainzel's estate in 1569. Reproduced from Dreyer, *TBOO* V, 1913–1929, p. 88

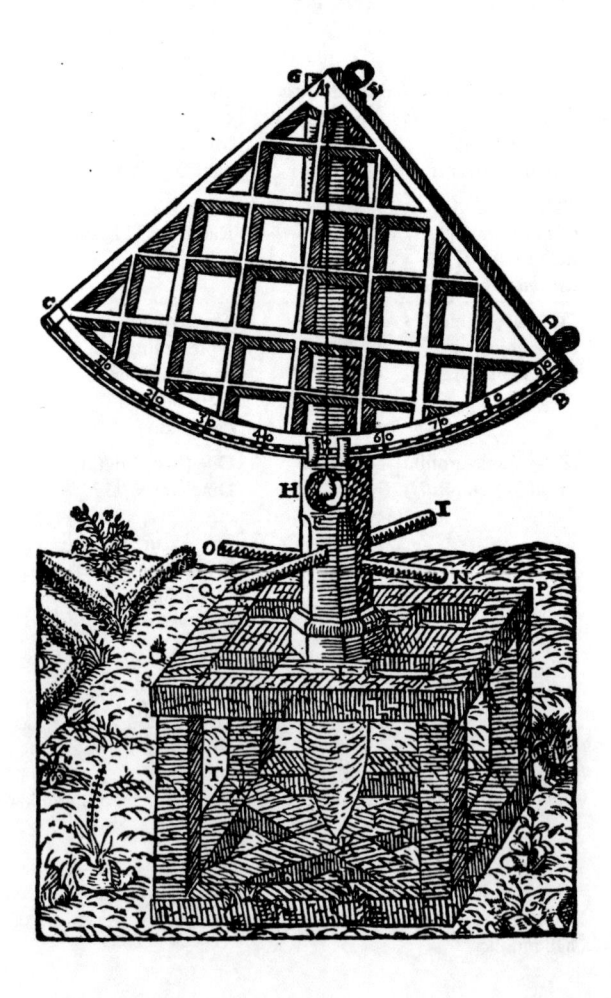

Jørgensdatter, a commoner from Knudstrup. Since it was forbidden by law for nobles to marry commoners, their union had to be morganatic. Given the unwarranted slander to which Tycho and Kirsten were in later years to be subjected, it is important to stress that, although Tycho's children could not inherit Tycho's estate or bear the Brahe name, they were considered legitimate in law.

The nature of Tycho's marriage prevented him from participating in the aristocratic life of the University of Copenhagen, so he decided, towards the end of 1571, to seek more congenial intellectual company at Herrevad Abbey, where his uncle Steen Bille resided. This former Cistercian abbey had been converted during the Reformation into a school and centre of humanist culture reflecting Melanchthon's influence on education. Tycho's main interest at Herrevad was alchemy. He could use the abbey's alchemical facilities to explore experimentally the astrological connection between the heavens and Earth. Tycho became completely immersed in these endeavours, his observing logs recording no astronomical observations until the appearance in the constellation Cassiopeia on 11 November of the New Star of 1572.

Cassiopeia is circumpolar from Herrevad. Tycho set up his sextant vertically at a northern window in order to measure the star's altitude at both upper and lower culmination (to allow for atmospheric refraction) in order to get its declination (which he found to be constant). He also had a cross-staff to measure the angular distance between the New Star and the star Schedir (α Cassiopeiae), which remained constant at $7°55'$ at both culminations. Tycho's observations have been lost, but he published the reduced angular distance measurements for the main stars in Cassiopeia and the New Star.[12] In line with his Paracelsian cosmological outlook, after giving the details of his astronomical observations and conclusions, Tycho devotes 17 pages to a discussion of the astrological significance of such an extraordinary celestial event. The separation in time of the New Star from the conjunction of Jupiter and Saturn of 1563 must surely presage a great upheaval of kingdoms. The star took on the redness of Mars as its light dimmed (yet a further ominous sign). The book also contains a prediction by Tycho of the lunar eclipse of 3 December 1573, together with a disquisition on the astrological significance of eclipses. Tycho noted that a 20 min adjustment would need to be made with respect to the *Prutenic Tables* for an accurate prediction of the timing of the eclipse. This indeed turned out to be the case.

News reached Tycho that the remuneration for the canonry of Roskilde Cathedral that had been promised to him when the canonry became vacant was to be used instead to increase the pay of the academic staff at Copenhagen University. Once again, Tycho was faced with the problem of securing an income that would enable him to continue his investigations and avoid the life of a courtier. By 1574, Tycho had moved his family to Copenhagen. Arrangements were made by Charles Dançay (1510–1589), envoy of France to Denmark and Sweden, for Tycho to give an oration in the reception hall of the French embassy to an audience that included King Frederick II (1534–1588, reigned 1559–1588) and academic staff from the University of Copenhagen.[13]

---

[12] Brahe (1573).

[13] Brahe (1621); Brahe (1923a), pp. 144–173.

The oration was well received and Tycho was granted permission to give a lecture course in astronomy at the University.

When news reached Tycho of an astronomical observatory at Hesse-Cassel, he abandoned the second half of his lecture course in order to pay a visit there. He arrived in 1575 to see for himself that Wilhelm IV, the Landgrave of Hesse-Cassel (1532–1592, reigned 1565–1592), did indeed have a lavishly equipped observatory and carried out his own observations, which were of far better quality than those of any of his contemporaries. It was at Cassel that Tycho met Christoph Rothmann (b. 1550–1560, d. after 1600), who had been appointed mathematician of Cassel in 1571 (he was responsible for computing and compiling the Cassel star catalogue). It was also at Cassel that Tycho became acquainted with Jost Bürgi (1552–1632), a gifted instrument and clock maker (he was to enter into Rudolf II's service in 1604 and became friends with Kepler). After leaving Cassel, Tycho's travels then took him to the Frankfurt Book Fair and then on to Venice. When he returned from Italy, he visited Regensburg to attend the coronation of the Holy Roman Emperor Rudolf II (1552–1612, reigned as Holy Roman Emperor 1576–1612). It was here that he met Thaddeus Hagecius (Tadeáš Hájek, 1525–1600) with whom he was to form a life-long friendship. After the coronation, Tycho returned to Denmark.

News of Tycho's outstanding qualities as an astronomical observer had reached King Frederick from the Landgrave, who strongly recommended Tycho for royal favour. Frederick communicated with Tycho to discuss offering him a choice of fiefdoms where he might build an observatory. Tycho, however, was unenthusiastic since a fief would entail court duties and his morganatic marriage would prevent him from integrating into courtly society. In any case he was planning to move to Basel, where, having already made his mark with *De nova stella*, he would be sure to find a congenial academic climate. He placed the matter of the king's generous offer in the hands of his uncle Steen and continued with his preparations for the move to Basel. Steen, who had direct access to the king, told him of Tycho's concerns over the duties involved with the fiefs offered and his fears that they would interfere with his astronomical observing and alchemical experiments. The king then astonished Tycho with an offer of the fief of Hven, an island (now belonging to Sweden) in the Øresund, whose distance from the mainland would release him from courtly duties and allow him to pursue his investigations undisturbed. The king asked Tycho to let him know his answer as soon as possible. Johannes Pratensis (1543–1576), a future professor of medicine who was friendly with Tycho, and Dançay both urged him to accept the king's offer, which Tycho did on 18 February 1576. The offer included a pension and an annual cash grant for setting up his residence, which, at the king's insistence, would house not only an astronomical observatory but also a fully equipped alchemical laboratory.[14]

---

[14] Brahe (1921), p. 109; Brahe (1996), p. 121.

3.2 Tycho's Arrival at Hven

On 23 May 1576 Frederick II conferred the fief of the island of Hven (Fig. 3.2) on Tycho. The inhabitants of the island were to provide tools for labour and two days' unpaid labour per week. Skilled builders and workmen could be ferried across from the king's castle in Helsingør that was undergoing renovation.[15]

The building was to be constructed in contemporary Italianate style. From a practical viewpoint, the castle had to house every aspect of Tycho's life on Hven, including residential areas, astronomical observations, alchemical investigations and the administration of the fief.

Work started immediately after Frederick signed the documents ceding the fief of the island to Tycho. The NS and EW axes of the building were marked out and excavation commenced, along with earthworks for the outer perimeter of the castle grounds. On 8 August 1576, noble visitors gathered at the laying of Uraniborg's cornerstone. Tycho made the first solar observations from Hven on 14 December. It would take a further four years for the castle to be made habitable.

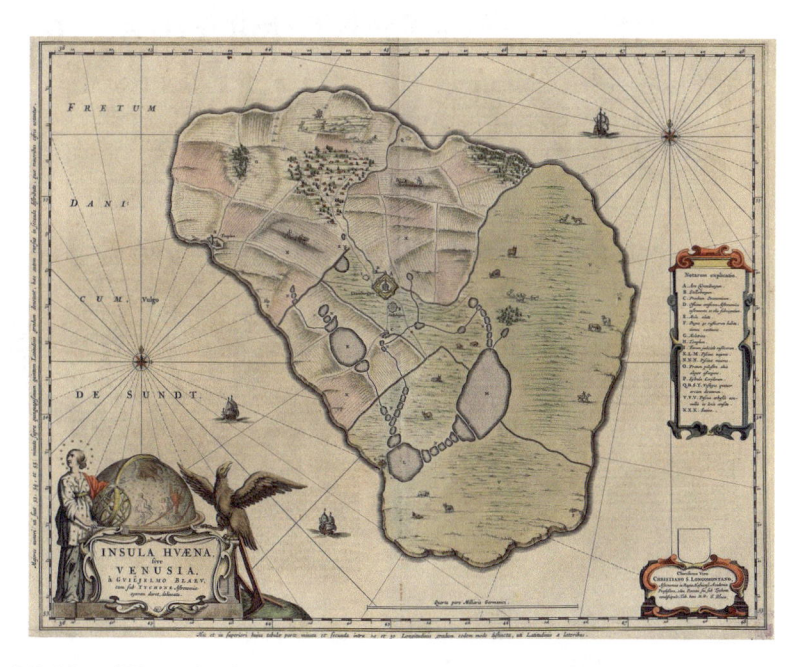

Fig. 3.2 Map of Hven, showing Uraniborg (near centre) and Stellaborg a little to the south-east. Reproduced from Joan Blaeu's *Atlas Maior*, Amsterdam, 1662–1672

---

[15] Thoren (1990), p. 113.

3.2.1 Uraniborg

Uraniborg (55°54′28″ N, 12°41′48″ E) is located more or less in the centre of the island of Hven. Tycho gives length measurements in cubits and digits. There are various definitions of the cubit. The lengths given in this chapter are taken from the Raeder et al. translation of *Mechanica*, where 1 cubit = 38.8 cm. Figure 3.3 shows the layout of the castle grounds. Two gates mark the east (*E*) and west (*D*) entrances to the grounds. Dogs were housed in kennels above the gates to announce the approach of visitors. A printing press was located at the south corner, and servants were housed in the north corner. Four pavilions were located inside the semicircular walls. Three hundred trees formed an avenue between the castle walls and the flowerbeds. Each wall was 91.4 m long, the inside diameter of the semicircular walls was 27.4 m, the tapered walls were 6.1 m thick at their base and 6.7 m high.

Figure 3.4 shows Uraniborg from the east (top) and in plan (bottom).[16] The elements of the building are labelled in the illustrations and described in the caption.

Fig. 3.3 Uraniborg Castle (A) and its grounds. Reproduced from Dreyer, *TBOO* V, 1921, p. 142

ARCIS VRANIBVRGI QVO AD TOTAM CAPACITATEM DESIGNATIO.

---

[16] Brahe (1921), pp. 138–145; Brahe (1996), pp. 145–151.

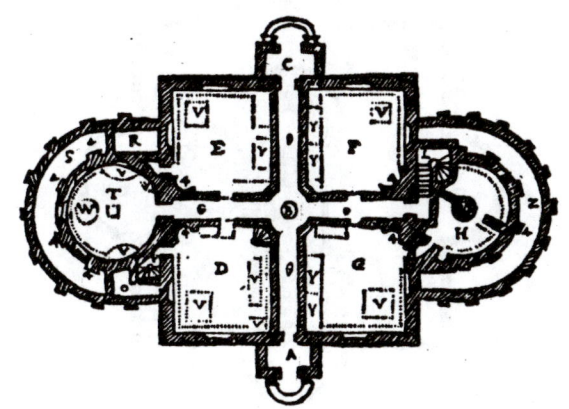

Fig. 3.4 *Top:* Castle of Uraniborg (the keys are better read from the Blaeu reproduction shown here). A: east door; C: west door; O: corridors; D: winter dining room; E, F, G: bedrooms; L: stairway; H: kitchen; T: library; W: globe; V: tables for assistants; Y: beds; G, H: basement windows; L: subterranean chemical laboratory; Z: wood cellar; D: red chamber; E: blue chamber; α: yellow octagonal chamber; X: upper storey windows; O: large south observatory, containing the azimuthal semicircle, Ptolemaic rulers, brass sextant, medium brass azimuthal quadrant; Q: octagonal gallery containing ball on which instruments to measure angular distances between stars were mounted; N: small observatory containing equatorial armillae; W: stairs leading to the basement laboratory and the observatory; R: large south observatory, containing rulers and the larger parallactic instrument, sextant, and bipartite arc, and Copernicus' parallactic instrument; S: small north observatory, containing an equatorial armillary; ε: uppermost octagonal gallery; β: chimneys; γ: octagonal structures bearing representations of the seasons; ν: clock (beneath belfry); λ: gilt Pegasus. Reproduced from Blaeu's *Atlas maior*, 1663. *Bottom:* plan of the castle. Reproduced from Dreyer, *TBOO* V, 1921, p. 142

ICHNOGRAPHIA STELLÆBVRGI.

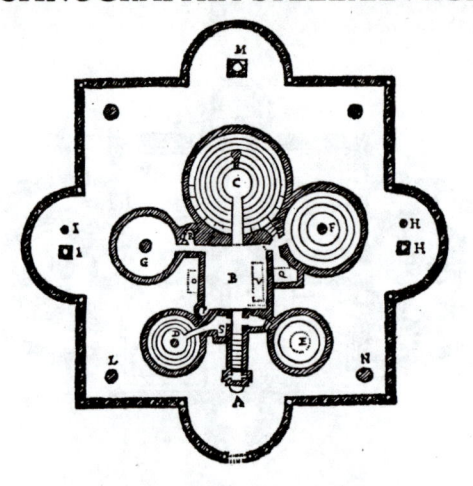

Fig. 3.5 *Top*: Stellaborg. Reproduced from Blaeu's *Atlas maior*, 1663. *Right*: A: portal to obser-vatory cellars; B: a round ceiling above the heating installation; C: cellar for the largest equatorial armillae; D: a cellar for the large revolving quadrant; E. cellar for zodiacal armillae; F. cellar for the large steel quadrant; G. cellar for the ball-mounted four-cubit sextant. *Bottom (view rotated 90 degrees anticlockwise)*: plan of Stellaborg. H, I: stone columns; K, L, N, T: ball mounts to support sextants; M: stone table; O: bed for Tycho; Q: larger bed for assistants; P: stove; V: table; S: entrance to underground passage to connect to castle. The square perimeter measures 70 feet (21 m) and the diameter of the semicircles is 24 feet (7.2 m). The entrance (A) faces north.[17] Reproduced from Dreyer, *TBOO* V, 1921, p. 146

---

[17] Dreyer (1890), p. 104 (Tycho does not state Stellaborg's orientation).

3.2.2 Stellaborg

After the completion of Uraniborg Castle Tycho saw the need to continue building new instruments. The main building had limited space, so in 1584 he built Stellaborg (Stjerneborg) close by (Fig. 3.5).[18]

3.3 Tycho's Instruments

Tycho described his observatories and instruments exhaustively in two works: *Astronomiae instauratae mechanica* (1598)—hereafter *Mechanica*—and *Astronomiae instauratae progymnasmata* (1602)—hereafter *Progymnasmata*.[19] The development of instrumentation at Uraniborg and Stellaborg proceeded at a hectic rate, 30 instruments being designed, built and upgraded over the 18 years of the observatory's existence.[20]

The instruments used by Tycho at Uraniborg and Stellaborg are listed in Table 3.1. Tycho continually modified their design to make observations carried out on them more accurate. He used such mechanical solutions as frameworks (variously of wood or steel) to avoid gravitational flexure in his larger instruments. This theme is recurrent in all of Tycho's design work:

> Only this I wish to state here with regard both to this instrument [(**B**) in Table 3.1] and to the others, namely, that all of it has to be as nearly perfect as is possible in every aspect and that, therefore, one should employ skilful craftsmen, who know how to carry out this sort of work artfully, or else can learn how to do it. And if they do not hit the nail on the head the first time (as they say), the constructor must not let himself be discouraged, but have the work repeated and improve the defects in every way, until none is left.[21]

3.3.1 Properties of the Eye

We summarize here some pertinent properties of the human eye, or rather the eye–brain mechanism. The eye's system of refractions, as Kepler discovered,[22] produces an inverted image on the retina. The brain compensates for this inversion and enables us to 'see' uninverted images.[23] The retina has two kinds of light-sensitive cells: cones (which react to colour) and rods (which do not). The most sensitive part of the retina is the *fovea centralis*. In low light conditions only the rods are activated (although

---

[18] Brahe (1921), pp. 146–149; Brahe (1996), pp. 152–155.

[19] Brahe (1602), published posthumously (see Sect. 3.8).

[20] Thoren (1973a).

[21] Brahe (1996), p. 20; Brahe (1921), pp. 18–19.

[22] See Donahue, 'Kepler's work on optics' (Chap. 7 of this volume).

[23] For a more detailed description of the eye consult Kitchen (1984), pp. 2–6.

colour is perceived for the brighter stars and planets). The iris controls the amount of light reaching the retina. At night, when the pupil is at its widest, the theoretical resolution of the eye would reach 20″, but resolution is degraded by atmospheric turbulence and the finite width of retinal cells. For two objects to be resolved, there must be at least one unexcited cell between the activated cells, making the actual resolution of the eye 1–2′. The final resolution of an observation, however, is not limited by the eye alone but may be improved with the aid of such mechanisms as transversals[24] and by making many observations of the same star. Tycho certainly had an intuitive understanding of the need for repeated observations of reference stars, as his observing logs[25] attest.

3.3.2 Tycho's Principal Instruments

Four instruments in particular enabled Tycho to produce his finest astronomical observations: the mural quadrant (**E**), the great steel quadrant (**G**), the great wooden quadrant (**F**) and the great steel armillary (**N**).

3.3.2.1 The Mural Quadrant (E)

The mural quadrant (Fig. 3.6) was the most important of Tycho's observing instruments. Accurately aligned with the local meridian (see Sect. 4.2), it enabled Tycho and his assistants to measure the culmination of stars across the plane of the quadrant arc. The observing team (Fig. 3.6) comprised the observer (*F*), a recorder (*G*) to log the transits, and a third member to call out the transit times.

The quadrant was fixed to the west wall of the south-west ground-floor room of the central part of the castle.[26] The brass arc of the quadrant had a radius of 205.74 cm, a width of 12.7 cm and a thickness of 5.08 cm.[27] There is a slit in the south wall of the room level with the 0° mark at the top of the arc.

There is a horizontal gilt brass cylinder whose axis is perpendicular to the plane of the arc. Its width is equal to the separation of vertical sides of the dioptres (to avoid parallax), which slide along the arc.

---

[24] See Sect. 3.3 of this chapter for a description of transversals.

[25] Brahe (1923b, 1924, 1925b, 1926).

[26] Dreyer (1890), p. 99.

[27] These values are taken from Dreyer (1890), p. 101, where the dimensions are given in Imperial measure (feet and inches).

Fig. 3.6 Observing with the mural quadrant. Three observers were assigned different tasks: the observer proper, the timekeeper and the amanuensis. Observations were made by sliding the dioptre to the correct altitude of the star about to transit, and the time recorded and duly noted. Reproduced from Brahe, T. (ed. Curtz, A.), *Historia coelestis*, 1666, p. 113

3.3.2.2 The Great Steel Quadrant (G) and the Revolving Azimuthal Quadrant (*F*)

The great steel quadrant (Fig. 3.7, *left*) was located in the south-west basement of Stellaborg. It was encased in a square steel frame and was more accurate than the revolving azimuthal quadrant referred to below. Strict verticality was obtained by means of plumb lines. The frame of the quadrant made contact with the azimuthal frame above the observer by means of two rods, one of which was used to measure azimuths. Single degrees are marked on the arc and each degree is further subdivided into single minutes, with transversal divisions allowing reading down to 1/6′. Even the supporting square frame has divisions that enable readings to be made to 1/6′.

The revolving azimuthal quadrant (Fig. 3.7, *right*) measures 'most accurately' altitudes and azimuths. Housed in the north-east basement of Stellaborg, it measured 155 cm in radius and the arc had transverse subdivisions that permitted the reading of angles as small as 15″. The pointer was equipped with peg-and-slit sights to minimize parallax error. At the top of the quadrant a pointer (*Q*) indicates the azimuth on the azimuth circle. The complex framework of the quadrant maintains rigidity. The perpendicular arm is aligned with the plumb line (*S*) to ensure verticality. The quadrant rotates about a strong iron axis. The azimuth circle zero point is aligned with the meridian to within 1′. The entire assembly is covered with a roof with observation windows.

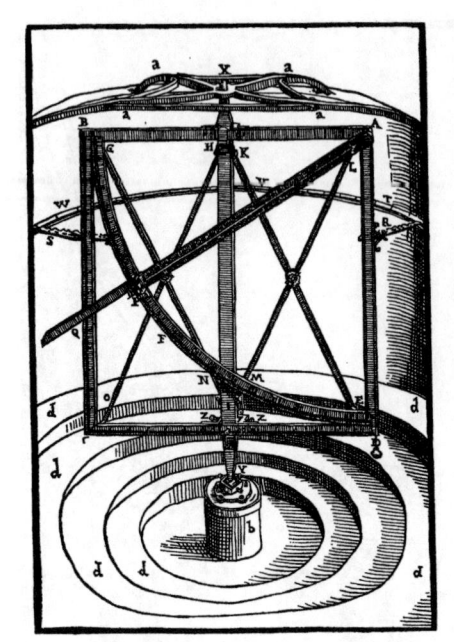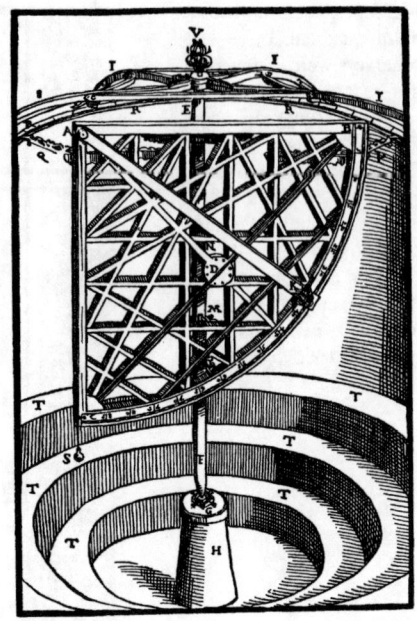

Fig. 3.7 *Left*: The great steel revolving quadrant. *Right*: The revolving azimuthal quadrant. Reproduced from Dreyer, *TBOO* V, 1921, p. 36 and p. 32

The two quadrants could also measure the altitudes and azimuths of stars and planets. They were also used to measure the declination of the Sun in order to establish the solar orbit (their zero-azimuth points were accurately aligned to within 1′ of the meridian).

3.3.2.3   The Great Armillary (N)

This giant (diameter = 272 cm) structure (Fig. 3.8) was quite revolutionary both for its unusual structure and because it is the first known instrument to have an equatorial mount. Again, Tycho has taken great care to avoid flexure by means of the circle's complex framework. The lower bearing is self-centring to ensure that the polar axis of the instrument remains correctly aligned with the Earth's polar axis. Single degrees are marked and readings can be made to 1/4′. This arrangement allowed declinations to be measured directly and accurately (by Tycho's mature standards). It had two alidades, enabling the observer to make a second observation by swinging the circle through 180°.

Fig. 3.8 The greatest equatorial armillary. Reproduced from Dreyer, *TBOO* V, 1921, p. 64

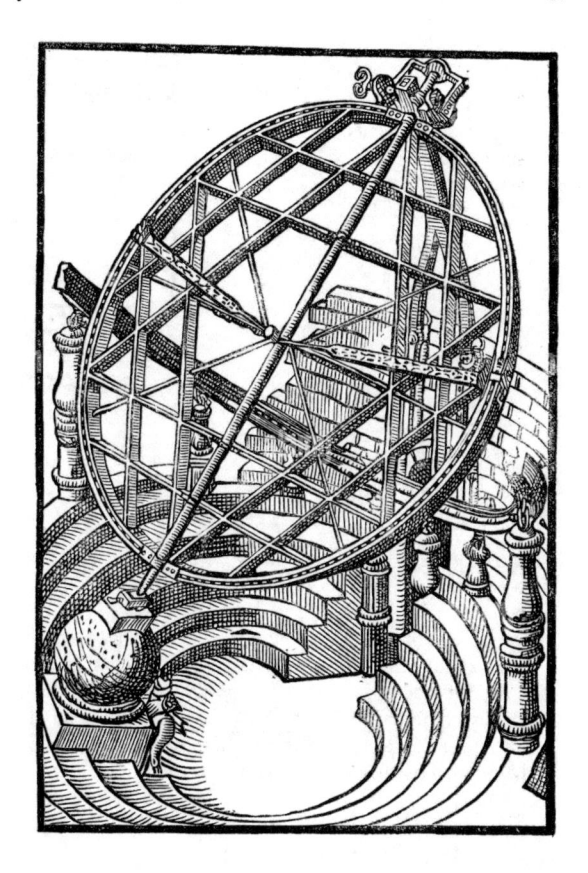

3.3.3   Transversal Subdivisions of the Arc

Tycho used a system of transversal subdivision of his arcs that enabled him to achieve a surprising degree of precision in his angular measurements. Figure 3.9[28] shows a 2-degree portion of an arc, with divisions corresponding to 1°, 1/2°, 10′, and 1′ (this last being represented by the distance between adjacent radial dots). Tycho explains further,

> Later on I adapted it conveniently to arcs on my instruments, as I stated ten years ago in my book on the comet of 1577 at the bottom of page 461. Here I say as follows: *For although the proof of the correctness of this method applies specially to rectilinear parallelograms, it may yet be maintained with good reason for curved lines also without appreciable error, provided the length is so small that the deviation from a straight line is imperceptible.*[29]

By interpolating between adjacent dots readings of 0.5′ or better could be made (depending on the width of the arc).

---

[28] Brahe (1921), p. 153.

[29] Brahe (1996), p. 161.

Fig. 3.9 A 2-degree section
of an arc showing how radial
dots gave enhanced precision
to angular measurements, as
described in the text.
Reproduced from Dreyer,
TBOO V, 1921, p. 153

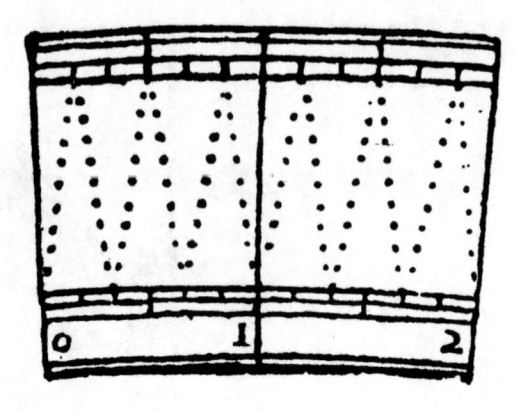

3.3.4 Dioptres and Other Sights

Prior to Tycho's introduction of the peg-and-slit dioptre, astronomers used pinhole
dioptres, which were prone to serious error. As Tycho explains,

> The method of observation through slits that have the same mutual distance as those on the
> other pinnule I invented driven by necessity. For when using the method which is otherwise
> ordinarily employed, it is extremely difficult to see stars through holes, especially through
> the pinnule farthest from the eye unless it is sufficiently large [i.e. bright]; and in that case
> one may err by a considerable fraction of a degree, since it is impossible to know whether
> the sighting has been made completely centrally.[30]

3.3.4.1 Two-Dimensional Sights

Tycho also used two-dimensional sights, which worked on the same principle as
their one-dimensional counterpart, but which could determine the position of a star
both horizontally and vertically (see Fig. 3.10). Tycho describes their construction
as follows:

---

[30] Brahe (1921), p. 155; Brahe (1996), p. 165.

The arrangement of pinnules or diopters which we have found to be the most suitable is such that the lower pinnule, or that closest to the eye, has slits on all four sides, exactly corresponding to the upper pinnule in such a way that they are at the same distance from its four sides with regard to the line of sight and correspond to them. This is indicated in the accompanying figure, as far as it was possible to do so on a plane surface. Here ABCD denote the pinnule that is held close to the eye of the observer, while EFGH is the other and more distant one which is located at the circumference of the instrument. Finally I denotes the alidade on to which these pinnules are fastened in a suitable way and at right angles to it. The pinnule FGHE must have exactly the same form as the other one, BCDA. The small springs, however, which are mounted on the lower pinnule on three sides and which are perfectly straight on the sides facing the pinnule, can be pressed towards the pinnule or removed a little from it. In this way the slits can be made perfectly equal, and it is also possible to widen or narrow the slits during the procedure, should this prove necessary. This can be done by means of an ingenious special arrangement on the other, that is the inner side of the pinnule. By turning one single screw, that is by one single manipulation, it is possible to widen or narrow all the slits simultaneously without any trouble or waste of time. The fourth slit which is carved on that side of the pinnule by which it is fastened to the alidade, remains unchanged all the time. It is seen a little above BA, and at the same distance from the plane of the alidade a second slit is seen in the upper pinnule at FE. This innermost slit, however, can be made adjustable in width in the same way as the others by a minor addition to the construction.[31]

Tycho then goes on to explain its use:

The use of the pinnules is for measuring altitudes of the stars. The alidade I is raised or lowered until the star is seen through the slit DA and in the slit HE at the side of the other pinnule, while at the same moment just as much of the star is seen through the slit BC at the other side GF. In that case there can be no doubt that this star has been sighted centrally and accurately. If it is desired to find azimuths as well, one has to look through another slit CD towards the forward side GH and simultaneously through the slit BA towards the other side FE; in this way the stars are observed most quickly. In making solar observations, however, circumstances are as follows. When the rays entering through a round hole in the upper pinnule in proportion to the amount of sunlight admitted by this hole in all directions fill a circle drawn on the inner side of the lower pinnule, then the required result is obtained. Further it should be noticed that in some instruments the pinnule farthest from the eye is of cylindrical form. The situation is the same as before, only it is now the shadow of the cylinder that has to be observed in the case of solar observations. Finally in the case of the armillae we make use of a round axis in order to make it possible to sight towards it from all sides. For both the cylinders and the round axes have the special advantage that they can be used not only by one but simultaneously by two observers.[32]

One should not underestimate the impact this innovation in sighting technique made on observational astronomy. Tycho recounts how the renowned mathematician Paul Wittich (*c.* 1546–1586), on examining Tycho's dioptres, 'uttered a cry of joy and assured me that he had now come to know something he had sighed for in vain for many years.'[33] Wittich would then go on to apply the same sighting technique to the instruments of Wilhelm IV, Landgrave of Hesse-Cassel (much to Tycho's annoyance).

---

[31] Brahe (1598), (no page numbers in original; Brahe (1921), pp. 154–155; Brahe (1996), pp. 163–164.

[32] Brahe (1598); Brahe (1921), p. 155; Brahe (1996), pp. 164–165.

[33] Brahe (1598); Brahe (1921), p. 155; Brahe (1996), p. 165.

Fig. 3.10 A typical
two-dimensional dioptre
used on Tycho's instruments.
Reproduced from Brahe, T.
(ed. Curtz, A.), *Historia
coelestis*, 1666, p. 111

The mural quadrant, with its fine angular gradations and parallax-free dioptres enabled declinations of unparalleled accuracy to be obtained.

3.4 The Accuracy of Tycho's Results

When Nicolaus Copernicus (1473–1543) published *De revolutionibus orbium coelestium* 1543, that work was greeted enthusiastically by Erasmus Reinhold (1511–1553), who then set himself the task of calculating ephemerides based on Copernicus' heliocentric model. The first edition of these *Prutenic* (Prussian) *Tables* appeared in

1551 and replaced the earlier *Alfonsine Tables* based on the Ptolemaic geocentric world system. They were reprinted in 1562, 1571 and 1585. We have seen (Sect. 3.1 of this chapter) Tycho's shock at the poor accuracy of predictions based on the *Alfonsine* and *Prutenic Tables*. We now examine how Tycho (and Kepler) set about reforming the predictive power of astronomy through ephemerides based on the *Rudolphine Tables* (see Chap. 9 of this volume), which were in turn based on Tycho's observations at Uraniborg and Stellaborg.

Even though the long-term reliability of the *Rudolphine Tables* was strong evidence for the accuracy of Tycho's observations, there has been a strange reluctance by some to accept Tycho's own assessment of his accuracies (plural because they depended on the instruments used and their state of repair). However, a number of studies, which we examine here, find in Tycho's favour. We shall see that his observations really were as good as he claimed.

When Tycho gives a value for the precision of a specific instrument, he refers to the smallest discernible graduation on the instrument's scale, not to the error in an estimated value when using that instrument. He distinguishes clearly between the resolution of his instruments and the observational errors that inevitably arise from their use. Tycho quantified the error introduced by adapting parallel transversals to the circular arc, and how he satisfied himself that the resulting instrumental error could safely be ignored.[34] The error introduced when observing with that instrument was of a different kind, arising from random errors in the reading of the scale. Tycho understood that this random error could be reduced by calculating the mean values of stellar altitudes from observations repeated over long periods of time such that the average random error would be smaller than the nominal resolution of the instrument itself.

Dreyer cites Tycho's mature assessment of the accuracy of his own work:

> [Tycho] divides his observations into "pueriles et dubitae" ['childish and doubtful'] (at Leipzig), "juveniles et mediocriter se habentes' ['juvenile and habitually mediocre'] (up to 1574) and "viriles, ratae et certissimae" ['virile, precise and absolutely certain'] (from 1576).[35]

But in spite of this severe judgement of his early work, even when still a callow youth Tycho set himself very high standards, and we shall see in what follows that, even on his worst nights, he was capable of producing useful results.

We now examine some examples of the closeness of Tycho's results to modern values.

---

[34] Brahe (1598); Brahe (1921), pp. 153–154; Brahe (1996), pp. 161–163.

[35] Dreyer (1890), p. 262, n. 2.

3.4.1 A Note on Predicting the Past ('Retrodiction')

Astronomical observations made from the surface of the Earth are susceptible to atmospheric refraction. There was no satisfactory mathematical model for atmospheric refraction in Tycho's day (indeed, he considered it to be negligible for altitudes greater than 20°). Observations made at widely different times ('epochs') must also take into account various secular variations in stellar positions (precession of the equinoxes is of primary concern in the case of Tycho's observations). In order to make proper assessments of the accuracy of Tycho's results, the historian must, besides taking precession into account, also apply a model of the atmospheric conditions that were likely to have prevailed at Uraniborg when Tycho made his observations. Once this has been done, the historian may then meaningfully compare Tycho's results with recent star catalogues. This process of retrodiction has been applied to Tycho's observations by several workers, as we shall now show.

3.4.2 Establishing the Meridian of Uraniborg

Prior to making any meaningful positional observations of the stars, it is necessary to determine as precisely as possible the local meridian of the site. An event occurred long after Tycho's death that led some mistakenly to question the accuracy of his observations. Tycho's observations were celebrated for their accuracy and his catalogue, if it could indeed be trusted, would be an essential database for astronomers working at other longitudes. Astronomers at Paris needed accurate assessments of the longitudes of both Paris and Uraniborg. Consequently, the leader of the Paris observers, Jean Picard (1620–1682), made a visit to the Uraniborg site in 1671[36] in order to calculate the accuracy of Tycho's measurement of the local meridian of Uraniborg. For this purpose he measured the azimuths of a couple of church spires on the mainland visible from Uraniborg. When comparing his calculations with Tycho's manuscript notes in Copenhagen, he found a difference of 14′ between his own measurements and those of Tycho. A 14′ discrepancy—almost the width of a half-moon (15′)—if real, would have seriously undermined confidence in Tycho. Picard was a reputed geodesist, but he had chosen church spires that were different from the ones used by Tycho so that most of the discrepancy between Tycho's and Picard's results could thereby be accounted for. Picard did not himself take the matter further; unfortunately, however, others wildly inferred from this difference that Tycho was a blunderer or that the meridian had somehow shifted by the time that Picard made his measurements. Dreyer[37] finally laid this myth to rest and put the matter to the test by examining the azimuth of a number of bright stars observed in 1582 when they were at, or close to, the prime vertical (in order better to highlight any noteworthy error in the azimuth circle of the instrument used). Dreyer concludes,

---

[36] Picard (1680).

[37] Dreyer (1890), pp. 358–360, 388–389.

> The observations of the comet of 1585 [...] prove conclusively that in that year the great armillae were in excellent alignment, so that Tycho cannot have made use of any badly placed meridian mark. I have also computed a number of observed altitudes and azimuths of stars from 1582, and from these it is evident that the zero line of the azimuth circle was within 1' of the meridian. [...] [As Tycho] frequently states that he verified his instruments by observations, it is impossible that he can, even before 1586, have made a mistake of 14' in azimuth in the adjustment of his numerous instruments.[38]

Regardless of Dreyer's debunking of the myth, the supposed misalignment of Tycho's meridian still occasionally surfaces.[39] Wesley, however, dismisses the myth entirely.[40]

3.4.3 Stellar Observations

To make accurate measurements of the motions of the Sun, Moon and planets it is first necessary to have accurate positions for a large number of reference stars distributed along the ecliptic; these positions are acquired through multiple observations over time of each star as it crosses the local meridian. Once a reference star has been observed in this way and its position accurately obtained, it may be used to measure accurately a standard star and a planet using other instruments away from the meridian. Such relative measurements are less accurate unless the individual positional errors of each measurement are complemented by additional simultaneous altitude measurements (in order to account for atmospheric refraction). The mural quadrant entered into operation in 1582 and by 1585, with the aid of transversals, was producing measurements with a precision of $\pm 1'$ even for unrepeatable observations—i.e. of objects moving against the background stars.[41]

Before discussing the quality of Tycho's stellar data in detail, let us visually compare the scattered residuals of Claudius Ptolemy's (*fl.* AD 129–141) star catalogue (Fig. 3.11, *left*) with those of Tycho (Fig. 3.11, *right*). With the exception of about a dozen stars, Tycho's data are very tightly clustered about the zero-residual line and immediately show an order of magnitude improvement over Ptolemy's. We shall discuss Rosa's analysis of Tycho's accuracy a little further on.

Tycho used the value 23°31'30" for the obliquity of the ecliptic,[42] which introduced what we would today call systematic error into the star positions of his catalogue that far exceeded the high precision of his many-times-repeated measurements of star positions (5" in the case of the mural quadrant). These large errors were further compounded by faulty assumptions concerning atmospheric refraction and the obliquity of the ecliptic, and were in no way the result of either inherent defects in Tycho's

---

[38] Dreyer (1890), p. 350.

[39] See, for example, Forbes, Meadows and Howse (1975), p. 6.

[40] Wesley (1978).

[41] Thoren (1973a).

[42] The obliquity value for 1 January 2000 stood at 23°26'21".448 (Lang, 2006, vol. II, p. 19).

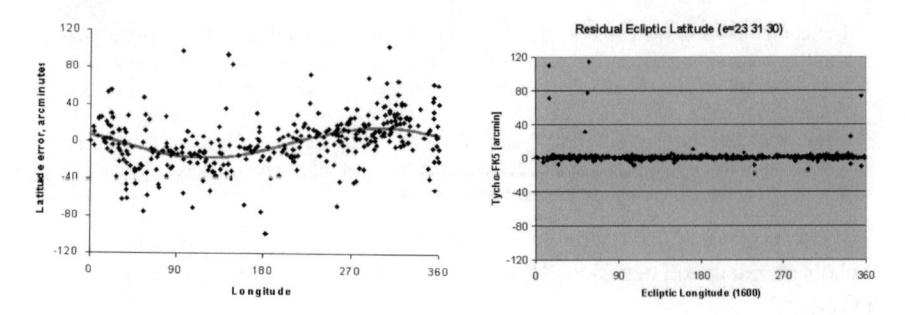

Fig. 3.11 *Left*: Residuals of Ptolemy's star catalogue. *Right*: Residuals of Tycho's star catalogue on an identical scale. The improvement over Ptolemy is immediately apparent) *Credit* Národní Technické Muzeum Praha

instruments or poor observing practices on his part. Use of a wrong value for the obliquity of the ecliptic does not affect the right ascension and declination, but it will systematically produce wrong values for ecliptic latitude and longitude in the conversion from equatorial to ecliptic coordinates.

A number of studies have been done to compare Tycho's results with more recent catalogues. Apart from such obvious factors as precession of the ecliptic, these studies include corrections for the slow changes in the obliquity of the ecliptic (along with other secular variations that need not concern us here) and atmospheric refraction. We shall see that these studies confirm the accuracy of Tycho's observations once these systematic errors are taken into account.

Friedrich Argelander (1799–1875) carried out a study of Tycho's half-sextant observations of the New Star of 1572[43] (Fig. 3.12). Even though the instrument (Fig. 3.13, *left*)[44] lacked many of the refinements of his later instruments (e.g. double-slit dioptres and/or transversals), Tycho managed to produce very good results for the position of the supernova. (The sextant in Fig. 3.13, *right*, is sometimes referred to as the instrument used for Tycho's observations of the New Star, but it is set up in this illustration to measure altitudes, not angular distances between celestial objects.) The familiar W-shape of the constellation is marked (in clockwise order) by the stars labelled *G, B, D, E* and *F*. The key (Fig. 3.12, *right*) identifies the stars used by Tycho and gives their equivalents in the stellar nomenclature system of Johann Bayer[45] (1572–1625). Notwithstanding the crudity of the instrument used, Argelander concluded that Tycho's half-sextant measurements of the distances between the stars of Cassiopeia were good to ± 41″. Argelander's investigations confirm that Tycho, even in his earliest astronomical endeavours, hankered obsessively after maximum accuracy in his observations.

---

[43] Argelander (1864).

[44] Instrument **S** in Table 3.1 (Fig. 3.13) is usually chosen to represent the actual sextant used.

[45] Bayer (1603).

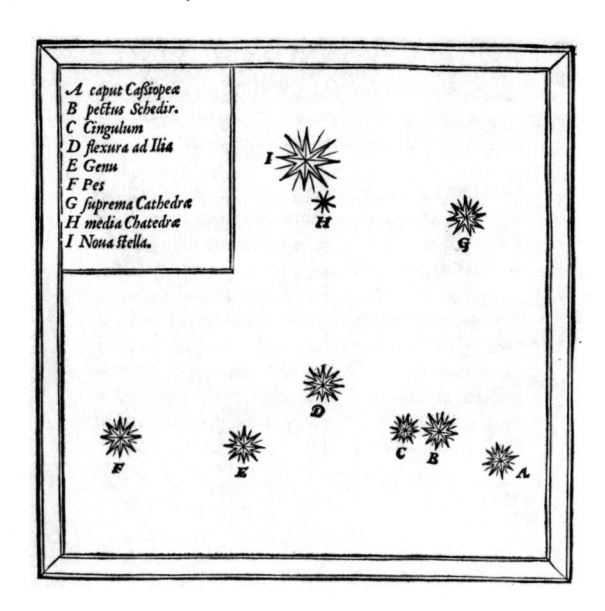

Fig. 3.12 Tycho's chart showing the New Star of 1572 (ι Cas not shown on map). *A*: Cassiopaia's head (ζ Cas); *B*: Chest (Schedir, α Cas); *C*: Waist (η Cas); *D*: Lower abdomen (γ Cas); *E*: Knee (δ Cas); *F*: Foot (ε Cas); *G*: Top of chair (β Cas); *H*: Base of chair (κ Cas); *I*: New star (SN 1572). Reproduced from Brahe, T., *Nova et nullius aevi memoria prius visa stella*, 1573, p. B1

Dreyer[46] compared the positions of nine standard stars used by Tycho (see Fig. 3.14, *left*) for the year 1586 with the positions of the same stars computed by James Bradley (1693–1762), Britain's third Astronomer Royal, for the year 1755 adjusted for proper motions accumulated over that time interval as calculated by A. J. G. F. von Auwers (1838–1915).[47] Dreyer corrected Tycho's mistaken assumption that atmospheric refraction is negligible for altitudes greater than 20° and tabulated the differences between the right ascensions and declinations as calculated by Tycho and Bradley. Figure 3.14 (*right*) reproduced from Dreyer,[48] shows how close Tycho's observations were (with atmospheric refraction duly taken into account) in comparison with those of Bradley (who was renowned for the exactitude of his astrometric work). Dreyer gives probable errors in Tycho's results of $\pm 24''.1$ for right ascension and $\pm 25''.9$ in declination—concordant with Tycho's[49] estimate of $\pm 25''$ (from observations with the mural quadrant and the great armillary).

Wesley[50] undertook a comparative study of the accuracy of the following of Tycho's instruments: the Mural Quadrant (**E**), the revolving azimuthal quadrant (**F**), the revolving great steel quadrant (**G**), the portable azimuthal quadrant (**C**),

---

[46] Dreyer (1890), pp. 387–388.

[47] von Auwers (1882).

[48] Dreyer (1890), p. 387.

[49] Dreyer (1890), pp. 351–352.

[50] Wesley (1978).

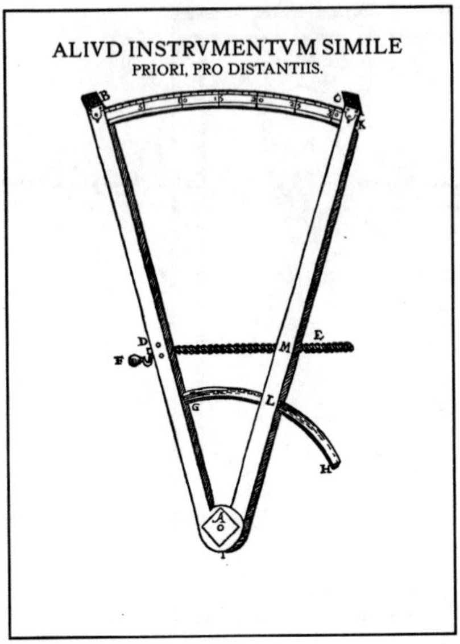
ALIVD INSTRVMENTVM SIMILE
PRIORI, PRO DISTANTIIS.

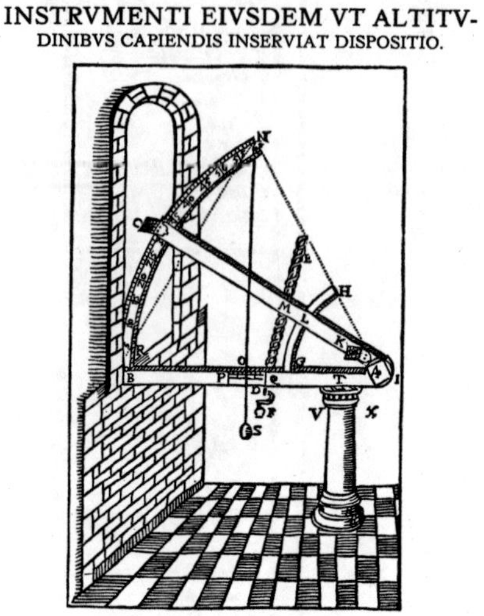
INSTRVMENTI EIVSDEM VT ALTITV-
DINIBVS CAPIENDIS INSERVIAT DISPOSITIO.

Fig. 3.13 *Left*: The half-sextant of the type with which Tycho observed the New Star of 1572. Reproduced from Dreyer, *TBOO* V, 1921, p. 80). The alidades (*AB* and *AC*) are tightly fastened together by a tenon (*A*) and the angle between the alidades is controlled by a long screw (*DE*) using the handle (*F*). The arc *GLH* serves to keep the alidade in the plane, and angles are read from the graduated 30-degree sector. When rested on a cross-bar the instrument's orientation can be controlled and measure made of angular distances between celestial objects. *Right:* A similar kind of instrument (this time a full sextant) mounted vertically for measuring altitudes. Note the plumb line for ensuring verticality. Reproduced from Dreyer, *TBOO* V, 1921, p. 84

Ascensiones Rectæ & Declinationes priùs adhibitarum Stellarum, idq; ad Annum 1585 completum.

Nomina Stellarum	Declinatio		Ascens. R.	
	P.	M.	P.	M.
Lucida ♈	21	$28\frac{1}{2}$ B.	26	$0\frac{1}{2}$
Oculus ♉	15	$36\frac{1}{4}$ B.	63	$3\frac{3}{4}$
Calx finiſtri pedis ♊	22	$38\frac{1}{2}$ B.	89	$29\frac{1}{6}$
Inferior Caput ♊	28	$57\frac{3}{4}$ B.	109	58
Cor Leonis	13	$57\frac{3}{4}$ B.	146	$32\frac{3}{4}$
Spica Virginis	8	$56\frac{1}{3}$ M.	195	$52\frac{1}{6}$
Borealis finiſt: man° ophiuchi	2	$33\frac{1}{4}$ M.	238	$11\frac{1}{3}$
Lucida Vulturis	7	$51\frac{1}{3}$ B.	292	$37\frac{1}{3}$
Prima Alæ Pegaſi.	13	$0\frac{2}{3}$ B.	341	$2\frac{1}{2}$

Ex his

	Bradley minus Tycho.			
	$\Delta\alpha$	$\Delta\alpha\cos\delta$	$\Delta\delta$	$\Delta\delta'$
α Arietis . .	+14″.7	+13″.7	−57″.0	−17″.4
α Tauri . .	+26.2	+25.2	−39.4	+ 9.6
μ Geminorum .	−29.0	−26.7	−41.5	− 3.6
β Geminorum .	−12.1	−10.6	−85.9	−56.5
α Leonis . .	+34.7	+33.7	−48.1	+ 4.4
α Virginis . .	+26.4	+26.1	−122.0	+ 0.6
δ Ophiuchi . .	−10.2	−10.2	− 6.5	+88.1
α Aquilæ . .	+76.2	+75.4	−44.2	+19.5
α Pegasi . .	+31.3	+30.4	−52.9	− 0.1

Fig. 3.14 *Left*: Tycho's standard stars. Reproduced from Brahe, T., *Progymnasmata*, 1602, p. 204. *Right*: Comparison between Tycho's and Bradley's calculations for Tycho's standard stars. The second column lists the difference in right ascension; the third column, the difference in right ascension (taking into account the convergence of RA meridians towards the north celestial pole); the fourth, the difference in declination without taking atmospheric refraction into account; and the last column, the difference in declination with due consideration for mean atmospheric refraction (reproduced from Dreyer, *Tycho Brahe*, 1890, p. 387)

the small brass quadrant (**B**), the astronomical sextant (**D**), the greatest equatorial armillary (**N**), the northern equatorial armillary (**L**) and the southern equatorial armillary (**M**). For the comparison he used the *Bright Star Catalogue*[51] (henceforth *SAO*, epoch B1950.0[52]) to calculate the effects of precession. Refraction tables were used to adjust for atmospheric refraction (although air temperature was ignored unless the stars were very close to the horizon). The *SAO* is a compendium of stellar positions derived from other star catalogues, each with its own positional errors. Wesley cites $0''.3$ as the typical standard deviation for stellar positions in the *SAO*. He studied nine stars from Tycho's catalogue (according to the number of measurements made with different instruments to get good statistics).

Wesley tabulates the mean errors and (where there were at least eight measurements) their standard deviations. His Table 1 lists the errors for meridian observations with the quadrants and the astronomical sextant (which, being vertically orientated, measured altitudes only). His Table 2 does the same for eight other stars. The armillaries are treated separately in his Table 3, where only the mean errors are given. In his Table 4, Wesley lists the average absolute error for measurements made with each instrument separately. We list these last results here:

Mural quadrant	$34''.6$
Revolving wooden azimuthal quadrant	$32''.3$
Revolving steel quadrant	$36''.3$
Portable quadrant	$40''.1$
Small brass quadrant (*Q. max*)	$48''.8$
Astronomical sextant	$33''.2$
Large equatorial armillary	$38''.6$

Wesley ignored all of Tycho's sextants except for the astronomical sextant (instrument **D**), owing to the problem of calculating the effect of atmospheric refraction for observations taken away from the meridian (these sextants, unless set up vertically, measured distances between stars, not their altitudes). A comparison of Wesley's Tables 1 and 2 reveals that the accuracy is better for Tycho's eight fundamental stars, the reason for this being that Tycho made more measurements of these stars. The errors listed for the armillaries are greater than those listed for the quadrants and the astronomical sextant. As for the other sextants, armillaries were used to measure distances between stars, so no altitude measurements were made, thus complicating the task of taking atmospheric refraction into account. Wesley concludes that none of the instruments was able to produce the $\pm 25''$ accuracy claimed by Dreyer, but that such a result might be possible by averaging the results of several instruments for at least some of the fundamental stars listed in Wesley's Table 1.

---

[51] Wesley does not specify which edition of the *Bright Star Catalog* he used. The third edition was published in 1964 and the fourth in 1982, so it is assumed here that he was using the third edition (Hoffleit, 1964).

[52] Star positions in catalogues are given for specific epochs. B1950.0 refers to Besselian epoch 1950.0. The Bessellian epoch was replaced by the Julian epoch (e.g. J2000.0) in 1984.

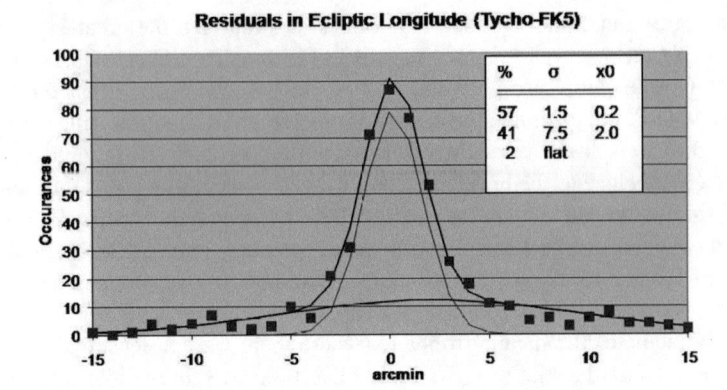

Fig. 3.15 Residuals (Tycho–FK5) in longitude. Thin line: narrow Gaussian distribution ($\sigma = 1.5'$) for 57% of Tycho's catalogue. Thick line: wider Gaussian distribution ($\sigma = 7.5'$) for the fainter 47% of the catalogue. Two per cent of the catalogue stars have an almost flat distribution

Maeyama[53] tested the accuracy of Tycho's observations for extreme altitudes of α Ursae Minoris (Polaris). For 125 of what Tycho considered to be his best observations of this star found an accuracy of 27 arc seconds. When Maeyama selected 29 of the best of these, the accuracy reached 13 arc seconds. He deduced that Tycho's measurements were quasi-normally distributed. Since α UMi was at high altitude when these measurements were made, atmospheric refraction did not distort the accuracy compared with stars measured at low altitude, which may have played an important part in the observations.

A more recent comparison of Tycho's results with modern values has been carried out by Rosa.[54] We have already seen his comparison of the catalogues of Ptolemy and Tycho in Fig. 3.11, where Tycho's latitude residuals (between the FK5 catalogue[55] and Tycho's is shown to be a factor of thirty times more precise than the latitude residuals of Ptolemy). It is interesting to note the sinusoidal zero-error line in the Ptolemy results. Rosa attributes the waviness of the zero-error line to Ptolemy's failure to perform a correct calculation for precession when using an earlier catalogue (whether this failure was a mere oversight or whether underhand motives on the part of Ptolemy were involved has been the subject of heated debate in recent decades). Figure 3.15 summarizes Rosa's analysis of Tycho's star catalogue. If we zoom in on Fig. 3.11 (*right*), the clustering of the Tycho–FK5 residuals is well within ± 4 arc minutes (see Rosa's Fig. 3.5). Rosa calculates a standard deviation (σ) of 10′ for Ptolemy (a figure normally attributed to Ptolemy's observations).

Of the 1024 stars in Tycho's catalogue Rosa finds that 57% (584) have a standard deviation of 1.5′; 41% (420) represent mainly fainter stars of considerably lesser quality ($\sigma = 7.5'$), and 2% (20) may be discounted entirely. Why such a discrepancy

---

[53] Maeyama (2002).

[54] Rosa (2010).

[55] Fricke et al. (1988).

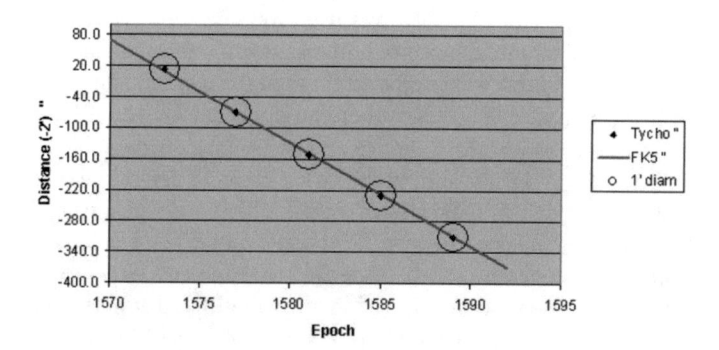

Fig. 3.16 Distance (in arc seconds) of Polaris from the North Celestial Pole plotted against observing epoch. Tycho's measurements are represented by black dots. The black line marks FK5 catalogue values for the range of epochs (spanning 20 years). The black empty circles of diameter 1 arc minute serve to highlight the extreme accuracy of Tycho's measurements (the greatest deviation of Tycho's measurements from the FK4 value is a mere 1″.2). The downward slope indicates the effect of precession over the 20-year period

in the quality of Tycho's results? Accurate star positions require many repetitions of the same measurements over very long periods of time to ensure high-precision mean values. Fainter stars are more difficult to measure and need more repetitions of observations, which takes longer. The lifetimes of the observatories (eighteen years for Uraniborg and less for Stellaborg) were not long enough for Tycho to accumulate accurate positions for many of the fainter stars.

Rosa ends with a discussion of Tycho's need for extreme accuracy of fundamental measurements (e.g. the altitude of the north celestial pole, which is equal to the observer's latitude) in order to maintain the 1′ accuracy of his other measurements. The height of the pole can be deduced from observations of a circumpolar star at upper and lower culmination. The mean of two such measurements will give the altitude of the pole and the angular distance of the star from the pole. Tycho employed this method by observing the star α Ursae Minoris (Polaris) some 1600 times over 20 years and selecting the 'best' value for a given night's observations. The difference between Tycho's results and modern values is in the range 0.2–1.2 arc seconds, so Tycho, for this particularly well-studied star, bordered on sub-arcsecond accuracy (see Fig. 3.16).

3.4.4   Meridian Observations of the Sun

The Sun is an important object in positional astronomy. Its annual path around the sky determines the ecliptic, a fundamental great circle that projects the actual orbit of the Earth onto the celestial sphere. The Sun's meridian altitude was measured on

890 days at Uraniborg from 1582 to 1590.[56] George Tupman (1838–1922) describes in detail how the dioptre in Fig. 3.10 was used for solar observations. Concentric circles were inscribed around the centre of the sunward side of the back plate of the dioptre and the Sun's image was projected through the central hole of the opposite plate. The observer would then judge when the projected image on the inner plate was concentric with the inscribed circles.[57] The projected image of the Sun was fuzzy around the edges, and the skill of the observer lay in determining how closely the Sun's image coincided with the circles. Around 100 observations of the meridian altitude of the Sun were made each year.[58] Tupman reported a mean error of $-47''$ in the observations for the years 1582–1585, $-38''$ for 1586 and $-21''$ for 1587–1590 (the number of solar observations falls of sharply after that date). Tupman's results were seriously affected by his use of Curtz's edition of Tycho's *Historia coelestis*[59] as his source. Dreyer, while conceding that *Historia coelestis* gave a good overall picture of Tycho's work, judged it not to be useful for serious scientific research.[60] Dreyer later produced what is now the standard edition of Tycho's observing logs.[61]

Wesley used Dreyer's edition of Tycho's observing logs.[62] Taking into account the longitude of Uraniborg and the equation of time, he used Tuckerman's tables[63] of the Sun's positions in the interval 1588–1591 (chosen because 1588 was the first year in which all of Tycho's main instruments were in operation)[64] and the year 1595 (the last year in which Tycho made solar observations at Uraniborg) to postdict the Sun's position for that interval to compare with Tycho's measurements. Wesley replaced the refraction estimates in Tycho's logs by the mean refraction, using mean temperatures for Copenhagen (the difference between mean and real refractions amounts to a difference of a couple of seconds of arc). The instruments used for solar observations during these years were the mural quadrant (**E**), the revolving quadrant (**F**), the revolving great steel quadrant (**G**), the astronomical sextant for altitudes (**D**) and the great equatorial armillary (**N**). Wesley reported the yearly average error for each instrument as shown below.

	1588	1589	1590	1591	1595
Mural quadrant	$-39''.2$	$-24''.8$	$-36''.0$	$+12''.1$	$-3''.7$
Revolving quadrant	$-48''.9$	$-21''.0$	$-42''.4$	$-1''.7$	$-40''.4$
Revolving steel quadrant	$-44''.4$	$-11''.1$	$-35''.5$	$-7''0.5$	$-41''.2$

(continued)

[56] Tupman (1900).

[57] Tupman (1900).

[58] Thoren (1990), p. 220.

[59] Brahe (1666).

[60] Dreyer (1890), pp. 371–372.

[61] Brahe (1923b, 1924, 1925b, 1926).

[62] Brahe (1923b, 1924, 1925b, 1926).

[63] Tuckerman (1964). The instructions for interpolation that Wesley refers to are to be found in Tuckerman (1962), pp. 4–7.

[64] Wesley (1979), p. 100.

(continued)

	1588	1589	1590	1591	1595
Astronomical sextant	− 34″.1	− 9″.5	−	−	−
Total yearly average for all four instruments	− 42″.8	− 17″.9	− 17″.9	− 3″.2	− 36″.3
Equatorial armillary	− 10″.4	− 9″.9	− 21″.9	− 5″.7	− 100″.2

In the case of declinations measured directly by the equatorial armillary, Wesley selected from the observing logs only those measurements made when the Sun was close to the meridian. Wesley's monthly error tables show no variation in the error with a maximum in June and minimum in November, as reported by Tupman.[65] Wesley attributes the fact that most of the errors in the solar observations are negative (as opposed to the stellar errors, which are predominantly positive) to a possible alignment error with the dioptre (used in projection mode for solar observations, as described earlier). Two readings were made of solar transits and there was an error of 10–15″, the second reading—according to Wesley—often being greater than the first. The armillary sphere errors for 1595 indicate a serious deterioration in the quality of observations made with that instrument in that year.

3.4.5 Planetary Observations

The *Prutenic Tables*, as we have seen, showed little improvement over the *Alfonsine Tables*[66] with regard to ephemerides based on them to predict planetary positions. It is characteristic of such tables that their reliability decreases over time as the models on which they are based become increasingly out of step with the heavens. In the case of Mars, for example, the *Prutenic Tables* did an extremely poor job of accurately predicting where in the sky that planet would be at any given time (see Fig. 3.17). Mars also strayed catastrophically from its predicted longitude for the greater part of its orbit, that deviation amounting to a huge 5° every 32 years (a situation dubbed 'the great Martian catastrophe' by one historian.[67]

Kepler was confident that Tycho's measurements of the longitude of Mars were accurate to within 2′, a belief that led him to consider as real the 8′ discrepancy in Mars' longitude at the orbital octants in *Astronomia nova*:

> Since divine benevolence has vouchsafed us Tycho Brahe, a most diligent observer, from whose observations the 8′ error in this Ptolemaic computation is shown, it is fitting that we with thankful minds both acknowledge and honour this benefit of God. ... Now, because

---

[65] Tupman (1900).

[66] A full account of the Alfonsine Tables is given by Chabas and Goldstein (2003) and Kremer (2023).

[67] Gingerich (2011).

Fig. 3.17 Graph showing the *Prutenic Tables'* poor ability to reproduce the ecliptic longitude of Mars. Reproduced from O. Gingerich, 'The great Martian catastrophe and how Kepler fixed it', *Physics Today*, 64(9), 2011 [https://doi.org/10.1063/PT.3.125]

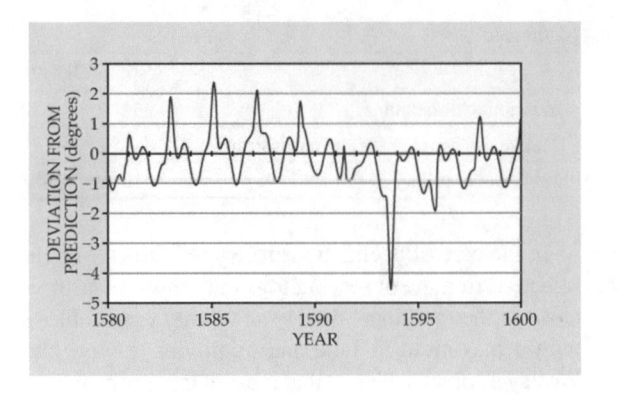

they could not have been ignored, these eight minutes alone will have led the way to the reformation of all of astronomy.[68]

Observations cannot be made repeatedly of a planet in a given position against the background stars since it will have moved to a new position in a matter of hours, so planetary observations cannot be subjected to the same statistical treatment as stars. Observational error will therefore be greater than for a star whose meridian transits are measured many times. Sextant measurements of the distance of a planet from reference stars were often made without reference to simultaneous altitude measurements, so it is difficult to make proper allowance for atmospheric refraction, and this would add to the overall uncertainty in the planet's true position. In short, it is to be expected that Tycho's measurements of the positions of Mars will not have the same quality as his measurements of his reference stars.

One of the main purposes of Tycho's observations of Mars at opposition was his attempt to prove that Mars was closer to the Earth than the Sun at such times, and that this distance could be established by determining the diurnal parallax of Mars at opposition.[69] In this way Tycho thought he could establish that the orbit of Mars could be explained by either Copernicus' heliocentric model or his own geoheliocentric model (see Sect. 3.5). Tycho's Mars campaign began with the 1582 opposition. According to the Copernican and Tychonic models, the distance of Mars at opposition would be about half the distance of the Earth from the Sun. Tycho accepted the Greek value of 3′ for the solar parallax and estimated a diurnal parallax for Mars at opposition of about 5′, which he would easily have detected. In reality, the diurnal parallax of Mars never exceeds 27″—well beyond the capabilities even of Tycho's best instruments. Although Tycho had failed to measure the yearned for diurnal parallax of Mars, his observations of that planet were fundamental to Kepler's discovery of the first two laws of planetary motion (see Chap. 4 of this volume).

---

[68] Kepler (1609), pp. 113–114; Kepler (1992), p. 286; Kepler (1990), p. 178 lines 1–12. See also Davis, 'Kepler's discovery of the planetary orbit' (Chap. 4 of this volume), where the matter is fully discussed.

[69] Gingerich and Voelkel (1998).

Kepler, after examining Tycho's manuscripts, gives a clear description of the problems involved in the use of diurnal parallax for this kind of measurement:

> In 1582, when Mars was opposite the sun in Cancer, I found incredible care in observing, with Tycho's manuscript title, 'For investigating the parallax of Mars', from which you will, however, deduce either no parallax at all, or one. exceedingly small. ... They compared the star Mars with nearby stars on the ecliptic, and frequently with ones at a great distance. Now it is usual to find the parallax of a mobile star (for Mars moves, with a retrograde motion when opposite the sun) by comparing morning and evening observations. It has thus happened that almost all the stars from which Mars's distance was observed in the morning are different from those by which it was observed in the evening. For a fixed star which is at hand in the morning (and higher than Mars), if it be near the ecliptic, has either set by evening (when Mars is in the west) or is rendered useless by refraction for this delicate procedure. Another star had to be substituted. But if the fixed stars are substituted for one another, there is always less trust in the procedure than if the same star had been retained.[70]

Once again, we see that atmospheric refraction complicates measurements made to determine the angular distance of a planet from nearby stars because refraction cannot be properly determined if measurements of the altitudes of the planet and stars in question are not made simultaneously with the angular distance measurements. Yet a further complication for Tycho was that Mars' closest approaches to Earth occur in July–August, when the nights at Hven are too short for diurnal parallax measurements (there is too little time for a sufficiently long baseline).[71]

We turn now to the errors arising when measuring angular distances of planets from background stars. Kepler was intimately involved with the interpretation of Tycho's observing logs during his labours to fit the planetary model most closely representing Tycho's accurate observations of Mars and the calculations he carried out many years later in drawing up the *Rudolphine Tables* (1627) based on that model. We have already seen how Kepler was confident that Tycho could not possibly have erred by so much as 8′ in his measurements of planetary positions. Thoren, along with Dreyer, accepts an error of approximately ± 1′ for Tycho's planetary observations made with the most accurate instruments (the mural quadrant, etc.) in the later years of Uraniborg's existence as a working observatory.[72] Unlike observations of stars, which are conveniently available in the form of a star catalogue, observed planetary positions are constantly changing and Tycho's observations have to be sought in the 1728 pages of his observing logs.[73]

Rosa,[74] after examining 191 observations of Mars made by Tycho from 1582 to 1601, confirms that Tycho consistently maintained an error of ± 1′ throughout his planetary work. Rosa used NASA JPL ephemerides to trace the orbit of Mars over this interval and plotted Tycho's Mars observations. There is a clustering of observations near various oppositions, especially those of 1585, 1587, 1589, 1593 and 1596. To ensure the best results, the distance of Mars from a number of stars was measured

---

[70] Kepler (1609); Kepler (1992), pp. 202–203.

[71] For a much fuller account see Gingerich and Voelkel (1998).

[72] Thoren (1973b); Dreyer (1890), p. 357.

[73] Brahe (1923b, 1924, 1925b, 1926).

[74] Rosa (2010). The rest of this section is based on the contents of this review.

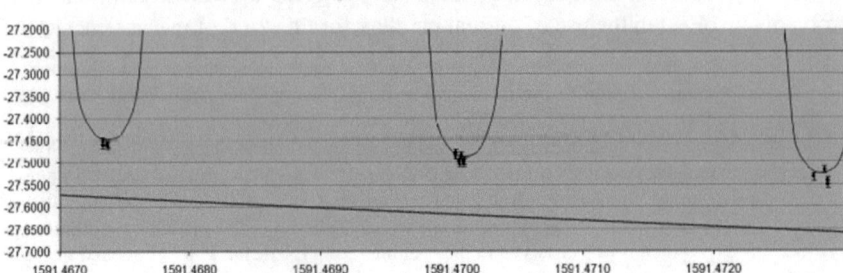

Fig. 3.18 Plot of three nights' observations of Mars at opposition as the planet crosses the meridian. The abscissa shows time in decimal years and the ordinate shows declination. See text for explanation

with at least two instruments, the mural quadrant being used to record meridian transits where possible. When comparing Tycho's recorded positions with postdictions derived from modern ephemerides, care must be taken to allow for refraction. A plot of three nights' observations of Mars during the opposition of 10–12 June 1591 is shown in Fig. 3.18. As Rosa explains. The downward sloping straight line represents the refraction-free locus, the U-shaped curves shows the path of Mars across the sky during the three nights from sunset to sunrise. Tycho's Mars observations are represented by black dots with 1' error bars (these are meridian-crossing observations). During summer months the ecliptic is low in the night-time sky in the northern hemisphere, so Mars is greatly refracted from its true position by atmospheric refraction. The figure shows clearly that Mars is refracted upwards by 8'—a huge systematic error that needed to be taken into account in order to appreciate the excellence of Tycho's measurements. Rosa's investigations, then, amply confirm the view held by Dreyer and Thoren that Tycho's planetary observations were accurate to approximately ± 1' once refraction is allowed for.

3.5 The Tychonic System

Tycho published his world system in a book bearing the title *On the Recent Phenomena in the Aetherial World* (1588).[75] A complete description of the system would take an entire chapter. The static Earth sits at the centre of the world. The Moon and the Sun (orbits 1 and 2 in Fig. 3.19) circle the Earth in an easterly direction. The stellar sphere rotates about the Earth once a day. The orbits of Mercury, Venus, Mars, Jupiter and Saturn (3, 4, 5, 6 and 7 respectively) are approximately centred on the Sun. Against this diurnal motion, each object in the planetary system has its own orbital period: $27^d.32$ (the Moon), $365^d.25$ (the Sun), 88^d (Mercury), 255^d (Venus), 684^d (Mars), 12^y (Jupiter) and $29^y.5$ (Saturn). Their orbital motion is

---

[75] Brahe (1588); Brahe (1922), pp. 1–378.

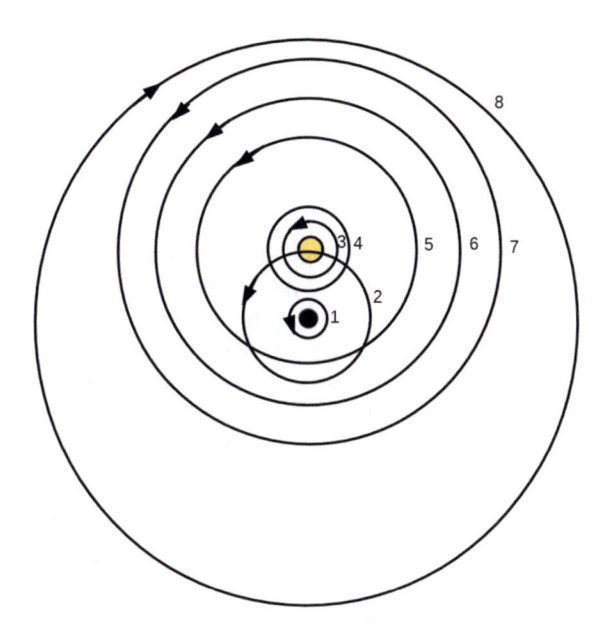

Fig. 3.19 The Tychonic system (simplified). Orbits of the Moon (1), the Sun (2), Mercury (3), Venus (4), Mars (5), Jupiter (6), Saturn (7) and the background stars (8). Not to scale

easterly. Comets, according to Tycho, followed orbits about the Sun 'like the figure commonly called an oval'.

In Tycho's world system, there were no crystalline spheres, and—as with all world systems built on circular motion—in this system Tycho also had to resort to an eccentric circle for the Sun and epicycles to explain various anomalies in the lunar orbit and those of the planets. As an example, for Saturn's orbit, Tycho used an anticlockwise deferent with period $29^y.5$ coupled with a clockwise epicycle (period $29^y.5$) centred on the circumference of the deferent and a further retrograde epicycle (period $14^y.75$) centred on the circumference of the first epicycle (he uses this mechanism for all the superior planets).

The orbit of Mars intersects that of the Sun. This is so to ensure that Mars when it is at opposition is closer to the Earth than it is to the Sun. As Dreyer[76] says of this intersection of orbits, 'As they are only imaginary lines and not impenetrable spheres, there is nothing absurd in this.' The Tychonic system is in reality an inversion of the Sun-centred Copernican system to a (static) Earth-centred system.

The question of Tycho's world system will resurface in Sect. 3.9 when we deal with the controversy with Nicolaus Reimarus Ursus (Nicolai Reimers Baer, 1551–1600).

---

[76] Dreyer (1953), p. 363.

3.6   Reception of Tycho's Observations

Tycho's observations of Mars, made with hitherto unattained precision and accuracy, were the bedrock on which Kepler built the edifice of his first two laws of planetary motion (see Chap. 4 of this volume). We have seen the high regard that Kepler had for Tycho's observing skills.

Some people, for example Galileo Galilei (1564–1642) writing in 1632, have interpreted the huge sums spent by Tycho on his instruments as sheer extravagance on Tycho's part,[77] or a ploy by Tycho to keep his workers in employment.[78] Others adopt the view that the high expenditure was the price necessary to fund the bold programme of instrumentation development that paid off magnificently in terms of the unprecedented accuracy of Tycho's scientific results. A careful examination of the evolution of the instruments at Uraniborg and Stellaborg shows that Tycho strove incessantly to improve his instruments to tackle new observational challenges. This could not be done cheaply.[79]

3.7   The Destruction of Uraniborg

Dreyer[80] and Thoren[81] provide detailed accounts of how Uraniborg's 18-year existence came to an abrupt end with the exile of Tycho from Denmark. When Frederick II died in 1588, four powerful noblemen took over the reins of state as Protectors (Frederick's eldest son Christian was eleven years old and deemed too young to reign). During this regency period, Tycho's pensions and funding for the observatories were secure, and Tycho was in the good graces of the Protectors.

The accession of Christian IV (1577–1648, reigned 1588–1648) to the throne on 29 August 1696, however, brought about a change in the way state finances were managed. Tycho, sensing an ill wind, attempted to ward off the coming fiscal storm by presenting a detailed case for the continuation of funding to the new king. Unfortunately, two powerful friends of Tycho at court had recently died, a circumstance that greatly weakened Tycho's representation at court. Christian IV made Christen Friis of Borreby (1581–1639) chancellor and Christoffer Valkendorf (1525–1601) high treasurer. These appointments spelt bad news for Tycho. His stewardship of the prebend of Roskilde came under close inspection after a series of incidents involving a dispute between Tycho and a tenant. Tycho's neglect of the upkeep of the 'Chapel of the Magi' in Roskilde Cathedral (his duty as steward of the prebend) also drew the attention of his enemies at court.

---

[77] Galilei (1632, 1953); Galilei (1632), pp. 387–389.

[78] Repsold (1908), p. 29.

[79] Thoren (1973b).

[80] Dreyer (1890), Chap. 9.

[81] Thoren (1990), Chaps. 11 and 12.

To compound Tycho's woes, when news of his fall from grace reached the population of Hven, the locals drew up a list of complaints of abuse by Tycho. In response to these complaints, Christian ordered Friis (the new high treasurer) and Axel Brahe (Tycho's brother, 1550–1616) to investigate the claims of mistreatment of tenants on Hven and also (ominously) to enquire whether Hven's priest, Jens Jensen Wensøsil, had firstly removed exorcism from the rite of baptism and secondly failed to censure Tycho for persistent failure to take the Sacrament. On April 14 privy councillor Ditlev Holk (1556–1633) was given the commission to pursue proceedings against the priest:

> Know you, that whereas a minister, by name Jens Jensen, has dared during the service of the church to act against the ritual, and he for such audacious conduct is to appear before our beloved the honourable and learned Dr. Peder Winstrup, superintendent of this diocese of Seeland, on the 22nd April: We order and command that you arrange to be present here in this town at the same time, and afterwards with the said Peder Winstrup in the said case to judge accordingly to what is Christian and right.[82]

Their findings and judgement are not known, but there is an entry in a diocesan record:

> [T]he minister of Hveen was dismissed in disgrace for not having kept to the ritual and prayer-book in the form of baptism ('I adjure thee') but acting differently; also for not having punished and admonished Tyge Brahe of Hveen, who for eighteen years had not been to the Sacrament, but lived in an evil manner with a concubine.[83]

This attack on the characters of Tycho and his common law wife can only have been politically motivated by Tycho's enemies at court. Not keeping to the baptismal ritual alone would have been ample grounds for Jensen's dismissal—or worse (such practice could only too easily have been interpreted as his having Calvinistic leanings). Tycho's common-law marriage with Kirsten was perfectly legal under Jutish law and did not warrant the scandalous description of it in the diocesan record.

One of Christian's first acts upon accession was to implement budgetary cuts, which included the withdrawal of Tycho's rents from a crown estate in Norway that had been granted to him by Frederick II. Tycho tried unsuccessfully to recover them. He then pleaded with Friis that he should at least be able to retain them until 1 May so that he could receive them when they fell due. He was denied both the rents and the requested extension. On 18th March 1597, the king ordered Valkendorf to withdraw Tycho's pension.

That was the last straw. Tycho realized that he needed to seek a new patron in order to continue his astronomical labours and provide security for his family. He and his assistants worked frantically to finalize the star catalogue. To the 777 well-observed stars, he added a number of less-well-observed stars to bring the catalogue to a grand total of 1017 stars. The last astronomical observations on Hven were made in March 1597. Tycho and his family left Hven for the last time on 29 March 1597, arriving in Copenhagen the following month.

---

[82] From Dreyer (1890), p. 236.

[83] Dreyer (1890), p. 236.

It had been made clear to Tycho that he was *persona non grata* at the Danish court. He left Copenhagen and travelled to Rostock, Wandsbeck, Wittenberg and finally Bohemia, where he would end his days. Meanwhile, the castle and observatory buildings on Hven were plundered and completely demolished. Did Christian ever live to regret having allowed Uraniborg and Stellaborg to be destroyed? In 1627 the publication of the *Rudolphine Tables*[84]—which, in other circumstances, might have borne the title *Christian* or *Danish Tables*—must surely have irked him. In 1625 Longomontanus (Christian Severin, 1562–1647), Tycho's successor as the King's astronomer and author of *Danish Astronomy,* published in Amsterdam in 1622[85]) suggested the building of an astronomical tower in Copenhagen. Christian acceded to Longomontanus' request and such an astronomical tower was built as part of a complex that included a students' church and university library. The tower, called the *Stellaeburgi Regii Hauniensis* (*Rundetårn* in Danish), was built in 1627 as part of the Trinitatis Complex, which still stands today.

3.8 Tycho at Benátky

The French astronomer and natural philosopher Pierre Gassendi (1592–1655)[86] says that Albert Curtz, SJ (1600–1671), future author of *Historia coelestis* (published posthumously in 1666 and consisting of a compilation of extracts from the works of Tycho) on a visit to Denmark, offered to negotiate with Emperor Rudolf II an invitation for Tycho to come to Prague, even offering his own residence to Tycho.[87] Hagecius, in correspondence with Tycho, had separately promised that he would be well received in Prague should he ever leave Hven.[88] On 2 January 1598 Tycho dedicated his star catalogue to Rudolf II and included a preface extolling its improvements over other catalogues.

Hagecius hinted to Tycho, in a letter dated 30 May 1598, that it would be wise for Tycho to present himself in Prague for an audience with Rudolf.[89] Tycho set out for Prague three months later with his entire entourage. He left his household at the residence of an official at Dresden before moving on to Prague, which was then in the throes of an epidemic. Rudolf had gone into isolation to avoid infection. An answer came from Rudolf a month later enjoining Tycho to wait for an invitation from him once the epidemic had abated. In spite of having received favourable signals from Rudolf, Tycho delayed his journey to Prague on account of his plans to have newly published works to present to the emperor. Since he had made changes to his lunar

---

[84] See Field, 'The long life of the *Rudolphine Tables*' (Chap. 9 of this volume). The story behind the Tables is described in Kepler (1969), 'Nachbericht' ('afterword').

[85] Longomontanus (1622), contains tables.

[86] Gassendi (1655); Thoren (1990) cites p. 131 of the Swedish translation.

[87] Curtz often used the pseudonym Lucius Barretus.

[88] Dreyer (1890), p. 223.

[89] Hagecius to Tycho, 30 May 1598, in Brahe (1925a), p. 68.

theory, he could not finish *Progymnasmata* in time for the audience so he decided on an ephemeris of the daily positions of the Sun and Moon for the year 1599. This entailed a large amount of calculation and was possibly the reason for his putting off the trip to Prague. After Rudolf's persistent enquiries whether Tycho had yet arrived, the Brahe entourage at last set out for Prague on 14 June 1598 and reached the city in July.

Tycho was gratified to discover that his fame had preceded him to the court of Rudolf II. He writes in a letter to Rozenkrantz, dated 30 August 1599:

> Immediately on my arrival (at court), I was received in a respectful and friendly way by many distinguished men, above all by the emperor's private secretary, Lord Johannes Barwitz.[90] … When I most humbly showed Barwitz the three writings I wanted to present to the emperor, and the letters of recommendation from the elector of Cologne and the duke of Mecklenburg, Barwitz said that he would speak with the emperor about them. … The following day he related that the emperor did not want to receive them from anyone other than myself, and that he would call me to the palace in a short time.[91]

(Barwitz is more usually known by his Latin name: Johann Anton Barvitius. He was born in 1555 and died in 1620.)

A few days later, Tycho was summoned to the emperor's residence outside Prague for a private audience in which he was warmly received by Rudolf. Shortly after, the imperial council settled an annual grant, which was to commence on 1 May 1598, and a suitable residence. Of three palaces offered to Tycho, he chose Benátky, a castle 38 km from Prague. By 28 August, Tycho had set up one of his instruments at the castle so, after the two-year hiatus since leaving Hven, Tycho was at last comfortably settled and back in business. He was now more concerned with the publication of his work, the protection of his reputation from attacks by Ursus and the fitting of his unprecedentedly accurate observations to his model of the world system. Longomontanus[92] was Tycho's most able assistant on Hven and had been tasked with developing Tycho's lunar theory and (initially) working on the orbit of Mars. When Tycho left Hven, they had parted company. Tycho later called Longomontanus to Benátky to help him with his publications and lunar theory. On completion of the lunar theory, however, Longomontanus returned to Denmark, where he obtained a professorship in astronomy. Earlier, while Longomontanus was still at Benátky, Tycho decided to take the Mars problem from him and hand it to a promising young newcomer who was about to be banished from Graz for his Lutheranism.

---

[90] For further information on Barvitius see Evans (1997).

[91] Tycho to Rozenkranz dated 30 August 1599 (OS) in Thoren (1990), pp. 410–413; Brahe (1925a), pp. 163–166.

[92] Hamel (2007).

3.9   Enter Kepler

Johannes Kepler,[93] a young graduate from the University of Tübingen, had published a book with the title *Mysterium cosmographicum* in 1596.[94] The publication of Kepler's book attracted the attention of Ursus, to whom Kepler had written on 15 November 1595[95] (with no reply from Ursus). Now that Kepler was a published author, Ursus deemed it to be not beneath his dignity to communicate with him. In his belated reply to Kepler, dated 29 May 1597, he requested a copy of *Mysterium cosmographicum*. Kepler's first letter had included an outline of his book and contained fulsome praise of Ursus,[96] who would later selectively cite Kepler's praise of him to his own advantage in his battle with Tycho over the geoheliocentric world system in his *Treatise on Astronomical Hypotheses*.[97] On 13 December 1597, Kepler wrote to Tycho to ask for his appraisal of *Mysterium cosmographicum*. Tycho replied on 1 April 1598 with a just appraisal of Kepler's hypothesis concerning the relative distances of the planets, noting that it was in approximate agreement with the Copernican hypothesis, but he took issue with Kepler on a number of points, particularly his a priori approach to constructing planetary models without recourse to accurate observational data and his belief in solid spheres (falsely attributed to Kepler by Tycho).[98]

Tycho had learned to be wary of all newcomers. On 21 April he wrote to Michael Maestlin (1550–1631) at Tübingen, a well-known astronomer and former teacher of Kepler, with forthright reservations concerning *Mysterium cosmographicum* and to ask him to set Kepler straight on the matter of Ursus.[99] Contact between Tycho and the young Kepler, then, had been tentatively established. And it marked the beginning of a collaboration that would change the course of astronomy.

Kepler's situation in Graz as a teacher of mathematics at the Protestant seminary had become untenable after Duke Ferdinand of Styria promised in 1598 to impose Catholicism on his region.[100] In January 1600, Kepler received a friendly invitation from Tycho (written in December 1599) to visit him in Prague. Baron Ferdinand Hofmann von Grünpichl und Strechau (1540–1607), a councillor to the emperor, offered to introduce him to Tycho. But Kepler had already begun the journey before Tycho's letter arrived. On 26 January, Tycho wrote to Kepler, 'You will come not so

---

[93] See Methuen 'Kepler, religion and natural philosophy: a theological biography' (Chap. 1 of this volume), for Kepler's biography.

[94] Kepler (1596, 1938); he published a second edition, with added notes, in 1621; see also Kepler (1981) and Kepler (1963). For a discussion of Kepler's cosmological views see Field, 'Kepler's cosmology' (Chap. 2 of this volume).

[95] Kepler to Ursus 15 November 1595; Kepler (1945), letter 26, pp. 48–49.

[96] Jardine (1984), p. 10.

[97] Ursus (1597).

[98] See Field (1988) and Field (2009). How Tycho's not believing in solid spheres plays a part is described by Field, 'Kepler and Galileo' (Chap. 8 of this volume).

[99] Tycho to Maestlin, 21 April 1598, in Kepler (1945), p. 204 ff.

[100] See Field, 'Kepler's cosmology (Sect. 1.2 of Chap. 1 of this volume).

much as a guest but as a very welcome friend and highly desirable participant and companion in our observations of the heavens.'[101] At long last, accompanied by one of Tycho's sons, on 4 February Kepler was taken to Benátky. Baron Hofmann, as good as his word, recommended that Kepler be taken under Tycho's wing.

Tycho wondered whether Kepler was secretly in league with Ursus (who had published Kepler's letter of praise in his *Treatise on Astronomical Hypotheses* (Prague, 1597), in which he had made a scurrilous attack on Tycho). Could Tycho really trust Kepler? As a precaution he limited Kepler's access to his observations and allowed him to use only those relating to Mars. As we have seen, Longomontanus had previously been handed the problem of Mars' orbit but had got nowhere with it. He resented having the problem taken from him and passed to an underling. This resentment was to cause difficulties for Kepler after Tycho's death. As a test of Kepler's loyalty, Tycho also set him the unenviable task of making a public defence of Tycho against the attacks of Ursus.

Kepler wrote of his being allowed access only to Mars observations:

> Tycho did not confer the multitude of them [planetary observations] upon me, except in as much as in an aside while dining he incidentally mentioned now the apogee of one, now the nodes of another. But when he saw that I had a daring mind, he decided perhaps the best way to deal with me was to give me my way with the observations of a single planet, namely Mars.[102]

To insure himself against any untoward diffusion of his proprietary data on Kepler's part (as had happened when Wittich reported his use of transversals to the Landrave of Hesse-Cassel), on 5 April 1600 Tycho made Kepler sign a written undertaking to 'keep everything that the well-informed Herr Tycho has communicated to me, or shall hereafter communicate to me, in whatever manner, top secret.'[103] Kepler was then placed under Longomontanus' supervision to work on the motion of Mars.

Kepler regarded the composition of *Apologia* as an imposition and in any case considered any attempt to reply to Ursus as beneath the dignity of Tycho; nevertheless he dutifully penned the defence of Tycho during the Christmas 1600 celebrations. The task took precious time from his astronomical labours. However, the defence gained Tycho's approval, and Kepler was now free to focus his full attention on the problem of Mars.

The first meeting of Kepler and Tycho had taken place on 4 February 1600. Kepler's original intention at Benátky had been to use Tycho's values for planetary eccentricities[104] and mean distances with which to test his planetary model described

---

[101] *Aduenies non tam hospes quam Amicus gratissimus nostrarum que in Coelestibus contemplationum, per ea quae nunc ad manus habeo, Instrumenta spectator et socius acceptissimus.* Tycho to Kepler 26 January 1600; Kepler (1949), letter 154, pp. 107–108.

[102] Kepler to Herwart, 12 July 1600, in Kepler (1949), letter 168, pp. 128–136, here p. 130.

[103] Kepler (1975), document 2.5, p. 48.

[104] In Tycho's day this term had two distinct meanings both of which differed from that used today for planetary orbits. Both meanings are given in the glossary of this volume.

in *Mysterium cosmographicum*, but Mars soon took all his attention. Kepler's difficult financial condition, however, now prompted him to act in a way that almost
wrecked this collaboration. In April 1600 his situation at Benátky *vis-à-vis* lodgings,
provisions and pay led him to draft a document[105] in which he made a long series
of demands in an insolent manner that infuriated Tycho. One demand particularly
irritated Tycho: 'When,' Tycho replied testily, 'had he ever made his assistants work
on Sundays or holidays?'[106] He angrily requested of Kepler 'to have more confidence
in him, and to conduct himself in future with more prudence and moderation towards
his benefactor, who had been very patient with him, and wished him and his well.'[107]

Tycho responded in writing to all of Kepler's grievances one by one.[108] One
particular demand by Kepler that he be granted time off to pursue his own projects
was flatly rejected by Tycho. Kepler, upon learning of this, left Benátky for Prague.
Tycho, knowing full well the value of having such a gifted assistant to work on his
geoheliocentric model, relented and let it be known that he would welcome Kepler
back to Benátky if he would apologize for his tantrums. Tycho was also worried that
Kepler might even decide to collaborate with Ursus, who had returned to Prague.[109]
But if Tycho agonized over the possibility of losing Kepler's theoretical expertise,
Kepler likewise realized that only Tycho's observations would be good enough to
decide which world model best fitted the motions of the planets. Kepler swallowed
his pride and penned a pathetic plea asking Tycho's forgiveness for his inexcusable
behaviour towards him.[110]

Rudolf had now decided to come out of quarantine and returned to Prague Castle.
He wanted Tycho to relocate nearer to him. Tycho, having just arranged everything to
suit his needs, found that he now had to quit Benátky. It is difficult to see how he could
have achieved any observational work of importance during all this toing and froing.
The emperor began to make increasing—mainly astrological—demands of Tycho.
By Christmas, Tycho was at last making observations from the castle belvedere.[111]

On Kepler's return, Tycho introduced him to Rudolf, who commissioned Kepler to
produce new tables of the planets based on Tycho's observations. Rudolf graciously
acceded to the tables' bearing the title *Tabulae Rudolphinae*. Shortly after this meeting
Tycho Brahe suddenly fell ill after a banquet and died of a bladder infection on 24
October 1601. On his deathbed he was heard to murmur repeatedly, 'Let me not be
seen to have lived in vain!' (*'ne frustra vixisse videar'*).[112] He begged Kepler, who

---

[105] Kepler (1975), pp. 40–42.

[106] Kepler (1975), p. 42.

[107] Quoted from Dreyer (1890), p. 301.

[108] Brahe (1925a), p. 272.

[109] Brahe (1925a), p. 299.

[110] Kepler to Tycho, April 1600, in Brahe (1925a), pp. 305–307; Kepler (1949), letter 162, pp. 114–
116.

[111] Brahe (1926), p. 241.

[112] Cited from Dreyer (1890), p. 309.

was to be his scholarly heir,[113] to continue with the plan to demonstrate through Tycho's planetary observations that the planets moved according to the Tychonic, not the Copernican, world system. Kepler did make a serious effort to justify Tycho's geoheliocentric theory; however, it was by finally settling on elliptical orbits for the planets that Kepler ensured that Tycho had indeed not lived in vain.[114] Tycho was buried in the Church of Our Lady before Týn in central Prague on 4 November 1601. Two days later Barvitius came to Kepler to inform him that the emperor had decided to name him Imperial Mathematician, and that he was to take responsibility for Tycho's papers and instruments.

That Kepler was indeed the most fitting candidate to take Tycho's place as Imperial Mathematician was amply confirmed some months after Tycho's death when Barvitius received a letter from Herwart, who wrote of Kepler:

> I, as one informed about these matters and having also some experience, know very well that at this time as far as one can judge from the works that have been published (*ex operibus editis*) no one can be found who can be compared both in intellectual power and in mathematics (*et ingenio, et fundamentis artis Matheseos*) with this Master Kepler, let alone be preferred to him, so that I have no doubt whatever that when it is brought to the attention of His Majesty most graciously and most humbly he will not let him go for any amount of money.[115]

Kepler saw himself as duty-bound and destined to bring Tycho's results to their full fruition. He writes to Maestlin:

> If, therefore, God has any interest in astronomy, a belief which demands piety, I hope that I shall achieve something in this field, as I see that God has united me with Tycho by an inexorable fate without having disunited us by the most serious misunderstanding.[116]

But Kepler's troubles were far from over. Tycho's heirs laid claim to Tycho's papers and instruments, on which they had set a value of 100,000 florins. Rudolf offered them 20,000 florins, to be paid from government (not his own) funds.[117] The family, while they waited for the sum to be paid in full, would receive 6 per cent of the principal per annum, dating from the death of Tycho.[118] Meanwhile, with respect to the observations, Kepler, in his role as Imperial Mathematician, took immediate action. As he explained in a letter to Christopher Heydon (1561–1623), a writer on astrology, in October 1605,

> I do not deny that upon Tycho's death, because of either the absence or insufficient expertise of the heirs, I boldly and perhaps arrogantly took charge of the observations he left behind,

---

[113] Longomontanus would have been Tycho's first choice as his successor as Imperial Mathematician but he had left Prague for Copenhagen.

[114] See Davis, 'Kepler's discovery of the planetary orbit' (Chap. 4 of this volume), for a detailed account of Kepler's discovery of the orbit of Mars.

[115] Cited from Dreyer (1890) p. 122; Herwart to Barwitz, 23 February 1602, Kepler (1949), letter 207, pp. 214–215.

[116] Kepler to Maestlin 10/12 December 1601, in Baumgardt (1952), pp. 65–68, here p. 66; Kepler (1949), letter 203, pp. 202–208, here. p. 203.

[117] Thoren (1990), pp. 462–463.

[118] Brahe (1928), p. 265.

against the will of the heirs, but nevertheless according to the not obscure command of the emperor. Since he had entrusted the care of the instruments to me, I interpreted the mandate broadly and took especially the observations to care for.[119]

Frans Tengnagel (1576–1622), Tycho's son-in-law, had his own plans to produce the *Rudolphine Tables* and, through his machinations, caused their publication to be delayed by twenty years (they did not appear in print until 1627). In 1602 Tengnagel demanded that Kepler return the observations to the heirs. Kepler refused to hand them over, in answer to which Tengnagel brought a suit against Kepler.[120] The family took control of the publication of the first volume of *Progymnasmata*, which contained an appendix written by Kepler but was not attributed to him. It also contained numerous errors that incensed Kepler, 'In this most miserable state of the book, which the author carried for twenty years, the heirs, hurrying to market, finally aborted it.'[121] Tengnagel did not stop there but decided to become directly involved in the production of the *Rudolphine Tables* on the grounds that Kepler was not honouring Tycho's wishes. The heirs prevailed in their suit against Kepler, whom they accused of unnecessarily delaying the publication of the *Tables*. As a further ignominy, Johannes Pistorius (1546–1608), Rudolf's father confessor, was to supervise Kepler's work.[122] Tengnagel's constant aim was to deny Kepler access to Tycho's observations, and squabbles between them persisted until an agreement was finally reached when both parties signed the Contract of 1604.[123] According to the terms of this contract, Kepler would be allowed free access to the observations, and Tengnagel would be allowed to use any results of Kepler deriving from his use of them. Neither would Kepler be allowed to publish anything without first seeking Tengnagel's permission. However, Tengnagel would not be allowed to alter any of Kepler's work without the latter's permission. The contract was to remain in force until the publication of the *Rudolphine Tables*, after which Kepler would be free to publish without restraint. One unfortunate product of this contract was Tengnagel's pompous preface to *Astronomia nova* (1609).[124]

Tengnagel's stranglehold on Kepler effectively ended when the latter left Prague for Linz to become mathematician for the Estates of Upper Austria in May 1612. The contract itself became null and void on Tengnagel's death in 1622. Kepler continued work on the *Tables* and saw them through to publication in 1627 (see Chap. 9 of this volume). Tycho and Kepler together had at last truly reformed astronomy.

---

[119] Kepler to Heydon, October 1505 in Kepler (1951), letter 357, pp. 22–27.

[120] Kepler to Longomontanus, beginning of 1605, in Kepler (1951), letter 323, pp. 223–228.

[121] Kepler to Herwart, 12 November 1602, in Kepler (1949), letter 232, pp. 337–339.

[122] Kepler (1969), p. 10*.

[123] Kepler (1975), docs 5.1 and 5.2, pp. 189–190.

[124] Kepler (1990), p. 17; Kepler (1992), pp. 43–44.

References

Alfonso, X. (1483). *Tabule astronomice illustrissimi Alfonsij regis*. Ioha[n]nis Hamman.
Alfonso, X. (1492). *Tabule astronomice illustrissimi Alfonsij regis*. Ioha[n]nis Hamman.
Argelander, F. W. A. (1864). Über den neuen Stern von Jahre 1572. *Astronomische Nachrichten, 63*, 273–276.
Baumgardt, C. (1952). *Johannes Kepler: Life and letters*. London: Victor Gollancz.
Bayer, J. (1603). *Uranometria omnium asterismorum, continens schemata, nova methodo delineata, aereis laminis expressa*. Augsburg: Christoph Mang.
Brahe, T. (1573). *De nova et nullius aevi memoria prius visa stella, …*. Copenhagen: Laurentius Benedicti.
Brahe, T. (1588). *De mundi aetherei recentioribus phaenomenis*. Uraniborg.
Brahe, T. (1598). *Astronomiae instauratae mechanica*. Wandsbek.
Brahe, T. (1602). *Astronomiae instauratae progymnasmata*. Prague.
Brahe, T. (1621). *De disciplinis mathematicis oratio*. Hamburg: Froben.
Brahe, T. (1666). *Historia coelestis* (A. von Curtz, Ed.). Augsburg: Simon Utzschneider.
Brahe, T. (1921). *Tychonis Brahe Opera Omnia, Tomus V: Astronomiae instauratae mechanica*. Copenhagen: Libreria Gyldendaliana.
Brahe, T. (1922). *Tychonis Brahe Dani Opera Ommia, Tomus IV: Scripta astronomica* (J. L. E. Dreyer & J. Raeder, Ed.). Copenhagen: Libreria Gyldendaliana.
Brahe, T. (1923a). *Tychonis Brahe Dani Opera Ommia, Tomus I: Scripta astronomica* (J. L. E. Dreyer & J. Raeder, Ed.). Copenhagen: Libreria Gyldendaliana.
Brahe, T. (1923b). *Tychonis Brahe Dani Opera Ommia, Tomus X: Thesaurus observationum … Tomus I* (J. L. E. Dreyer, Ed.). Copenhagen: Libreria Gyldendaliana.
Brahe, T. (1924). *Tychonis Brahe Dani Opera Ommia, Tomus XI: Thesaurus observationum … Tomus II* (J. L. E. Dreyer & J. Raeder, Ed.). Copenhagen: Libreria Gyldendaliana.
Brahe, T. (1925a). *Tychonis Brahe Dani Opera Ommia, Tomus VIII: Epistolae astronomicae, Tomus III* (J. L. E. Dreyer, Ed.). Copenhagen: Libreria Gyldendaliana.
Brahe, T. (1925b). *Tychonis Brahe Dani Opera Ommia, Tomus XII: Thesaurus observationum … Tomus III* (J. L. E. Dreyer & J. Raeder, Ed.). Copenhagen: Libreria Gyldendaliana.
Brahe, T. (1926). *Tychonis Brahe Dani Opera Ommia, Tomus XIII: Thesaurus observationum … Tomus IV* (J. L. E. Dreyer & J. Raeder, Ed.). Copenhagen: Libreria Gyldendaliana.
Brahe, T. (1928). *Tychonis Brahe Dani Opera Ommia, Tomus XIV: Epistolae et acta ad vitam Tychonis Brahe* (E. Nystrøm, Ed.). Copenhagen: Libreria Gyldendaliana.
Brahe, T. (1996). *Instruments of the renewed astronomy* (Translated from the Latin by H. Raeder, E. Strömgren and B. Strömgren, and revised with commentary A. Hadravová, P. Hadrava and J. R. Shackelford). Prague: Koniasch Latin Press.
Chabás, J., & Goldstein, B. R. (2003). *The Alfonsine Tables of Toledo* (*Archimedes*, vol. 8). Dordrecht: Kluwer.
Dreyer, J. L. E. (1890) *Tycho Brahe: A picture of scientific life and work in the sixteenth century*. Edinburgh: Adam and Charles Black.
Dreyer, J. L. E. (1953). *A history of astronomy from Thales to Kepler*. New York: Dover.
Dürer, A. (1515). Two star charts, for the northern and southern hemispheres. Nuremberg.
Dürer, A. (1555). Two star charts, for the northern and southern hemispheres. Nuremberg.
Evans, R. J. W. (1997). *Rudolf II and his world* (2nd ed.). London: Thames and Hudson.
Field, J. V. (1988). *Kepler's geometrical cosmology*. London and Chicago: Athlone Press and Chicago University Press (Reprinted by Bloomsbury, 2014).
Field, J. V. (2009). The realism of Copernicus and Kepler. *Oriens-Occidens, 7*(2009), 239–267.
Forbes, E. G., Meadows, A. J., & Howse, D. (1975). *Greenwich observatory* (Vol. 1). London: Taylor & Francis.
Fricke, W., et al. (1988). Fifth fundamental catalogue. Part 1. The basic fundamental stars. *Veroeff. Astron. Rechen-Inst., 32*, 1–106.
Galilei, G. (1632). *Dialogo sopra i due massimi sistemi del mondo*. Florence: Landini.

Galilei, G. (1953). *Dialogue concerning the two chief world systems—Ptolemaic & Coperican* (Translated from the Italian by Stillman Drake and with Foreword by Albert Einstein). Berkeley and Los Angeles: University of California Press.

Garcaeus, J. (1556). *Tractatus brevis et utilis, de erigendis figuris coeli, verificationibus, revolutionibus et directionibus.* Wittenberg: Heirs of Georg Rhaus.

Gassendi, P. (1655). *Tychonis Brahei, Equitis Dani, Astronomorum coryphaei, Vita.* The Hague: Adrian Vlacq.

Gingerich, O. (2011). The great Martian catastrophe and how Kepler fixed it. *Physics Today, 64*(9), 50–54.

Gingerich, O., & Voelkel, J. R. (1998). Tycho Brahe's Copernican campaign. *Journal for the History of Astronomy, 29*(1998), 1–34.

Hamel, J. (2007). Severin, Christian. In T. Hockey et al. (Eds.), *Biographical encyclopedia of astronomers, Vol. II: M–Z* (pp. 1041–1042). New York: Springer.

Hoffleit, D. (Ed.). (1964). *Catalogue of bright stars.* New Haven: Yale University Observatory.

Jardine, N. (1984). *The birth of history and philosophy of science.* Cambridge: Cambridge University Press.

Kepler, J. (1596). *Mysterium cosmographicum.* Tübingen: G. Gruppenbach.

Kepler, J. (1609). *Astronomia nova aitiologêtos, seu physica coelestis, tradita commentariis de motibus Stellae Martis.* Heidelberg: E. Vogelin.

Kepler, J. (1938). *Johannes Kepler Gesammelte Werke. Band I: Mysterium cosmographicum/De stella nova* (M. Caspar, Ed.). Munich: C. H. Beck.

Kepler, J. (1945). *Johannes Kepler Gesammelte Werke. Band XIII: Briefe 1590–1599* (M. Caspar, Ed.). Munich: C. H. Beck.

Kepler, J. (1949). *Johannes Kepler Gesammelte Werke. Band XIV: Briefe 1599–1603* (M. Caspar, Ed.). Munich: C. H. Beck.

Kepler, J. (1951). *Johannes Kepler Gesammelte Werke. Band XV: Briefe 1604–1607* (M. Caspar, Ed.). Munich: C. H. Beck.

Kepler, J. (1963). *Johannes Kepler Gesammelte Werke. Band VIII: Mysterium cosmographicum (editio altera cum notis) / De cometis, Hyperaspides* (F. Hammer, Ed.). Munich: C. H. Beck.

Kepler, J. (1969). *Johannes Kepler Gesammelte Werke. Band X: Tabulae Rudolphinae* (F. Hammer, Ed.). Munich: C. H. Beck.

Kepler, J. (1975). *Johannes Kepler Gesammelte Werke. Band XIX: Dokumente zu Leben un Werk* (M. List, Ed.). Munich: C. H. Beck.

Kepler, J. (1981). *The secret of the universe: Mysterium cosmographicum* (Translated from the Latin by A. M. Duncan with introduction and commentary by E. J. Aiton). Norfolk CT: Abaris Books.

Kepler, J. (1990). *Johannes Kepler Gesammelte Werke. Band III: Astronomia nova* (2nd ed.) (M. Caspar & Kepler-Kommission, Ed.). Munich: C. H. Beck.

Kepler, J. (1992). *New astronomy* (Translated from the Latin with and Introduction, Glossary and Notes by W. H. Donahue). Cambridge: Cambridge University Press.

Kitchen, C. R. (1984). *Astrophysical techniques.* Bristol: Hilger.

Kremer, R. L. (2023). *Alfonsine astronomers at work: Computing planetary positions in Europe, 1350–1560.* Turnhout: Brepols.

Lang, K. R. (Ed.). (2006). *Astrophysical formulae* (vol. II, 3rd ed.). Heidelberg: Springer.

Longomontanus, C. (1622) *Astronomia Danica.* Amsterdam: Johan and Cornelius Blaeu.

Maeyama, Y. (2002). Tycho Brahe's stellar observations. An accuracy test. *Acta Historica Astronomiae, 16*, 113–119.

Picard, J. (1680). *Voyage d'Ouranibourg, ou Observations Astronomiques Faites en Dannemarck.* Paris: Imprimerie royale.

Reinhold, E. (1551). *Prutenicae tabulae coelestium motuum* (reprinted 1562, 1571 and 1585). Tübingen.

Repsold, J. A. (1908). *Zur Geschichte der Astronomischen Messerzeuge von Pürbach bis Reichenbach 1450 bis 1830.* Leipzig: Wilhelm Engelmann.

Rosa, M. R. (2010). How really precise and accurate are Tycho Brahe's data? In *Kepler's heritage in the space* age (A. Hadravová, T. J. Mahoney & P. Hadrava, Eds., pp. 102–113). Prague: Národní Technické Muzeum.

Thoren, V. E. (1973a). Tycho Brahe: Past and future research. *History of Science, 11*, 270–282.

Thoren, V. E. (1973b). New light on Tycho's instruments. *Journal for the History of Astronomy, 4*, 25–45.

Thoren, V. E. (1990). *The Lord of Uraniborg: A biography of Tycho Brahe.* Cambridge: Cambridge University Press.

Thorndike, L. (1949). *The sphere of Sacrobosco and its commentators.* Chicago: Chicago University Press.

Tuckerman, B. (1962). *Planetary, lunar, and solar positions, 601 B.C. to A.D. 1, at five-day and ten-day intervals.* Memoirs of the American Philosophical Society (Vol. 56).

Tuckerman, B. (1964). *Planetary, lunar, and solar positions, A.D. 2 to A.D. 1649, at five-day and ten-day intervals.* Memoirs of the American Philosophical Society (Vol. 59).

Tupman, G. L. (1900). A comparison of Tycho Brahe's meridian observations of the Sun with Leverrier's Solar Tables. *The Observatory, 23*, 132–135, 165–171.

Ursus, N. R. (1597). *De astronomicis hypotesibus tractatus.* Prague.

von Auwers, A. J. G. F. (1882) *Neue Reduction der Bradley'schen beobachtungen aus den Jahren 1750 bis 1762.* St. Petersburg: Commissionare der K. Akademie der wissenschaften.

Wesley, W. G. (1978). The accuracy of Tycho Brahe's instruments. *Journal for the History of Astronomy, 9*, 42–53.

Wesley, W. G. (1979). Tycho Brahe's solar observations. *Journal for the History of Astronomy, 10*, 96–101.

Chapter 4
Kepler's Discovery of the Planetary Orbit: The Goldilocks Solution

A. E. L. Davis

4.1 Kepler's Discoveries: An Overview

The *Kepler* mission launched in 2009 (so named to celebrate the quatercentenary of the publication of *Astronomia nova*) lived up to its name. This was chosen by a group of NASA astronomers that included Carl Sagan (1934–1996), because the spacecraft was to investigate exoplanets (planets orbiting other stars); its success in identifying them has been measured in thousands. The particular exoplanets it was searching for were those that might support life. The main condition is that the planet must orbit within the *habitable zone*, defined by the possible existence of liquid water, so neither too hot nor too cold—and hence the distance of the planet from its star had to be not too large, nor too small, but just right. For this reason, astronomers refer to this distance range 'the Goldilocks zone': and there is a connection with Kepler's procedure, as we shall now see.

For more than four centuries almost all commentators have been content to accept that Kepler's solution to the problem of planetary motion was merely an approximation (or even a lucky guess). The belief has been that, using the best available observations, he was able to show that the planetary positions seemed (near enough) to lie on an ellipse, as the curve is defined by Apollonius of Perga (*c*. 240–*c*. 190 BC), with the Sun at one focus—that is, an entirely ad hoc (non-theoretical) result. Only since 1981 has it been demonstrated to the contrary—that Kepler approached the problem entirely differently, relying on Euclid's straightedge-and-compasses geometry, he used the precision of those tools to construct exact curves against which he could successively test the observed (approximate) positions of the planet concerned. He naturally started from a circle (Stage I) but this was too large. Then he constructed what turned out to be an oval (Stage II), but that was too small. For Stage III he invented a way to construct a curve that lay in the middle, which turned out to be just right. Even then Kepler had not finished: he did not immediately recognize that

A. E. L. Davis (Deceased) (✉)
University College London, London, UK

© Springer Nature B.V. 2024
A. E. L. Davis et al. (eds.), *Reading the Mind of God*, Springer Praxis Books,
https://doi.org/10.1007/978-94-024-2250-4_4

the unknown curve he had constructed was precisely an ellipse as that curve was defined by Archimedes (287–212 BC). This finally turned out to be the basis for a geometrically perfect theory (which can be proved with a rigour exacting enough to match modern standards). How he did it will be the subject of this chapter.

Kepler did not use algebra for his proofs: he believed that the geometry of Euclid alone was appropriate for the analysis of the heavens, the realm of God. Nowadays, since people are not accustomed to reasoning by geometry, we shall not expect our readers to follow each step of Kepler's procedures. Instead, we shall start by setting out the procedure by which Kepler constructed the planetary orbit, and provide an outline of his background in ancient astronomy; then in subsequent sections we shall track in detail how he got there. In the concluding section, we shall express the orbit in algebraic notation that modern readers are more likely to be familiar with.

Figure 4.1 shows the diagram that Kepler invariably used as a basis for finding the orbit of Mars, based on a strictly heliocentric configuration in which the Sun A is fixed. Kepler followed tradition in starting from the (simplified) assumption that the planet will move in the circle centre B, diameter CD, eccentrically placed with respect to A, and known as the eccentric circle. Q is a typical position of the planet on that circle at Stage I; the eccentric circle provides the geometrical framework for illustration of the planetary laws. This is because at the culmination of Kepler's investigations, in Stage III, the position of the planet P that satisfies observations (to the limit of precision attainable by Tycho's observations) lies on the ordinate QH of that circle defined by Q.

Now we can state what Kepler discovered:

Ellipse Law (Law I)—the path of the planet is the particular ellipse $CPFD$ [with the Sun at one focus];

Area Law (Law II)—the time in orbit is measured by the area CAP swept out by the radius vector.

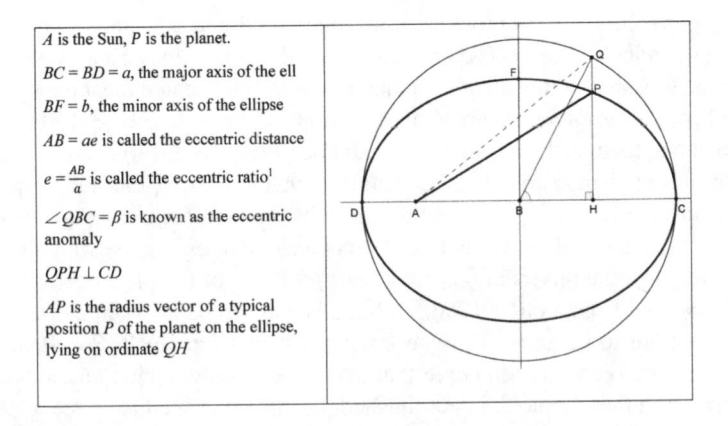

A is the Sun, P is the planet.

$BC = BD = a$, the major axis of the ell

$BF = b$, the minor axis of the ellipse

$AB = ae$ is called the eccentric distance

$e = \frac{AB}{a}$ is called the eccentric ratio[1]

$\angle QBC = \beta$ is known as the eccentric anomaly

$QPH \perp CD$

AP is the radius vector of a typical position P of the planet on the ellipse, lying on ordinate QH

Fig. 4.1 The geometry of the orbital ellipse that Kepler discovered

(The last phrase of the statement of Law I has been parenthesized because no reference to the focus occurs anywhere in the text of *Astronomia nova*, so it is an anachronism in that context. Kepler defined the particular ellipses in the manner of Archimedes rather than that of Apollonius, as we shall confirm in Sect. 4.5 below, where the distinction will be explained).

Astronomia nova (1609) is the work that contains Kepler's major astronomical discoveries, and therefore it is the main subject of the present analysis.[1] In Chap. 1, Kepler applied the term *orbit* in an astronomical context. The new term incorporated the two constituents listed above: first the planetary path, determined by the radial distance from the fixed source of motion, and then the measure of time, which described how the planet moved ('timewise') along that path, starting from aphelion *C*. (Kepler always followed the ancient practice of measuring from aphelion, the greatest distance from the Sun.) These constituents provide the pair of *orbital coordinates* of a typical point *P* on the ellipse that are required to specify where the planet is, and when it is there. (Though these relations are nowadays known as Kepler's laws, Kepler himself did not so describe them; yet he would have regarded them as part of God's governing plan for the Universe.) It is notable that the mathematical rigour of the geometrical structure involved will entail an exact solution—though reality itself is not so precise, as Isaac Newton (1642–1727), and then Albert Einstein (1879–1955), subsequently demonstrated.

The third and last of the Keplerian planetary laws was formulated independently, and provided a mathematical synthesis for the planetary system. It was stated later, in *Harmonice mundi* (1619).[2]

Kepler had enormous respect for his predecessors; though evidently his traditional approach did not inhibit a new interpretation where necessary, and it was his good fortune that the exact way the Universe works kinematically can be derived by applying the simplest geometrical principles—as Figs. 4.1 and 4.8 indicate.

Kepler received a thorough grounding in the work of the Greek classical authors at the University of Tübingen in preparation for his intended career as a Lutheran minister: both Plato (428/427 or 424/423–348/347 BC) and Aristotle (384–322 BC) greatly influenced his approach to astronomy. Ancient astronomers believed that uniform circular motion alone could be identified with celestial perfection, so it must have come as rather a shock when they noticed, quite early on, that certain bodies (the 'wanderers'[3]) moved in a less ordered way against the background of the stars. Nevertheless, with great ingenuity, the Greeks managed to minimize the disruptive effect of such irregularities on the perfection of the heavens; by permitting a celestial body to possess more than one circular motion. Thus they ensured that the principle of uniform circular motion remained inviolate, applying what we shall

---

[1] So this book is cited here only by chapter number, and by page number in the modern Latin edition (Kepler, 1990). Translations of the short extracts quoted in this paper are my own, but I have also benefited from access to the revised edition of the translation by Donahue (Kepler, 2015).

[2] See also Field, 'Kepler's cosmology' (Chap. 2 of this volume) and Field (1988).

[3] The five naked-eye planets, together with the Sun and the Moon.

refer to as the Platonic precept,[4] by requiring each of the multiple motions possessed by a body to take place in its own circle, uniformly about its own centre.

Up to Kepler's era, indeed, this fundamental tenet was never challenged, and in particular it was always believed that circular motion—rotation—was the only motion natural to the heavens, requiring no cause other than its status as a provision of God. However, Kepler became aware that the ancient astronomy had never managed to account for the motion of the planets as accurately as the observations now gave him the potential to do—so he was stimulated to speculate about the possibility of physical causes in the heavens. Unfortunately, the physics of Kepler's day was inadequate, and though previous commentators have stated that Kepler used physical causes to determine the orbit precisely, this was not in fact mathematically feasible because the suggested causes were bound to be unsound. Rather, Kepler found the orbit by traditional geometrical reasoning, and only later was he able to propose a pair of causes to 'justify' it (see Sect. 4.7). Accordingly, it seems that the translation into English and German of the Greek word *AITIOLOGETOS* in the title of *Astronomia nova* to mean 'based upon causes', is a misrepresentation of Kepler's intention. One instead should adopt the translation 'New Astronomy with an account of reasons …', as set out by Gregory in Chap. 5 of this volume.

Nevertheless, as a convinced Copernican (see Sect. 4.2.1), Kepler regarded the Sun as the overriding physical meta-cause, because of its association with the (infinite) power of God. In relation to a single planet, it played two particular roles:

• The Sun is fixed in position at the hub of the world, as the origin of coordinates;
• The Sun is the source responsible for generating all celestial motion.

The first of these expresses Kepler's most fundamental astronomical belief; the second was stated as the title to chapter 33 of *Astronomia nova*: 'the power that moves the planets is located in the body of the Sun'. In fact, this second statement was expanded in the title to chapter 38: 'the motion of each [planet] is compounded from two causes'. So undoubtedly Kepler realized that it made sense that the Sun, as a rotating celestial body, would be capable of generating the motion of a planet in just two ways: round itself in a circle (motion of revolution) or towards and away from itself (radial motion). Taken with the assumption that the position of the Sun is fixed, and noticing that rotation about a fixed axis is implied in the title to *Astronomia nova* (Chap. 34), the conclusion is that the orbit produced will be restricted to a plane—precisely a one-body problem, in fact. (One may doubt whether Kepler appreciated the full implications at this early stage of his investigations, but he did in fact demonstrate the same thing from observations, as we shall see in Sect. 4.2.3)

Hence in a modern sense the situation must be described as strictly kinematical. It is of course consistent with the essential duality of the two laws, of their two constituents and the two coordinates, circular and rectilinear, as well as, now, of the component motions, circular and radial (following Aristotle). Moreover all

---

[4] This name has been chosen for specificity because the precept was frequently attributed to Plato himself.

these pairs were composed of two quantities that were mutually perpendicular—Euclid's term for this was *orthogonal*—so they behaved in accordance with a generally accepted principle of orthogonal independence, which states that a pair of individual components should be treated independently if and only if their directions are mutually perpendicular. It is interesting that, almost contemporaneously, Galileo Galilei (1564–1642) applied this principle with advantage in treating the motion of a projectile by considering its vertical and horizontal motions independently in the Third and Fourth Days of *Discourses concerning Two New Sciences*.[5] Certainly Kepler put the principle into practice frequently, to supply implied justification for his handling of circumsolar and radial motion separately—as we shall so treat them in what follows.

4.2 Astronomical Considerations—The Observational Phase (*Astronomia nova,* Chaps. 1–31)

Overall, Kepler's work in *Astronomia nova* fell into two phases. The first phase dealt with practical astronomical matters, by organizing the raw observations and reducing them to usable form, as outlined in the present section. The second phase was tackled in Kepler's chapters 32–60, applying the framework set out in Sect. 4.3 which Kepler used to determine the theoretical orbit, discussed in Sect. 4.4 onwards of the present chapter.

Kepler proposed his orbit for Mars as representing the actual path of the planet in space; that is, not merely its apparent changes in position against the pattern of the fixed stars. Moreover, as a Copernican he believed no division should be made between the behaviour of celestial and terrestrial bodies—after all, Copernicanism made the Earth a planet. This allowed Kepler to apply notions and insights gained from observing terrestrial bodies when he came to consider the behaviour of bodies in the heavens.

4.2.1 Application of Strict Heliocentricity (**Astronomia nova,** Chaps. 3–6)

Greek astronomers had progressively developed techniques for 'saving the appearances'—that is, for providing a sufficiently accurate representation of the motions of the heavenly bodies to agree with what was seen from the Earth. This work culminated in the geocentric arrangement perfected by Claudius Ptolemy (*fl.* AD 129–140): Kepler always expressed great admiration for his skill and thoroughness. The Ptolemaic configuration was modified only minimally in the succeeding centuries and remained dominant until the ideas of Nicolaus Copernicus (1473–1543)

---

[5] Galilei (1638, 1913).

transformed astronomy, by setting the Earth in motion. Copernicus' book made an enormous impact on Kepler, who was introduced to it at university by his teacher Michael Maestlin (1550–1631), and he at once became a convinced Copernican. However, Copernicus had merely exchanged the relative positions of the Earth and the Sun in the Ptolemaic configuration, putting the Sun at rest, but at the centre of the Earth's orbit. Thus the planet–Sun distances were measured from the centre of the Earth's orbit as origin, rather than from the actual Sun. Obviously, Kepler was obliged to recalculate every planetary distance at the outset, in order to represent the path of the planet round the Sun as the single, plane closed curve that he envisaged as his goal. As to the practicalities, the calculations involved solving a series of individual plane triangles Sun–planet–Earth by applying the triangulation method (used by terrestrial surveyors, then and now, to find the distance of an inaccessible object) to the splendid stock of observations made by Tycho Brahe (1546–1601).[6] Hence it was Kepler who was responsible for turning the Copernican planetary arrangement into a system that was strictly heliocentric, thus enabling Kepler, through his first law, to make the Sun truly the centre of planetary motion.

It is interesting that the many benefits of heliocentrism not mentioned by Copernicus (and reaped by Kepler, as we shall demonstrate in this chapter) led Kepler to remark (apropos of another topic, in chapter 14), that Copernicus was ignorant of his own riches.[7]

4.2.2   Why Mars? (Astronomia nova, Chap. 7)

Kepler commonly referred to his main astronomical work, which we know as *Astronomia nova*, by the less formal name 'Commentaries on Mars'. It is true that Kepler originally investigated the orbit of Mars because that was the task allocated to him when he joined Tycho Brahe in Prague in 1600. However, as he explained in *AN* chapter 7, he came to regard it rather as a disposition of Providence,[8] when he recognized that Mars is the unique planet whose orbit Tycho Brahe's observations could hope to determine. (It is not generally realized that Kepler's belief that all the planets obey the laws he eventually proved for Mars must have been an assumption by analogy: the accuracy of Brahe's observations was sufficient only to distinguish the lengths of the axes of the orbits of the remaining planets, and not the shape of their paths themselves.) Consequently, this particular planet remained the principal concern throughout his work. Four factors contributed to the observability of Mars:

• Mars is an outer planet (and therefore seldom too close to the Sun for favourable viewing conditions).

---

[6] For a more detailed account, see Davis (1992); see also Field, 'The long life of the *Rudolphine Tables*' (chapter 9 of this volume) and Mahoney, 'Measuring the heavens: how Tycho Brahe revolutionized observational astronomy' (chapter 3 of this volume).

[7] Kepler (1609), ch. 14, p. 81; Kepler (1990), p. 141, line 3.

[8] Kepler (1609), ch. 7, p. 53; Kepler (1990), p. 109 lines 7–8.

- Mars is the only one of the outer planets whose path is sufficiently non-circular for its shape to be determinable by measurements made using open sights.
- Mars is the nearest to the Earth of the outer planets (and therefore changes in its position appear larger, allowing more accurate observations).
- Mars is the nearest to the Sun of the outer planets (and therefore makes more frequent circuits, allowing more observations).[9]

4.2.3 Kepler's 'Zeroth Law' (Astronomia nova, Chaps. 13–14)

In the next part of *Astronomia nova* Kepler continued to work exhaustively on analysing Tycho's data. In early astronomy, the *aphelion* and *perihelion* of each planet were among the first points whose positions were established with any certainty, which emphasized the existence of one axis of symmetry (known to astronomers as the *line of apsides*). But at intermediate points of the path, the observations badly needed systematization. We now recognize that the motion of a planet does take place about the Sun and can therefore be expressed straightforwardly in terms of orthogonal components, with the Sun as origin; but this motion inevitably appeared irregular, sometimes random, when viewed from the Earth, or even (as Copernicus viewed it) from the centre of the Earth's orbit.

In the course of chapters AN13–14, Kepler invented new ways of testing the observations and was able to satisfy himself that the orbit of Mars lay in a fixed plane through the Sun, and to determine the constant inclination of that plane to the plane of the ecliptic (numerous determinations gave the angle as $1°50'$). Then he inferred that the orbits of all the planets would be planar, and thus was able to conclude that all the orbital planes would pass through the Sun. This observational confirmation that the orbit was a plane curve was of the greatest importance, because it supported the theoretical soundness of Kepler's two-dimensional kinematical treatment of the orbit referred to in Sect. 4.1 (and Kepler showed that he appreciated its significance by demonstrating it again in chapter 52). Dreyer[10] drew attention to this 'important discovery' and Gingerich[11] has emphasized it by suggesting that it should be named 'Kepler's zeroth law'. Looked at another way, Kepler had managed to solve the 'problem' of the latitudes (which had caused difficulty for all earlier astronomers), by establishing that, in the kinematical context, latitude would be associated with a constant angle because there was no concept of mass to introduce any variation.

---

[9] I am indebted to J. V. Field for pointing out this additional feature in discussion some years ago.

[10] Dreyer (1906); see revised version: Dreyer (1953), p. 383.

[11] Gingerich (1975), p.263.

4.2.4   The 'Vicarious Hypothesis' (**Astronomia nova,** Chaps. 16, 18–20)

Kepler deeply appreciated his good fortune in the circumstances in which his life became intertwined with that of Tycho Brahe. He was able to use Tycho's splendid store of observations, whose precision achieved (or, as we have seen in Chap. 3 of this volume, even exceeded) the limit of resolution achievable with the human eye.[12] In chapter 19 of *Astronomia nova*,[13] Kepler affirmed his debt to Tycho in a situation in which he had felt obliged to abandon an early attempt, which he later named the 'Vicarious Hypothesis' (*hypothesis vicaria*), simply because of its failure adequately to satisfy observations:

> Since divine benevolence has vouchsafed us Tycho Brahe, a most diligent observer, from whose observations the 8′ error in this Ptolemaic computation is shown, it is fitting that we with thankful minds both acknowledge and honour this benefit of God. ... Now, because they could not have been ignored, these eight minutes alone will have led the way to the reformation of all of astronomy.[14]

Kepler relied on this unprecedented level of accuracy throughout both phases of his work—in the first phase for assessing the accuracy of observations, and in the second phase for checking his trial orbits against observations (to be demonstrated in Sect. 4.4 below).

We shall refer to Fig. 4.2—reiterating the standard notation employed throughout this chapter—in order to explain Kepler's abortive attempt in *AN* chapter 16 to introduce some theoretical improvement. In ancient astronomy Ptolemy had taken the important step of moving the centre of the proposed circular path away from the point at which he measured time (the equant E) and Kepler decided to investigate whether Ptolemy had actually found the optimum position for it. In chapter 16, Kepler therefore proposed an alternative position for the centre of the supposed orbital circle (between B and E). However, this involved him in tiresome and long-winded calculations, all to no purpose as it turned out, since the new theory, which at first seemed promising because it satisfied one group of observations, notably failed when presented with a different set.

In the event, Kepler was able to salvage something useful from this investigation, because the arrangement possessed the quirk that it was accurate enough to use as a surrogate for some purposes, so he was occasionally able to employ it as a check in his

---

[12] For a discussion on the acuity of the eye and Tycho's observing techniques see Mahoney, 'Measuring the heavens: how Tycho Brahe revolutionized observational astronomy', in Chap. 3 of this volume.

[13] Kepler (1990), p. 178, lines 1–12.

[14] For a discussion of the quality of Tycho's observations see Mahoney, 'Measuring the heavens: How Tycho Brahe revolutionized observational astronomy' (chapter 3 of this volume). See also the Wikipedia entry for the Vicarious Hypothesis: https://en.wikipedia.org/wiki/Vicarious_Hyp othesis#:~:text=Kepler%20used%20these%20four%20observations%20to%20determine%20t he,time%20and%20location%20of%20the%20observation%20would%20match. (Accessed 26/ 06/2022).

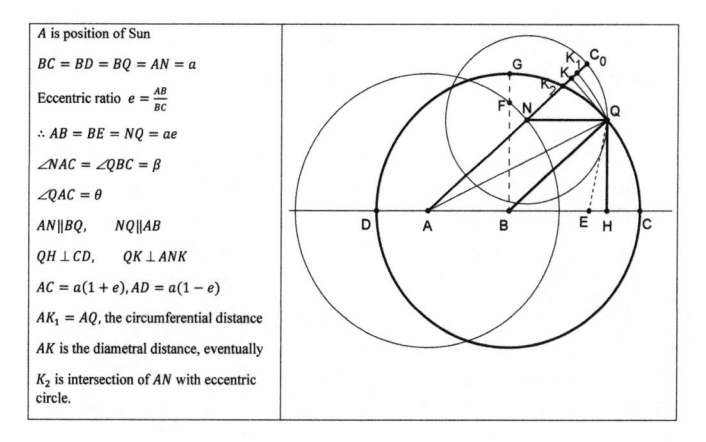

Fig. 4.2   The invariant framework and the Ptolemaic parallelogram

subsequent work (thus the reason for its name). But Kepler was not yet ready even to envisage the two major breakthroughs by which he eventually reformed astronomy.

4.2.5   Increasing the Density of Observations (Astronomia nova, Chapter 17)

Kepler was aware of various secular (long-term) changes and periodic perturbations that affected the orbits of both Mars and the Earth (due, we now recognize, to the presence of other bodies in the system—and the existence of mass as a feature of the real world). While searching for a simple solution, Kepler found a way (in *AN* Chap. 17) to compensate for the secular changes and the perturbations by tabulating the cumulative effects using values recorded over many centuries, from which he obtained tiny annual differences that were then applied to 'reduce' (that is, to adjust) every one of Tycho's observations to a common standard. Thus he successfully combined all the appropriate observations into a single dense set. It has not been previously remarked that this procedure would have enormously increased the size of the database available to him, and it certainly must have helped to counter some observational deficiencies (by providing nearby alternatives, when he sensed that a particular value looked questionable). We can certainly argue that this procedure may well have significantly helped to account for the fact (often mentioned but so far not adequately explained) that Kepler's fairly frequent individual errors never seemed to affect the overall picture as his results converged to exactitude.

4.2.6 More Accurate Representation of the Earth's Orbit (Astronomia nova, *Chapters 22–31*)

One of Kepler's major problems was to determine the motion of Mars from an observing platform, the Earth, that was itself in motion, and so the orbit of the Earth had to be soundly established first. Kepler felt that its orbit should be represented more accurately by a Ptolemaic eccentric circle, just like all the other planets (see Sect. 4.3.1 below), though that had not previously been thought necessary because the path of the Earth is so nearly a circle. In chapters 22–28 Kepler used an innovative method to confirm the validity of this improvement, by viewing the Earth at various points of its orbit from a fixed position of Mars. It was a notable consequence that in chapter 30 Kepler tabulated the first-ever systematic set of associated distance–time (that is, orbital) coordinate values for a planet (the Earth) at intervals of $1°$, for its half-orbit from *aphelion* to *perihelion*. (A complete table for Mars was not produced until Kepler published the *Rudolphine Tables* in 1627.[15])

As the result of the prolonged astronomical investigations of the first phase, Kepler had now prepared the necessary astronomical groundwork and had achieved a considerable simplification of the structures assumed by his predecessors. Only then was he able to move on to the second, theoretical phase.

4.3   The Fundamental Geometrical Framework for the Second Phase (and Standard Notation)

Having thoroughly cleansed the Augean stables,[16] Kepler set out to discover the orbit of Mars. The common features of the framework within which he examined both constituents of the orbit consisted of a closed curve, repeating (and thus periodic), having one axis of symmetry, and lying in a plane. We shall first compare typical points of different proposed orbits, occurring at near enough the same time, to determine the distance of the planet from the Sun geometrically, and then we shall look separately at how Kepler handled the measurement of time.

---

[15] Kepler, J. (1627, 1969); see also Field, 'The long life of the *Rudolphine Tables*' (Chap. 9, this volume).

[16] Kepler so described the situation in a letter to Longomontanus in early 1605 (Kepler, 1951, letter 323). He was referring to one of the Labours of Hercules.

4.3.1 The Invariant Geometrical Framework Derived from Ptolemy

Since the representation of both constituents of the orbit—the time as well as the path—has to be geometrical, we begin (see Fig. 4.2) by identifying the structure within which the two can be related. (Its final form has been anticipated in Fig. 4.1 in order to state Kepler's results.) Partly out of respect for his predecessors, and also because he was by nature a traditionalist, Kepler based his construction on the simplest possible Ptolemaic framework, from *Almagest* III, 3, representing the supposed orbit of the Sun around the Earth (but Kepler transposed it to heliocentricity for his purpose, of course). The motions produced were evidently in accordance with the Platonic precept in its simplest manifestation (see Sect. 4.1 above), consisting of only three uniform motions in perfect circles about their respective centres. Its basis depended on the geometrical equivalence of the epicycle-with-deferent[17] to the eccentric circle: the fact was said by Ptolemy (in *Almagest* XII, 1), to have been discovered by Apollonius (which is a bit ironic, as we shall see in Sect. 4.5 below). This structure will provide the foundation for all Kepler's orbit constructions, and the notation and symbols we introduce here will be adopted as standard throughout the remainder of this account.[18] The centres of the deferent circle (A) and of the eccentric circle (B) lie on CD, known in astronomy as the *line of apsides*. The point Q initially represents the proposed typical position of the planet, and Q moves on the epicycle (centre N) that revolves on the deferent whose own centre lies at the fixed point A, which marks the position of the Sun. When we fix the three uniform angular motions, round B, A, N, to be equal (but that at N in the opposite direction), we obtain the simple situation that was illustrated in Fig. 4.1; the consequent constructed parallels ensure that the angles of rotation about the respective centres at A, B, N are all equal. Then the resultant motion of the planet at Q takes place in the eccentric circle centre B diameter CD. We set QH as the ordinate of Q, and name C_0 as the point where the epicycle cuts AN extended.

The following quantifications will be used throughout: the radii of the deferent circle and the eccentric circle are of equal length, and are denoted by a, the standard or norm-length for an individual planet, while the radius of the epicycle is designated by $AB = ae$, where e is here defined as the eccentric ratio having the value $e < 0.1$ for every planet except Mercury.[19] (Evidently, the adjective 'eccentric' is used here in reciprocal senses, the place of the Sun at A being known as the eccentric position, and the circle centre B is described in astronomy as the eccentric circle. Kepler probably regarded AB as the eccentric distance. In any case we avoid the term *eccentricity*, which will not be employed anywhere in this chapter because of the way it inhibits

---

[17] This term means 'carrying circle' (it carries the epicycle).

[18] This standard notation was initiated by Kepler himself in *AN* Chap. 40 (though unfortunately he did not continue it thereafter).

[19] Mercury was excluded because its nearness to the Sun and low altitude as seen from Tycho's observatories made observations unreliable.

a sound assessment of Kepler's method, by suggesting to the reader a conic section well before any such idea had entered Kepler's mind.)

Most importantly, $\angle QBC$, the angle measured at the centre B, specifying the position of Q, will throughout this account be designated by the typical value β known in astronomy, both ancient and modern, as the *eccentric anomaly*. This angle must be distinguished from the polar angle (or true anomaly, which Kepler called the equated anomaly) θ, always measured at A, which became the natural basis of analysis immediately after Kepler's time. However, Kepler followed his predecessors in adopting the eccentric anomaly β as the foundation for his successful interpretation. He avoided the theoretical use of the equated anomaly in all circumstances, though he occasionally used it in practice.

The point E on the line of apsides, so placed that B bisects AE, is called the *equant*[20] (introduced by Ptolemy to provide a representation of the planet's uniform motion in orbit by $\angle QEC$). This feature was one of Ptolemy's greatest contributions to planetary theory: ironically, having insisted on its application to the orbit of the Earth (see Sect. 4.2.6 above), Kepler almost immediately abandoned it in favour of his own theory of time (see Sect. 4.6 below).

Two lengths, which turn out to be of the greatest significance, are defined here: the *circumferential distance AK_1*, and the *diametral distance AK*. They were highlighted and named by Kepler himself, but not until chapter 57,[21] though they had already been illustrated and employed at the first stage, in chapter 39,[22] and played a vital role in chapter 56, as we shall see. In our standard notation we have $AQ = AK_1$ by construction, while QK is the half-chord of the epicycle, perpendicular to the diameter ANK_1, which defines AK, also on that diameter. Kepler always drew the epicycle and the eccentric circle separately, right up to Chap. 58, but the evidence shows that he was aware, right from the start, of their common properties and how they fitted together, so the two diagrams have been conflated here, for the convenience of the reader.

4.3.2   Grading the Eggs: The Mathematical Basis

We here make a temporary digression into modern algebra to reveal the pattern of Kepler's progress in constructing the correct path. While following in outline the treatment initiated by Whiteside (1932–2008) in 1974,[23] we adopt the Keplerian choice of taking the eccentric anomaly $\angle QBC = \beta$ as the independent variable. Then the general form of an expression for the Sun–planet distance r, which will encapsulate the circle and all the various curves that Kepler successively proposed,

---

[20] The term 'equant' was not coined until the mediaeval period.

[21] Kepler (1609), ch. 57, first paragraph, p. 269; Kepler (1990), p. 348, first marginal note.

[22] Kepler (1609), ch. 39, pp. 186–87; Kepler (1990), p. 257.

[23] Whiteside (1974), note 23.

is found by Davis[24] to be:

$$r = a\left(1 + e\cos\beta + \lambda_0 e^2 \sin^2\beta + \lambda_1 e^3 \sin^2\beta\cos\beta\right) + O\left(e^4\right),$$

where $O(e^n)$ indicates terms involving e to the fourth and higher powers.

The mathematical properties of this encapsulating equation ensure that it represents a simple closed curve (its origin of coordinates being the Sun at A) passing through the apsides C and D. Then, the value of the leading coefficient λ_0 in the equation will operate to classify the mathematical 'curve-finding' phase of Kepler's work into three distinct stages:

Stage 1 (chapters 39–44): the large-grade curve ($\lambda_0 = \frac{1}{2}$ gives the eccentric circle)
Stage 2 (chapters 45–50): the small-grade curve ($\lambda_0 = -\frac{1}{2}$ gives all the ovoids)
Stage 3 (chapters 51–60): the medial-grade curve ($\lambda_0 = 0$ gives the size of the correct orbit).

When with great reluctance Kepler abandoned the circle as unsatisfactory (as described in Sect. 4.4 below), he was obliged to adopt curves that he envisaged generically as egg-shaped: he called them 'oval' or 'ovoid' rather at random. In countries where eggs are sold by size (rather than by weight) the modern sizing process is called grading. Kepler's eggs were actually all the same length, CD, and we decide their grades just by their width along the 'lesser' or perpendicular axis. Every Keplerian curve can be assigned to one of the grades listed above. Then the values of subsequent coefficients λ_1, λ_2, ... account for variations in shape of the range of curves within each grade.[25] And because this algebraic classification correlates perfectly with the geometrical structure, the above terminology is very useful in keeping track of progress in detailed discussions of *Astronomia nova*.

We recall from Sect. 4.2.4 that Kepler had already had occasion to celebrate the splendid accuracy of Tycho Brahe's observations, which justified him in abandoning a proposed hypothesis (a model orbit) because it exhibited a discrepancy of 8'. For our present purpose of curve-finding it was necessary for Kepler to consider the question of accuracy also in relation to the orbits of the planets in the equations above, which evidently involved considering the various powers of e also as numerical values. It has seemed to me that Kepler had a built-in instinct for accuracy and that he must have carried out this process automatically. Readers may find it helpful to refer to the figures set out in Table 4.1. (Incidentally, it is interesting that radians seem to crop up in this context without comment or naming, but according to their modern definition,[26] as if confirming that circular measure was entirely natural, at least in astronomy.)

The standard set by Tycho will be described in what follows as the *stipulated level of observational accuracy*; it underpinned the discovery of the planetary laws.

---

[24] Davis (1981).

[25] Discussion of variations in shape were irrelevant in the frequent cases when the geometrical structure was unsatisfactory.

[26] One radian (1^c) = $180°/\pi$.

Table 4.1 Equivalences for Mars and the Earth (for use to assess comparative levels of observational accuracy)

Mars Orbital radius $a = 1.00000$	Earth $(T)^a$orbital radius $a_T = 1.00000$	Value in Keplerian units (KU) [1 KU = 0.00001 rad (^{c})]	Sexagesimal measure equivalent
e		0.09265	5° 18½′
e^2	e_T	0.01800	1° 2′
		0.00858	30′
½e^2		0.00429	15′
¼e^2		0.00215	7½′–8′
e^3		0.00080	3′
	$e_T{}^2$	0.00029	1′

^{a}The subscript 'T' is used to distinguish quantities that refer to the Earth. The orbits of the two planets are generally treated separately but in certain circumstances, when they are treated together in the same (triangulation) diagram, it is essential to remember that $a = 1.52350$ for Mars, in terms of $a_T = 1.00000$ for the Earth—or, in illustrations, $a_T \approx 2/3a$.

Of course, paths of different sizes cannot be compared unless one considers typical positions that occur simultaneously. Kepler could not assess this because in his early work he did not have a precise geometrical way of expressing elapsed time as the planet moved round its orbit—but it evidently happened that his sound mathematical instinct led him to consider typical points controlled by the eccentric anomaly β in such a way that the difference in time between them is negligible. The question of time measurement will be considered in Sect. 4.6 below.

4.4 The Curve-Detection Phase: Construction of the Satisfactory Distance

4.4.1 Kepler's Characteristic Construction

Next, we shall demonstrate that the classification into grades, set out as a modern mathematical scheme in Sect. 4.3.2 above, correlates precisely with Kepler's method of geometrical construction clearly conceived within the Ptolemaic framework (see Sect. 4.3.1 above), as developed here in Fig. 4.3.

In the Europe of Kepler's day, it was taken for granted that astronomy was geometrical. Indeed, in the quadrivium of the four mathematical sciences taught in universities astronomy was seen as a real-world instantiation of geometry. And the basis of geometry was to be found in Euclid's *Elements*. In constructing a proposed planetary path, Kepler invariably followed the Euclidean method, which demanded that the only instruments used should be a pair of compasses and a straightedge (an unmarked ruler). Kepler's procedure entailed defining typical points of successive

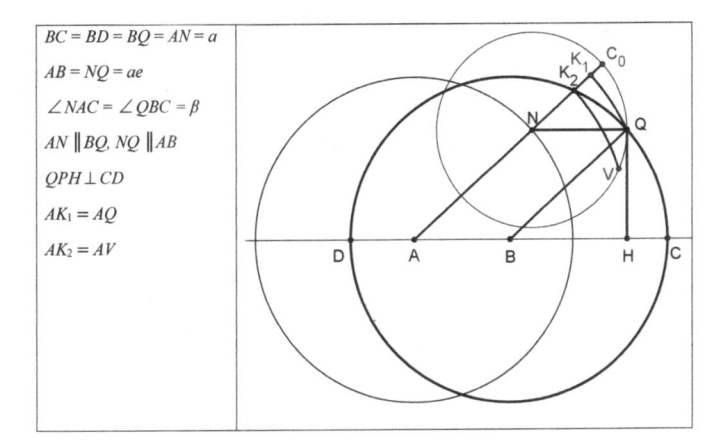

Fig. 4.3 Kepler's characteristic construction of the planetary orbit

proposed paths from the Ptolemaic framework by selecting a typical value of eccentric anomaly (β), and then drawing arcs with centre A by taking various lengths along AN extended. This procedure will be named *Kepler's characteristic construction*— though unfortunately Kepler did not always draw the arcs clearly enough to explain his method. Now there was a convention of Euclidean geometry, long accepted by scholars, which demanded that, before one can construct an arc, the length of its radius must be specified, as a line segment, terminated by a specified point at each end. Therefore, it was more than just convenient that there were pre-existing points on AN available to supply Kepler with radii of appropriate lengths, and hence we can be sure that the characteristic arcs are authentically Keplerian. The kernel of this representation is the 'Ptolemaic parallelogram' $ABQN$, which is part of the common structure of each diagram, while the line ANC_0 which controls the turning of the epicycle about A, also functions as a direction indicator ($\angle NAC = \beta$) along which the distances of all the typical points will be measured in turn.

4.4.2   Stage I: The Large-Grade Path (**Astronomia nova**, Chaps. 39–44)

Kepler naturally chose to start from the simplest possible situation—the eccentric circle itself, directly generated from the epicycle moving on the deferent circle as described and illustrated in Sect. 4.3.1 above. However—though Kepler did not use this approach—the typical point Q of the eccentric can be drawn in a way that is consistent with the characteristic construction used for the subsequent non-circular curves, if we start from the point K_1 and then draw the arc K_1Q of radius $AK_1 =$

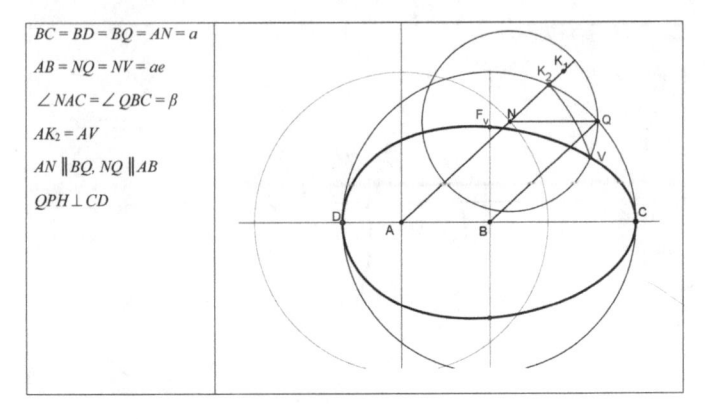

Fig. 4.4 The composite ovoid of the second stage

AQ: the point K_1 actually appears in Kepler's own epicycle diagram of chapter 39.[27] Kepler was extremely reluctant to reject a circular path, because of the weight of ancient tradition that lay behind it. Nonetheless, he eventually became convinced that Tycho Brahe's observations produced distances that would be consistent only with a path that lay within the circle, except at the apsides.

4.4.3 Stage II: The Small-Grade Path (Astronomia nova, Chap. 45–50)

For the second stage, a suitable point K_2 has already appeared in Fig. 4.2 as the point of intersection of the eccentric circle and *AN* extended; this produced the arc K_2V having radius AK_2 ($< AK_1$) which cuts the epicycle at *V*, applying Kepler's characteristic construction; and that arc appeared in Kepler's own diagram in chapter 46.[28] Thus *V* became the typical point of the small-grade curve, as illustrated in Fig. 4.4. (This illustration has of course been computer generated; but the equation of the curve can in fact be found by simple algebra, so we can identify it as being a small-grade ovoid, in the mathematical scheme of Sect. 4.3.2 above.)

At the second stage Kepler ended up, as he announced in chapter 50, with a 'composite oval path' (*via ovalis composita*)—so described in chapter 49,[29] which consisted of a combination of the ovoid of chapter 49 with ovoid VI of chapter 50. This 'composite oval path' *CVD* was analysed in detail for the first time in 2009,[30] and its geometrical construction has been rigorously established; so here it is only necessary to set out the result. When the typical point *V* was identified as lying on

---

[27] Kepler (1609), p. 186; Kepler (1990), p. 257.

[28] Kepler (1609), p. 218; Kepler (1990), p. 291.

[29] Kepler (1609), p. 236; Kepler (1990), p. 311, line 13.

[30] Davis (2009).

the epicycle, a number of geometrical equalities unexpectedly turned up—though Kepler himself had no inkling of the exactness that this brought to the analysis. In the end, however, these properties were irrelevant as this curve turned out to be merely an intermediate stage (because the distances, when tested against observations, turned out to be too small).

In the course of these six chapters, with enormous persistence, Kepler considered as many as twelve ovoids, but, apart from the two that formed the 'composite ovoid' just discussed, the remainder were quickly abandoned because they were either inadequately specified, or internally inconsistent. Eventually, Kepler relied on the distance determinations that he carried out, mostly in chapters 41–44; special cases were tabulated in chapter 47.[31] As observational uncertainties steadily reduced, these values were genuinely able to establish that the distances in the eccentric circle were too great, and those in the small-grade curve too small—both by a maximum of around $7'$–$8'$ at the first and third octants. And this defect was of a size that Kepler could not ignore, because of his confidence in the accuracy of Tycho Brahe's observations.

4.4.4   Stage III: The Medial-Grade path: The Goldilocks Solution

Even more calculations, in the course of chapters 51–53, finally convinced Kepler that the distances from the first stage were too long, and those from the second stage were too short, more or less by the same amount. However, he was unable to move on until he had invented a geometrical way of constructing a distance that lay in the middle. Kepler said this idea came to him as a numerical coincidence in Chap. 56,[32] but it would hardly have provided the 'flash of enlightenment' he described had it not been underpinned by geometry. Kepler found the essential point K 'in the middle' by constructing the line through Q, perpendicular to AN extended, to meet it at K, once more conforming to the Euclidean convention explained in Sect. 4.1 above. It was a notable coincidence (which I suspect Kepler did not fully appreciate until later) that the radius of the characteristic arc KJP he invented for the third stage (see Fig. 4.5) on which the typical point of the medial-grade curve would lie, was identical to the already-defined *diametral distance AK*.[33] Hence these diametral distances specifying the third stage almost stealthily replaced the *circumferential distances* from the first stage. (Because of their importance, the original Keplerian definition of these two lengths is displayed in the original framework diagram of chapters 39 and 40, in Sect. 4.3.1 above.)

Thus, via two distinct constructions, we have quantified a length AK (the *diametral distance*) that satisfies observations: its length is just right. However, its direction is not yet determined: the typical point could be taken either at J (whose locus Kepler

---

[31] Kepler (1609), p. 228; Kepler (1990), p. 302.

[32] Kepler (1609), pp. 267–68; Kepler (1990), pp. 345–346.

[33] Kepler (1609), ch. 53, p. 263; Kepler (1990), p. 347, penultimate line.

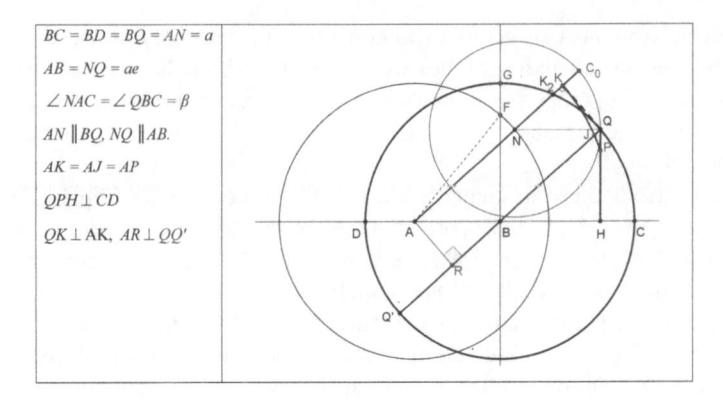

Fig. 4.5 The correct distance at the third stage

called *via buccosa*), or at *P*, because at first he had no way to decide which was the right choice. Details of the sophisticated misapprehension, involving some erroneous reasoning in Chap. 58, have been explained in a ground-breaking paper by Whiteside in 1974.[34]

Clarification only arrived with the theory set out in Chap. 59, where the geometrical reason that made point *P* the unique solution was revealed. Until then, Kepler did not have the faintest idea what curve his characteristic construction had produced.

However, from Fig. 4.5 we can derive the algebraic expression for the length *AK* (note that it is identical to the modern formula for the medial curve set out in Sect. 4.3.2 above):

$$AK = AN + NK = a + ae\cos\beta.$$

(This will turn out to be the equation that will identify the Keplerian (Goldilocks) solution: but it is fair to say that very few modern mathematicians would recognize it as an ellipse—even when they were made aware that β represented the eccentric anomaly.)

---

[34] Whiteside (1974).

4.5 Discovery of the Ellipse (*Astronomia nova,* Chap. 59)

4.5.1 Specification of the Elliptic Path

This section of our analysis is especially important because it is necessary to repudiate a well-entrenched error that used to be widespread among commentators (it should no longer be current)—the error being to claim that, in his astronomical work, Kepler relied, explicitly or implicitly, on the work of Apollonius.[35]

While Kepler had long been hoping to find a curve whose geometry he was familiar with, it was not until the beginning of Chap. 59, as recorded in Protheorema I, that what probably came to him as a supplementary 'flash of enlightenment' reminded him how to find the definition of the unique curve that would fit the observations. He identified the curve as the ellipse found in the work of Archimedes, *On Conoids and Spheroids* Proposition 4[36]: Kepler had access to this in a good Latin translation from the Greek by Federico Commandino (1509–1575), published in Venice in 1558.[37]

The Archimedean ellipse could supply the solution Kepler required, because it was defined by the major and minor semi-axes—and these elements (*a, b*—see Fig. 4.6) had already been evaluated by Kepler in connection with his unknown curve (as we confirm shortly). Its defining property was stated without proof by Kepler in Protheorema I of chapter 59. In our standard notation it is:

$$\frac{PH}{QH} = \frac{BF}{BC} = \frac{b}{a}.$$

This relationship is known nowadays as the *ratio property of the ordinates*. It produces an ellipse which is aptly described in the present context as a *compressed circle*.

It is moreover of interest in determining the underlying source of Kepler's astronomy to confirm that the basic ratio property is derived from the cone, simply by taking a (finite) plane section as confirmed by Davis (2007, Sects. 3–6).[38] In his introduction to his edition of *Apollonius* of 1896 T. L. Heath (1861–1940)[39] identified it as a property known from earlier writers (that is, before Archimedes), and

---

[35] Field (2010) offers a possible explanation of the origin of the error.

[36] This is so numbered in Heath's English translation (Archimedes, 1897) but Kepler follows Commandino's edition (Archimedes, 1558), where it appeared as Proposition 5. That is how it is referred to, without clarification, in Donahue's revised translation of *Astronomia nova* (Kepler, 2015).

[37] Archimedes (1558).

[38] Davis (2007).

[39] See Apollonius (1896), Archimedes (1897) and Euclid (1908).

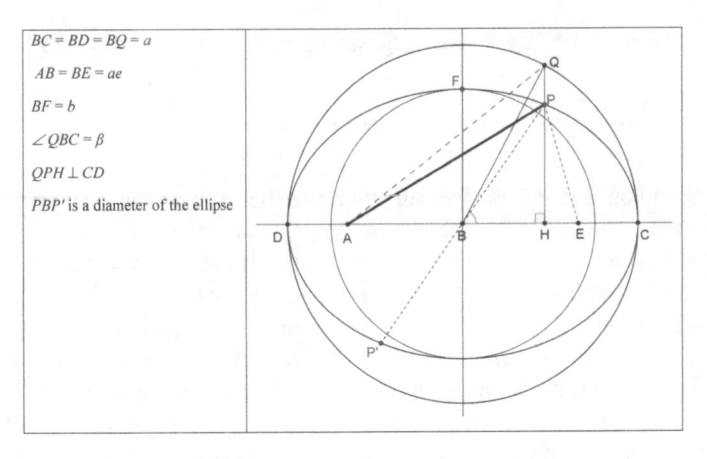

Fig. 4.6 Geometrical correlation of the orbit from Archimedes: *On Conoids and Spheroids*

this agrees with what Kepler wrote in Chap. 47.[40] Earlier proofs of the property may have been lost quite early on, perhaps because it was regarded as elementary.

The value of the minor semi-axis b had been determined earlier, from a short, previously unnoticed section in Chap. 53,[41] which has been analysed only by Davis[42]: it used an observer's trick of getting as close as possible to the object. Kepler considered observations of points on or near the quadrant F of Mars viewed from the Earth when it happened to be in favourable positions (rather than from the Sun as usual). From such observations he was able to determine a value for the quadrant distance (whose potential exactness could of course be checked as a special case of Kepler's characteristic construction), to find:

$$AF = a.$$

Thus, F lies on the deferent circle (see Fig. 4.2), and then application of Pythagoras' theorem to $\triangle AFB$ (using the eccentric distance AB, which was well observed) gave the 'lesser' semi-axis $BF = b$ of the curve. Hence, with its semi-axes known, the elliptic path is precisely specified.

4.5.2 Determination of the Associated Area

In Protheorema II of chapter 59, Kepler went on to state the result of the Archimedean proposition mentioned above based on that ratio-property of the ordinates: that the area of the ellipse is in the same known ordinate ratio to the area of the circle. It was

[40] Kepler (1609), ch. 57, p. 226; Kepler (1990), p. 300, lines 30–36.

[41] Kepler (1609), p. 260; Kepler (1990), p. 338, line 21–p. 339, line 2.

[42] Davis (1992).

in Protheorema III that Kepler himself extended that result to partial (sector) areas. He deduced from Archimedes' proof[43]:

$$\text{Ellipse segment } PHC = \frac{b}{a}\text{Circle segment } QHC.$$

Then he applied *Elements* bk 4, 1 (since the two triangles have the same base), to state that[44]:

$$\triangle PAH = \frac{b}{a}\triangle QAH.$$

Hence, as Kepler said, 'by composition' of these two results—that is, by combining the two pairs of pieces (both visually and mathematically) using his diagram above—he obtained[45]:

$$\text{Ellipse sector } PAC = \frac{b}{a}\text{circle sector } QAC.$$

4.5.3 Absence of the Focus in Astronomia nova: Dichotomy of Traditions

The Archimedean ellipse is wholly and uniquely defined without reference to a focus. Kepler had always assumed that the Sun was positioned at some fixed point A eccentric to the centre of the original circular path, whose distance ($AB = ae$) from the centre was very exactly known (and frequently revised) from astronomical observations; this observed length was the radius of the epicycle which was part of the fundamental framework described in Sect. 4.3 above. Thus the position of the Sun was never associated with any theoretical (geometrical) property. In Protheorema VII Kepler restated the Pythagorean relationship (illustrated in Kepler's own diagram in the Euclidean way as the difference of squares) using the right-angled triangle $\triangle$ ABF from which he had previously determined the minor semi-axis (see Sect. 4.5.1 above), here to formalize the position of the Sun in terms of the known major and minor semi-axes:

$$AB^2 = AF^2 - BF^2, \text{ or } BC^2 - BF^2 \left(\text{or } a^2 - b^2\right).$$

In Chap. 11 of this volume, Knobloch mentions Kepler's motivation to study the *Conics* of Apollonius (in the edition of Commandino) initially for his book

---

[43] Kepler (1609), ch. 59, p. 287; Kepler (1990), p. 368, lines 3–4.

[44] Kepler (1609), ch. 59, p. 287; Kepler (1990), p. 368, lines 6–7.

[45] Kepler (1609), ch. 59, p. 287; Kepler (1990), p. 368, lines 7–8.

Astronomiae pars optica[46] and describes how Kepler made good use of Apollonius' work in both his optical and his stereometrical writings. Intriguingly, in Chap. 4, Sect. 4 of Kepler's book, we find the evidence that Kepler himself invented the term *focus*,[47,48] to describe the pair of points (foci) associated with the reflection property of a central conic (ellipse or hyperbola) first proved by Apollonius (*Conics* III, 48). Yet, in *Astronomia nova* (even when Kepler had finally determined that the curve was an ellipse), the term focus was not mentioned in the text,[49] nor were any propositions from Apollonius cited there. One can suggest a practical reason for this: Kepler may have believed that an ellipse potentially possessed a number of 'centres', of which the focus was only one (of a pair, naturally)—a situation analogous with that of a triangle, which he knew had a number of different 'centres'.[50] Hence, Kepler could have regarded the position of the Sun as one of several interesting points in the ellipse. However, there is also a compelling theoretical reason arising from the dichotomy between the Archimedean tradition, which, for central conics, depended on double-axis symmetry and orthogonal treatment; and the Apollonian tradition, with its panoply of tangents and oblique axes. We emphasize that Kepler, trained in the geometry of Euclid, viewed orthogonality as associated with perfection and would therefore initially have regarded the Apollonian treatment as inappropriate for the purpose of astronomy (because of the connection of the heavens with the Deity).

Subsequent evidence shows that it was Kepler himself who eventually introduced the concept of the focus into astronomy, over a decade later, in 1621. The term appeared in *Epitome* V, I, 3.[51] It is clear that fresh thinking was involved then, because, before setting out the Apollonian property: $AP + EP = 2a$ (from *Conics* III, 52), Kepler stated a version of his own, which I have not seen elsewhere: $AP + AP' = 2a$ where PBP' is a diameter of the ellipse (so P, P' are diametrically opposite points). This permitted Kepler to place the position of the Sun finally and firmly at the focus, and he was thus able to state Law I in the form in which it appears in textbooks today:

The path of the planet is an ellipse with the Sun at one focus.

Hence it can be argued that it was Kepler himself who successfully merged the Archimedean tradition with the Apollonian tradition—indeed, so effectively that nowadays nobody is aware that a dichotomy ever existed.

---

[46] Kepler (1604, 1939). According to a letter he wrote in May 1603 to Herwart von Hohenburg, in Kepler (1949), letter 256.

[47] Kepler (1604), ch.4, Sect. 4, p. 93; Kepler (1939), p. 91, line, 17.

[48] More details are given by Knobloch, 'Kepler's contributions to mathematics' (this volume, Chap. 11).

[49] The word occurred once in a prefatory poem (Kepler, 1990, p. 15; Kepler, 2015, p. 41), with the non-mathematical meaning 'hearth'.

[50] Four were known to the Greeks: circumcentre; incentre; orthocentre; median centre.

[51] Kepler (1618), part I, Sect. 3, p. 659; Kepler (1991), p. 372, lines 16–19.

4.6 Development of the Theory of the Time-Measure (*Astronomia nova*, Chaps. 32–34, 38–40, 59)

4.6.1 The Stipulated Level of Analytical Exactitude

In Kepler's day—and for long after that—angles were the sole feature that could be measured directly by astronomers; so it is entirely fortuitous that those same angles provided the means to express intervals of time (as do analogue timepieces in terrestrial terms, even nowadays). It was Kepler who took a great leap forward by introducing the idea of small angles to lead to an alternative theoretical representation.

For the analysis that follows, we need to notice that the numerical value that Kepler used throughout his astronomical work as his standard a (the radius of the original circle): was invariably $a = 10^5 = 100,000$ (while one radian, mentioned in Table I in connection with observational accuracy, bore the same relation to 'one Keplerian unit'). In his day, it was usual for every practitioner to select his own standard value (and by chance Kepler's choice turned out to be perfect for his astronomical purpose[52]). The reason for making one's own selection, which may seem odd to present-day readers, lay in the lack of generally agreed notation for the representation of decimals in that era (as well as some unfamiliarity with their use), so people instead calculated with whole numbers only. Of course, this is exactly the same as working nowadays correct to five decimal places (in Kepler's case): anything smaller was treated as negligible, and simply disregarded. This level 'correct to five decimal places' will be described, in relation to Kepler's astronomy, as the *stipulated level of analytical exactitude*. It is of great importance in our forthcoming discussion of small quantities.

The mathematics of small quantities is a topic mentioned also in Chap. 11 of the present volume in connection with Kepler's work on mensuration in *Stereometria doliorum* (1615).[53] Neither in that context nor in his astronomy did Kepler consciously initiate any theory or introduce technical terms such as were developed merely a decade or so later: the small parts under consideration could be made as small as necessary for the particular purpose, while remaining always discrete. It seems as if the purpose was simply to achieve smoothness of the particular curve (or surface in the case of stereometry), by ensuring that the irregularities were undetectably small, so that progression from one value to the next would appear to be continuous.

In the mensuration calculations of *Stereometria doliorum*, Kepler gave no indication of the size of the small sections he was investigating. By contrast, in his astronomy, Kepler's purpose was to ensure he had a sound foundation for his theoretical work, and he chose to divide the eccentric circle into 360 parts at the centre B (as mentioned in Sect. 4.6.3 below), in effect selecting $1°$ as his small interval. Now

---

[52] He used 10^7 as unit in his work on logarithms; see Kepler (1960).

[53] Kepler (1615, 1960).

it can easily be checked from a scientific calculator (or by reference to Table 4.I—which sets out the equivalences between degree and radian measures) that, correct to five decimal places,

$$1° = 0.01745^c = \sin 1°,$$

is strictly true, and this will ensure that a circular arc and its corresponding chord (and consequently a circular sector and its corresponding isosceles triangle), both subtended by a small angle of 1°, will become indistinguishable at just that particular level. Thus, it happened (and it is one of several unexplained but fortunate coincidences that crop up throughout this account) that Kepler satisfied this requirement in practice, simply as a result of his *stipulated level of exactitude*. Accordingly, these intervals ≤ 1° were sufficiently small for any results based on them to be regarded as exact in Kepler's terms.

4.6.2   From Ptolemaic Equant to Keplerian Innovation: The Measure of Uniformity (Astronomia nova, Chaps. 32 and 33)

Kepler generally used the Latin word *mora*, which means 'delay', or more appropriately, 'duration', to denote 'time spent'. The word has frequently been translated as 'elapsed time', regardless of whether it was *micro* or *macro*. This is not helpful, and in order to understand Kepler's intentions it is essential to take care, whenever time is mentioned, to distinguish a micro-amount—a small interval, or increment, of time (δt)[54]—from a macro-amount. We shall refer to the macro-amount t as the *time in orbit* (or orbital time)—defined as the time taken to reach a typical position of the planet in its (initially circular) orbit, measured from aphelion. We must further distinguish these two amounts from the *periodic* (that is, total) *time T* (though Kepler frequently considered the half-period instead, because the symmetry of the orbit about the major axis was always assumed).

The Ptolemaic equant theory had supplied a preliminary macro-representation of time, for any position of the planet, as an angle ($\angle QEC$ in Fig. 4.2) that increased uniformly. Kepler was initially very pleased when he managed to satisfy himself that the Earth also possessed an equant (see Sect. 4.2.6 above), because this meant that any new theory would apply to all six planets. However, he then became aware that this Ptolemaic way of representing time was not accurate enough for Mars, and had to be rejected, because that planet's eccentric ratio is more than five times as large as that of the Earth (see Table 4.1). It was in Chap. 32 that he began to develop a theoretical foundation for time that was entirely new to astronomy, justified

---

[54] The author (aware that it is an anachronism) is consciously using the calculus notation δt. Calculus was discovered by Isaac Newton in 1665–1667 and independently by Gottfried Wilhelm Leibnitz (1646–1716).

initially, nevertheless, by an argument from the Ptolemaic configuration (transposed to heliocentricity, of course), in its simplest possible manifestation, illustrated in Fig. 4.2.

In Chap. 32, Kepler started to consider the orbit of Mars by carrying out a geometrical investigation of the situation at the apsides. He found that the small intervals of time around C and D, as determined by Ptolemaic equant theory, were proportional to the apsidal distances AC and AD (see Fig. 4.2). This is the first occasion on which time was assessed other than by an angle, and is of enormous significance in the technical progress of astronomy.

Yet in Chap. 33, Kepler abandoned the equant mechanism. This may well mark the moment he made the changeover from the first to the second phase of investigation, when his astronomy became entirely new. He went on to generalize his insight to a typical point Q of the circular path, by stating[55]:

> ... the [small amounts of] time taken by a planet, in equal parts of the eccentric circle, ... are in the same ratio as the distances of those parts from the point whence the eccentric distance [or the eccentric ratio] is reckoned.[56]

This introduces a general consideration, seldom specifically mentioned as an essential ingredient. Underpinning all mathematics, there is a fundamental requirement that quantifiable entities should possess some *measure of uniformity* (explicit or implicit), defined as the standard of reference, suitably chosen to provide a basis of comparison. The size of this measure is generally irrelevant—in everyday life it is often merely a matter of common consent—but its presence is a primary prerequisite in mathematics; it started in arithmetic as the counting unit, and appeared in the early geometry of the Greeks. When mathematics developed to involve the more sophisticated notion of a function in algebra, and subsequently in calculus, the *measure of uniformity* took the role of the independent variable, while the function was subsumed into the dependent variable.

Here we quantify the *measure of uniformity* to be used for astronomical time. Kepler chose this to be a small part of the eccentric circle round which, by definition, motion was uniform about its centre: this part was specified, as occasion arose, either by a small arc $a.\delta\beta$ of its circumference, or by a small angle $\delta\beta$ at its centre. Then, expressing the statement just quoted from chapter 33 in modern notation (but involving no distortion of Kepler's intention), we obtain:

$$\frac{\delta t}{(a)\delta\beta} \propto r.$$

On the other hand, the same Keplerian relationship in its inverse form will express strictly circular motion:

[55] Kepler (1609), ch. 33, p. 170; Kepler (1990), p. 236, lines 8–11.

[56] *Moras planetae, in aequalibus partibus circuli eccentrici ... esse in proportione ea in qua sunt ad invicem eorundem spaciorum abscessus a puncto, unde computatur eccentritas.* The point mentioned is of course the position of the Sun, as Kepler confirmed later.

$$\frac{(a)\delta\beta}{\delta t} \propto \frac{1}{r}.$$

Kepler realized that this representation was realistic because it was compatible with what was known about planetary motion in general: when the planet is far from the Sun, it moves more slowly; while when the planet is near the Sun, at a lesser distance, it moves faster. He put it pithily, again in Chap. 33: *major minorque distantia, majoris minorisque morae*[57] 'the further [distance] the longer [time], the nearer the shorter'.

4.6.3   Illustration to Represent the Increment of Time (Astronomia nova, Chap. 40)[58]

In Chap. 40 Kepler set out the details of the sectioning he proposed. His intention is less easy to follow than it should be, because the two statements were separated in the text:

> I began by dividing the eccentric circle into 360 parts [by lines from B], as if these were the smallest [possible] parts, and supposed that, within each such part, the distance does not change. … I cut the area [*planum*] of the eccentric circle into 360 parts by lines drawn from the point [A] whence the eccentric distance [or the eccentric ratio] is reckoned.[59]

This needs illustration to aid understanding, but Kepler did not supply any diagrams. However, since he was one of the first mathematicians to introduce the analysis of small quantities into mathematics, it is not surprising that the habit of illustrating them individually had not yet developed. (There were also practical reasons which might have caused problems—for instance the cost of expert draughtsmen, or perhaps technical difficulties in representing new mathematical ideas.)

So we provide a diagram on Kepler's behalf, in Fig. 4.7—but this should not be regarded as an extreme anachronism, since within half a century it had become normal practice to illustrate small intervals.

Kepler directed (as quoted) that the eccentric circle should be sectioned from its centre into sectors (or equivalently isosceles triangles), typically q_1Bq_2, having equal angles $\delta\beta = 1°$ at B, so they subtended equal arcs (or equivalently bases of the triangles) of length $a.\delta\beta$ (where a is the radius of the circle) at the circumference. Then the ends of these equal arcs were to be joined to A to create 360 sectors or triangles, typically q_1Aq_2, where the angles at A were also small, but obviously no longer equal. Thus, Kepler could employ q_1Bq_2 (being of common size all round the orbit) to represent the measure of uniform time, while the area q_1Aq_2 represented the increment of time that varied as the planet moved round its (circular) orbit.

---

[57] Kepler (1609), ch. 33, p. 168; Kepler (1990), p. 236, lines 32–33.

[58] This section, and the next, have been explained in more detail in Davis (2015).

[59] Kepler (1609), ch. 40, pp. 192–193; Kepler (1990), p. 263, lines 26–27 and p. 264, lines 8–10).

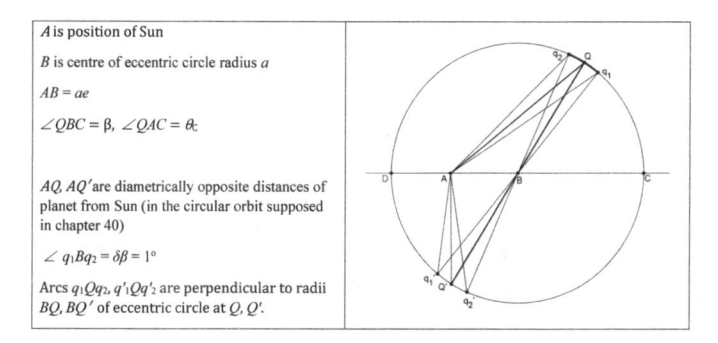

Fig. 4.7 Original increment of time illustrated from *AN* Chap. 40

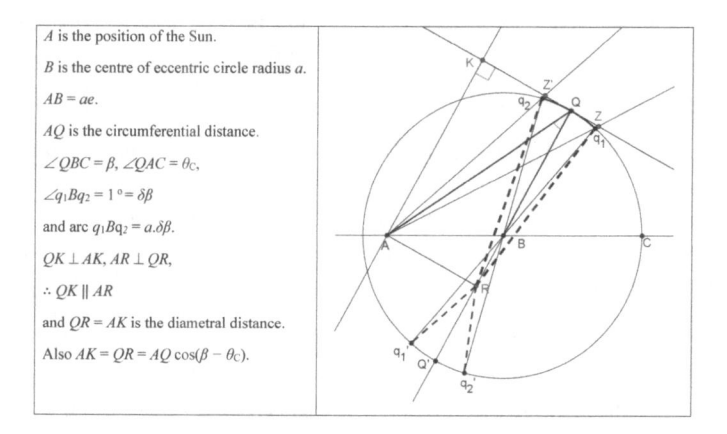

Fig. 4.8 The quantified increment of time from *AN* Chaps. 56 and 59

This idea was so new that Kepler invoked the authority of Archimedes from his work *Measurement of the Circle* Proposition 3 to confirm the suggested involvement of area:

> … it occurred to me that all these distances are contained in the area [*planum*] of the eccentric circle. For I recalled that Archimedes, in seeking the ratio of the circumference to the diameter, once similarly divided a circle into an infinity [a very large number] of triangles ….[60]

The fact that Kepler envisaged the distances as contained in the area is emphasized by Knobloch in Chap. 11 of this volume, by reference to erroneous translations which, by implication, mis-describe Kepler's intentions and have led to serious misinterpretations. (In fact, Kepler invariably used the distance AQ, typically denoting the mid-value of the distance within each interval, as a stand-in to refer to the area q_1Aq_2, which commentators have found extremely confusing and indeed misleading.) Nevertheless, the representation could not yet be regarded as satisfactory because the

---

[60] Kepler (1609), ch. 40, p. 193; Kepler (1990), p. 264, lines 4–7.

evaluation of micro-area q_1Aq_2 was not straightforward—it could not be expressed geometrically. Kepler himself pointed out the source of the difficulty—that each small arc q_1q_2 was oblique (rather than perpendicular) to its radius vector AQ[61]: this difficulty remained unresolved from chapter 40 until late in the third stage.

4.6.4 Quantification of the Increment of Time Achieved (Astronomia nova, Chap. 59)

Kepler did not initially adapt his theory of time to a non-circular orbit while he was struggling to find the right curve. As we have seen in Sect. 4.4, it was in Chap. 56[62] that he pinpointed the curve of correct grade; but theoretical confirmation did not occur until chapter 59 Protheorema IX, where he formally stated:

> the diametral distances are [to be] taken in place of the circumferential ones.

So Kepler did just that: he replaced the *circumferential distance AQ* with the corresponding *diametral distance QR* (as noted previously, the importance of this pair of lengths is such that they have been defined as part of the underlying structure, in Sect. 4.3.1 above). When this substitution is carried out, the length involved in the measure of the increment of time is altered to $QR = AQ \cos AQR$. In the original representation of Chap. 40, that length AQ was oblique to the small arc q_1Qq_2 of the orbit, as Kepler had noted there, while QBR is now perpendicular to that small arc as required. Moreover, from Fig. 4.8—a development of Fig. 4.7, again on Kepler's behalf—we see that QK and AR are parallel (this is precisely true when we stipulate the level of exactitude set out in Sect. 4.6.1 above—or one could always select a more restrictive level if necessary). Accordingly, by geometry, the original micro-area remains unaltered by the substitution, and (since the two triangles are on the common base q_1q_2) we can share Kepler's conviction that their areas are the same, as established with a reference to Euclid (*Elements* bk 1, 37):

$$\Delta q_1 Aq_2 = \Delta q_1 Rq_2 = (^1/_2)QR(a.\delta\beta).$$

(It happens that Newton used the identical Euclidean proposition in the general proof of the area–time law in *Philosophiae naturalis principia mathematica* (*Mathematical Principles of Natural Philosophy*, London, 1687, Book 1), Proposition I—but it would be too complicated to pursue this here, since the two approaches were methodologically distinct.)

By this change of formulation, Kepler demonstrated that the diametral distance $AK = QR$ that he had already employed successfully to specify the elliptical path, would provide the geometrical basis for a quantified theory of micro-time as well:

---

[61] Kepler (1609), ch. 40, p. 196; Kepler (1990), p. 267, lines 4–9.

[62] Kepler (1609), ch. 55, p. 167; Kepler (1990), p. 345, line 35–p. 346, line 13.

$$\text{Increment of time } \delta t = (\textstyle\frac{1}{2}) Q R(a.\delta\beta).$$

4.6.5   The Problems of Determining the Macro-Time

What astronomers needed in practice, however, was the time at a given position of the planet, rather than to know how it got there. Obviously, that macro-time was determined by adding all the increments (micro-times) up to that point. (Kepler got plenty of practice in carrying out numerical summations, by tackling numerous examples throughout *Astronomia nova*. The aids to calculation available in Kepler's day were extremely limited and he could seldom afford to employ anyone to help him with this drudgery.[63])

Thus, the macro-time t taken from C to Q (giving the time and position in the eccentric) could evidently be obtained by summation, and in fact could be easily determined by inspection (for instance from Fig. 4.7), merely by noticing that the small areas were strictly contiguous, with a common vertex at A, to give $t \propto QAC$. Moreover, in chapter 40,[64] Kepler had already introduced a sectioning of the circle sector that allows the macro-time to be expressed in a way familiar to modern readers:

$$t \propto \text{circle sector } QAC$$
$$= \text{sector } QBC + \triangle QAB$$
$$= \text{sector } QBC + \textstyle\frac{1}{2}AB.QH$$
$$= \textstyle\frac{1}{2}\, a^2(\beta + e\sin\beta).$$

Nevertheless, at the final step two problems arise. There is understandable disagreement between commentators (pending further research), as to what extent Kepler was justified, or whether he had jumped to a premature conclusion, when he stated that he had found the (macro-)time at every position, not only for the orbit as a whole. (Modern readers are easily convinced of the theoretical result by a simple integration, but this technique lay several decades in the future.) However, in Protheorema XV (the last one), he applied laborious term-by-term summations of increments of time,[65] in a numerical method that would generate a value for the macro-time, at any whole degree of eccentric anomaly. Moreover, the resulting sums agreed, 'even

---

[63] The logarithms that would have speeded up these calculations were not invented by Napier until 1614; and Kepler was subsequently motivated to invent a system of his own: see Knobloch, 'Kepler's contributions to mathematics' (this volume, Chap. 11). Kepler later helped to invent a calculating machine, his colleague being Wilhlem Schickard, see article by Prager (1975).

[64] Kepler (1609), ch. 49, p. 192; Kepler (1990), p. 265, lines 8–9.

[65] Kepler (1609), ch. 60, p. 297; Kepler (1990), p. 375, lines 13–23.

to the second'[66] with the formula for the macro-time set out above.[67] It seems fairly certain that Kepler satisfied himself as to the end, but he was evidently not happy about the means, which he described as $\dot{\alpha}\gamma\varepsilon\omega\mu\acute{\varepsilon}\tau\rho\iota\kappa o\varsigma$ ('ungeometrical').

Additionally, Kepler got involved in some misdirected investigation in Protheore-mata XIII and XIV, which created unnecessary confusion. Probably the trouble was that Kepler had determined the increments of time in relation to the circular orbit, but had not recognized how this solution could also apply to the ellipse, by increments. However, this minor gap in the account is of no consequence, as it turns out, since we have shown in Sect. 4.5.2 that in Protheorema III he had already established the required result for the macro-time by proving[68]:

$$\text{ellipse sector } PAC = \frac{b}{a}\text{circle sector } QAC.$$

Thus Kepler could now validly apply the above proportionality of time in relation to the circle (derived from chapter 40) to deduce for the ellipse:

$$t \propto \frac{b}{a}\text{circle sector } QAC = \frac{1}{2}ab(\beta + e\sin\beta).$$

Thus Law II has been established convincingly from the observations.[69]

4.7   Chasing Causes

4.7.1   The Background of Orthogonality

After fairly extensive investigation, the best succinct account that I have come across, correctly expressing Kepler's final views on causes, has been given by the historian of astronomy Agnes Clerke (1842–1907) in 1905. She stated that Kepler 'was driven to the twofold expedient of creating a whirling medium for maintaining the revolutions of the planets, and of supposing the Sun to exercise a "magnetic influence", by which they were drawn into closed orbits.'[70] This account clearly distinguishes the two causes, as Kepler did, in accordance with the Aristotelian Principle of Economy, which demanded one cause per motion.

---

[66] Kepler (1609), ch. 60, p. 297; Kepler (1990), p. 375, lines 19–23.

[67] This formula, as stated by Kepler (1990, p. 375, lines 21–22) was defective, but has been corrected by an insertion in the text made by Donahue (Kepler, 2015) in each edition of his translation, explained by a footnote.

[68] Kepler (1990), p. 368, lines 7–8.

[69] The time taken in orbit is measured by the area swept out.

[70] Clerke (1905), p. 10.

The Aristotelian view about motion was not adequately superseded until some decades after Kepler's death. According to Aristotle, every celestial motion was directly associated with its cause, both in amount, and in direction. An Aristotelian 'force' was held to act only in its own direction, and through contact alone: when these conditions were satisfied, the cause was proportional to the motion, and supplied a measure of it.

Kepler did not specify the physical mechanisms that he believed would account for the elliptical orbit he had discovered, though a decade or so later he put forward some speculative analogies in his *Epitome of Copernican astronomy*[71]; in *Astronomia nova* he merely suggested how supposed properties of the Sun might produce the expected effects. Initially he found it difficult to disentangle the various strands, but this problem resolved itself as he came to appreciate the essential orthogonality of every aspect of the orbit: coordinates, constituents, components of motion, causes. And in *Epitome* Book V, Part 1, Sect. 4, published in 1621, well after the orbit had been established, he unequivocally distinguished the pair of causes, while their independence was ensured by the principle of orthogonal independence, as we have described in Sect. 4.1 above:

> Because the orbit of the planet is eccentric, two components of motion are combined; … one is the revolution round the Sun produced by one power of the Sun, the other is the radial motion [libration] resulting from a power of the Sun distinct from the first.[72]

4.7.2   The Precedent Cause, Associated with the Circumsolar Motion in Time

The general revolution of the bodies in the planetary system about its centre, the Sun, in which every planet took part, was obviously the most outstanding feature that had to be accounted for, as Kepler realized. Very near the beginning of his theoretical investigations (in part of the title to chapter 34 of the *Astronomia nova*) he speculated that the mechanism responsible was the rotation of the Sun; he stated 'the Sun rotates on its axis'. Indeed, he attempted to confirm this idea by determining the period of rotation—he selected three days 'for archetypal reasons'; while the period was actually independently recorded from observations of sunspots quite shortly afterwards (1609–1610) by Galileo, Christopher Scheiner (1573/5–1650) and Thomas Harriot (1560–1621). Kepler was content to accept the (quite different) value (25 or 26 days) found by these observers.[73] Within this scheme, each planet individually moved round its own circuit centred on the Sun, supposedly driven by the rays of the Sun, which Kepler envisaged as hitting the planet perpendicularly, moving it 'instantaneously' in a circle as they swept past it: thus Kepler's account of the circumsolar motion was in accord with the Aristotelian tenets concerning 'force'.

---

[71] Kepler (1618–21, 1991).

[72] Kepler (1621), Book V, Part 1, Sect. 4, p. 668; Kepler (1991), p. 377, lines 33–36.

[73] Kepler (1990), Chap. 34, p. 245, line 2, and Kepler (1991), IV, II, 1, p. 296, lines 16–17 respectively.

Now in modern notation, the circumsolar motion (discussed in Sect. 4.6.3) is determined by:

$$\frac{a\delta\beta}{\delta t} \propto \frac{1}{r}.$$

When Kepler came to examine the cause of circumsolar motion in *Epitome* IV, III, Sect. 2, by matching it, as Aristotle required, to the quantity of motion itself, he became aware that this particular cause also decreased linearly with distance from the Sun. The action of the lever (which was a familiar artefact in Kepler's day—indeed an Archimedean device) provided him with a practical analogy, which he illustrated in that section, in whose title he associated the cause with 'motion in longitude' (*in longum*).

4.7.3   The Lesser Cause, Associated with the Radial Motion Producing an Ellipse

As Kepler pointed out in *Astronomia nova* (chapter 38), not only was there a common 'force' that acts on all of the planets, but each planet must also possess some individual power that enabled it to vary its distance from the Sun—Kepler called it *vis insita* ('innate' or 'inherent' force). Searching for fresh stimulation, Kepler came across *De Magnete* (*On the Loadstone*), a book on magnetism published in London in 1600 by William Gilbert (1544–1603). Kepler immediately took up Gilbert's suggestion that the Earth itself was a huge magnet, and extended the idea to all six planets and even to the Sun: he announced, in the title to *Astronomia nova* (Chap. 34) that 'the Sun is a magnetic body'. Kepler discussed magnetism in more detail in *Epitome* Book IV, Part III, Sect. 3, whose title referred to 'motion in altitude' (*in altum*). By this, of course, he meant the changing (radial) distance from the Sun; I suspect that this term was invented by Kepler partly for its propaganda value, because only in a strictly heliocentric situation is it proper to consider 'height above the Sun' as equivalent to radial distance (Even in Kepler's day, it was commonplace for geocentric astronomers to describe heavenly bodies as being at a certain 'height' or 'altitude' above the Earth, which was not only supposed central but also considered to be in the lowest position in the Universe). Only Stephenson[74] seems to have noticed Kepler's idiosyncratic usage of 'altitude' in connection with Chap. 39 of *Astronomia nova*. Indeed, this appears to be significant with respect also to Chap. 38, but it is a matter that will require more detailed investigation.

The application of magnetism to the heavens required adaptation of Gilbert's ideas. Kepler had to suppose that the Sun had the same polarity all over its surface (the other pole being at its centre) and that it exerted a magnetic influence with power enough to affect every planet similarly, however far from the Sun. Moreover, the

---

[74] Stephenson (1987), p. 76.

Sun's magnetism was essentially passive, simply capable of controlling the direction of motion. Kepler further supposed that each planet possessed an internal structure consisting of several disparate sets of fibres sensitive to magnetism, each proposed for a specific purpose. The particular set of fibres of interest here was arranged in or parallel to the plane of each orbit, fixed in the direction of its ordinates—that is, perpendicular to its line of apsides (the major axis of its ellipse), so that the radial motion would depend on an angle that varied as the planet moved round its orbit.[75] This involves what we now call a sine function (because the motion was associated with the cosine of the complement of the angle made with the axis), which Kepler and his contemporaries were quite familiar with in practical applications, and which he illustrated by a selection of physical analogies (not all sound, in fact) discussed in *Epitome* Book V, Part I, Sect. 4.1. Consequently, Kepler envisaged that the magnetism of the Sun would work by activating the magnetism of the set of fibres which would then produce the observed radial motion.

Incidentally, Kepler found that he could, at last, account for the randomness of the eccentric ratios (ellipticities or now, in modern terms, eccentricities) of the various planets (it had always seemed an odd circumstance that they were not ordered according to the distance of each planet from the Sun). In *Epitome* Book IV, Part III, Sect. 3[76] he stated that the [instrumental] cause of the individual eccentric ratios of a planet would be due to the strength or potency of its fibres, in accordance with the fact that the radial motion depends on the eccentric ratio of the particular ellipse. (However, Kepler undoubtedly believed that the final cause of the eccentric ratios in the planetary system was the demonstration of the harmonies through the motions of all the planets—which he had dealt with separately, because of its overriding importance, in *Harmonice mundi*, Book V.[77])

4.7.4 Assessment of Accuracy

We conclude by pointing out that though Kepler (obviously) lacked any under-standing of the later treatment of orbits based on the Newtonian formulation in terms of acceleration rather than speed, he managed sound estimates of the first derivatives, the motions, because he had developed a good informal understanding of incremental change. Thus, he could explain the motions as above, by causes that seem on the whole amazingly sensible, but were of course quite wrong. However, we remark that the correct modern velocity in an elliptical orbit can be anachronistically derived simply by applying modern algebra to find the resultant, by composition, of Kepler's components of motion (subject to adjustment of the constants, since Kepler

---

[75] Though Kepler subsequently introduced a minor deflection of the fibres, its effect is negated when modern mathematical methods are applied to the analysis.

[76] Kepler (1620), p. 592; Kepler (1991), p. 338, lines 36–44.

[77] Kepler (1619), Book V; Kepler (1940).

stated motions as proportions, in the Euclidean way). This should surely be regarded as a very satisfactory achievement.

To complete the picture, it must be said that, by modern standards, 'motion in latitude' (in the title to *Epitome* Book III, Part III, Sect. 4, following the two sections just discussed) cannot exist within the Keplerian synthesis, because we have assumed a kinematical situation in which orbital motion takes place in a plane. This has been accounted for in Sect. 4.1 above (and Kepler established it observationally, as confirmed in Sect. 4.2.3 above). However, Kepler believed that he had to identify a mechanism to retain each planet in its orbital plane, so just for that purpose he invented another distinct set of magnetically sensitive fibres, similarly activated by the Sun, supposedly lying within the body of the planet, whose direction was rigidly fixed in the plane of the orbit of each planet. These fibres were passive, so they had no other effect.

4.8 Conclusion: An Almost-Complete Synthesis

The laws of planetary motion are generally regarded as Kepler's most notable discoveries. We conclude by presenting a modern assessment of the status of Kepler's achievements in formulating the two laws that govern the motion of an individual planet, while avoiding the inclusion of contributions made by those who followed him (however eminent). It is a matter of regret that those laws are frequently stated, and always illustrated, as if they were Newtonian, omitting the circular framework which synthesized them (see Figs. 4.1, 4.3 and 4.8), which appeared in all Kepler's relevant diagrams. We are aiming here, by contrast, to ensure that Kepler receives credit for the distinctive features he invented, as well as for the conclusions he reached.

We have noted that in chapter 59, protheorema XI, Kepler provided a proof of the soundness of his method for Law I by showing that $AP = AK$, where AP is the radius vector of the elliptical path, and AK is the medial distance produced by his characteristic construction (see Sect. 4.4 above), that he named the *diametral distance*. This proof is sound according to modern standards, using elementary propositions of Euclid (Pythagoras' theorem and similar triangles) to demonstrate that the end P of the constructed arc lay on the ordinate QH and satisfied the Archimedean ratio property defined by the semi-axes, so that P was a typical point of the ellipse.[78] Separately, Kepler had verified that the actual path of the planet approximated to that geometrical ellipse at the stipulated level of observational accuracy. Hence Law I may properly be described as established observationally by Kepler in agreement with geometrical theory.

Meanwhile, we have shown (see Sect. 4.6.4 above) that in chapter 59, protheorema IX, Kepler represented each increment of time by an increment of area depending on the length $QR = AK$. This was a theoretical formulation, with no observational input, which formally established the area–time equivalence for micro-quantities,

---

[78] This Keplerian proof has been set out in full in Davis (1992), pp. 156–157.

enabling time to be represented geometrically. Nevertheless, on this foundation of theory, Law II, giving the macro-time of the planet in terms of an area (here the area swept out by the line joining the planet to the Sun—the radius vector—was again established observationally).

Thus Kepler had discovered the pair of mathematical laws which together gave the position of the planet (distance and time coordinates), where the confirmation of that orbit lay in its agreement with the best observations. While it cannot therefore be said that Kepler's determination of the planetary laws was other than observation-based, nevertheless the planetary orbit itself was a theoretical construct, its exactitude founded on geometry alone.

The notable geometrical feature linking the two laws is the length AK, which Kepler called the *diametral distance*—illustrated in both its manifestations in Fig. 4.9 as implied in Chapter 59, Protheorema XII. It is evident—proved by the geometry of Euclid—that the two planetary laws are completely and exactly defined (in the Ptolemaic framework) by two equal lengths AP, QR having AK as their common measure, where:

$$\text{diametral distance} = AK = AP = QR.$$

Neither Kepler's contemporaries nor most of the astronomers who followed him were aware that these laws were exact, and indeed before the end of the century the Keplerian treatment was superseded by the comprehensive dynamical approach initiated by Newton. Hence it is rare to encounter the two laws interpreted geometrically, and even rarer to find them in combination, expressed with precision in algebraic notation, their relationship established by the presence of a common variable. We set them out here, for the benefit of readers with a background in modern science:

$$\text{Law I} : r = a(1 + e\cos\beta)$$

$$\text{Law II} : t \propto \tfrac{1}{2}a^2(\beta + e\sin\beta) \text{ or } t = \beta + e\sin\beta$$

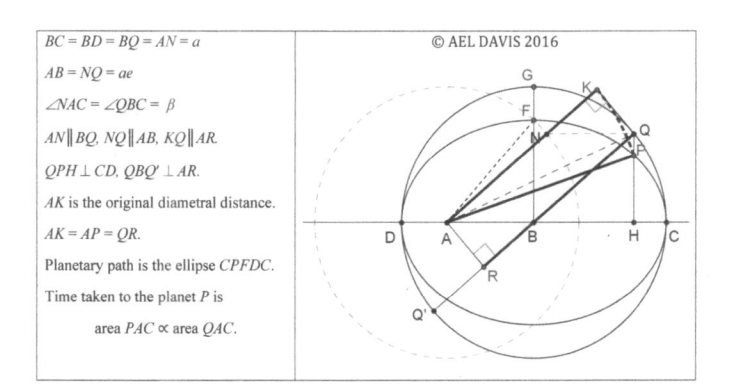

Fig. 4.9 Correlation of the two laws through the diametral distance

And we include, as a distinctive feature, a unique version of the orbital equation of a planet expressed algebraically, explicitly in terms of its constituent coordinates, distance and time:

$$t = \cos^{-1}\left(\frac{r-a}{ae}\right) + \frac{\sqrt{a^2e^2 - (r-a)^2}}{a}.$$

Kepler's aim was to discover the planetary orbit as the mathematical (geometrical) entity which would represent the Universe created by God. That was inevitably an idealized solution, and modern astronomers have hastened to point out, rightly, that such an orbit does not represent reality, and therefore success belongs to the person (Newton, of course) who introduced the consideration of matter (in *Principia* Book I, item XI). Alternatively, for the purpose of the present discussion, we can choose to assess Kepler's success in his terms, in achieving mathematical exactitude. Moreover, it is important to appreciate that, if modern physics is understood as the convergence of succeeding approximations by which one hopes to arrive at reality, it is essential to start from an existing exact structure. That is what Kepler provided, and Newton, beginning from his mould-breaking synthesis set out in *Principia* Book I, propositions II and III, gave proofs that Kepler's area law implied the existence of a central force.

References

Apollonius (1896). *Apollonius of Perga: Treatise on conic sections*. Edited and translated from the Greek by T. L. Heath. Cambridge: Cambridge University Press.

Archimedes (1558). *Archimedis Opera non nulla à Federico Commandino Vrbinate nuper in Latinum conuersa, et commentariis illustrata*. Translated from the Greek by F. Commandino. Venice: Pauus Manutius.

Archimedes (1897). *Archimedes: Works*. Edited by T. L. Heath and translated from the Greek by J. L. Heuberg. Cambridge: Cambridge University Press.

Clerke, A. M. (1905). *Modern cosmogonies*. London: Adam and Charles Black.

Davis, A. E. L. (1981). *A mathematical elucidation of the bases of Kepler's laws*. London: Printed on demand by University Microfilms International.

Davis, A. E. L. (1992). Grading the eggs (Kepler's sizing-procedure …). *Centaurus, 35*(2), 97–191.

Davis, A. E. L. (2009). Kepler's "via ovalis composita": Unity from diversity. *Journal for the History of Astronomy, 40*, 55–69.

Davis, A. E. L. (2007). Some plane geometry from a cone. *Mathematical Gazette, 91*(521), 235–245 (sections 3–6).

Davis, A. E. L. (2015). The geometrical root of the area-measure of time. *Journal for the History of Astronomy, 46*, 297–324.

Dreyer, J. L. E. (1906). *History of the planetary systems*. Cambridge: Cambridge University Press.

Dreyer, J. L. E. (1953). *A history of astronomy from Thales to Kepler*. Revised with Foreword by W. H. Stahl. New York: Dover.

Field, J. V. (1988). *Kepler's geometrical cosmology*. London: Athlone Press (reprinted by Bloomsbury in 2014).

Field, J. V. (1988a). What is scientific about a scientific instrument?, *Nuncius, 3.2*, pp. 3–26.

Field, J. V. (2010). Kepler's place in the history of science. In A. Hadravová, T. J. Mahoney, & P. Hadrava (Eds.), *Kepler's heritage in the space age, Acta historiae rerum naturalium necnon technicarum* (Vol. 10, pp. 11–16). Prague: Národní Technické Muzeum.

Galilei, G. (1638) *Discorsi e Dimostrazioni Matematiche intorno à due nuovi Scienze Attenenti alla Mecanica & i Movimenti Locali.* Leiden.

Galilei, G. (1913). *Discourses on two new sciences.* Translated from the Italian by H. Crew and A. de Salvio, with an Introduction by Antonio Favaro. New York: Dover.

Gingerich, O. (1975). Kepler's place in astronomy. *Vistas in Astronomy, 18*, 261–278.

Kepler, J. (1604). *Ad Vitellonium paralipomena, astronomia pars optica.* Frankfurt: Claudius Marnius and Heirs of Johann Aubrius.

Kepler, J. (1609). *Astronomia nova aitiologêtos, seu physica coelestis, tradita commentariis de motibus Stellae Martis.* Heidelberg: E. Vogelin.

Kepler, J. (1615). *Stereometria doliorum vinariiorum ....* Linz: Johann Planck.

Kepler, J. (1620). *Epitome astronomiae Copernicanae*, Book IV. Linz: Johann Planck.

Kepler, J. (1939). *Johannes Kepler Gesammelte Werke. Band II: Astronomia pars optica.* F. Hammer (Ed.). Munich: C. H. Beck.

Kepler, J. (1940). *Johannes Kepler Gesammelte Werke. Band VI: Harmonice mundi.* M. Caspar (Ed.). Munich: C. H. Beck.

Kepler, J. (1949). *Johannes Kepler Gesammelte Werke. Band XIV: Briefe 1599–1603.* M. Caspar (Ed.). Munich: C. H. Beck.

Kepler, J. (1951). *Johannes Kepler Gesammelte Werke. Band XV: Briefe 1604–1607.* M. Caspar (Ed.). Munich: C. H. Beck.

Kepler, J. (1960). *Johannes Kepler Gesammelte Werke. Band IX: Mathematische Schriften.* F. Hammer (Ed.). Munich: C. H. Beck.

Kepler, J. (1990). *Johannes Kepler Gesammelte Werke. Band III: Astronomia nova.* M. Caspar, M. and Kepler-Kommission (Ed.). Munich: C. H. Beck.

Kepler, J. (1991). *Johannes Kepler Gesammelte Werke. Band VII: Epitome astronomiae Coperni-canae.* M. Caspar and Kepler-Kommission. Munich: C. H. Beck.

Kepler, J. (2015). *Astronomia nova* (W. H. Donahue, Trans.). Santa Fe: Green Lion Press.

Prager, F. D. (1973, 1975). Kepler als Erfinder, in Krafft, F., Meyer, K., and Sticker, B., (eds) (1973), *Internationales Kepler-Symposium Weil der Stadt 1971*, Hildesheim, pp. 385–405. English translation: Kepler as inventor, in Beer, A. (ed.), *Vistas in Astronomy*, 18, 887–89.

Stephenson, B. (1987). *Kepler's physical astronomy.* New York: Springer.

Whiteside, D. T. (1974). Keplerian planetary eggs, laid and unlaid, 1600–1605. *Journal for the History of Astronomy, 5*, 1–21.

Chapter 5
The Translation of the Title of Kepler's *Astronomia nova*

Andrew Gregory

The full title of Kepler's *Astronomia nova*, as given on the title page, is shown in Fig. 5.1.

What I want to look at here are some issues of translation and of historiography, of how we approach Kepler's texts. Most of the translation here is uncontroversial, but the Greek *AITIOΛOΓHTOΣ* (*aitiologêtos*) is problematic, as is SEU PHYSICA COELESTIS. The uncontroversial part runs 'New Astronomy… Treated by Means of Commentaries on the Motions of the Star Mars, from the Observations of Tycho Brahe, Gentleman.' The question here is whether *aitiologêtos, seu physica coelestis* means something like 'based upon causes, or celestial physics' or means something like 'with explanations, or the natural philosophy of the heavens.'

5.1 *Aitiologêtos*: Cause or Explanation?

On the original title page, *AITIOΛOΓHTOΣ* (*aitiologêtos*) is in Greek characters in distinction to the rest of the title. The specific form *aitiologêtos* is not found in ancient texts though some cognate forms are.[1] The second part of the word is straightforward enough, meaning the study of or talk about. The initial part of this word though brings us to our first crux. How do we translate the *aitio*-part of this word? Here it is critical to recognize that the Greek word *aitia* and its cognates have several meanings, one of which is cause, so translation here involves a choice. If we go to Liddell, Scott and Jones's Greek Lexicon (LSJ), then *aitia* may mean blame or culpability and it may also mean reason, explanation or cause. Some advances in scholarship on Aristotle

---

[1] One can find some contraries, meaning 'hard to account for'.

A. Gregory (✉)
University College London, London, UK
e-mail: andrew.gregory@ucl.ac.uk

© Springer Nature B.V. 2024
A. E. L. Davis et al. (eds.), *Reading the Mind of God*, Springer Praxis Books,
https://doi.org/10.1007/978-94-024-2250-4_5

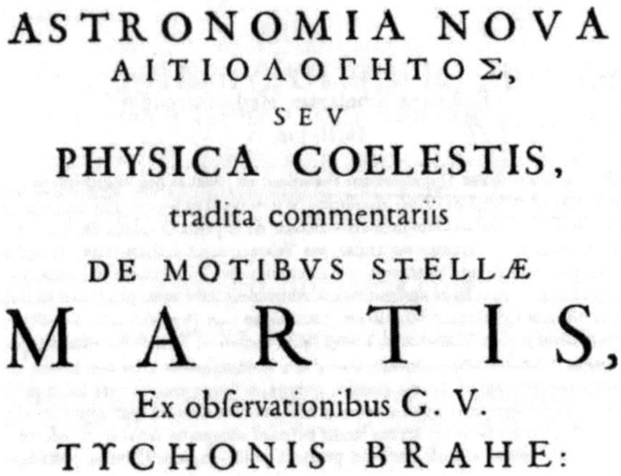

ASTRONOMIA NOVA
ΑΙΤΙΟΛΟΓΗΤΟΣ,
SEV
PHYSICA COELESTIS,
tradita commentariis
DE MOTIBVS STELLÆ
MARTIS,
Ex obfervationibus G. V.
TYCHONIS BRAHE:

Fig. 5.1 Title page of *Astronomia nova* (Reproduced from Kepler, *Astronomia nova*, Heidelberg, 1609, large folio)

(384–322 BC) and Plato (428/427 or 424/423–348/347 BC) are also relevant here. It is well known that Aristotle had a scheme of four *aitiai*, material, efficient, formal and teleological ways of describing a phenomenon. These were often known as Aristotle's four causes. However, since Hocutt's seminal 1974 paper, 'Aristotle's Four Becauses', it has generally been recognized that Aristotle offered a scheme of four becauses, four types of reason or four types of explanation.[2] Hocutt argued that Aristotle's efficient explanations were the only ones which reasonably matched the modern conception of cause. I would argue that it is best to say that Aristotle had four types of explanation, one of which loosely resembles the modern conception of cause, rather than anachronistically impose a modern conception of cause on Aristotle.

There has also been a similar discussion about the meaning of *aitia* in Plato. The seminal paper here is Vlastos's *Reasons and Causes in the Phaedo*.[3] Again, it is now considered unwise and too crude simply to translate all instances of *aitia* and its cognates as 'cause' in Plato. Some of Plato's *aitiai* may resemble modern causes but by no means all do. Again, it is better to think in terms of Plato on explanations, some of which may resemble modern causes, than to impose a modern theory of cause on him. It is also important to note that the Latin *'causa'* displays similar ambiguity to the Greek *aitia*. Lewis and Short in their Latin Lexicon (L&S) give *'that by, on account of*, or *through which any thing takes place* or *is done*; *a cause, reason, motive, inducement.'*

So to translate *aitiologêtos* as 'cause' then is not the only, or indeed the primary option.[4] Which translation we choose is effectively a historiographical decision,

---

[2] Hocutt (1974).

[3] Vlastos (1969); cf. Sedley (1998).

[4] Here I disagree with some commentators, such as Pavel Gabor, who think that to translate as 'cause' is obvious. See Gabor (2020).

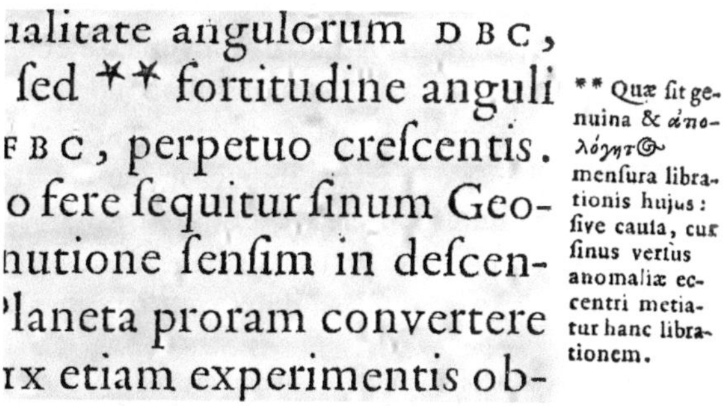

ialitate angulorum D B C,
fed ✱✱ fortitudine anguli
F B C, perpetuo crefcentis.
o fere fequitur finum Geo-
nutione fenfim in defcen-
'laneta proram convertere
ix etiam experimentis ob-

✱✱ Quæ fit ge-
nuina & ἀπο-
λόγητ☉
menfura libra-
tionis hujus:
five cauſa, cur
finus verſus
anomaliæ ec-
centri metia-
tur hanc libra-
tionem.

Fig. 5.2 Marginal note from *Astronomia nova* (Reproduced from Kepler, J., *Astronomia nova*, Heidelberg, 1609, Chap. 57, p. 269)

determined by, or perhaps helping to determine, our broader picture of Kepler, or if we view this more locally, our picture of the *Astronomia nova*. Should we see Kepler (or just *Astronomia nova*) as still part of an ancient tradition where causal explanations are part of a wider array of acceptable explanations, or should we see Kepler (or just *Astronomia nova*) as moving towards a more modern conception of cause as it relates to astronomy? One important consideration here is that *aitiologêtos* is spelt in Greek lettering, *ΑΙΤΙΟΛΟΓΗΤΟΣ*, in distinction to the rest of the title, which may indicate orientation to an existing, Greek derived system of explanation.

5.2   Other Uses of *Aitiologêtos* in Kepler

I agree with Ernst Kühn that *aitiologêtos* occurs in a marginal note in Chap. 57 of *Astronomia nova*[5] (see Fig. 5.2).

What is printed in modern editions is 'Quae sit genuine et ἀπολόγητος mensura librationis hujus'. However, *apologêtos* makes little or no sense here.[6] In the 1609 edition what we arguably find is ἀτιολόγητος with a tau (τ) and an iota (ι) and not a pi (π).[7] It is then easy to see how ἀτιολόγητος became ἀπολόγητος due to a misreading of τι for π. In my experience of dealing with ancient Greek texts, such transmission errors are not unusual and Kühn gives some other examples of this error.[8] The error of ἀτιολόγητος instead of αἰτιολόγητος, omitting the first iota, is also an easy one to make. This I can testify from my own experience of writing on *aitiai*. One has to be careful to check spellings in both transliterated and Greek versions of the word. So two small, very plausible errors have resulted in *apologêtos*

---

[5] Kühn (2010).

[6] Donahue (1992) considers ἀπολόγητος to be enigmatic, and conjectures ἀπολογέτικος instead. His translation of the passage is: 'This is the genuine measure of this reciprocation, supported by reason; in other words, it is the reason.'

[7] See Kühn (2010), pp. 71 and 73, where his reproductions of the 1609 page make this clear.

[8] Kühn (2010).

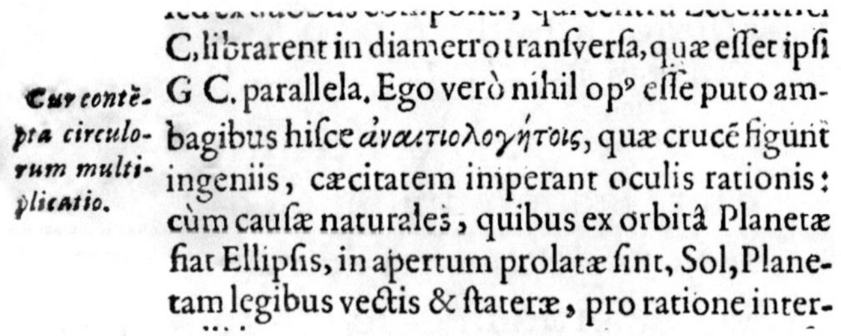

C, librarent in diametro tranſverſa, quæ eſſet ipſi
Cur contē. G C. parallela. Ego verò nihil opᵖ eſſe puto am-
pta circulo- bagibus hiſce *ἀναιτιολογήτοις*, quæ crucē figunt
rum multi- ingeniis, cæcitatem imperant oculis rationis:
plicatio. cùm cauſæ naturales, quibus ex orbitâ Planetæ
fiat Ellipſis, in apertum prolatæ ſint, Sol, Plane-
tam legibus vectis & ſtateræ, pro ratione inter-

Fig. 5.3 Marginal note and the use of the Greek word *anaitiologêtos* in the text (Reproduced from Kepler, J., *Tabulae Rudolphineae*, Ulm, 1627, Praecepta, cap. XX, p. 59)

where $αἰτιολόγητος$ was in all likelihood the intended original. I would translate this text:

Quae sit genuine et αἰτιολόγητος mensura librationis hujus ('So this is the natural and explanatory measure of this libration').

I agree with Kühn that there is a 'tight brace' between the title of *Astronomia nova* and Chap. 57 which is brought to the fore by the use of *aitiologêtos* written in Greek characters in both.[9] So this is not just a matter of a word used in the title, but is fundamental to the project of the book.

There is a further use of *aitiologêtos* by Kepler, this time in its negative form in the *Rudolphine Tables* (Ulm, 1627), *Precepts*, Chap. 20, p. 57 (see Fig. 5.3).[10]

Here we have a marginal note of 'Why disregard the multiplication of circles?' The text it refers to runs:

I believe there is no need for these non-explanatory (*anaitiologêtois*) windings…

when the natural causes by which ellipses are generated from the planetary orbits are brought into the open.

Again, in the Latin text we have a word spelt out in Greek characters, *anaitiologêtois*.

This is an alpha-privative form, so this is a negation of *aitiologêtois*. As with *aitiologêtos*, the meaning is reasonably clear. We do find the alpha-privative from of *aitios, anaitios*. Primarily in Greek literature this means 'not being the fault of, guiltless' but in Greek philosophy it means 'not being the explanation'.[11] This fits very well with Kepler's sense here. The multiplication of circles is non-explanatory and we can bring the reason why planetary orbits are ellipses into the open.

We might push this a little further in relation to Plato, *Laws* X, 898a, where Plato argues that ideal celestial motion is regular and uniform around the same point. and is 'according to a single scheme and a single order (*hena logon kai taxin mian*)'. While Kepler's proposal of ellipses in one sense goes against the Platonic scheme

---

[9] Kühn (2010).

[10] Kepler (1627), *Praecepta*, ch. XX, p. 57; Kepler (1969), p. 132.

[11] See Aristotle (1989), 65b16; Aristotle (2020) 1401b30.

of regular circular motion, there is also a sense in which it accords very strongly with this passage from the Laws. Unlike epicycles, eccentrics and equants, which all have multiple centres for motion and the planet does not travel in a regular manner around the proper centre, elliptical motion is around one point and each planet has one ellipse (one scheme and one order) rather than each planet having a different and largely arbitrary construction of circles. So for Kepler there may be a stronger sense in which the multiplication of circles is *anaitiologêtos*, non-explanatory.

It is important then that *aitiologêtos* is not a hapax in Kepler, a word which occurs only once. This gives us a wider context for the application of *aitiologêtos*. It does not definitively solve the problem of whether to translate as 'cause' or 'explanation', though explanation works perfectly well in both of these contexts and certainly in our second example here, there is a strong case for explanation rather than cause.

5.3 Explanation and the Mind of God

A key historiographical question is to what extent it is proper to describe Kepler in some sense as a neo-Platonist? If Kepler did place himself within a Platonic tradition, then it is important to be aware of some very famous passages in Plato concerning *aitiai*. Let us start with *Timaeus* 46de:

> All of these are *sunaitiai* (auxiliary explanations) which the god uses as tools to instantiate the form of the good. However, they are thought by most men to be not the *sunaitiai* but the *aitiai* (explanations) of all things, cooling and heating, packing together and dispersing and all such actions… we must speak of both types of *aitiai*, but keep separate those which with the aid of mind generate that which is beautiful and good, from those which are devoid of understanding and in each case produce chance, unordered results.[12]

The import here is clear enough. For Plato, the sorts of explanations which we would reckon to be causal are *sunaitiai*, 'auxiliary explanations', while in the *Timaeus* at least, the cosmogonical actions of Plato's craftsman god, the Demiurge, are guided by intelligence and aim at the good, and are a different type of explanation. This sort of view is also clear in the *Phaedo*, in the passage known as 'Socrates' Autobiography'. Socrates (*c*. 470–399 BC) says that the reason he is in prison is not to do with an analysis of his physical constituents, but that his mind has chosen this with the good in mind. At *Phaedo* 99b he is famously disparaging of those who are:

> Unable to distinguish between the real reason (*aition*) for something and that without which the reason (*aition*) could ever be a reason (*aition*).

Socrates rejects physical explanations as inadequate in general and specifically in cosmology rejects explanations in terms of some physical support for the Earth (a vortex, or air supporting the Earth) in favour of explanations which state why it is good for the Earth to be where it is and be stable.

---

[12] Cf. Plato (1989), 76d.

So one might argue that Kepler writes *aitiologêtos* in Greek in the title, invoking a specific tradition of explanation using more than causes, but not *sunaitiologêtos*, which would indicate a break with that tradition in favour of physical or causal explanations.[13] The strong line here would be that *aitiologêtos* deals only with the intelligent explanations of the heavens, relating to how a Christian god has put the heavens together, the more moderate view being that in line with the *Timaeus* passage, Kepler intends to declare both sorts of *aitiai*. In either case, 'cause' would not be an appropriate rendering of *aitiologêtos*. Kepler, as a reader of Plato, would be well aware of these issues.

There are also some important parallels between some of Kepler's work and Plato's *Timaeus*. At *Timaeus* 34a ff., the Demiurge, Plato's god who brings order from primordial chaos, sets the ratios for the paths of the planets, does so for the best and does so with considerations derived from musical theory. As is well known, Kepler in *Harmonice mundi* has the spacing of the planets set in terms of geometrical considerations and the parameters of the ellipses set in terms of musical consider-ations.[14] Ultimately for Kepler it is a Christian god who will have put this cosmos together.[15] There is a deeper strategic affinity here though. For Plato, there are no accidental features of the cosmos and the Demiurge has a reason for each of the choices he makes in setting up the best possible cosmos. So Plato's *Timaeus* is an attempt to read the mind of the Demiurge, to divine how this god put the cosmos together in the best possible fashion. Kepler too seems to think that the number of planets, the ratios of their orbits and the parameters of the orbital ellipses are in need of explanation, but not a causal one and he attempts to read the mind of a Christian god. This approach is quite different from that of modern cosmology where these features would be thought accidental and not in need of any special explanation.[16] So unless *Astronomia nova* marks a significant break for Kepler 'explanation' is probably better than 'cause' for *aitiologêtos*.

5.4   Cause and Explanation

One might ask, do we need to translate *aitiologêtos*? Could we not use the modern English derivative, aetiology? So the first part of the title might be New Astronomy, Aetiologically. That might initially look like a neutral way of treating the matter, but the modern word aetiology has very strong causal connotations. It is used mainly in medical contexts to describe the causes of a disease. So to refuse to translate in

---

[13] Here I disagree with Gabor (2020), who believes that Kepler used the new term *aitiologêtos* to indicate a new approach. If this was his intention, then *sunaitiologêtos* would have made a much stronger, clearer statement and one might ask why the Greek characters if he intended a break.

[14] See Field (1988) and Field, 'Kepler's cosmology' (this volume, Chap. 2).

[15] In 1689 Leibniz made the interesting (unpublished) comment: 'The angels had watched over that he might be the first among mortals to publish the laws of the heavens, the truth of things, and the principles of the gods' (Clark, 1992, p. 102).

[16] On the similarities of the programmes of Plato and Kepler, see Gregory (2022).

favour of transliteration is also a significant historiographical decision, and one which may introduce an anachronism.[17] It certainly favours the causal/physics approach to Kepler. Let us look at a translation of one of the further instances of *aitiologêtos*. Kühn, in 2010, translates 'Quae sit genuine et αἰτιολόγητος mensura librationis hujus' as:

> Was das naturgegebene und ursächliche (äitiologische) Maß dieser Schwankung ist ('What the natural and causal (aetiological) measure of this variation is').[18]

So we have 'causal (aetiological)' for *aitiologêtos*. Kühn also talks of Kepler's 'natural, causal magnet like forces' in relation to planetary motion, which would seem to support this sort of translation.[19] An important further consideration though is how some phenomena were understood in Kepler's time. Were they understood using something like a modern conception of cause, or were they understood in another manner? Magnetism is clearly a critical instance here. As magnetism is now part of electromagnetic theory, which can be understood in a straightforwardly causal manner, it is tempting to assume that Kepler had a causal understanding of magnetism. That is not the case though. In Kepler's period magnetism was often treated as part of natural magic.[20] Even if we look at the foundational text for the scientific study of magnetism, which was influential on Kepler, William Gilbert (1540–1603), *De Magnete, Magnetisque Corporoibus, et de Magno Magnete Tellure: Physiologia noua, Plurimis & Argumentis, & Experimentis* (*On the Loadstone and Magnetic Bodies, and on That Great Magnet the Earth: A New Physiology, Demonstrated with Many arguments and Experiments*), magnetism is not treated in a straightforward causal manner. If we look at how Kepler treated magnetism, with the idea of *anima motrix*, 'motive soul', it is clear that there will need to be further explanation than straightforward causality here. Here too there can be a historiographical translation issue. 'Motive soul' is the literal translation of *anima motrix* and fits with the understanding of magnetism of the time. Translations such as 'moving power' provide a much more modern feel more in line with our current understanding of magnetism. So for this passage, I hold to my earlier translation of 'So this is the natural and explanatory measure of this libration'.

The word 'physiologia' in Gilbert's book title might be better rendered 'account of its nature', which is its literal meaning from the Greek, than physiology, which brings us to the next issue in the title of Kepler's *Astronomia nova*.

---

[17] Gabor (2020).

[18] Kühn (2010), my translation from the German.

[19] Kühn (2020).

[20] In his *Magia naturalis*, Giovanni Battista della Porta (1535–1615) has Chap. 7 on 'The Wonders of the Loadstone'.
(della Porta, 1588).

5.5 *Physica*: Physics or Natural Philosophy?

The second, and related issue of translation in the title of *Astronomia nova* is what we do with *physica* in *aitiologêtos seu physica coelestis*. Again we have options and some similar historiographical considerations will be in play. A possible translation for *physica* is 'physics' but L&S give 'natural science, natural philosophy, physics'. Standard usage in texts before Kepler would indicate that 'natural philosophy' is a better rendering for *physica* than 'physics'. The Latin *physica* and other cognate terms derive from the Greek word *phusis*. There are several reasons why it is a clear error to translate this term as 'physics' rather than 'nature'. Firstly, to study the *phusis* of something was to study its origin, development and current constitution. That could be the nature of something specific or the nature of the cosmos as a whole. LSJ have *'origin... the natural form* or *constitution* of a person or thing *as the result of growth'* for *phusis*. Secondly, *phusis* derives from *phuein*, 'to grow' and so can carry a strong organic sense to it. Thirdly, as Mourelatos has argued, *phuein*, 'to grow', the verb from which *phusis* derives, can have a sense of dynamic being, of coming into being where, *einai* expresses a more static sense of being.[21] In ancient Greece, the *phusiologoi* were those who spoke about nature in this broad sense, often in an organic manner, and were by no means exclusively physical philosophers or any correlate of physicists. Nature, *phusis*, was often conceived of as organic and its explanation often used biological rather than physical or mechanical analogues. Their study of *phusis* certainly included a whole range of disciplines beyond physics as is clear both from the extant writings of the *physiologoi* and the report of Plato and Aristotle.[22]

5.6 Conclusion

Clearly the questions of how we translate *aitiologêtos* and how we translate *seu physica coelestis* are related, 'cause' going with 'physics of the heavens' and 'explanation' going with 'nature of the heavens'. Equally clearly, which we choose is a significant historiographical decision. I would say that the evidence (etymology, contemporary usage, use of the Greek form) slightly favour the latter view. However, the most important things here are to be aware that we are making a significant historiographical decision with this translation and that 'cause' is not the only, the obvious or even the primary translation of *aitiologêtos* given recent advances in scholarship.

My thanks to A. E. L. Davis and J. V. Field for their comments and suggestions in relation to this chapter. For further consideration of translation problems see W. H. Donahue ('Translating Kepler') in Chap. 13 of this volume.

---

[21] Mourelatos, A. P. D. (2018) *Bryn Mawr Classical Review* 2018, review of Laks and Most, *Early Greek Philosophy*. https://bmcr.brynmawr.edu/2018/2018.03.15/.

[22] For more on *phusis*, see Gregory (2021a,b).

References

Aristotle. (1989). *Categories. On interpretation. Prior analytics.* Edited and translated from the Greek by H. P. Cooke and H. Tredenck. Loeb Classical Library 325. London: Heinemann.

Aristotle. (2020). *Rhetoric.* Edited and translated by J. H. Freese and revised by G. Striker. Loeb Classical Library 193. London: Heinemann.

Clark, W. (1992). The scientific revolution in the German Nation. In R. Porter, & M. Teich (Eds.), *The scientific revolution in national context* (p. 102). Cambridge: Cambridge University Press.

della Porta, G. B. (1558). *Magia naturalis.* Naples: Orazio Salviani.

Donahue, W. H. (1992). *New astronomy.* Cambridge: Cambridge University Press.

Field, J. V. (1988). *Kepler's geometrical cosmology.* London and Chicago: Athlone Press and University of Chicago Press.

Gabor, P. (2020). It's all Greek: About three of Kepler's book titles. Part II Astronomia Nova Aitiologetos. https://www.vaticanobservatory.org/sacred-space-astronomy/its-all-greek-about-three-of-keplers-book-titles-part-ii-astronomia-nova-aitiologetos/

Gregory, A. (2021a). Plato's reception of presocratic natural philosophy. In C. C. Harry, & J. Habash (Eds.), *The reception of presocratic natural philosophy in later classical thought.* Leiden: Brill, pp. 44–80.

Gregory, A. (2021b). *Early Greek philosophies of nature.* London: Bloomsbury.

Gregory, A. (2022). Mathematics and cosmology in Plato's Timaeus, *Apeiron* 2022.

Hocutt, M. (1974). Aristotle's four becauses. *Philosophy, 49*(90), 385–399.

Kepler, J. (1627). *Tabulae Rudolphinae.* Ulm: Jonas Saur.

Kepler, J. (1969).*Johannes Kepler Gesammelte Werke. Vol. X: Tabulae Rudolphinae.* F. Hammer. Munich: C. H. Beck.

Kühn, E. (2010). Das ἀπολόγητος – Rätsel in Keplers >>Astronomia Nova<<. In K. Gaulke, & J. Hamel (Eds.), *Kepler, Galilei, das Fernrohr und die Folgen* (pp. 66–88). J. Verlag Harri Deutsch.

Plato. (1989). *Timaeus. Critias. Cleitophon. Menexenus. Epistles.* Edited and translated from the Greek by R. B. Bury. Loeb Classical Library 234. London: Heinemann.

Sedley, D. N. (1998). Platonic causes. *Phronesis, 43*, 114–132.

Vlastos, G. L. (1969). Reasons and causes in the Phaedo. *The Philosophical Review, 78*, 291–325.

Chapter 6
Kepler and the Reform of Astrology

Sheila J. Rabin

Kepler wrote a considerable amount on astrology. Though he lambasted what he considered excessive reliance on astrology and tried to reform the way it was practised and the theory behind it, he accepted it as a valid part of the study of the heavens.

6.1 Renaissance Astrology

During the Renaissance the *Tetrabiblos* of the second-century AD astronomer Claudius Ptolemy became the leading textbook of astrology.[1] Ptolemy referred to two kinds of prediction by means of the heavenly bodies: the first concerned 'the movements of Sun, Moon, and stars in relation to each other and to the Earth', which he had written about in his earlier work on astronomy, the *Almagest*, and the second to 'investigate the changes' brought about by the motions of the heavenly bodies',[2] which came to called astrology. Ptolemy formulated the basic elements of traditional astrology, for example, establishing special characteristics for what he considered the planets in his geocentric Universe, which included the Sun and the Moon, and for the signs of the zodiac; deeming aspects or the angular distances between two planets or lines drawn between those planets converge on Earth, which he established as conjunction (0°), sextile (60°), square (90°), trine (120°) and opposition (180°), good or bad; dividing the sky into twelve houses that represent various facets of a person's life; using for prediction a method called progressions or directions, the

---

[1] For general histories of astrology, see Thorndike (1923–1958); North (1986); Tester (1987); Campion (2008–2009); Rutkin (2019), vols. 2 and 3 forthcoming. For astrology in the Renaissance; see also Garin (1983); Westman (2011); Rutkin (2006); Dooley (2014).

[2] Ptolemy (1980), p. 3.

Sheila J. Rabin (✉)
Emerita, Saint Peter's University, Jersey City, NJ, USA
e-mail: rabinhist@gmail.com

© Springer Nature B.V. 2024
A. E. L. Davis et al. (eds.), *Reading the Mind of God*, Springer Praxis Books,
https://doi.org/10.1007/978-94-024-2250-4_6

claim that the location of a planet at birth portended the character of the year in a person's life corresponding to the degree the planet has advanced from the subject's nativity.

The seventh-century writer Isidore of Seville (*c.* 560–636) called the two parts of the study of the heavens *astronomia* and *astrologia*; he also divided astrology into an acceptable 'natural' and unacceptable 'judicial' form that practised divination.[3] Nevertheless, the terms 'astronomy' and 'astrology' were often interchangeable through the medieval and early modern periods. For example, the fourteenth-century author Nicole of Oresme (*c.* 1320/1325–1382) titled a polemic against astrologers *Tractatus contra astronomos*, though he used the term *astrologia* and alternatively called the practitioners *astrologi*.[4]

Isidore of Seville notwithstanding, the study of the heavenly bodies, whether in the form of astronomy or astrology, was not significantly pursued in the early Middle Ages. It grew in the twelfth century when the Latin West was flooded with translations of books on philosophy, mathematics and medicine from the Islamic world. Particularly important for the history of astrology were the Aristotelian corpus, both the *Almagest* and the *Tetrabiblos* by Ptolemy, and works by Islamic authors, such as the ninth-century writer Abu Mashar (787–886), whose *Greater Introduction to Astronomy* in Latin translation was frequently used.[5] The Islamic tradition bequeathed an almost indissoluble relationship between astronomy and astrology and Aristotelian philosophy as the foundation of natural philosophy, including the study of the heavenly bodies. Though mathematical astrology had not entered Greece during the classical age and consequently was not part of Aristotle's philosophy, he did suggest that the Universe from the Moon and beyond was made of a non-physical fifth element that is immutable and moves the lower world, and this was often seen as supporting astrology. For example, the fifteenth-century physician Lucio Bellanti (d. 1499) cited this to defend astrology against the attack by Giovanni Pico della Mirandola (1463–1494).[6]

6.2   Astronomy and Astrology in Universities and Princely Courts

The combination of astronomy and astrology within the framework of an Aristotelian natural philosophy was reinforced by the universities that were coming into being in the twelfth century. Astrology was taught together with astronomy in the mathematical, philosophical and medical curricula.[7] University-educated physicians, like Bellanti, used astrology to locate the causes of diseases and to help determine their

---

[3] Isidore of seville (2006), p. 99.

[4] Oresme (1952), pp. 123–141.

[5] Lemay (1987).

[6] Bellanti (1553), p. 7; Pico (1946, 1952); see also Akopyan (2020).

[7] For astrology in the universities, see, for example, Grendler (2002), pp. 408–429.

treatment.[8] In a sense an astrological chart was the medieval and Renaissance version of the patient's medical history. Matters of health and the weather were part of natural astrology and consequently theologically sound. Judicial astrology involved the reading of personal characters and the prediction of human events from the heavenly configurations and was suspect, but it could be difficult to ascertain where natural astrology ended and judicial astrology began, especially in the matter of health. More and more frequently astrologers/astronomers were also found in princely courts, especially from the fifteenth century, both as advisers to princes and as proof that a prince's court was culturally and intellectually chic.[9]

The Protestant Reformation of the sixteenth century did nothing to dislodge astrology from its place in the Lutheran universities despite biblical prohibitions against divination. Though Martin Luther rejected astrology as against the Bible, the university curriculum was set up by Philipp Melanchthon, and Melanchthon continued the study of astrology as inseparable from astronomy in the Lutheran institutions.

He regarded astronomy as the study of 'celestial motions' and astrology as that of 'celestial effects' and he considered both necessary for understanding divine governance of the heavens.[10] Ptolemy's *Tetrabiblos* was the major textbook, and astrology was still taught in the context of Aristotle's natural philosophy. Thus, when Kepler studied mathematics under Michael Maestlin in Tübingen, he learned astrology along with astronomy as the study of the heavens and how together they were a means of understanding how the divine worked in the Universe.[11]

6.3 The Status of Astronomy and Astrology

Despite their established status, Kepler repeatedly criticized both astrology and astrologers and differentiated astrology from astronomy. In his treatise about the comets of 1607 and 1618 he reproved astrologers who claimed to be able to predict future events from these comets and asserted that he 'did not condemn them for impiety but for stupidity and vanity'.[12] He did see in comets a divine warning of future, but undefined, disturbances if human beings were lax in their faith and let

---

[8] Siraisi (1990), pp. 134–136.

[9] Moran (2006). Deimann and Juste (2015); for specific courts, see Azzolini (2013); Hayton (2015).

[10] Kusukawa (1995), p. 131 (on Melanchthon's attitudes toward astrology and its implementation in the curriculum, see Chaps. 4 and 5; see also Methuen (1998), Chap. 3; Brosseder (2004); on the relationship of astrology to the Protestant Reformation, see Barnes (2016).

[11] On Maestlin's support of astrology, see Methuen (1998), pp. 129–132; See also Westman (2011), pp. 262–264, which emphasizes his doubts.

[12] Kepler (1619a), p. 122; Kepler (1963), p. 248.

themselves be led into evil ways.[13] He also wondered if the comet of 1607 caused the warmth and drought that year, but he did not assert it.[14]

Kepler repeatedly deplored the fact that working in astrology was a financial necessity for him. He wrote in *Third Man in the Middle*[15] (*Tertius interveniens*) that astrology

> benefits the study of astronomy … It is true that this astrology is a foolish little daughter … but, dear God, where would her mother, the highly rational astronomy be if she had not had this foolish daughter?[16]

In 1618 he also wrote Matthäus Wacker von Wackenfels (1550–1619), the imperial councillor and long-time associate, regarding his wish to work on the *Rudolphine Tables* (*Tabulae Rudolphinae*) and the *Ephemerides*, 'In order to raise the money for the *Ephemerides* of two years I wrote a cheap calendar with two prognostications; this seems at least a bit more decent than begging.'[17]

6.4   Kepler's Practice as an Astrologer

Kepler did not merely practise astrology for pecuniary reasons; as that 1610 defence of astrology *Third Man in the Middle* clearly shows, he accepted the validity of natural astrology.[18] Indeed, it would have been difficult for a sixteenth century thinker to reject it entirely apart from the theological arguments of Melanchthon and his followers. As has been pointed out, 'no general refutation of astrology was a reasonable proposition while the influence of the Sun upon the weather (in determining the seasons) and that of the Moon upon the sea (in causing the tides) were regarded as examples of astrological 'force' in action'.[19] On the other hand, Kepler did not accept astrology as he had been taught it. Already in his calendar for 1598, he declared his intent to reform it.[20]

His critique of traditional astrology was so strong that he found it necessary to affirm his belief in astrology to his former professor Maestlin.[21] In his 1606 treatise *On the New Star* (*De stella nova*) he went so far as to state that 'by right, I should have been able to be seen as conceding everything to [Giovanni Pico della Mirandola's] judgement about the worthlessness of astrology'. But he could not reject it entirely

---

[13] Kepler (1619a), p. 127; Kepler (1963), p. 252.

[14] Kepler (1619a), pp. 110–111; Kepler (1963), p. 238.

[15] Rosen coined this translation of the phrase Tertius interveniens. See Rosen (1984), p. 257.

[16] Kepler (1941), p. 161.

[17] Kepler to Wacker, beginning of 1618, in Kepler (1955), letter 783; Baumgardt (1951), p. 130; see also Bauer (2015); Deimann and Juste (2015), pp. 205–219.

[18] For general treatments of Kepler's astrology, see Simon (1979); Field (1984); Boner (2013).

[19] Field (1984), p. 220.

[20] Kepler (1597); Kepler (1993), pp. 37–39.

[21] Kepler to Maestlin, 15 March 1598, in Kepler (1945), letter 89, p. 183, lines 142–146; see also Field (1984), p. 196.

because he had had successful experiences using astrology, and so he added, 'I will not deny ... that there is great vanity of experience vaunted by astrologers ..., but I will not on that account concede that experience has been nothing'.[22] Other astrologers failed, but not Kepler, because his was a reformed and superior astrology. Or as he so charmingly put it in *Third Man in the Middle*,

> No one should consider it unbelievable that out of astrological foolishness and godlessness a useful sense and holiness could not also be scraped out and found, in unclean slime not also a snail, mussel, oyster, or eel useful for eating, out of the big heap of caterpillar egg droppings not also a silk spinner, and finally out of an evil-smelling dung heap not also a good granule from a busy hen, a peach, or a gold nugget'.[23]

The most important facet of traditional, Ptolemaic astrology that Kepler rejected was giving any astrological significance to the signs of the zodiac. In his 1606 treatise about the New Star that had appeared in 1604, *On the New Star*, he asserted that the 'images [of the zodiac] were not formed by nature'.[24] He further claimed that

> a human image might have earned the name of certain individuals who, to be sure, could be historical, or real, or mythical. ... And indeed not dissimilarly, the peasants in turn have assigned names of animals to certain constellations whose origin is mythical, using poetic practice.[25]

Kepler maintained that the images in the zodiac were not natural but were created by the imagination of human observers. Therefore, they could not cause the effects generally ascribed to them because they could not have the characteristics attributed to them. How could someone born under Taurus be bull-headed if those stars only formed a bull in the imaginations of peasants who looked at them? Such a claim about the cause of a person's personality could be considered an example of sympathetic magic, just like the belief that a plant shaped like the liver could promote the health of the liver.[26] Sympathetic magic was an example of an occult cause in the period when Kepler wrote. As Henry Cornelius Agrippa (1486–1535), whose influential work *Three Books of Occult Philosophy* had been printed in 1533, wrote about 'occult qualities', 'their causes lie hid, and man's intellect cannot in any way reach and find them out'.[27] Thus, by rejecting the astrological significance of the signs of the zodiac, Kepler was rejecting occult causation in astrology.

Likewise Kepler rejected other astrological ideas involving the zodiac, for example, the distribution of the signs of the zodiac among the seven planets.[28] He rejected the possibility of speaking about the fortune and misfortune of the whole world, a country, a city, and so on, for he insisted that no one could establish a

---

[22] Kepler (1606), Chap. 8, pp. 30–31; Kepler (1938), p. 184.

[23] Kepler (1941), p. 161.

[24] Kepler (1606), Chap. 5, p. 20; Kepler (1938), p. 174, see also p. 184; on Kepler's use of images in practice, see Greenbaum (2015).

[25] Kepler (1606), Chap. 5, p. 21; Kepler (1938), p. 175.

[26] Kieckhefer (1989), p. 13.

[27] Agrippa (2005), p. 60.

[28] Kepler (1610); Kepler (1941), p. 185.

nativity for such entities; he also rejected a division of countries among the signs of the zodiac.[29] He did not, however, reject the astrological usefulness of the zodiac as such, without reference to the images:

> nature itself indeed does not divide [the zodiac] into twelve precise parts but only displays the occasions for their receiving these divisions, as when the Moon conjuncts with the Sun during every year in all twelve zodiacal places.[30]

Thus, the zodiac was a convenient method of dividing the visible sky and of indicating place and season.

6.5 Planetary Aspects

The true gold nugget for Kepler's astrology was the planetary aspects, the angular distance between two planets as measured from the Earth; he called them 'a pearl of nobility from astrology'.[31] In addition to the Ptolemaic aspects, Kepler added three new ones: the quintile, 72°; the sesquiquadrate, 135°; and the biquintile, 144°.[32] Kepler used the aspects to forecast the weather, and he kept records to test the accuracy of his predictions. In *Third Man in the Middle* he mentioned that he had been recording and testing weather predictions for sixteen years.[33] He also listed the weather of each case from 1592 to 1609, when a conjunction of Saturn and the Sun occurred in Capricorn and Aquarius, to demonstrate the validity of his use of aspects as opposed to those astrologers who relied on the zodiacal sign and thereby were led astray in their predictions.[34] Kepler believed that these aspects would affect the Earth's soul 'which is prompted and, as it were, excited by the aspect and stirs up the weather and events in the sky'.[35] Just as the body of a human being had a soul, Kepler believed that the Earth had a soul, though the Earth soul was not connected to a mind as the human soul was.[36]

Just as he believed that the Earth soul was moved by aspects, which allowed the astrologer to predict the weather, so he believed that the aspects would stir the soul of human beings, which gave the astrologer insight into people. He suggested in *Third Man in the Middle*,

---

[29] Kepler (1941), pp. 241–242.

[30] Kepler (1606), Chap. 7, p. 27; Kepler (1938), p. 181.

[31] Kepler (1941), p. 209.

[32] Kepler (1602), Thesis 38; Kepler (1941), p. 22; 'On Giving Astrology Sounder Foundations', translated from the Latin by Field (1984), pp. 250–251; Kepler (1619b); Kepler (1940), Book 4, Chap. 5, pp. 250–251; Kepler (1997), p. 340.

[33] Kepler (1941), p. 205.

[34] Kepler (1610); Kepler (1941), pp. 254–256.

[35] Kepler (1619b), Book 4, Chap. 7; Kepler (1940), p. 268; Kepler (1997), p. 362; on the Earth in Kepler's astrology, see Boner (2005); see also Rabin (1997), esp. p. 764.

[36] Kepler (1619b), p. 269; Kepler (1997). p. 364.

The human being in the first igniting of his life, when he first lives for himself and cannot remain any more in his mother's body, receives a character and image of all the configurations of the heavenly bodies, or of the shape of the rays streaming toward earth, and retains it until he is in his grave.[37]

Kepler maintained that this image of the heavens left an impression on both the physical body of the person and on the character and personality, and it influenced the person's relationship with other people as well. By these means, Kepler declared,

a very big difference between people will be produced, that one will become good, lively, joyful, trusting, another sleepy, indolent, careless, obscurantist, forgetful, timid, and what are such general qualities that are compared to the beautiful and exact or extensive, unsightly configurations and to the colours and movements of the planets.[38]

For Kepler, the geometrical configuration of the heavens was imprinted on the human soul at birth. The relationship of the planets to each other through their aspects caused different traits to develop; thus, the birth chart would give a description of a person. In this way geometry became the archetype of the human being, just as Kepler believed it was for the natural world.[39]

The belief in the efficacy of aspects for astrological prediction was consistent throughout Kepler's career as an astronomer and astrologer. He wrote about aspects in his 1596 work, *The Secret of the Universe* (*Mysterium cosmographicum*), in Chap. 12,[40] in a book that otherwise did not deal with astrology. He went into much more detail in *On Giving Astrology Sounder Foundations* (*De fundamentis astrologiae certioribus*) from 1602, where he began explaining the importance of aspects in his astrological thinking in thesis 38 and then from thesis 52 gave a series of predictions based on the various aspects.[41] In his 1606 work *On the New Star*, Chaps. 8 and 9 are devoted to defending the astrology of aspects against Pico's attack.[42] He dealt with aspects in many different parts of *Third Man in the Middle,* which appeared in 1610, but he particularly devoted theses 59 through 63 to developing his ideas about them.[43] And finally in book 4 of *The Harmony of the World* (*Harmonice mundi*), published in 1619, he again described them as part of his picture of that universe.[44] He summarized these ideas in *Third Man in the Middle*:

[N]ature does not take pleasure from any proportion that would be taken from such rejected figures whether it be in voices or in rays of stars. And, on the other hand, all proportions of

---

[37] Kepler (1610); Kepler (1941), p. 209.

[38] Kepler (1610); Kepler (1941), pp. 209–210.

[39] On the archetypes in his astronomy, see Martens (2000).

[40] Kepler (1596), Chap. 12; Kepler (1938), pp. 42–43; Kepler (1981), pp. 135, 137.

[41] Kepler (1602); Kepler (1941), pp. 22–35; Field (1984), pp. 250–268.

[42] Kepler (1606), Chaps. 8 and 9; Kepler (1938), pp. 184–194.

[43] Kepler (1941), pp. 201–209.

[44] Kepler (1619b), Book 4; Kepler (1940), pp. 207–286; Kepler (1997), pp. 287–385. The most complete commentary on this is in Field (1988), pp. 127–142.

voices and chords which are taken from the knowable figures[45] give in music its harmonies and in planetary rays all proportions that appear when two light rays strike together (as far as they are noted in the daily experience and recording of the weather itself in nature's drive toward violent weather), such [harmony] is also found under the knowable figures and not one under the unknowable. And so a wonderful secret follows from this, that nature is God's image and geometry is the archetype of universal beauty. So much was put into the work through creation; so much could be known in geometry through finitude and equations. And what falls outside the limits, comparison, and knowledge would also remain unformed and uncreated in the world. That which is given no special beauty or shape but of corporeality, fortune, and accident, which in themselves are boundless, would be abandoned as, for example, individual fruits and flowers are, indeed, found which have seven, nine, or eleven branches or leaves when the species commonly varies in the individuals. But no species is found which does not regularly contain this number, as five, six, four, three, ten, twelve, etc.[46]

For Kepler, here again geometry was the key for understanding all that is beautiful and true as the divine created the Universe according to knowable geometric patterns. Violation of the natural geometric proportions resulted in discord, anomaly, ugliness. Kepler believed this to be true in music; he believed it true in nature. In astrology, an aspect would produce harmony if it belonged to one of the regular polygons that can be constructed with straightedge and compasses, discord if not. This was Kepler's view of the natural world: ordered, comprehensible, reducible to a single principle, as the divine Creator's world ought to be. Fortunately for Kepler's formulation, the seven-toed ichthyostega was not discovered by the West until the twentieth century.[47]

6.6   The Uncertainty of Astrological Prediction

Despite his trust in aspects Kepler did not believe that astrology could produce certain knowledge the way astronomy could, either concerning the weather or human beings. With the weather there was variability because of geography and season. A conjunction between Mars and the Sun was supposed to cause heat, but

> in winter instead of the heat it is mild, with thunder and rain, as in December 1598 and in February 1601. In spring such a conjunction drives off what it finds, namely much still rough air, as in April 1603...[48]

And when it came to the issue of producing good wine, the importance of which to Kepler may be surmised by his writing a book in 1615 titled *The Stereometrics of*

---

[45] A knowable geometrical figure is one that can be constructed by the means allowed in Euclid's *Elements*, that is straightedge and compasses. See also Field, 'Kepler's cosmology' (in this volume, Chap. 2).

[46] Kepler (1941), p. 204.

[47] See Gould (1993).

[48] Kepler (1941), p. 188.

Wine Casks (*Stereometria dolorium vinariorum*) on how to measure the volume of wine the casks,[49] geography played a big role:

> In Italy there is good, spirited wine, for the countryside faces the midday Sun. Along the Rhine there is also much wine but gentler, for the countryside faces north and yet has deep valleys to retain the heat. Along the upper Danube there is no wine because the countryside is not protected against the harsh winds from the snowy mountains. But down below in Austria and Hungary there is good, strong wine because the land faces east and south and starts to become deep between very high mountains. The Elbe produces little wine, for the countryside faces north and is more level than other regions.[50]

Kepler's weather predictions were very general and would not be adequate in the age of hourly predictions available on the internet.

His astrological statements regarding human beings were even more general and their accuracy even more limited. He was not an astrological determinist; he did not believe that astrologers could predict future contingencies by the configurations of the heavenly bodies. 'The stars incline; they do not compel', Kepler asserted.[51] He explained that there are many matters that he took into account when trying to understand human beings and their actions. 'For example,' he wrote,

> when I see that there are many beautiful aspects in a birth chart, and it is so provided that there is no melancholy or lack of reason but rather an inner joyous nature appears, and if the person is already at an appropriate age for marriage, a bachelor, and is in a land where one does not vow eternal chastity, then I may well say on the issue of marriage if such a one will not be situated in a lowly station, so he will acquire a rich wife.

He went on to explain that he

> predicted nothing particular here, and as regards the marriage it must also remain in doubt whether or not it will take place. But my unfailing principle is to be general, that it is of a good reasonable nature that one is wise to seek. The rest that concerns such particular points is only probable.[52]

As examples of 'utterly worthless' predictions, he proffered.

> that the newly-born's wife will be born in this or that land, will have a hidden defect on the body, that she will not remain faithful to her husband, will have so and so many children, and the newly-born will have two, three, or more wives.[53]

Kepler used his own background as a particularly poignant example of the limitations of dependence on heavenly configurations:

> First, then, there was added to the aspects of the planets the daily imagination of my mother during her pregnancy, whose mother-in-law, my grandmother, an enthusiast for popular medicine, which was also practised by my father, was an object of admiration; secondly,

---

[49] On the mathematics of Kepler's treatment of this matter, and its historical significance, see Knobloch, 'Kepler's contributions to mathematics' (in this volume, Chap. 11).

[50] Kepler (1941), p. 235.

[51] Kepler (1941), p. 243.

[52] Kepler (1941), p. 232.

[53] Kepler (1941), p. 232.

there was added the fact that I was born a man, not a woman, a difference in sex which the astrologers seek in vain in the heaven. Thirdly, I take from my mother my bodily constitution, which is more suited to study than to other kinds of life. Fourthly, my parents' means were limited, that is to say there was no land for me to be born to and to cling to. Fifthly, there were schools available, there were examples available of the liberality of the magistrates to boys who were suited to study.[54]

Here Kepler describes facets of his upbringing that affected his development and had nothing to do with the sky. The belief that a mother's imagination affected the development of the foetus was commonly accepted during the Renaissance. But the fact that he was male meant that despite his mother's intelligence, he got a formal education and she did not. His family did not have the means to enable him to be a man of leisure so that he had to earn his living, but the poverty of his parents was offset by the fact that the dukes of Württemberg, the duchy where he grew up, provided scholarships to students who (like him) lacked financial means but were academically promising. And as he grew older, he continued to be affected by developments that were outside the influence of the sky:

Yet in this my stars were not Mercury as morning star in the angle of the seventh house, in quartile with Mars, but they were Copernicus, they were Tycho Brahe, without whose books of observations everything which [has] now been brought by me into the brightest daylight would lie buried in darkness; not Saturn the overlord of Mercury, but Rudolph and Matthias, each a Caesar Augustus, my overlords; not the lodging of the planets, Capricorn for Saturn, but Upper Austria the home of Caesar, and the ready liberality of its nobles, on an unusual pattern, in answer to my petition.[55]

Again, Kepler was influenced by the books he read, like Copernicus's *On the Revolutions*; the people he worked with, like Tycho Brahe, whose observations were necessary to his astronomical discoveries; and those who were his patrons, Emperors Rudolf II and Matthias and the nobles who supported him. These made possible his particular mathematical and astronomical accomplishments, not the sky.

On the other hand, Kepler pointed out that celestial occurrences could have a non-astrological effect on events on Earth. He used the example from Herodotus (*c.* 484–*c.* 425) of the battle between the Medes and the Lydians in which a solar eclipse forced them to stop fighting and make peace. Kepler explained, 'the eclipse … had not alone made the peace but only frightened the people and gave them guidance so that they would be eager for peace'.[56] Both as a principle of his astrology and as a testament to his Christian beliefs, Kepler also asserted, 'I do not mean to defend the prediction of future events that are contingent on the particular, for they depend on human free will'.[57] And he used a widespread prejudice to illustrate the futility of assuming that astrology alone can predict all human actions:

The conjunction of Saturn and the Moon should be the cause that someone is going to be cheated by a Jew. But if this conjunction takes place on the Sabbath, then no one in Prague

---

[54] Kepler (1619b), Book 4, Chap. 7, p. 170; Kepler (1940), p. 279; Kepler (1997), p. 376.

[55] Kepler (1619b); Kepler (1940), p. 280; Kepler (1997), p. 377.

[56] Kepler (1610); Kepler (1941), p. 199.

[57] Kepler (1941), p. 198.

will be cheated by any Jew, and, on the other hand, several hundred Christians will daily be cheated by Jews and *vice versa*, and yet the Moon runs below Saturn only once a month.[58]

Human free will could always intervene in human actions: an otherwise eligible man could decide not to marry; a Jew was determined not to engage in business on the Sabbath. This made prediction through astrology highly unreliable, and that lack of reliability reinforced a very real scepticism Kepler had about dependence on astrological judgements.

6.7   The Court Astrologer

As Imperial Mathematician, Kepler was court astrologer to his patron, the Holy Roman Emperor Rudolf II, Holy Roman Emperor (1552–1612, reigned from 1576). While Rudolf made Prague into an intellectual and cultural centre and promoted religious peace in a period of great religious tensions, he was weak politically and militarily, and his brother Matthias, who had been encroaching on Rudolf's territory, sought to usurp the imperial throne as well. In 1611 agents of Matthias approached Kepler and requested that he write a bogus horoscope predicting Rudolf's fall. The assumption was that if Rudolf received such a prediction from Kepler, he would take it as an absolute prediction, that he would lose and would simply let Matthias take over without a fight. Kepler described this situation to one of the emperor's advisers and concluded with the recommendation,

> I am of the opinion that astrology has to be withdrawn not only from the Senate but also from the heads of those who want to advise the Emperor today to the best of their abilities; one must keep astrology entirely from the emperor's mind.[59]

Court astrologers were common among the Renaissance rulers as advisers, and Kepler fulfilled this function as well, but he knew that astrology had its limitations as a source of knowledge, and it could lead to disaster if not handled properly. Astrologers were not any less prone to the influence of corruption than any other type of government functionary.

6.8   Qualities of Individual Planets

We have noted that among his astrological principles Kepler rejected the images of the zodiac. He complained that the images were created within the human imagination and did not exist in reality. Kepler likewise rejected the belief that the planets affect the Earth or its inhabitants because of personal qualities inherent in the planets; for

---

[58] Kepler (1941), p. 163.

[59] Kepler to an Anonymous Nobleman, 3 April 1611, in Kepler (1954), letter 612, p. 375, lines 79–81; Baumgardt (1951), pp. 99–100.

example, he did not accept that Mars could make someone aggressive because Mars was an aggressive planet. Just as the claim that the images of the zodiac provided similar personality characteristics to the human being, the idea that such qualities could come from the planets could also be considered examples of sympathetic magic. Instead, Kepler asserted that the effects of the individual planet resulted from its relationship to the Sun, which provided heat, and the Moon, which provided moisture. He then assigned varying degrees of heat and moisture to each of the planets; each could be excessive, average, or deficient in heat and moisture.[60] Kepler denied that planets could be inherently good or bad, but from these physical characteristics good or harm could come to the human being. Mars and Saturn involved excess, which could cause harm; Jupiter involved temperance, which could bring good.[61] Not only did Kepler reinforce a physical conception of the influence of the planets, but as a good Christian, this idea counteracted the accusation that astrology was inherently pagan because of its association with pagan mythology. As far back as the fifth century, Saint Augustine rejected the study of astronomy because of pagan associations.[62]

6.9   Progressions and Wallenstein

Kepler also accepted the theory of progressions, by which events in a particular year could be foretold from the degree to which a planet had advanced since the nativity: 'The doctrine of progressions will earn fine consideration from me', he wrote in *Third Man in the Middle*,

> If I would allow, with Copernicus, that the Earth revolves, then the proportion naturally embedded between a day and the year turns out to be one to 365, whether we will be carried around in the Universe with a field, a house, or a ship for our dwelling. And it is, therefore, more believable that in progressions and the nativities of human beings who are the inhabitants of this ship, this proportion should also rule.[63]

Kepler found in the doctrine of progressions a natural geometric proportion that did not exist in other astrological doctrines. The year was a virtual circle and each day a degree in that circle. Once again, a basis in geometry helped Kepler decide which theories to accept and which to reject. And he also liked the doctrine of progressions because he saw it as amenable to the Copernican system of a rotating Earth.

The doctrine of progressions was what made possible the predictions in Kepler's codicil to the famous horoscope for Count Albrecht von Wallenstein. In 1608 Kepler had been asked by an intermediary to chart the nativity of an unnamed Bohemian lord.

---

[60] Kepler (1602), Thesis 24; Kepler (1941), pp. 1617; Field (1984), pp. 239–242; Kepler (1610); Kepler (1941), pp. 172–175.

[61] Kepler (1610); Kepler (1941), p. 176.

[62] Augustine (1958), pp. 65–66.

[63] Kepler (1610); Kepler (1941), p. 185; see also Kepler (1619b), Book IV, Chap. 7; Kepler (1940), pp. 284–285; Kepler (1997) pp. 383–384.

It appears, however, that Kepler did, in fact, know that his client was Wallenstein, for Kepler wrote his name in code on the document.[64] Martha List has suggested that the reason for this secrecy was that 'Wallenstein, in fact, did not want his plans, which were already ambitious at that time, to be undermined by his rivals through knowledge of his horoscope'.[65] Wallenstein was constantly seeking astrologers' advice, and he would not have wanted his enemies to have access to their findings about him. Kepler's interpretation of the birth chart was quite revealing of the count's character:

> Thus may I in truth say about the lord, that he has an alert, excited, industrious, restless temperament, eager for all kinds of novelties, not liking common human pursuits but seeking new, untried, solitary paths, yet for all that has much more in his thoughts than he lets outwardly be seen or felt. Saturn on the ascendant makes for deep, melancholic, constantly alert thoughts, alchemy, magic, sorcery, communion with spirits, scorn and disregard for human law and custom, also all religions, makes everything suspicious and distrustful that God or men do as if it were pure fraud and it were underneath much different from what one pretends.[66]

Kepler met Wallenstein again in 1624. The count offered him a position as astrologer in his fief of Sagan in Silesia and requested that Kepler further elaborate the horoscope. Kepler used progressions to extend the chart ten years, to 1634, when Kepler foresaw 'terrifying chaos in the land' with respect to Wallenstein in March of that year.[67] Wallenstein, the imperial general who apparently was engaged in secret negotiations with the enemy, was assassinated by imperial agents on 25 February 1634.

It must be noted that Kepler's attitude toward astrological prediction would make it difficult to assume that Kepler was, in fact, predicting Wallenstein's death. As we have seen, Kepler eschewed such exact predictions, and given that it was the middle of the Thirty Years War and Wallenstein was a major player in that war, he was continually faced with 'terrifying chaos'. Furthermore, Kepler did not end his prediction with 1634 because he foresaw Wallenstein's death; he ended with 1634 because his task was to provide predictions for ten years.

6.10   Heliocentrism and Astrology

Kepler's acceptance of the heliocentric system was in no way an argument against acceptance of astrology. Astrology centres on the location of its subjects. If the subjects were on the Earth, then the astrologer would only be concerned with how the celestial configurations appear from the Earth, and astrology would appear to be geocentric; if those subjects were on the Sun, then the astrologer would be concerned with how the celestial configurations appear from the Sun, and astrology would

---

[64] List (1971), p. 130.

[65] List (1971), p. 13.

[66] Kepler (2009), pp. 449–450.

[67] Kepler (2009) pp. 469–470.

appear to be heliocentric. More important were the physical inferences from the Copernican system. This adoption of a 'unified physics' was far more of an obstacle to his acceptance of astrology.[68] Kepler rejected the distinction between a physical sublunar world and a non-physical, immutable heaven that moved the lower world, on which traditional astrology had been based. Thus, he also rejected elements of sympathetic magic in astrology and justified his acceptance of the effects of planetary configurations with physical explanations.

Moreover, astrology had to fit his geometric conception of the universe. 'Geometry', he declared in *The Harmony of the World*, 'which before the origin of things was coeternal with the divine mind and is God himself (for what could there be in God which would not be God himself?), supplied God with patterns for the creation of the world'.[69] Just as geometrical archetypes were crucial to his astronomy, so they were the foundation of his astrology. Otherwise astrology could not be effective in the divinely-created Universe.

Astrological aspects did not fit into Kepler's harmonic conception of the Universe as he would have liked, because the last necessary element in the formation of his ideas, observation, showed imperfect agreement between musical intervals and astrological aspects.[70] Nevertheless, Kepler's reformed astrology fitted into his worldview; he accepted those elements used for astrology that were compatible with his idea of a physical Universe. Aspects were based in geometry, and they were observable and measurable.

As an astronomer trained in the sixteenth century, Kepler was taught to accept astrology as the 'practical' side of the study of the heavens and a means of understanding the divine plan, with the same validity as astronomy, and his practice of astrology reinforced his acceptance of its place in that study. But he could not integrate traditional astrology with his Copernican astronomy, particularly with his concept of celestial physics, and so he set out to reform astrology. He rejected many traditional astrological ideas, including the belief that the images of the zodiac are formed by nature and effect changes. In keeping with his idea of the geometrical divine plan, Kepler emphasized aspects, the element of astrology that not only occurs in nature but is also measurable. Thus, Kepler tried to make his astrology consistent with his astronomy.

---

[68] On the physical issues, see also Simon (1979), pp. 42–43.

[69] Kepler (1619b), Book 4, Chap. 1, p. 119; Kepler (1940), p. 223; Kepler (1997), p. 304.

[70] Kepler (1610); Kepler (1941), p. 205.

References

Agrippa, H. C. (2005). In transl. anonymous, Whitefish, MT.

Akopyan, O. (2020). *Debating the stars in the Italian Renaissance: Giovanni Pico della Mirandola's Disputationes Adversus Astrologiam Divinatricem and its reception*. Brill's studies in intellectual history (Vol. 325). Leiden: Brill.

Augustine. (1958). *On Christian doctrine* (Translated from the Latin by D. W. Robertson, June). New York: Macmillan.

Azzolini, M. (2013). *The Duke and the stars: Astrology and politics in Renaissance Milan*. Cambridge, MA: Harvard University Press.

Barnes, R. B. (2016). *Astrology and Reformation*. Oxford: Oxford University Press.

Bauer, K. (2015). Johannes Kepler between two emperors. *Astrologers and their clients in Medieval and Early Modern Europe* (Deimann & Juste, Ed., pp. 205–219). Cologne: Bohlau.

Baumgardt, C. (Ed. & Trans.). (1951). *Johannes Kepler: Life and letters*. London: Victor Gollancz.

Bellanti, L. (1553). *De Astrologica veritate* (3rd ed.). Basel: Jakob Kundig.

Boner, P. J. (2005). Soul-searching with Kepler: An analysis of anima in his astrology. *Journal for the History of Astronomy, 36*, 7–20.

Boner, P. J. (2013). *Kepler's cosmological synthesis: Astrology, mechanism and the soul*. Leiden: Brill.

Brosseder, C. (2004). *Im Bann der Sterne: Caspar Peucer, Philipp Melanchthon und andere Wittenberger Astrologen*. Berlin: De Gruyter.

Campion, N. (2008–2009). *A cultural history of western astrology* (2 Vols.). London: Hambledon Press.

Deimann, W., & Juste, D. (Eds.). (2015). *Astrologers and their clients in Medieval and Early Modern Europe*. Cologne: Bohlau Verlag.

Dooley, B. (Ed.). (2014). *A companion to astrology in the Renaissance*. Leiden: Brill.

Field, J. V. (1984). A Lutheran astrologer: Johannes Kepler. *Archive for History of Exact Sciences, 31*, 189–272.

Field, J. V. (1988). *Kepler's geometrical cosmology*. London and Chicago: The Athlone Press and Chicago University Press.

Garin, E. (1983). *Astrology in the Renaissance: The zodiac of life* (Translated by C. Jackson, J. Allen, C. Robertson & E. Garin). London and Boston: Routledge & Kegan Paul.

Gould, S. J. (1993). Eight little piggies. In *Eight little piggies: Reflections in natural history* (pp. 63–78). New York: Vintage.

Greenbaum, D. G. (2015). Kepler's personal astrology: Two letters to Michael Maestlin. In C. Burnett & D.G. Greenbaum (Eds.), *From Masah'allah to Kepler: Theory and practice in Medieval and Renaissance astrology*. Ceredigion, Wales: Sophia Centre Press.

Grendler, P. F. (2002). *The universities of the Italian Renaissance*. Baltimore: Johns Hopkins University Press.

Hayton, D. (2015). *The crown and the cosmos: Astrology and the politics of Maximilian I*. Pittsburgh: University of Pittsburgh Press.

Isidore of Seville. (2006). *Etymologies* (Translated from the Greek by Stephen A. Barney, W. J. Lewis, J. A. Beach, & O. Berghof). Cambridge: Cambridge University Press.

Kepler, J. (1596). *Mysterium cosmographicum*. Tübingen: Georg Gruppenbach.

Kepler, J. (1597). *Screibkalender und Practica auf 1598, 1597*. Graz: Hansl Schmidt.

Kepler, J. (1602). *De fundamentis astrologiae certioribus*. Prague: Schuman Press.

Kepler, J. (1606). *De stella nova*. Prague: Paulus Sessius.

Kepler, J. (1610). *Tertius interveniens*. Frankfurt am Main: Georg Tampach.

Kepler, J. (1619a) *De cometis*. Augsburg: Andreas Apergen.

Kepler, J. (1619b). *Harmonice mundi*. Linz: Joannes Plank.

Kepler, J. (1938). *Johannes Kepler Gesammelte Werke. Band I: Mysterium cosmographicum / De stella nova* (M. Caspar, Ed.). Munich: C. H. Beck.

Kepler, J. (1940). *Johannes Kepler Gesammelte Werke. Band VI: Harmonice mundi* (M. Caspar, Ed.). Munich: C. H. Beck.

Kepler, J. (1941). *Johannes Kepler Gesammelte Werke. Band IV: Kleinere Schriften 1602–1611 / Dioptrice* (M. Caspar & F. Hammer, Eds.). Munich: C. H. Beck.

Kepler, J. (1945). *Johannes Kepler Gesammelte Werke. Band XIII: Briefe 1590–1599* (M. Caspar, Ed.). Munich: C. H. Beck.

Kepler, J. (1954). *Johannes Kepler Gesammelte Werke. Band XVI: Briefe 1607–1611* (M. Caspar, Ed.). Munich: C. H. Beck.

Kepler, J. (1955). *Johannes Kepler Gesammelte Werke. Band XVII: Briefe 1612–1620* (M. Caspar, Ed.). Munich: C. H. Beck.

Kepler, J. (1963). *Johannes Kepler Gesammelte Werke. Band VIII: Mysterium cosmographicum (editio altera cum notis) / De cometis / Hyperaspides* (F. Hammer, Ed.). Munich: C. H. Beck.

Kepler, J. (1981). *Mysterium cosmographicum: The Secret of the Universe* (Translated from the Latin by A. M. Duncan). New York: Abaris Books.

Kepler, J. (1993). *Johannes Kepler Gesammelte Werke. Band XI.II: Calendaria et Prognostica / Astronomica minoria / Somnium* (V. Bialas & H. Grössing, Eds.). Munich: C. H. Beck.

Kepler, J. (1997). *Memoirs of the American Philosophical Society, Vol. 209: Harmony of the world* (Translated from the Latin with and Introduction and Notes by E. J. Aiton, A. M. Duncan, & J. V. Field). Philadelphia: American Philosophical Society.

Kepler, J. (2009). *Johannes Kepler Gesammelte Werke. Band XXI.II:II: Manuscripta astrologica / Manuscripta pneumatica* (F. Boockmann et al., Eds.). Munich: C. H. Beck.

Kieckhefer, R. (1989). *Magic in the Middle Ages*. Cambridge: Cambridge University Press.

Kusukawa, S. (1995). *The transformation of natural philosophy: The case of Philip Melanchthon*. Cambridge: Cambridge University Press.

Lemay, R. (1987). The true place of astrology in Medieval science and philosophy: Towards a definition. In P. Curry (Ed.), *Astrology, Science and Society* (pp. 57–73). Woodbridge: The Boydell Press.

List, M. (1971). Das Wallenstein Horoskop von Johannes Kepler. Zur Geschichte seiner Entstehung. In G. Maar (Ed.), *Johannes Kepler—Werk und Leistung* (p. 130). Linz: Gutenberg.

Martens, R. (2000). *Kepler's philosophy and the New Astronomy*. Princeton: Princeton University Press.

Methuen, C. (1998). *Kepler's Tübingen: Stimulus to a theological mathematics*. Aldershot: Routledge.

Moran, B. T. (2006). Courts and academies. In K. Park & L. Daston (Eds.), *The Cambridge history of science, vol. 3, Early modern science* (pp. 253–263). Cambridge: Cambridge University Press.

North, J. D. (1986). *Horoscopes and history*. london: Warburg Institute.

Oresme, N. (1952). Tractatus contra astronomos. In G. W. Coopland (Ed.), *Nicole Oresme and the astrologers: A study of his Livre de divinacions*. Liverpool: Liverpool University Press.

Pico, G. (1946, 1952). *Disputationes adversus astrologiam divinitricem* (E. Garin, Ed., 2 Vols.). Florence: Vallecchi Editore.

Ptolemy, C. (1980). *Tetrabiblos* (Translated from the Greek by F. E. Robbins). Loeb Classical Library 435. New York and London: Heinemann.

Rabin, S. (1997). Kepler's attitude toward Pico and the anti-astrology polemic. *Renaissance Quarterly, 50*, 750–770.

Rosen, E. (1984). Kepler's attitude toward astrology and mysticism. In B. Vickers (Ed.), *Occult and scientific mentalities in the Renaissance* (pp. 253–272). Cambridge: Cambridge University Press.

Rutkin, H. D. (2019). *Sapientia astrologica: Astrology, magic and natural knowledge, ca. 1250–1800, vol. 1, Medieval structures (1250–1500): Conceptual, Institutional, Socio-Political, Theologico-Religious and Cultural*.

Rutkin, H. D. (2006). Astrology. In K. Park & L. Daston (Eds.), *The Cambridge history of science. Vol. 3: Early modern science* (pp. 541–561). Cambridge: Cambridge University Press.

Simon, G. (1979). *Kepler: Astronome astrologue*. Paris: Gallimard.

Siraisi, N. G. (1990). *Medieval and Early Renaissance medicine: An introduction to knowledge and practice*. Chicago: Chicago University Press.

Tester, S. J. (1987). *A history of western astrology*. Woodbridge: The Boydell Press.

Thorndike, L. (1923–1958). *A history of magic and experimental science* (8 Vols.). New York: Cambridge University Press.

Westman, R. S. (2011). *The Copernican question: Prognostication, skepticism, and celestial Order*. Berkeley: University of California Press.

Chapter 7
Kepler's Work on Optics

W. H. Donahue

Kepler's work in optics was extraordinarily far-reaching, rising from its foundations in the role of light in God's creation and extending to the geometrical theory of light rays, the physical theory of reflection and refraction (including the focal properties of lenses and mirrors and the the the design of telescopes), the functioning of the eye in vision, the nature of the light of the stars, Sun, Moon, and planets, and the techniques to be used in the observation of eclipses, especially solar eclipses. In the course of these widely varied investigations, he also made excursions into a critique of Aristotle's account of light and vision, a debate on how levers work, the principle of the balance (expressed dynamically), a novel and succinct exposition of the theory of conic sections, a provocative argment that Euclid's optical treatise shows that Euclid was a heliocentrist, a historical account of all the known records of solar eclipses, and a compendium of pranks one can play with telescopes.

Despite this remarkable breadth of inquiry, Kepler's interest in optics sprang from particular problems that arose in his astronomical work, specifically in his early observations of solar eclipses. He was fortunate in having studied astronomy with Michael Maestlin (1550–1631) at the University of Tübingen. Maestlin was one of the few astronomers of that time who fully accepted the Sun-centred astronomy of Nicolaus Copernicus (1473–1543), and he was also notably skilled in observation, unusual among university professors, most of whom limited their interest to models of planetary motion and the casting of horoscopes for medical diagnosis. Kepler reports that, while he was an undergraduate, Maestlin had invited him to observe and measure the time and extent of a solar eclipse,[1] showing how to project the sunlight through a small opening (which might just be a chink in a roof) to obtain an image of the eclipsed Sun.

---

[1] Eclipse of 21/31 July 1590; see Kepler (1604), p. 396; Kepler (2000), p. 399.

W. H. Donahue (✉)
St. John's College, Santa Fe, NM, USA
e-mail: william.donahue@sjc.edu

© Springer Nature B.V. 2024
A. E. L. Davis et al. (eds.), *Reading the Mind of God*, Springer Praxis Books,
https://doi.org/10.1007/978-94-024-2250-4_7

After Kepler took up a teaching position in Graz, in Austria, Maestlin kept him informed about astronomical matters and, in particular, told him of a letter he had received from the Danish astronomer Tycho Brahe (1546–1601) regarding the solar eclipse of 1598.[2] Brahe said that his observations showed that the apparent diameter of the Moon, when it eclipsed the Sun, was smaller than it appeared at similar positions in the sky outside of eclipses. This, he believed, could not be the result of the Moon adopting a greater distance from the Earth in solar eclipses: there was no other evidence of such a gratuitous jump. He therefore concluded that the Moon must become smaller, at least in optical terms, when interposed between the Sun and Earth.[3] Brahe went so far as to assert that there had never been, and could never be, a total eclipse of the Sun. Initially, after reading of Brahe's observations and conclusion, Kepler thought it possible that the Moon has its own atmosphere that reflects light when the Moon is full (adding to its apparent diameter), but allows light to pass during solar eclipses (resulting in its showing a smaller diameter). However, in view of the many historical reports and observations of total solar eclipses, Kepler doubted Brahe's conclusions, particularly relating to totality of eclipses. This led him to begin a study of ray optics, initially through a reading of the *Perspectiva Communis* of John Peckham, Archbishop of Canterbury (c. 1230–1292),[4] though he also consulted the pseudo-Aristotelian *Problems*.[5]

Apparently, at this time (1590), Kepler had not yet seen the treatise on optics by Ibn al-Haytham (c. 965–1040), known in the west as Alhacen or Alhazen. This massive work, printed in Latin translation in 1572, is remarkable for its methodical, experimental approach, and for the strict application of the principle of the rectilinear propagation of light in uniform transparent media.[6] However, Peckham was familiar with the book,[7] through a Latin translation of the late twelfth or early thirteenth century, and Alhazen's optical work helped form the ray-tracing development of perspective drawing by Piero della Francesca (c. 1412–1492) and Albrecht Dürer (1471–1528). Kepler would later make a careful study of Alhazen and frequently refers to the book in his *Optics*. He had also read Dürer's *Underweysung* (see below and Fig. 7.1), which provided an important clue, although it was not an optical treatise. However, at the time when he was struggling to understand the behaviour of light passing through apertures, he had to depend mainly upon Peckham.

---

[2] Brahe to Maestlin, 21 April 1598 (O.S.). The complete letter is published in Brahe (1925), pp. 52–55; also Brahe (1858). An extract is published in Brahe (1945).

[3] 'This, however, must be noted: that the Moon, at new moons belonging to eclipses, does not appear with that magnitude which it otherwise has at full moons, although it is at the same distance from the earth: it is, as it were, squeezed in by about a fifth part, owing to certain causes that will be discussed elsewhere.' Tycho Brahe to Michael Maestlin, 21 April 1598 (O.S.), Brahe (1925), p. 55 ll. 9–12; also in Brahe (1858), p. 46.

[4] The *Perspectiva* was first published in Milan in 1482 or 1483. See Lindberg (1970), pp. 56–57.

[5] See Kepler (1604), p. 38; Kepler (1939), p. 47; Kepler (2000), pp. 55–56.

[6] For Ibn al-Haytham's theory of light, see Sabra (A. I. Sabra, 'Ibn al-Haytham', in Gillespie, C. (1970–1980), vol. 6, pp. 191–192; reprinted in Sabra (1994), Sect. II.

[7] Lindberg (1970), p. 20.

Fig. 7.1 Dürer engraving of ray-tracing. Reproduced from *Underweysung der Messung*, 1525, p. 181

Kepler found Peckham's account confused and his two-dimensional diagram unhelpful. But, crucially, Kepler had also studied Dürer's methods of tracking visual rays in making perspective drawings, using threads to represent rays.[8] In pondering what Peckham had written, Kepler realized that the rays of light from luminous bodies could likewise be traced using threads. He wrote,

> Since I was unable to understand the very obscure sense of the words from a diagram drawn in a plane, I had recourse to seeing with my own eyes in space. I set a book in a high place, which was to stand for a luminous body. Between this and the pavement a tablet with a polygonal hole was set up. Next, a thread was sent down from one corner of the book through the hole to the pavement, falling upon the pavement in such a way as to graze the edges of the hole, the image of which I traced with chalk. In this way a figure was created upon the pavement similar to the hole. The same thing occurred when an additional thread was added from the second, third, and fourth corner of the book, as well as from the infinite points of the edges. In this way, a narrow row of infinite figures of the hole outlined the large quadrangular figure of the book on the pavement.

Kepler then extends the proof to apply to a luminous surface:

> It was thus obvious that this was in agreement with the demonstration of the problem, that the round shape is not that of the visual ray but of the sun itself, not because this is the most perfect shape, but because this is generally the shape of a luminous body. This is the first success in this work.[9]

---

[8] Dürer (1525), esp. p. 181.

[9] Kepler (1604), Chap. 2, p. 39; Kepler (1939), pp. 47–48; Kepler (2000), p. 56.

And what a success! In this one 'magnificent account', as historian Stephen Straker aptly described it,[10] Kepler realized three things important for understanding light and the formation of images. First, when undisturbed by changes of medium and by obstructions, light really does travel in straight lines, just as ancient geometrical optics states.[11] Second, every one of the infinite number of rays of light passing through an opening (or, as he would later write, a lens) participates equally in forming an *image*, shaped jointly by the form of the light source and the form of the opening or lens, in a geometrically precise and determinate way. Third, the light source is fully and accurately understood as an array of single luminous points, each of which acts as the origin of a single ray of light. Kepler achieved this conceptualization through a consistent application of linear transmission of light together with his adoption of Dürer's treatment of artistic perspective, replacing the eye in Dürer's model with one of the innumerable points of the luminous body.

The procedure that was first outlined in this way as an experience, Kepler sets forth more formally in the opening series of propositions in Chap. 2. Proposition 1 posits a single luminous point shining light through a small window, and shows that the rays from that point form an image on the opposite wall that is similar to the shape of the window. In Proposition 2, Kepler imagines the luminous point of Proposition 1 to be at 'an incalculably great distance', and shows that the image will be quantitatively the same as the window. Proposition 3 inverts the situation, now supposing a luminous surface of determinate shape, while the window has shrunk to a single point. In this case, a low point on the luminous body sends a ray through the tiny window and illuminates a more elevated point on the wall, while a ray from a higher point illuminates a lower point on the wall (see Fig. 7.2).

In Propositions 4 and 5, Kepler combines the conclusions of the first three propositions. Proposition 4 begins by supposing a single luminous point:

> For if, on the one hand, we pretend that it is a single point that shines, the rays transmitted through the boundaries of the window, since they meet at their origin, are proportionally farther apart as they go forth, and thus take up more space on a more remote wall than they do at the closer window, by Prop. 1 of this chapter.

Kepler then extends the proof to apply to a luminous body (see Fig. 7.3):

> Let *PNQ* be the luminous surface, whose center is *E*, and let *FGHO* be the window. Thus ... the center *E* of the luminous surface [since it is a luminous point] will create the figure *IKLM* on the wall, similar to the window *FGHO*, [but not smaller].[12] Now ... through the individual points of the window, individual inverted images of the luminous surface are transmitted, such as you see at *M* ..., transmitted through point *O* .... And since *EOM* is the ray from the center of the luminous body, and the middle of all those that intersect each other at the point *O*, the remaining ones are either beyond or this side of it, and the one that descends from the point *Q*, which is on the inside with respect to the window, is now made to be on the outside

---

[10] Straker (1970), p. 391.

[11] The philosophical side of ancient optics (Aristotle's view, in particular) contrasted markedly with geometrical optics. Kepler criticizes Aristotle's light theory in Appendix to Chap. 1 of Kepler (1604), pp. 29–37; Kepler (2000), pp. 43–54.

[12] The bracketed words are editorial, either clarifying or summarizing points in Kepler's text.

Fig. 7.2 Projection through
a small aperture. Reproduced
from *Optics*, Chap. 2 Prop. 3
diagram, ed. 1604, p. 43

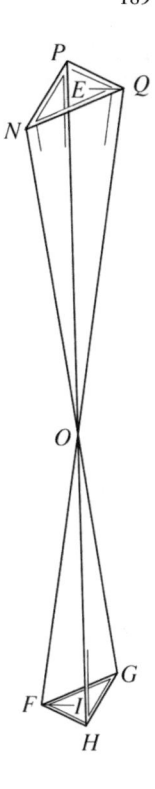

by the intersection [of rays] that takes place at *O*. The same description can be applied to all
the points. In this manner, a perimeter will be created that is greater than *IKLM*.

Kepler soon got a chance to test this new understanding of the formation of
images behind openings. On his return to Graz from Prague, in late June of 1600, he
set up a large pinhole-type instrument in the town square; he had previously built the
instrument for this purpose (see Fig. 7.4).[13] A solar eclipse was predicted for June
30/July 10 1600, and he hoped that his measurements of the Sun's apparent diameter
and of the size of the eclipsed portion would vindicate his view that the anomaly that
had troubled Brahe was the effect of projecting the eclipse through a small opening,
which had the effect of adding an illuminated region (whose breadth was equal to the
radius of the opening) to the edges of the Sun's projected image. His measurements
(later confirmed by other solar observations) were consistent with his optical theory,[14]
and in the enthusiasm that followed upon this success, he spent the next few weeks
writing the series of proofs described above that later was incorporated into the *Optics*
as Chap. 2. In a letter to Maestlin (September 9, 1600) he wrote,

---

[13] The instrument and its use are described at the beginning of Chap. 11 of Kepler (1604).

[14] See Kepler's presentation of his observations in Kepler (1604), Chap. 11, pp. 422–430; Kepler
(2000), pp. 423–429.

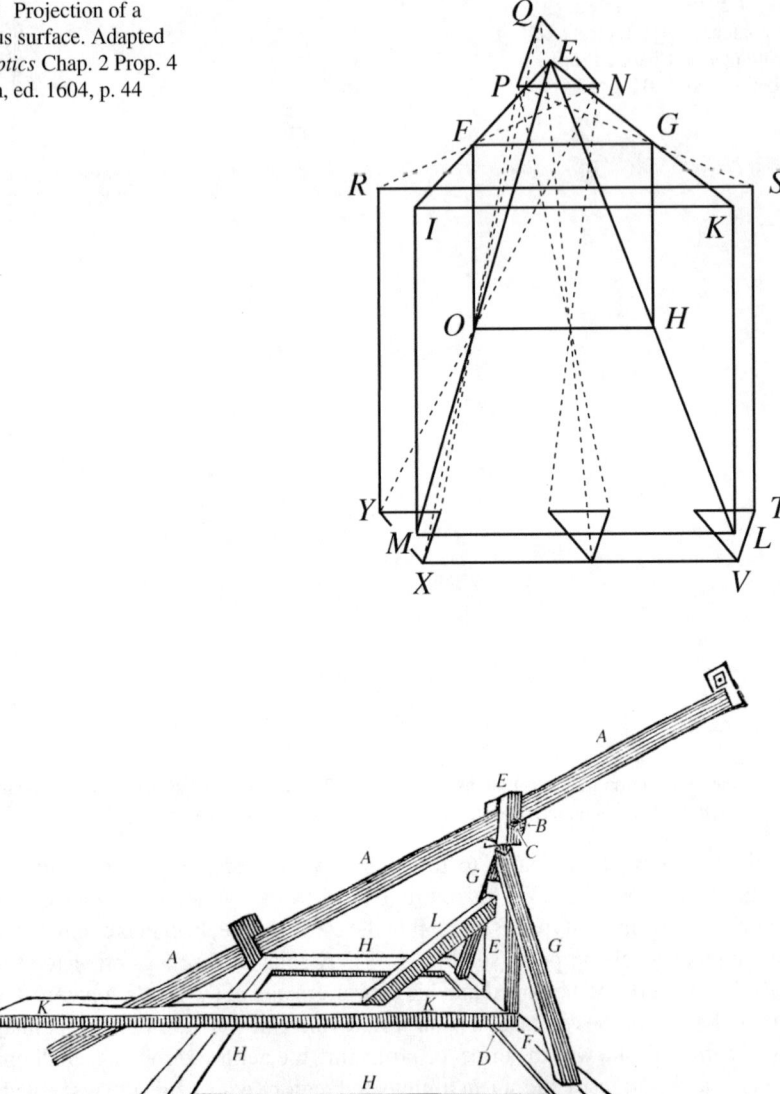

Fig. 7.3 Projection of a luminous surface. Adapted from *Optics* Chap. 2 Prop. 4 diagram, ed. 1604, p. 44

Fig. 7.4 Eclipse instrument. Reproduced from *Optics* Chap. 11, ed. 1604, p. 338

It was indeed a costly eclipse, and one nonetheless from which I learned the cause of the Moon's displaying such a small diameter at new Moon on the ecliptic. As a consequence, during the rest of the month of July I wrote "Paralipomena" to Book II of Witelo's *Optics.*[15]

---

[15] Kepler to Maestlin, 9 September, 1600 (O.S.), Kepler (1949), p. 150.

Kepler's words here make it clear that from the beginning he intended to characterize his contribution to optical theory with a word derived from ancient Greek. *Paralipomena* literally means 'things left along the side': evidently, he intended to situate his theory in the tradition of ray optics begun by Euclid and Ptolemy, while acknowledging the advances made by the thirteenth century Silesian optical writer Witelo, at the same time implying that he, Kepler, has added some things that Witelo (and his Arabic source) had 'left on the side' (or omitted).

The manuscript was ready for a prospective printer by 16 December 1600. And here it was itself left on the side, owing to unfavourable circumstances: it was becoming increasingly clear that Kepler, a Lutheran, could not continue to live in Catholic Graz.[16] The question of where to go was rather neatly solved by an invitation from Tycho Brahe to become part of his team of assistants. The invitation came more as a consequence of Brahe's wish to have Kepler on his side (and under his thumb) in a plagiarism quarrel than as an acknowledgement of Kepler's abilities[17]; however, it had profound consequences for Kepler's career and for the future of his optical work.[18]

The story of Kepler's moving his household to Prague and his rocky relationship with Brahe has little direct bearing on Kepler's optical work. But two events brought Kepler's attention back to his shelved *Paralipomena*. First, Brahe's primary assistant, Longomontanus (Christian Severin, 1562–1647), had left Prague permanently in August 1600,[19] and second, Brahe died suddenly in October 1601. Kepler was, largely by default, Brahe's evident successor as Imperial Mathematician. He was officially appointed two days later.

This abrupt change of fortune presented a problem for Kepler. As Imperial Mathematician, he was expected to produce (in addition to calendars and prognostica) major works that would do credit to his patron. Since Kepler was, at the time, working on a new theory of the motion of Mars, his *Commentaries on Mars* would seem to be an obvious choice. However, in early April 1602, in a brilliant two-page analysis of Mars' positions compared with the positions required by a circular orbit, Kepler concluded that the orbit could not be circular, but had to be, in Kepler's words, squeezed inwards slightly at the sides.[20] At this point, he realized that he knew hardly anything about the exact shape of the orbit, or how it could be generated. Since he had already announced that he would publish his optical researches by the following Christmas, leaving the Mars book for the following Easter,[21] he set the Mars work aside and concentrated primarily on the *Optics*.

---

[16] See Charlotte Methuen, 'Kepler, religion and natural philosophy: a theological biography' (this volume, Chap. 1).

[17] The tangled sequence of events leading to Brahe's invitation to Kepler and Kepler's arrival in Prague is thoroughly presented by Thoren (1990), pp. 432–438.

[18] Brahe also corresponded with Maestlin about Kepler. For details see Mahoney, 'Measuring the heavens' (this volume, Chap. 3).

[19] Christianson (2000), p. 316.

[20] Donahue (1996).

[21] Letter to Longomontanus, early 1605; letter no. 323 in Kepler (1951), p. 140.

Inevitably, the project turned out not to be as simple as it had seemed at first. As Kepler wrote in his letter of dedication to the Emperor Rudolf II, he thought a book on astronomical optics had to contain an account of atmospheric refraction. But that required an understanding of refraction of light in general, which in turn demanded a study of the nature of light. He was still working on these problems in the summer of 1603, as he wrote to his friend and patron Herwart von Hohenburg (1522–1611): 'Measuring refractions: here I got stuck. Good God! What a hidden ratio! All the *Conics* of Apollonius had to be devoured first, a job which I have now nearly finished.'[22] The completed manuscript was presented to the Emperor as a New Year's gift in January 1604. Delays in getting the manuscript back, so he could send it to the printer, and delays in the printing itself, further bedevilled the publishing process. The finished book was finally ready for sale in time for the Frankfurt Book Fair in the autumn of 1604, with many appended endnotes and a lengthy errata sheet.

However, as Kepler himself makes clear in the introductory chapter of the *Optics*, he believed light—the result of God's first creative act—to be fundamental to the structure and drama of the Universe. Here we see Kepler's Christian faith and his training for ministry in the Lutheran Church. The understanding of light, both physical and divine, was for Kepler an essential assignment that God had given to human beings. All of Kepler's works may be viewed in one way or another as expressions of this directive. Indeed, it may not be wide of the mark to see Kepler's example of the luminous book shining through a small window (in Chap. 2 of the *Optics*)[23] as a metaphor for Biblical illumination interpreted by limited human understanding. So we should step back, both literally and figuratively, to *Optics* Chap. 1. Here Kepler sketches out the origin of the Universe as an image of the Trinity, in the form of an immense sphere. Continuing, he writes,

> This, then, is the authentic, this is the most fitting, image of the corporeal world, which anything that aspires to the highest perfection among corporeal created things takes on, either simply or in some respect. The bodies themselves were confined separately within the limits of their surfaces and could not by themselves have multiplied themselves into an orb. For this reason, they were endowed with various powers, which, though they do have their nests in the bodies, nevertheless, being somewhat freer than the bodies themselves and lacking corporeal matter (though they do consist of their own kind of matter which is subject to geometrical dimensions), may proceed forth and might try to achieve an orb, as appears chiefly in the magnet, but shows plainly in many other instances. What wonder, then, if that principle of all adornment in the world, which the divine Moses introduced immediately on the first day into barely created matter, as a sort of instrument of the Creator, for giving form and growth to everything—if, I say, this principle, the most excellent thing in the whole corporeal world, the matrix of the animate faculties, and the chain linking the corporeal and spiritual world, has passed over into the same laws by which the world was to be furnished. The Sun is accordingly a particular body, in it is this faculty of communicating itself to all things, which we call light; to which, on this account at least, is due the middle place in the whole world, and the centre, so that it might perpetually pour itself forth equably into the whole orb. All other things that have a share in light imitate the Sun. From this consideration

---

[22] Kepler to Herwart von Hohenburg, May 1603, Kepler (1949), p. 396.

[23] Kepler (1604), Chap. 2, p. 39; Kepler (2000), p. 56.

there arise, in a way, certain propositions, which are among the principles in Euclid, Witelo, and others.[24]

This final sentence, which may strike us as merely a transition to the series of propositions that follow, should be understood instead as a challenge to the students of Nature (often followers of one or another sect of Aristotelian philosophy) to take seriously, as physical explanations, the mathematical ray optics of Euclid, Ptolemy, Alhazen, Witelo, and others. This bold challenge is of a piece with Kepler's insistence that mathematical astronomy and philosophical speculation about the heavens must be brought together into a coherent science, an explanation that makes physical sense and that also produces a true and accurate mathematical account of the positions of real celestial bodies.[25]

The same striving for reconciliation is evident in Kepler's optical theory, perhaps most vividly in his physico-mathematical account of refraction. Chapter 4 of the *Optics* is devoted to determining the mathematical law of refractions; however, the physical foundation for this, as well as many other properties of light, is laid in a series of propositions in Chap. 1, which is title 'On the Nature of Light'. And in general, as in other Keplerian works, ideas tend to overflow the topical boundaries within which they originate. Accordingly, to avoid confusion, or at least to attenuate it, it may be helpful to give a systematic summary of the book, chapter by chapter.

7.1 The Overall Structure of Kepler's *Optics* of 1604

The account of this work presented so far has traced the particular origins and remarkable development of Kepler's investigation of geometrical optics and its application to small 'windows', especially in the observation of solar eclipses.

The *Optics* as a whole divides neatly into two parts, as its full title (*Paralipomena to Witelo, by Means of which the Optical Part of Astronomy is presented*) indeed suggests.[26] The first part, the 'Paralipomena to Witelo', consists of five chapters on light and vision, while the second, the 'Optical Part of Astronomy', uses the conclusions of the first part to consider a variety of astronomical topics, especially those relating to solar and lunar eclipses. The running heads in the two parts make this division explicit.

In the first part, the extremely interesting **Chap. 1, 'On the Nature of Light'**, is followed by four more chapters, each devoted to a particular way in which light is modified by material things.

---

[24] Kepler (1604), Chap. 1, p. 7, Kepler (2000), pp. 19–20.

[25] See the Translator's Introduction to Kepler (2015), p. xxiii.

[26] This division is also made clear in the running heads: Chaps. 1–5 are headed 'Paralipomena to Witelo', while Chaps. 6–10 are headed, 'Optical Part of Astronomy'.

The introductory parts of Chap. 1, on the centrality of light in Creation, have already been noted. The rhapsodic image quoted above, of the divinely created tripartite cosmos, all linked by the light of the central Sun, is followed by a tightly organized sequence of 38 propositions, grouped according to the different aspects of light.

> Propositions 1–5: the infinite instantaneous outflowing of light, and the formation of linear rays.
>
> Propositions 6–9: the quantification of light, its rarefaction in breadth and area and constancy along each ray.
>
> Propositions 10–14: light and bodies. Light interacts only with surfaces (since it is a surface). It passes through or is stopped according to the density or rarity of the surface that it encounters.
>
> Propositions 15–17: Colour is the result of light encountering a body. It is 'light entombed in a pellucid body'.
>
> Propositions 18–19: Rebounding or reflection of light.
>
> Propositions 20–21: Refraction or 'breaking' of light.
>
> Propositions 22–31: Coloured light.
>
> Propositions 32–35: Heating properties of light.
>
> Propositions 36–38: Burning and bleaching power of light

The apparent orderliness of this summary is breached by two surprising interludes. The first comes in the midst of Proposition 20. Why, Kepler asks, should light be deflected towards the perpendicular when entering a denser body? 'This whole matter', he writes, 'depends upon the principle of the balance, and should be derived from its source'. This introduces several pages of arguments about the balance[27] directed against Aristotle, Jordanus de Nemore (*fl.* thirteenth century), Guidobaldo del Monte (1545–1607), and Girolamo Cardano (1501–1576).[28] It would be superfluous to review the arguments; what is relevant in the present case is that writers on mechanics at that time often thought about levers, pulleys, and so on, in terms of moving rather than static forces.

It is undoubtedly this approach to the laws of the balance that Kepler found useful to his analysis of the refraction of light. His explanation is rather obscure, but he considers light as if it were a body, infinitely thin but extended across its direction of travel. It is driven forward as if by oars, and the force of an oar is diminished when

---

[27] Kepler (1604), pp. 17–21, Kepler (2000), pp. 29–34.

[28] It may come as a surprise that Archimedes' treatment of the lever is not mentioned in this discussion. According to Brown (1978), p. 187, 'The static tradition [represented by Archimedes], with its superior mathematics, would make its impact [on the Latin Schoolmen] through the dynamic tradition rather than as an independent entity. This fact is illustrated by the relative obscurity that fell upon the translations of Archimedes' works. Latin versions of Archimedes' *Equilibrium of Planes* and his hydrostatic *On Floating Bodies* were made in 1269, but the inherent difficulty of understanding the unfamiliar mathematics, augmented by unsatisfactory texts, severely restricted their circulation. Consequently, these treatises did not find a place with the standard works that collectively comprised the science of weights.'

Fig. 7.5 Refraction
diagram. Reproduced from
Optics Chap. 1, ed. 1604,
p. 20

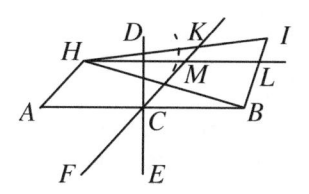

deployed in a denser medium.[29] The result is that when a ray of light encounters the surface of a denser medium (*HL*, in Fig. 7.5), the oar on the more open side of the angle of contact (*CML*) has more power than the oar on the acute side of the angle (where the oar must dip into the denser medium). As a result, the moving 'spark of light' (*lucula*) swerves towards a line perpendicular to the point of contact with the refracting surface, along the broken line. The resemblance of the impulse theory of light proposed by Christiaan Huygens (1629–1695) in Huygens (1690) to Kepler is striking.[30]

The other interlude, or rather postlude, is an analysis of Aristotle's theory of light and vision in *On the Soul*, which Kepler treats in a formal scholastic manner. This addition to Chap. 1, which Kepler calls an appendix, is referred to only once, very obliquely, in the rest of the *Optics*.[31] The problem that Kepler was addressing was the incompatibility between the mathematical science of optics (as proposed by Euclid, Ptolemy, Alhazen, Peckham and Witelo) and the philosophical view of the nature of light and vision, as presented by Aristotle in a short section of *On the Soul*.[32] In a prefatory paragraph to the Appendix, Kepler writes,

> Aristotle reigns everywhere, while the optical writers turn a blind eye and privately remain content with their liberty. Therefore, in order to make opposites illuminate opposites by placing them together, and to lure the Aristotelians at last into the school of the opticians for the aim of either learning or refuting, it seemed right here to discuss explicitly Aristotle's comments on vision.[33]

Evidently, Kepler intended to do for optics what he was at the same time doing for astronomy: to forge an accommodation between mathematical modelling and physical explanation. Henceforth, natural philosophy would be required to produce a mathematically accurate account of observations, while mathematical models would in turn have to make physical sense. It was a deeply radical programme.

Chapter 2, 'On the Shaping of Light', is about light passing through apertures in opaque walls: the theory of the *camera obscura*. Here Kepler presents formally the realization that arose from his empirical approach described above, prompted by his reading of Peckham and Dürer. This purely geometrical exposition, based strictly on

---

[29] The attempted demonstration is on pp. 33–34 of Kepler (2000); Kepler (1604), pp. 20–21. It is especially interesting that when Kepler tried to explain the motions of planets he likewise imagined them equipped with oars. See Kepler (1609), Chap. 38, pp. 184–185; Kepler (2015), pp. 299–300.

[30] Huygens (1912).

[31] Kepler (1604, Chap. 5, p. 210; Kepler (2000), p. 225.

[32] Aristotle (1935), 418a 29–419a 24.

[33] Kepler (1604), Chap. 1 Appendix, p. 29; Kepler (2000). p. 43.

the classical ray optics of Euclid and Ptolemy, solved the problem of the 'shrinking Moon' that had bothered Brahe.[34] It also showed that the various modifications of ray optics proposed by Aristotle and others were distracting and unnecessary.

Chapter 3, 'On the Foundations of Catoptrics and the Place of the Image', deals with the optics of mirrors (that is, catoptrics proper), but also veers into the subject of refraction. Here, Kepler is especially concerned with defining an 'image', and showing how ray optics can account for image formation.

Chapter 4, 'On the Measure of Refractions', now takes on the perplexing topic of how refractions are quantitatively determined. The question is explicitly posed within the purview of astronomy: that is, Kepler limits his treatment to refractions in the Earth's atmosphere. However, he does attempt, with limited success, to develop a general formula for refraction.

The inquiry is long and involved. This can be seen even in the section headings:

1. On the debate between Tycho and Rothmann[35] upon the matter of refractions;
2. Refutation of various authors' various ways of measuring refractions;
3. Preparation for the true measurement of refractions;
4. On the sections of a cone;
5. What kind of quantity measures refractions?
6. Causes of the quantity of refractions (followed by eleven lengthy propositions);
7. Consideration of those things that Witelo advised were necessary for astronomy;
8. Whether the refractions are the same in all times and places;
9. On the observation of the Dutch in the far north;
10. Conjectures from antiquity concerning refractions;

I shall make just a few comments on this difficult chapter.

Section 3, the 'preparation', expresses the hope that, just as curved mirrors display focal properties deriving from mathematical properties of conic sections, there might be some connection between conic sections and the measure of refractions. What Kepler realizes at this point is that most of his readers will not have the least inkling about the properties of conic sections. So he dashes off Sect. 4, a wonderful and brilliant four-page compendium of all the sorts of lines that can arise from the cutting of a cone, from the straight line through to the hyperbola, parabola, and ellipse, to the circle, all based upon focal properties. He shows, in a single diagram, how these curves transform into one another, and presents readily understood ways of constructing conic curves with string and tacks.[36] And then, in Sect. 5, with a very involved algebraic analysis, he shows that, in the matter of refractions, none of these curves will work![37]

---

[34] Kepler (1604), Chap. 2 Proposition 9, p. 54; Kepler (2000), p. 70.

[35] Christoph Rothmann (1550–c. 1605), Court Mathematician to William, Landgrave of Hesse.

[36] Kepler (1604), Chap. 4 Sect. 4, pp. 92–96; Kepler (2000), pp. 106–110. See also Davis (1975).

[37] Kepler (1604), Chap. 4 Sect. 5, pp. 96–123; Kepler (2000), pp. 110–123. The equations are all quadratic, with tangent functions as coefficients. See, for example, Kepler (1604), pp. 100–101; Kepler (2000), pp. 114–115.

Kepler begins Sect. 6, where he thrashes out an iterative rule for approximating atmospheric refractions, as follows:

> Forsooth, reader, I have kept you and myself hanging long enough now, while I tried to gather the measures of different refractions in a single packet, meanwhile acknowledging that the cause is not in this measure. For what do refractions, which we have established to be fundamentally in the plane surfaces of transparent media, have in common with conic sections, which are mixed lines? For that reason—may God look kindly upon us—we shall now also busy ourselves with the causes of this measure. For even if we shall perhaps still stray somewhat from the goal, it is nonetheless preferable to show our industry in looking around, rather than our lassitude in inaction. If among the optical propositions above we have explained the cause of refractions correctly in general, the specifics must also be correctly derived from the same source. But in prop. 20 above, we proposed as a cause the resistance of the medium, by which the spreading of light is hindered, by material necessity. Now it must be seen whether we are able to arrive by following these tracks.[38]

In other words, he gives up hope of finding a neat geometrical or algebraic expression for refractions, though we can see in hindsight that he has assembled nearly all the pieces that would be needed. His final tables of atmospheric refractions show that his iterative method has come very close to the empirically established quantities of atmospheric refractions at different altitudes. The maximum discrepancy amounts to less than half a degree. Of this he writes,

> This tiny discrepancy should not move you; believe me: below such a degree of precision, experience does not go in this not very well-fitted business. You see that there is a large inequality in the differences of my figures and Witelo's. But my refractions progress from uniformity and in order. Therefore, the fault lies in Witelo's refractions.[39]

Kepler concludes this chapter with several sections showing the inconstancy of atmospheric refractions (most notably, the large anomaly reported by the Willem Barentsz (c. 1550–1597) expedition to Nova Zembla (1596–1597),[40] and finally, with a historical account of astronomers' recognition of the facts of refraction throughout all ages.

It is remarkable that in the whole of this long chapter, considering refraction at various surfaces, there is no treatment of glass lenses, which had for centuries been used to correct defective vision. The reason for this becomes evident in **Chap. 5, 'On the Means of Vision'**, which is Kepler's game-changing treatment of the way the eyes work. The first section of this chapter is a detailed account of the eye, proceeding

---

[38] Kepler (1604), Chap. 4 Sect. 6, pp. 109–110; Kepler (2000), p. 123.

[39] Kepler (1604), Chap. 4 Sect. 6, p. 116; Kepler (2000), pp. 128–129.

[40] The Barents expedition of 1596–1597 was attempting to find an ice-free route to the Orient by sailing east across Siberia. They reached the northern end of Nova Zembla (77° north latitude), where they became icebound and eventually lost their ship. They managed to build a hut and survived the winter, returning to Kola (near Murmansk) in small boats in the spring, when they arranged passage back to Amsterdam.

In late winter of 1597, they were astonished to see the edge of the Sun appear above the horizon on January 24, two weeks before the date shown by calculations. It was later determined that this was the result of an anomalous refraction created by a thermal inversion layer stretching for hundreds of miles, and is now known as the Novaya Zemlya effect, after the Russian name for Nova Zembla. See Pitzer (2021), pp. 185–186.

from the place of the eye in the world and in animals, to the way the head relates to the body and the eyes relate to the other parts of the head, to the importance of binocular vision, and finally, to the parts of the head whose function is to assist the eyes, and to the detailed anatomy of the eye itself. Kepler states at the beginning of the section that he will be disagreeing with a number of authorities whose accounts are unclear and 'at risk of uncertitude'. He therefore proposes to gather together, 'as if in the role of principle', descriptions from the testimony of 'the most reputable anatomists'.

The anatomists he calls upon are, first, the Swiss physician Felix Platter (1536–1614), author of *De Partium Corporis Humani Structura et Usu* (Basel, 1583) and, second, Johannes Jessenius a Jessen (1566–1621), professor of medicine at Prague, author of *Anatomiae Pragae ... Historia* (Wittenberg, 1601), and a friend of Kepler. In introducing these authorities, Kepler explains that he had never performed any dissections and would therefore not be qualified to earn the trust of his readers on his own merits. Although he gave a remarkably detailed description of ocular anatomy, he decided later, at the urging of 'friends' (presumably Jessenius), to include an extra unnumbered sheet in the book reproducing Platter's detailed plate of the anatomy of the human eye.[41] He accordingly urged readers to avoid getting mired in the verbal description, but to refer directly to the plate.

Kepler appears to have taken some care and advice in choosing to follow Platter, who disagreed with the received view (which Jessenius held) that the 'crystalline humour' (now known as the 'lens') is the sensitive part of the eye and is located in the eye's geometrical centre. (It is remarkable that Platter's diagrams were included in the *Optics* on Jessenius's recommendation, despite his disagreement with Platter's conclusions.) On the basis of his own anatomical studies, Platter placed the lens near the front of the eye, and located the faculty of discerning in the retina. This revision opened the way for Kepler to realize the analogy between the *camera obscura* (the principles of which he established in Chap. 2) and the eye: the iris and the lens together constitute the opening, while the retina becomes the back wall upon which the image is cast.

Although the detailed geometrical analysis of the paths of light rays from the source through the pupil and lens to the retina is given later in the chapter, Kepler did not wish to leave his readers waiting for the astonishing conclusion. In the style of the mathematicians, near the beginning of Chap. 5 Sect. 2 (which is titled, 'On the Means of Vision') he summarizes what he is going to demonstrate later in the chapter. Kepler's succinct statement of the eye's operation is as follows:

> And this vision, finally, is the most distinct, when all the light of the same point, howsoever much it is spread over the breadth of the cone admitted through the opening of the uvea [the pupil], is brought together by two refractions, one at the cornea, the other at the posterior surface of the crystalline humor [the lens], and illuminates most strongly a single point of the retina, namely, the orifice itself of the nerve bearing the visual faculty or spirit; and no other rays from any other lucid point can fall upon that point, because of the beneficial action

---

[41] The plate and its numbered key are inserted following p. 177 of Kepler (1604); Kepler (2000), pp. 188–191.

of the blackness and opacity of the uvea, of the narrowness of the opening, of the ciliary processes [the iris], and of the rest, which will be described shortly.[42]

The revolutionary insight behind this description is an application of the same ray-tracing principle, learned originally from Dürer, that gave Kepler the key to the secrets of light passing through openings. Only here the rays are deflected by refraction, in such a way that each point of the retina is illuminated by one and only one point in the luminous external world. The two pyramids[43] of light and vision that had been proposed by different strains of classical ray optics[44] were now separated into sequential actions, one diverging pyramid of rays spreading from a point of a luminous body, and one converging pyramid of the same rays brought together at a single point at the back of the eye.

In describing this, Kepler writes:

I say that vision occurs when an image of the whole hemisphere of the world that is before the eye, and a little more, is set up at the white wall, tinged with red, of the concave surface of the retina. How this image or picture is joined together with the visual spirits that reside in the retina and in the nerve, and whether it is arraigned within by the spirits into the caverns of the cerebrum to the tribunal of the soul or of the visual faculty; whether the visual nerve itself and the retina, as to lower courts, might go forth to meet this image—this, I say, I leave to the natural philosophers to argue about. For the arsenal of the optical writers does not extend beyond this opaque wall, which in fact occurs first in the eye.[45]

The most counterintuitive aspect of Kepler's new form of ray-tracing was that the image, or 'picture' (Kepler's word)[46] is inverted. He wrote,

Those things that are on the right outside, are depicted at the left side of the wall, the left at the right, the top at the bottom, the bottom at the top.

As with other matters in Sect. 2, he postpones discussion of this startling conclusion until after his mathematical demonstrations.

Kepler concludes Sect. 2 with a detailed account of the anatomy of the eye, describing what each part does. It is remarkable that, while he was a powerful theorist in numerous areas, Kepler began with the anatomical facts when he undertook to explain how vision occurs. In doing so, he distinguished himself from medical authors

---

[42] Kepler (1604), Chap. 5 Sect. 2, p. 172; Kepler (2000), p. 183.

[43] Lindberg (1970, p. 243, note 8) writes, 'In works translated from Arabic, the term "*pyramis*" is used even when the figure has a round base and hence could aptly be designated by the term "*conus*"'.

[44] Euclid and Ptolemy's optical geometry considered rays of vision coming forth from the eye. In contrast, Ibn al-Haytham argued that optics should be about light emanating from luminous bodies. Each used the pyramid model, originating in the first case at the eye, and in the second, at each point of the luminous body. In practice, later optical authors used whichever model suited the problems they were considering. See Lindberg (1976), pp. 11–13 and 21–23; Straker (1970), pp. 76–77.

[45] Kepler (1604), Chap. 5 Sect. 2, p. 168; Kepler (2000), p. 180.

[46] Kepler (1604), p. 193; Kepler (2000), p. 210. For the revolutionary implications of this term see Alpers (1983), Chap. 2.

Fig. 7.6 'Caustic' diagram.
Reproduced from *Optics*,
Chap. 5, ed. 1604, p. 194

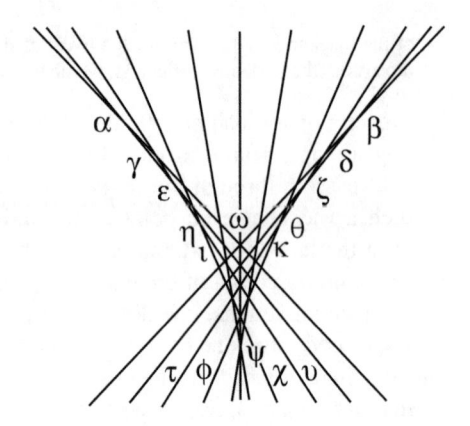

such as Andreas Vesalius (1514–1564) and Realdo Colombo (1516–1559),[47] whose anatomical descriptions were tainted by their previous ideas of how the eye works.

This anatomical and physiological preview serves as an introduction to Sect. 3, which demonstrates formally the claims made in Sect. 2. And it is in this place, not the chapter on refraction, that the behaviour of light passing through spherical bodies is shown. Kepler is concerned with the eye, and in particular its lens, for which an aqueous globe is a good analogy. Even though he lacked a mathematical expression for the relation between angles of incidence and refraction, Kepler was able to provide a comprehensive account of how convex spherical lenses function. For example, in Propositions 9 and 15, he demonstrates what is now known as spherical aberration, and draws an accurate illustration of the way the rays cross each other, forming a ray pattern now called the 'caustic' (Fig. 7.6). In short, he has demonstrated clearly all the properties of lenses that would be required in order to design and build a telescope. We will return to this matter below.

In contrast, what Kepler was rightly excited about was what his study of lenses implied for the theory of vision. In his Sect. 4, in which he summarizes the conclusions of the arguments of Sect. 3, he writes,

> What confirms me is the most universal line of argument, used by Witelo himself. The effect of vision follows upon the action of illumination, in manner and proportion. But the retina is illuminated distinctly, point by point, by the individual points of the objects, and most strongly through the individual points. Therefore, it is at the retina, not elsewhere, that the most distinct and most evident vision can take place.[48]

About the inverted picture he writes,

> As for the inversion of my picture, that might be raised against me in objection, which Witelo avoided with great care: he first attributed flatness to the crystalline, contrary to obvious experience, in order to maintain that opinion. And since the present opinion affirms that that surface is bulging, by the testimony of Witelo there occurs an inversion of the

---

[47] Lindberg (1976, pp. 169–175) lists more than a dozen anatomists following the generally accepted central position of the lens.

[48] Kepler (1604), Chap. 5 Sect. 4, p. 205; Kepler (2000), pp. 220–221.

likeness. And I, for my part, tied myself in knots for the longest time, trying to show that left cones that are made of the right ones in entering the opening of the uvea [the pupil], are again cut beyond the crystalline in the middle of the vitreous humor, and that another inversion occurs, and that the parts that were made to be left again became right before they reach the retina. Nor was there an end of this useless anxiety until I hit upon Prop. 11 and 12 among those preceding, by which this opinion was most evidently refuted. And even if I had upheld what was proposed, there was still to be a remaining complaint: the hemisphere was going to be reversed. For those now stand facing us on the outside, judging these parts to be right, those left, have images directly opposite whose right parts will be taken to the left, as is to be seen in mirrors. For the eye which to you is right becomes the left one for your image. I shall say nothing about how the picture's concavity was verging inwards towards the head, while the concavity of the object was verging in the opposite direction.[49]

In the subsequent discussion, Kepler does what he can to convince the reader that the inversion of the image is real, and will not result in the world appearing upside down. The chapter concludes with critiques (often witty) of a number of other writers on optics, most notably, Giovanni Battista della Porta (1535–1615), and a section on how errors of vision can influence astronomical observations.

7.2 The Optical Part of Astronomy

This concludes the 'Paralipomena' half of the book. The remainder, on 'The Optical Part of Astronomy', contains fewer arresting discoveries, but Kepler presents a great deal of valuable historical research. The chapter titles are:

6. On the varied light of the stars.
7. On the shadow of the Earth.
8. On the shadow of the Moon and daytime darkness.
9. On parallaxes.
10. Optical foundations of motions of the heavenly bodies.
11. On the observations of the diameters of the Sun and Moon, and eclipses of the two, following the principles of the art.

Much of this part is based on historical observations, in an attempt to discover, for example, whether a truly total solar eclipse is possible, and, on the contrary, whether the Moon's apparent diameter is always great enough to cover the entire disc of the Sun (Chap. 8). This chapter lists 28 historical reports of eclipses (or possible eclipses) that appear to have been total. The other sort of eclipse, when a ring of the Sun's disc is visible around the Moon (an 'annular eclipse') was generally thought to be impossible. However, there was just one credible instance. This was on 9 April 1567, in Rome, observed by Christoph Clavius SJ (1538–1612), an experienced observer. Kepler considers this eclipse carefully, mentioning observations that he had made himself that might be mistaken for seeing the Sun surrounding the Moon.

---

[49] Kepler (1604), Chap. 5 Sect. 4, pp. 205–206; Kepler (2000), p. 221.

Chapter 11 contains a wealth of information about techniques of naked-eye astronomy, of possible value for reconstructions of historical equipment and methods.

Chapter 10 deserves special attention because of the wry and perhaps playful claims it puts forward. At the outset, Kepler presents the idea that many features of the standard planetary models are based on optical illusions—a claim that deserves serious consideration. In the first half of the chapter, Kepler gives examples of apparent motions in the heavens that could be understood as motions of the observer. But then Kepler presses the point farther. After a brief account of how motions that are real in the Ptolemaic system become optical phenomena for Copernicus, Kepler writes,

> As regards Copernicus, however, this whole illusion of standing still and retracing of steps is demonstrated most beautifully from optics. And although these things are more appropriately learned from the author himself, nevertheless, so that nothing might there be said which would affect the reader negatively, I shall repeat the fundamentals in three words from Euclid himself. It is indeed my judgement that if we had not had other arguments by which antiquity had tested this Copernican opinion, this passage alone would have been sufficient to vindicate Copernicus from the truth of Pythagoras. First, it is evident, not only in itself but also from the commentary of Proclus, that all of Euclid's geometry is Pythagorean and aims at the knowledge of the five regular figures which are called 'cosmic': Euclid was therefore a Pythagorean. Next consider for me the bundle of Euclidean propositions in his *Optics*, namely, 53, 54, 55, 56, 57, 58, which Witelo carried over into his Book IV Propositions 134, 135, 136, 128, 132, 133, 129. In these propositions, Euclid propounded pure, unadulterated Copernican astronomy.[50]

Then, as if to counter the incredulous objection of the reader, Kepler rehearses, step by step, the arguments of those six propositions from Euclid, showing how each, properly understood (according to Kepler's view), is a thinly veiled presentation of heliocentric astronomy, which Euclid presumably learned from his Pythagorean teachers.[51]

Can Kepler be speaking seriously here, or is this a kind of rhetorical stunt? Kepler offers no hints, but I suspect that this is an instance of what Nicholas Jardine calls a 'serious joke'.[52]

7.3   Kepler, Galileo, and the Telescope

Among the copies of his first published book, *Mysterium cosmographicum* (Tübingen, 1596), that Kepler sent forth to those who he hoped would appreciate it was one that made its way to the mathematics professor at the University of Padua,

---

[50] Kepler (1604), Chap. 10, pp. 301–302; Kepler (2000), p. 342.

[51] Copernicus (1543, I. 10) cites Euclid's demonstrations that Kepler cites above; Kepler takes the hint and enlists Euclid as a crypto-heliocentrist. This is of a piece with the Renaissance myth of a *prisca philosophia* of which only a few traces survived. Pythagoras and his school were thought to have inherited much of that tradition. See Copernicus (1543, fol. 4^A); Copernicus (1992, pp. 4–5).

[52] Jardine (2009).

Galileo Galilei (1564–1642). In return, Kepler received an elegant and non-committal letter of thanks, along with Galileo's revelation that he, too, was a follower of Copernicus.[53] Kepler responded, urging Galileo to express his views publicly, but heard no more—that is, not until Kepler's friend the courtier Johann Matthäus Wacker von Wackenfels (1550–1619),[54] came by Kepler's house in 1610 all excited about the news that a certain Italian had discovered four new planets, and many other things, by means of an optical instrument.[55]

It is hard to imagine what Kepler must have thought upon hearing that this same Galileo from whom he had previously heard had mounted a combination of lenses in a tube and had turned it to the heavens, with spectacular results. In the *Optics*, Kepler had worked out the way to use ray optics to account for image formation by spherical lenses. He had constructed a large pinhole instrument for observing eclipses, and had all the information necessary to project enlarged images with a single lens. He had set this out in detail, but the part about the action of lenses was in Chap. 5, Sect. 3, which was aimed at showing how the eye functions. It is even possible that Kepler already had some experience with the use of a lens in observing the Sun: in his response to Galileo, he writes:

> Let Galileo stand next to Kepler, the former observing the Moon with face turned to heaven, while the latter observes the Sun, turned away towards a tablet (so that the lens should not burn his eye), each man using his own instrument.

This is from p. 12 of Kepler's published response, *Conversations with the Sidereal Messenger* (*Dissertatio cum Nuncio Sidereo*, Prague, 1610). The word Kepler uses for 'lens' is *specillum*. What is suggested here is that a lens should be installed at the pinhole at the upper end of his 360 cm long instrument (see Fig. 7.7).[56]

The caution about burning the eye suggests personal experience; two years later he wrote of using his instrument with a lens and a dark glass, which nevertheless resulted in a near-blinding.[57] Could he have already tried this combination before learning of the telescope? Following the passage quoted above, Kepler writes,

---

[53] Letter from Galileo to Kepler, 4 August 1597, in Kepler (1945), pp. 130–131. Kepler replied to this letter in October: Kepler to Galileo, 13 October 1597, in Kepler (1945), pp. 144–146. See J. V. Field, 'Kepler and Galileo' (this volume, Chap. 8). English translations of both of these letters are in the Appendix to that chapter.

[54] For more on Wacker von Wackenfels, see Edward Rosen's Note 44 to his translation of Kepler's *Dissertatio cum Nuncio sidereo* (Kepler, 1965, pp. 60–61).

[55] Kepler told of his friend's excited visit in his letter to Galileo of 19 April, which he published, with modifications, as *Dissertatio cum Nuncio Sidereo* (Prague, 1610); see also Kepler (1610, 1941). His account appears on pp. 1–2 of the 1610 edition. The *Dissertatio* was translated by Rosen (Kepler, 1965) with more than 400 highly informative notes. Kepler's account of Wacker's news appears on p. 10 of the translation.

[56] Kepler's illustration of this instrument was printed in Kepler (1604), p. 338; Kepler (2000), p. 349.

[57] Kepler's letter to Wacker von Wackenfels, Kepler (1955), p. 8; translated by Edward Rosen in Kepler (1965) note 179, p. 97.

Fig. 7.7 How a single
convex lens creates an
inverted 'picture' (*pictura*, in
Latin) of a luminous source,
just as Kepler proposes to
Galileo on p. 12 of the
Dissertatio. Reproduced
from Item XLV of the
Dioptrice, ed. 1611, p. 17

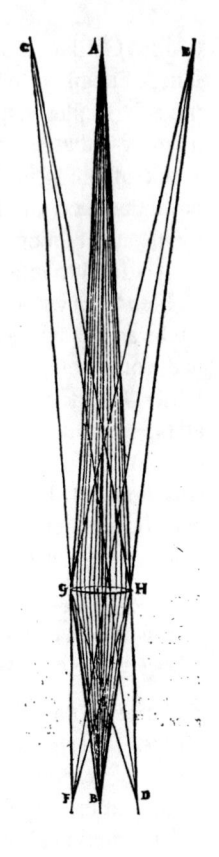

First of all, it is extremely fortunate that I myself have been involved in observing the spots
of the same Moon, not (like you) with face turned towards it, but turned away. You have a
diagram for this on p. 247 of my book.[58]

Indeed, there is a picture of the Moon on this page, but it contains very little detail.[59]
Kepler makes no mention of a lens, and there is no reason to suppose that by this
time he had added one to his instrument.

He continues just below,[60]

From this I get the idea of competing with you in carefully observing those small spots
that you first noticed in the brighter part. But I hope I will complete this using my way of
observing, facing away from the Moon, in this way: I shall let the Moon's light enter through
an opening and fall upon a tablet built around the pole[61] [running lengthwise along the
instrument], but with a crystalline lens with a spherical convexity of a great circle, and with
the tablet [at the lower end of the instrument] adjusted to the location of the confluence of

---

[58] Kepler (1610), p. 13; Kepler (1965), p. 23.

[59] Kepler (1604), Chap. 6 Sect. 9, p. 247; Kepler (2000), p. 259.

[60] Kepler (1610), p. 13; Kepler (1965), p. 23.

[61] Latin, 'pertica'.

> the rays. Thus with a pole twelve feet long,[62] the body of the Moon will be perfectly pictured with the size of a large silver coin. I have demonstrated the device in Prop. 23, p. 196, and on p. 211, of my book.

The two passages he cites[63] do indeed describe what happens when rays of light pass through a narrow opening and are then refracted by an aqueous sphere: Kepler clearly has the eye, not his instrument, in mind in these passages from the *Optics*, but in the *Conversation* he remarks that the argument is exactly the same when applied to a lens placed below the opening.

The clarity of Kepler's description may seem to hint that he had already constructed this instrument. Perhaps he had done so, but was disinclined to initiate a priority dispute with Galileo. What is certain, however, is that Kepler already had the optical theory needed to have constructed a *camera obscura* with a lens at its opening.

If Kepler had had all he needed to have invented a telescope suitable for astronomical use six years before Galileo, why did it not occur to him to do so? One factor that has already been mentioned is that his study of lenses was very narrowly conceived as leading to an understanding of how the eye works. Nonetheless, a few pages before the passages quoted above,[64] Kepler describes (with numerous references to specific pages in the *Optics*) how he would design a telescope with an objective lens that would be shaped to avoid spherical aberration and a concave eyepiece that would bring the rays together on the retina of the viewer's eye. Although he had never seen one of Galileo's telescopes, he describes its construction with impressive accuracy—except that Kepler would add more lenses!

But immediately afterwards he compliments Galileo on contriving a way of precisely measuring very small angles by means of the telescope. He writes,[65]

> Since your achievement along these lines vies with Tycho Brahe's highly precise accuracy of observation, it may not be amiss to digress somewhat.
>
> That master of all the sciences, Johannes Pistorius,[66] asked me more than once, I recall, whether Brahe's observations were so refined that in my opinion absolutely nothing could be lacking in them. I vigorously maintained that the pinnacle had been reached, and that nothing further was left to human enterprise, because the eye would not permit greater precision, nor would the effect of refraction, which alters the position of the stars with reference to the horizon. In rebuttal, he steadfastly declared that some day somebody would come along who would devise a more exact procedure with the help of lenses. I objected on the ground their

---

[62] About 360 cm.

[63] Kepler (1604), Chap. 5 Sect. 3, p. 196; Kepler (2000), pp. 212–213; Kepler (1604), Chap. 5 Sect. 4, p. 211; Kepler (2000), pp. 225–226.

[64] Kepler (1610), pp. 9–11; Kepler (1965), pp. 19–21.

[65] Kepler (1610), p. 11; Kepler (1965), pp. 21–22.

[66] Johannes Pistorius (1546–1608) was a fervent and zealous Catholic convert (from Lutheranism via Calvinism) who became the confessor to the Emperor Rudolf II. As Kepler says, he was a very learned polymath and author of a wide variety of books. As an intimate of the Emperor he was able to be of great assistance to Kepler in obtaining funds and assisting in negotiations with Tycho Brahe's heirs. He and Kepler became friends, as well as (mostly) friendly antagonists in religious matters. See Kepler (1965), Rosen's Note 167, pp. 92–94.

refractive properties made lenses unsuitable for reliable observations. But now at last I see that Pistorius was in part a true prophet. To be sure, Brahe's observations speak for themselves and need no praise. For what an arc of 60° is in the heavens, or 34′, is known through Brahe's instruments by themselves. But whereas Brahe in this way measured celestial degrees in the heavens ... now your telescope, Galileo, surpasses these attainments.

Brahe's instruments incorporated design improvements that increased the precision of naked-eye instruments by a factor of ten, from plus or minus 10 arc minutes down to plus or minus one arc minute. From his experience as an observer, Kepler thought that at the one-arc-minute level, the limit of resolution of the human eye had been reached. Moreover, he was not confident that magnification by lenses could be made linear, so that larger intervals as viewed in the telescope would be accurately mappable onto smaller intervals in the heavens. Thus before reading Galileo's account, he felt no impulse to try using lenses, which he understood well, to perform observations that required fine distinctions.

Kepler's mention of the arcs of '60° ... or 34″' deserves a comment. Tycho Brahe's instruments were designed to measure both large angles and small ones down to a few arc minutes, and they did this superbly well. What Kepler realized about the telescope is that by magnifying objects it was capable of discerning and measuring much smaller angles, such as the apparent dimensions of the orbits of Jupiter's moons. When the larger angles were involved, Kepler could see no advantage in using the telescope.

7.4 Dioptrice

Whatever emotions Kepler might have felt in realizing that his advocacy for Tycho Brahe and his own doubts about a future for lenses in astronomy had led him to miss the chance to invent the astronomical telescope, he was clearly spurred into action by the experience. In a burst of activity that was extraordinary even for him, he wrote a magisterial treatise, titled *Dioptrice*, on the functioning of lenses, both individually and in combination of up to three lenses. This concise book, organized into 141 numbered proofs or statements, leads to a final 'concealment' (an account of which, respecting Kepler's coyness, I defer until later in this chapter). He started work on the book in August, not long after the printing of the *Dissertatio*, and it was finished by the end of September.

The contrast with the *Optics* could hardly be greater, despite its relation to the earlier work. Presentation of the *Dioptrice* is tight and spare. The body of the book takes up only 80 pages in a small quarto format, with a 28-page introduction much of which discusses recent letters from Galileo. While the *Optics* is a journey of discovery, a far-reaching investigation, the *Dioptrice* has a more narrowly defined aim, as stated in its full title:

Dioptrics

or

Demonstration of the Effects upon Vision and Visibles Resulting from "Conspicilla",

that is,

Glasses or Pellucid Crystals

(The word 'dioptrics' was Kepler's invention (see his explanation below). 'Conspicilla' was at that time being used as a name for eyeglasses.)

Duke Ernst of Bavaria, Archbishop Elector of Cologne (1554–1612), brought a telescope, made by Galileo himself, to Prague in mid-August, 1610, thus giving Kepler his first opportunity to see for himself what Galileo had described (see Kepler, 1611b, fol. *3r; Kepler, 1941, p. 318). In his dedication to Duke Ernst,[67] he explains that his *Optical Part of Astronomy*, published six years before, had never been impugned, despite its having explained for the first time many things about how vision and how lenses function. 'Therefore', he wrote,

> it was fitting that I show that these same fundamentals, by which I had built up the functioning of vision, and the actions of simple lenses, also suffice for supporting the composition of several viewing lenses into a single fishing rod.[68] This has indeed succeeded so well (and this is an indication of truth) that it is impossible to carry out this demonstration upon any principles whatever, other than those that I have used. And since Euclid had made Catoptrics, which is about reflected rays, into the image of Optics, with a name derived from the chief instrument of my book, *Dioptrics*, was born.[69] It is mostly concerned with a ray refracted by dense pellucid media, both natural, as in the human eye, and artificial, in the variety of lenses. By this, it is distinguished from Catoptrics, as one species against another. However, it is Dioptrics that is prior, and Catoptrics posterior, because Catoptrics is concerned with images, which, whatever they are in general, cannot be understood without knowledge of the eye, which must be sought from Dioptrics.[70]

Kepler goes on to explain that despite having presented the fundamentals of vision and refraction in the *Optics*, he does so again here. This is partly because the human eye is itself an assembly of lenses, so that neither one can be understood without the other. But also, he says, some readers of the *Optics* had complained that parts of it were rather obscure, so he will use the *Dioptrice* as an opportunity to clarify and make explicit things that were unclear in the earlier work. He promises explicit definitions, continuous numbering of propositions, and more diagrams. He concludes by writing,

> If I have not entirely removed the obscurity from this work, I hope those who are students of [natural] philosophy will grant me a degree of pardon for my weakness, and will consult this work with good spirit.[71]

---

[67] Kepler's dedication to the *Dioptrice* is on four unnumbered sheets following the title page of the 1611 edition; also in Kepler (1941), pp. 331–333.

[68] The Latin is *arundo,* more usually spelled *harundo.* Generally, it means 'reed', but it can also denote a fishing rod or pole. In the opening sentence of the dedication, Kepler had mentioned, as a great invention of the age, the *Arundo dioptrica.* Why would Kepler choose *arundo* rather than the more literally accurate *tubus*? I suspect he is alluding to the great 'fishes' that Galileo had caught with his rod-like instrument.

[69] Euclid had named his book *Catoptrics* (κατοπτρικός) because it dealt with mirrors (κάτοπτροι). Kepler played on this name, replacing κατα with δια, 'through', because it deals with lenses.

[70] Kepler (1611a), fol.) (2 r–) (2 v; in Kepler (1941), p. 331 l. 36–332, l. 1–2.

[71] Kepler (1611a), fol.) (2 v; in Kepler (1941), p. 332, l. 14–16.

Having said this, he partly takes it back in the Preface that follows (which was written a year later, just before publication):

> I am showing you a book, dear reader, that is mathematical; that is, one that is not very easy to understand. It not only requires much intelligence in the reader, but also, chiefly, attention of the mind, and an incredible desire to come to know the causes of things.[72]

Despite the promise to satisfy this 'incredible desire', the reader working through the first twenty numbered items in the *Dioptrice*, which presents the fundamentals of refraction on flat surfaces, will find little in the way of causes. No mention is made of why light behaves as it does, and no attempt is made to derive the mathematical theorems from physical principles. This is in marked contrast with Kepler's aim in both the *Optics* and *Astronomia nova*, and, indeed, with Kepler's warning near the beginning of the Preface. Describing the way the ancients adopted purely geometrical models for planetary motions, he wrote,

> We should beware lest what happened to the ancients might happen to us: that, being too securely trustful in the one eye of Optics in this examining of the planets' orbit, they closed the other eye of Physics. Thus, what should bave been equally attributed to reasoning of both Optics and Physics we would have attributed to Optics alone, and so again would have missed the target. On this subject see my *Optical Part of Astronomy* [i.e. the *Optics*], and *Commentary on the Motions of Mars* [i.e. *Astronomia nova*].[73]

Having given the reader this warning, Kepler presents his subject matter in a formally mathematical style, with very little in the way of Physics.[74] It is striking, however, that he omits many of the formal features of Euclidean geometry.[75] Instead he gives us 141 numbered short statements (only some of which are followed by longer proofs or explanations), each referred to only by a Roman numeral and not by any general name. Most of them are given headings, which include Definitions, Problems, Axioms, Postulates, Propositions, Optical Axiom, Porism, and at least one Note. Some have no headings, but appear to be of the same kind as the one with the most recently used heading. Since Kepler chose to identify these elements only by Roman numerals,[76] using only the numerals when referring to them elsewhere, I shall refer to them as 'items'.

The items are presented in twelve unnumbered groups; all but the first group bear headings. The structure of the book is made evident by the list of headings[77]:

---

[72] Kepler (1611a), fp. 1; in Kepler (1941), p. 334, l. 5–8.

[73] Kepler (1611a), p. 3; Kepler (1941), p. 335, l. 36–41; Kepler (1937) 'Introduction', p. 20; Kepler (2015), p. 19.

[74] Although in Kepler's time the Latin word *physica* did not mean what it does today (nor what it meant when Aristotle coined it), Kepler was very clear about the distinction between what he called 'mathematics' and what he called 'physics'.

  For reflections on Kepler's use of the word *physica*, see Andrew Gregory, 'The Translation of the Title of Kepler's *Astronomia Nova*' (this volume, chapter 5).

[75] For the formal features of a classical geometrical proof, see Proclus (1992), p. 159.

[76] Kepler chose to write IIX for VIII, XIIX for XVIII, and so on. I have substituted the current notation for these numerals.

[77] Except for the numerals, the section names are those given by Kepler.

1. [No heading, these items are an introduction to refraction on flat surfaces] items I–XX

    So much for the plane crystal: now, on curvilinear crystals:

2. First, on light: items XXI–XXIV

3. On the lens: items XXV–XXXIII

4. Convergence [i.e. focusing] of a [half] lens: items XXXIV–XXXVII

    So much for a single, solitary, convex lens surface; now,

5. The complete lens: items XXXVIII–XLI

6. The effects of a lens by itself: items XLII–LVI

    So much for the convex lens and its uses, other than with respect to the eye. Now for its usefulness in aiding vision:

7. And first, on vision itself: items LVII–LXV

    So much for the eye and vision. There follows:

8. The uses of the lens with respect to the eye: items LXVI–LXXXV

    So much for a single convex lens; now:

9. On convex lenses combined with one another: items LXXXVI–LXXXIX

    So much for convex lenses:

10. On hollow lenses: items XC–C

    So much for convex lenses by themselves, and hollow lenses by themselves:

11. On hollow and convex lenses combined: items CI–CXXIV

    So much for the simple instrument:

12. Next, the "concealment" [κρύψις]: items CXXV–CXLI

The order of presentation follows what was promised in the Dedication: the fundamentals of refraction are presented first, and then, at item LVII, we are given an account of vision. The laborious procedure for computing refractions that Kepler laid out in Chap. 4 Sect. 6 of the *Optics* is set aside for practical reasons. Even in the *Optics*, he despaired of finding the true law of refraction, writing 'even if we shall perhaps still stray somewhat from the goal, it is nonetheless preferable to show our industry in looking around, rather than our lassitude in inaction.'[78] Now, in the *Dioptrics*, he simplifies his account by restricting it to the air-to-glass interface, and limiting the angle of incidence to a maximum of thirty degrees. With these constraints, he can rely on two items that he presents as axioms:

VII. Axiom: Refractions in crystal up to thirty degrees of inclination are perceptibly proportional to the inclinations.

VIII. Axiom: The angle of refraction in crystal, up to the stated limit, is approximately one third of the inclination in air.

Clearly, Kepler's use of the word 'axiom' is different from ours. His usage is closer to Aristotle's, for whom it denoted statements that are assumed without proof in formal logic. They may or may not be true, but are assumed so for the sake of argument. In the present case, these 'Axioms' are rules that are approximations for an as yet unknown law of refraction.

---

[78] Kepler (1604), p. 110; Kepler (1939), p. 104; Kepler (2000), p. 123.

It should be noted that Kepler measured the angle of refraction with respect to the direction of the incident ray. Thus, if the incident ray were not at all bent, the angle of refraction would be zero. At an angle of incidence of 30° (measured from the perpendicular), the angle of refraction would be 10°, which is the amount by which the refracted ray is deflected from its previous course. Therefore, the angle of the refracted ray, measured from the perpendicular to the surface (as is the convention now), would be approximately 20°.

Snel's law of refraction—named after Willibrord Snel (1581–1626) and recorded in a manuscript dated 1621[79]—states that

$$\frac{\sin(i)}{\sin(r)} = k,$$

where i is the angle of incidence, r is the angle of refraction (both measured from the perpendicular to the surface, and k is the index of refraction. For glass, a typical value of k is 1.52.

Kepler's approximation, stated in modern terms, is

$$r = \frac{2}{3}i.$$

Snel's and Kepler's results are shown visually in Fig. 7.8 and quantitatively in Table 7.1.

Evidently, Kepler's approximation works reasonably well.

Here we are clearly in a realm different from what we experienced in the *Optics*. Kepler is evidently playing the part of the optical engineer, where 'pretty close' is good enough. Compare this with our account of refraction in *Optics*, Chap. 1 Proposition 20, where Kepler goes boldly forth into wonderful and strange realms to understand what the true cause of refraction is. There, his ray of light has zero thickness in the direction of travel, but seems to have a tiny bit of breadth and encounters an increase in the density of the medium with differential torque, as if it were propelled by oars. The resulting angle of refraction, or better, deflection,[80] is to be determined by the kind of dynamic principle of the lever proposed by Renaissance mechanics, based upon moving action rather than the static approach of Archimedes (c. 287–212 BC).[81] One can see Kepler's imagination jumping way ahead of anything

---

[79] The law was known in Baghdad in the tenth century and seems to have been discovered by Ibn Sahl (c. 940–1000) in about 984. There is evidence it was rediscovered by Thomas Harriot (1560–1621), who did not publish his results, and then independently by René Descartes (1596–1651), who published the law in his *Optics* (Descartes, 1637).

[80] At the beginning of Chap. 1 of the *Optics* (Kepler, 1604, pp. 5–6; Kepler, 2000, pp. 17–18), Kepler considers the terminology for refraction and reflection. He proposed 'deflection', but rejected it because it could apply to both mirrors and transitions to different media. He proposed *repercussus* for mirrors and *infractus* for rays entering water or glass.

[81] See the note on Kepler (1604), Chap. 2, Prop. 20, above.

Fig. 7.8 Comparison of Snel's Law with Kepler's refraction computation in the *Dioptrice*

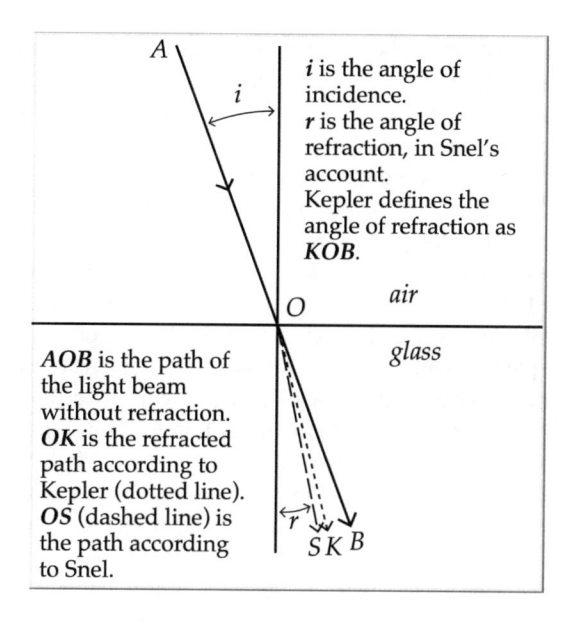

i is the angle of incidence.
r is the angle of refraction, in Snel's account.
Kepler defines the angle of refraction as **KOB**.

AOB is the path of the light beam without refraction. **OK** is the refracted path according to Kepler (dotted line). **OS** (dashed line) is the path according to Snel.

Table 7.1 The refraction laws of Snel and Kepler compared

Angle of incidence: i	Angle of refraction (by Snel's Law): r	Angle of refraction (by Kepler's approximation, in modern terms)
0°	0°	0°
5°	3.3°	3.3°
10°	6.6°	6.7°
15°	9.8°	10.0°
20°	13.0°	13.3°
25°	16.1°	16.7°
30°	19.2°	20.0°

available in contemporary mathematics. He was trying to do for optics what he later did for astronomy in *Astronomia nova* (1609).

Contrast this with what we have seen in the *Dioptrics*, where there is no pretence of a physical explanation: Kepler deliberately looks at the problem with only his Optical eye open. This makes his presentation feel surprisingly modern. There is no attempt, for example, to define light, or rays: these are taken as undefined elements, having certain mathematical properties such as convergence and divergence. The introductory section on lenses (Sect. 3, in the above list of the groups of items) is just a series of terms relating to the forms of lenses. With these things set forth, and with his simple rules governing refraction (for glass and crystal only), Kepler jumps right into his mathematical exposition.

Because of the very different aims of the *Optics* and the *Dioptrics*, Kepler could not simply carry over conclusions from the former to prove theorems in the latter. In the *Optics*, Kepler's interest in refraction was confined to two special topics: atmospheric refraction, which had to be considered when making astronomical observations; and refraction in the eye, by which it creates images or pictures of the outside world. Neither of these involved lenses to any great extent, and even in the eye, Kepler modelled the path of light rays using glass spheres filled with water. This enable him to use Euclid's propositions on circles brilliantly: he established many characteristics of lenses, such as the focusing of rays, spherical aberration, limitations on light paths, and location of images, using single elements—but only as applied to complete spheres, which of course would apply directly to vision. When it came to thin lenses, both convex and concave, he had an entirely different set of problems.

The trick that initially opened up the realm of lenses was (as mentioned above) simplification of the mathematics of refraction by restricting it to air-to-glass interfaces and to a limited range of angles. His strategy in developing his account of the action of lenses was to consider air-to-glass refractions (Item XXXIV) and glass-to-air refraction (Item XXXV) separately, then to combine the two effects to show the action of a spherical convex lens as a whole (Items XXXVIII and XXXIX). His treatment in Item XXXIX is especially ingenious; he considers the paths of only two rays, one passing through the centre of the lens (perpendicularly to the surface, therefore unrefracted) and the other passing through the exact edge of the lens, thus (for the sake of analysis) being refracted by both surfaces of the lens at once. This approach sidesteps the more difficult question of the angles of the rays inside the lens.

Kepler's approach in this proof anticipates a technique of the differential calculus. His argument depends implicitly on the idea of examining the two transitions of the ray through the surfaces at an infinitesimal distance from the edge of the lens, so that the passage of the ray through the glass would not have to be considered. The infinitesimal distance (in the calculus as developed later) would then be allowed to approach zero as a limit, leaving the two refractive surfaces but eliminating the glass in between. Kepler simply ignored the paradox of two refractions occurring at once, and jumped directly to the conclusion.[82]

Once Kepler has established the basic focal properties of lenses (in Items XXXIV, XXXV, XXXVIII and XXXIX), he can proceed to establish many of the properties that he had already demonstrated in the *Optics*, such as how and where lenses form visible images, or pictures on paper. These are beautifully captured in the concept of a 'pencil' (*penicillum*) of rays, introduced in Item XLV.[83] The classical meaning of *penicillum* is a painting brush with diverging bristles, and in repurposing it as an optical term Kepler imagines rays of light spreading out from a luminous source. The accompanying figure, reproduced here (Fig. 7.9), clearly illustrates the idea. Once can easily visualize the relation between object and picture: how the lens mediates

---

[82] For more on Kepler's use of infinitesimals and indivisibles, see Knobloch, 'Kepler's contributions to mathematics' (this volume, Chap. 11).

[83] Kepler (1611a), p. 17; Kepler (1941), p. 368.

Fig. 7.9 Pattern of rays passing through a convex lens. Reproduced from *Dioptrice* LXXV, ed. 1611, p. 32

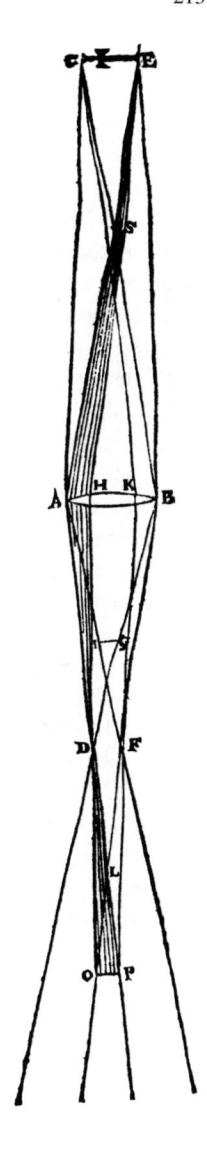

between them (Item XLVI), and why the single convex lens creates an inverted picture (Item XLIV).

The remaining 95 items, though they hold to the geometrical presentation of the first part of the book, cover a wide variety of phenomena, much more than might be guessed from the section headings given above; more, indeed, than can be adequately covered in the present chapter. Unfortunately, there is no English translation of the *Dioptrics* as yet,[84] and I am unaware of any extensive study of the work. The best

---

[84] There is a German translation (Kepler, 1904).

that can be done here is to cover a few of the most important theorems and topics in the remainder of the book.

The eighth section, 'on the uses of the lens with respect to the eye', covers much more than the description suggests. Ray optics is just lines in space; like Dürer's strings, the pencils of rays provide a transition between the luminous points of the object and the received points of the image. They do not get one to the main topic, which is the act of seeing. Also, there is a complication: there are two different kinds of image, namely, what can be seen directly with the eye, and what can be projected onto a sheet of paper or a wall. And they are somewhat mysteriously found in different places.

Kepler works out many of the complexities of lenses and seeing in Items LXX to LXXVI. One governing principle is that vision cannot be clear when rays from a point of the visible object converge as they approach the eye (this is shown in Item LXV). On the other hand, as emerges in Items XL to XLVI, divergent or parallel rays will not project a picture: the rays must be converging, in such a way that rays from one point of the object all converge to form a single point of the picture.

So the places where direct vision of the rays can occur must be found in between the places where the rays converge to a point. In Item LXX, Kepler says that such a place can be found beyond the convex lens, but before what he calls the *punctum concursus*, the 'point of convergence'. The location of this point depends upon the angle of divergence or convergence of the rays entering the lens. The point we would call the 'focal point of the lens' is the point of convergence of rays that are parallel before entering the lens. In the adjacent diagram (Fig. 7.9) (which accompanies Item LXXV), the meeting place is along the line *DF*, where *D* is the image point corresponding to *E* of the object.

But somewhere above *DF*, in the region of the line marked *IG*, the rays are converging, but perhaps not too much so. Interestingly, Kepler says nothing about this convergence in Item LXX: he is primarily interested in showing that the image that an eye in this region sees is not reversed. In Item LXXI, he draws attention to this, noting that the rays are converging, and will therefore not give clear vision, by Item LXV. However, in Item LXXIV, he places the eye at the point of meeting—at *D* or *F* or some point in between—and notes that since the rays are converging here most of all, vision will be completely confused. Then in Item LXXV, when the eye is placed farther out, beyond the point of meeting, as at *OP*, the eye can see the image, but it is inverted. This he demonstrates geometrically at some length. He chose this point because the rays from the two ends of the object overlap, and form an area of intersecting rays that is comparable to the size of the eye's pupil.

This sequence, and especially the citation of Item LXV, which describes what things look like but is not logically required, suggests that (much to his credit) Kepler is writing this with a lens in hand, watching what happens when he moves his eye gradually farther from the lens. Some of his statements and assumptions are questionable, but he has worked out the ray diagram clearly, and at the same time is faithfully reporting what he sees.

When Kepler comes to the point of adding a second convex lens, the first thing he does is to invent what we now know as the 'Keplerian telescope' (Item LXXXVI). It has been said that, lacking necessary resources, Kepler was never able to make such an instrument. While it may be true that he did not assemble a pair of appropriately matched lenses in a tube of a suitable length, it seems implausible, given what we have just read, that he would not have had a couple of convex lenses to play with, and that it would not have occurred to him to line them up with his eye and look at things through them.

His reasoning, in explaining this combination of lenses, is careful, with a diagram (Fig. 7.10) showing the paths of important rays. His argument is that the lenses should be placed so that the rays through the lens nearer the object diverge excessively (this would occur beyond, but close to, the point of meeting), while the lens near the eye has a contrary convergence, which 'remedies the excessive divergence of the former one'. For this reason, the combination makes the light 'approach the eye so as to present distinct vision'. In other words, the rays entering the eye are either parallel or very slightly diverging. He says that the magnification of the pair depends on the proportion of the lenses between themselves 'which depends on the choice of the maker', but he claims that (by Item LXXX) the lens near the eye, which he shows does not invert the image, must magnify the image at least to some extent.

Kepler's proof is difficult to follow, especially because when he introduces the second lens (the eyepiece) he begins by interposing it between the objective lens and the eye, but then without warning describes what this second lens would do if it were used to view the object by itself, at a specified position, independently of the objective lens. He then describes, qualitatively, how the lenses in combination, appropriately spaced, produce a compound instrument: that is, when looked through, it produces a clear inverted image. This logically tangled argument suggests that it arose from an experimental procedure rather than isolated reasoning. A few minutes of looking out of the window with a pair of lenses would result in an inverted, magnified image. The difficult part would be to figure out what is happening to the rays as they pass through the two lenses to the eye. It would have been nearly impossible to consider all the possible placements of lenses and their points of meeting to find the exact conditions that would produce an in-focus image.

Kepler's achievement here is extraordinary, and not just because he created a new kind of instrument. His more important accomplishment was to show how to use the ancient science of ray optics, together with his one extension of it into the phenomena of refraction, to analyse compound systems of lenses.

7.5 Analysis of the Galilean Telescope

In view of Galileo's lens combination (concave eyepiece, convex objective), Kepler's section on this type of telescope (Sect. 11, Items CI–CXXIV) is fascinating. Galileo provided a rudimentary account of how he designed the instrument in *The Assayer* (*Il Saggiatore*, Rome, 1623), which contains no optical theory, stating only how he

Fig. 7.10 Convex lenses combined to form a telescope. Reproduced from *Dioptrice* LXXXVI, ed. 1611, p. 43

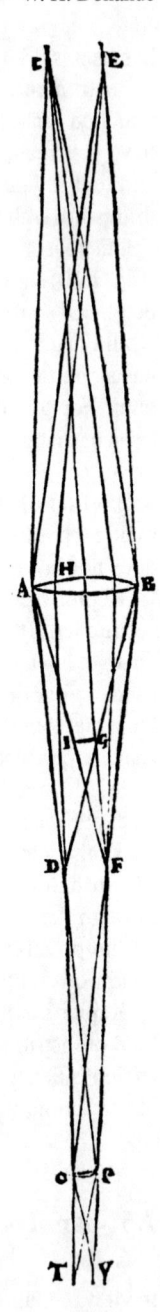

arrived at that combination of lenses.[85] In the *Sidereus Nuncius* (Venice, 1610), the work in which he announces the discoveries he has made with the telescope, he says even less about the instrument, although he wrote, 'On another occasion we shall publish a complete theory of this instrument.'[86] Kepler's masterly development of optical theory, in contrast, allowed him to give a thorough ray-optical treatment of the Galilean instrument. Although Galileo surely knew more than he let on, Kepler comes through as the consummate optical theoretician.

Because we know that, through the assistance of the Archbishop Elector of Cologne (to whom the Dioptice was dedicated), Kepler had been able to examine and use one of Galileo's telescopes,[87] we do not have to speculate about how much of what he writes is based on experience. He clearly knows the instrument and has experimented with a variety of lenses. His ray tracing gives an accurate description of how the instrument works.

Near the beginning of the eleventh section, Kepler notes (Item CIV) a striking difference between the Galilean telescope and those using convex lenses only. In the latter, the secondary lens (*OP*, in the diagram for LXXXVI) is placed closer to the eye (in the neighbourhood of *TV*) than is the meeting point *DF* of the rays coming from the first lens, while in Galileo's instrument, the meeting point is beyond the concave secondary lens—indeed, beyond the eye itself. The concave lens receives the converging rays and redirects them through refraction into either parallel or diverging rays. This results in an upright image when viewed directly though the telescope.

When, on the contrary, the instrument is readjusted to project a picture of the visible object onto a screen such as a wall or sheet of paper (Item CV), the secondary lens must be repositioned so as to redirect the rays into more elongated pencils that remain convergent. Kepler had already used a single lens with a long focal length in conjunction with his eclipse instrument (Fig. 7.4). He described this extended pinhole instrument, twelve feet long (about 3.6 m), at the beginning of Chap. 11 of the *Optics*; in the *Disssertatio cum Nuncio Sidereo*, he proposed the addition of the lens to assist in observation of the Moon[88] Here in Item CV, he replaces the lens with a Galilean telescope, placing the concave lens farther from the objective, nearly at the point of meeting. Kepler notes that this arrangement produces an enlarged and inverted image on the paper, which he calls a 'picture', equivalent to a 'real image' in today's optics.

This form of projection telescope was soon being used for observing the Sun, obviating the need for smoked glass or other means of attenuating the sunlight. The Jesuit astronomer Christopher Scheiner (1573/5–1650), for example, built an elegant

---

[85] Translated as *The Assayer* by Drake and O'Malley in Galileo (1960), pp. 151–336. The outline of Galileo's reasoning process is this: more than one lens will be required; two kinds of lenses exist; therefore, the instrument must consist of some combination of convex and concave lenses; and the combination of the convex and the concave gave the desired result (Galileo, 1960, p. 213).

[86] Galilei (1610), fol. 7r; Galilei (1989), p. 39. The promised theory was never published.

[87] This is Ernst, Duke of Bavaria, mentioned at the beginning of Sect. 7.4. See Kepler (1611b), fol. *3ʳ; Kepler (1941), p. 318 l. 37–40.

[88] Kepler (1604), pp. 335–339, Kepler (2000), pp. 347–350; Kepler (1610), p. 13; Kepler (1965), p. 46; see Sect. 7.1 above.

Fig. 7.11 Scheiner's
projection telescope.
Reproduced from *Rosa
Ursina sive Sol* Book II,
1630, p. 77

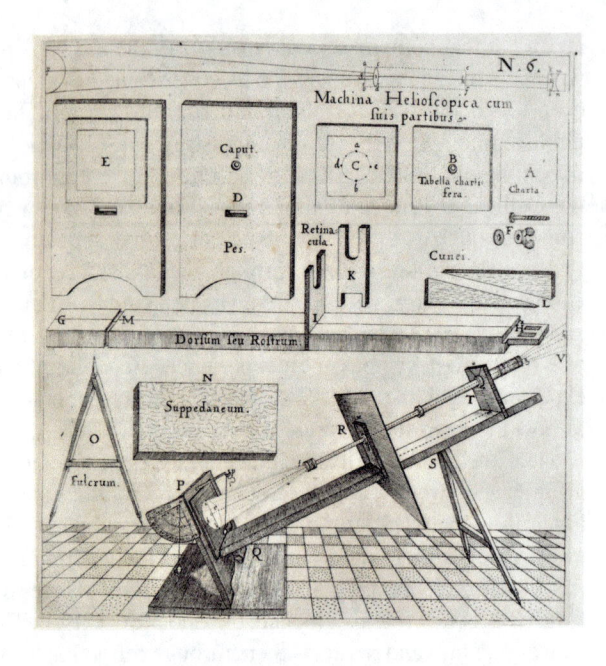

instrument incorporating the convex/concave optics, and used it for his extensive study of sunspots. Figure 7.11 shows his engraving of the instrument, constructed in the 1620s.[89] A cross-section of the telescope, showing the two lenses, appears at the upper right.

Items CVII and CVIII are converses of one another, and show that any combination of concave eyepiece and convex objective may be used, provided the objective has a greater radius of curvature than the eyepiece. These two propositions provide a general account of the functioning of the Galilean telescope (Fig. 7.12). The next three (Items CIV–CXI) show how to position the two lenses of a Galilean telescope in relation to the meeting point of the rays coming from the convex objective lens.

Many of these items are related to what is now called the 'focal length' of the lens. Although Kepler did not use this term, the concept was clearly stated in the enunciation of Item XXXIX:

> The meeting place (of incoming parallel rays) after the lens occurs at the point which is close to a semidiameter from the opposite surface of the convex lens—that is, in its centre.

'Meeting place' by itself, without the parallel-rays specification, is a problematic term because (as Kepler suggests here) it is not a fixed length, but is dependent upon the law of refraction and the convergence or divergence of the rays entering the lens. Thus, the meeting of the actual rays does not occur at a single well-defined point (as he proved in the *Optics*).[90] He prefers to use the radius of curvature of

---

[89] Scheiner (1626–1630), Book II p. 77.

[90] Kepler (1604) Chap. 5 Prop. 19, pp. 193–194; Kepler (2000), pp. 210–211.

Fig. 7.12 Galilean
telescope. Reproduced from
Dioptrice CV, ed. 1611, p. 55

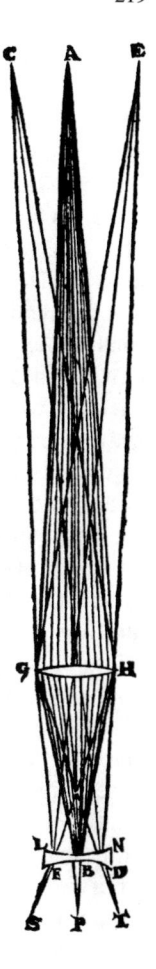

the lens, which is a well-defined measurable quantity, rather than the 'distance of
convergence', as he would say. The focal length of a lens, in modern optics, may
be determined approximately by shining a distant light source through a lens and
measuring the distance from the optical centre of the lens to the projected image
of the light source.[91] As a result of the absence of generally accepted terminology,
Kepler uses several separate items to state what we would now gather into a single
statement or equation. Thus, where he says (Item CXII) 'Convex lenses with a large
circle need a long distance from the concave lens and the eye; with a small circle,
a short distance', this connects a long focal length with a convex lens having a
long radius of curvature, and vice versa. And then Items CXIII and CXV–CXVIII
correspond to the single equation

---

[91] The exact formula for more general optical systems involves at least four variables. See Wikipedia,
article 'Focal Length'. (Accessed 31 October 2022).

$$f_1/f_2 = m,$$

where f_1 and f_2 are the focal lengths of the objective lens and the eyepiece respectively, and m is the magnification.

Items CXIX and CXX are about lens apertures. A large lens gathers more light and so makes objects 'clearer and stronger'. In item CXXII Kepler considers the effect of a stop, restricting the passage of light rays to the middle part of the objective lens. Galileo had proposed the use of such stops in the *Sidereus Nuncius*, in hopes that the diameter of the stop would restrict the field of view, and would thus allow small differences in angular positions to be measured accurately.[92] All efforts to turn this idea into a practical measuring instrument failed.[93] Kepler tried using the telescope with and without the stop, and got its effect wrong in an interesting way. He had noticed that an open objective lens can create rainbow colours around luminous objects, and that putting a stop over the lens or inside the tube lessens this effect. But he attributed it to 'spirits' in the retina, and saw the diaphragm as being like the iris in the eye, applied to the telescope.[94] Of course, what he has discovered is the chromatic aberration of the lens: the wedge-shaped edges of the lens act as a prism, refracting the different colour components of light at slightly different angles. However, to notice an effect is not to explain it: the correct account was provided by Isaac Newton (1642–1727) in the 1660s.[95] Kepler had already mentioned this phenomenon in his little book on his observations of Jupiter's satellites.[96] He attributed it there to 'weakness of the faculty of sight' (*imbecillitas visus*), not to the heavenly bodies themselves, since he had observed the same effect, the 'colours of the rainbow', in daytime observations. This is very likely the first published account of the phenomenon. At the end of Sect. 11, in Items CXXIII and CXXIV, Kepler writes briefly about the use of the telescope. Item CXXIII is a Problem: 'To view the visible up high, down low, from the right or left—wherever you wish'. Anyone who has used a telescope of the Galilean design will know what Kepler is describing. The concave eyepiece has an extraordinarily wide exit pupil, which spreads out the pencils of rays coming from different points of the object. However, the circular opening of the objective end of the tube, viewed through the concave eyepiece, seems tiny. One has to look *through* that small circle, and beyond it one finds a magnified piece of the luminous field. It seems rather claustrophobic until one realizes that, without redirecting the tube, one can move the eye up, down, left right—whatever—and previously unseen parts of the visible object will come into view.

Item CXXIV, which concludes the section, shows a way of estimating the magnification of a telescope by comparing the size of an object (see with the unaided eye) with the size as viewed in the telescope. It is surprising that, having read Galileo's

[92] Galilei (1610), fol. 7r; Galilei (1989), p. 39.

[93] Galilei (1989), p. 39, note 32.

[94] Kepler (1611a), Item CXXII, pp. 64–65; Kepler (1941), p. 403.

[95] Newton invented the reflecting telescope to avoid the chromatic aberration of lenses that he had explained by his study of prisms. See Westfall (1980), pp. 156–174.

[96] Kepler (1611b), fol. 4^R; Kepler (1941), p. 320, ll. 8–12.

more elegant and precise method, Kepler proposes this procedure that depends on visual estimating rather than measurement. Galileo described drawing two circles, each on a separate sheet of paper, one having a diameter twenty times that of the other, placing both at a distance and viewing the smaller circle through the telescope with one eye while viewing the large circle with the other eye, unaided. If the circles appear equal, the magnification is 20×. A precise comparison can be obtained by adjusting the distance of one of the circles.[97]

7.6 The Final Section: *Κρύψις*

Now we come to the mysterious part. The Greek title of this last section can mean 'concealment' or 'mystery, secret'. Is Kepler trying to hide something? Is he about to reveal secrets of Optics? He does not explain, so we shall have to dig out his meaning from the seventeen items in this section.

The items fall into two distinct groups, introduced by three preparatory propositions (Items CXXV–CXXVII) showing the results of using some simple compound lenses. Kepler then considers a new class of lenses, of which one surface is convex and the other is concave. In Item CXXVIII, he considers the case where the surfaces have the same curvature, and concludes that such a lens leaves the rays unaffected, though slightly displaced. In the following items, Kepler allows the second surface to have a curvature different from the first surface, and concentrates mainly on whether, and under what conditions, the rays exiting the mixed lens converge or diverge. In items CXXX, he says that this kind of lens is called a 'meniscus'. This part concludes with Item CXXXIV, which argues that every compound lens or meniscus lens is equivalent to some simple convex or compound lens. Then we get the final group of items, which seem more along the line of 'tricks'. One such trick involves hiding an extra lens inside the tube of a telescope, invisible to the user. In Item CXXXV, for example, he installs a second convex objective lens behind the first one, relying upon Item CXXV, which shows that the doubled convex lens is equivalent to a single lens of a shorter focal length. This shorter focal length cuts the length of the instrument in half, which might be a puzzle for someone who has some experience with the telescope. A similar trick (Item CXXXVI) involves doubling the concave eyepiece lens, which will increase the magnification of the instrument. In Item CXXXIX, the trick telescope has a concave lens at each end: hidden behind the concave objective lens is a convex lens, making a long-focal-length telescope that (by CXIII and CXV–CXVIII) increases the magnification. Item CXXXVII plays a similar trick, three different ways. In CXL, both lenses appear convex, but the image is upright (the eyepiece has a contiguous concave lens with strong curvature), and finally, in Item CXLI, the whole instrument is contrived to look inverted end to end (again with hidden compound lenses, in two different ways).

---

[97] Galilei (1610), fol. 6v; Galilei (1989), p. 38.

So the mystery title, κρύψις, might best be translated 'concealment'. Kepler reveals mysteries in the earlier items, and then delights in practical jokes to finish the book on a prankish note.

Kepler has tended to be characterized either as a powerfully imaginative theorist, or alternatively (and perhaps incongruously) as a computational drudge. In the *Dioptrice*, however, we see quite a different Kepler. He steps onto the stage in the formal costume of the mathematician, presenting his subject through the use of mathematically organized proofs. Perhaps largely for this reason, the book has been described as an epoch-making exposition of optical theory. But hidden beneath the formality is a wonderful combination of diagrammatic intuition and practical experience. He clearly knows this stuff, in a way that bespeaks much tinkering with lenses, both by themselves and in combinations. This is where the earlier items begin, and this is where we end up. There is, to be sure, some fine mathematical thinking in the book, most notably in Item XXXIX, where Kepler anticipates the differential calculus by considering a light ray at the very edge of a lens as if passing through both surfaces at once. But the surprising ending, with Kepler playing the role of the stage magician (with perhaps and appreciative nod in the direction of Porta and his 'Natural Magick')[98] seems on reflection to be in keeping with the more earthy subtext of this surprising treatise.

References

Alpers, S. (1983). *The art of describing*. Chicago: University of Chicago Press.

Aristotle. (1935). *On the soul* (Greek text edited and translated by W. S. Hett). Cambridge, Mass.: Heinemann and Harvard University Press (Loeb Classical Library 288).

Brahe, T. (1858). In Kepler, J. (1858), *Kepleri Astronomi Opera Omnia, Volumen I* (Ch. Frisch, Ed., pp. 44–46). Frankfurt: Heyder & Zimmer.

Brahe, T. (1925). *Tychonis Brahe Dani Opera Omnia*, Tomus VIII (J. L. E. Dreyer, Ed.). Copenhagen: Libraria Gyldendaniana, pp. 52–55.

Brahe, T. (1945). *Johannes Kepler Gesammelte Werke, Band XIII: Briefe 1590–1599* (M. Caspar, Ed., pp. 204–205). Munich: C. H. Beck.

Brown, J. E. (1978). The science of weights. In *Science in the Middle Ages* (D. C. Lindberg, Ed.). Chicago: University of Chicago Press.

Christianson, J. R. (2000). *On Tycho's Island*. Cambridge: Cambridge University Press.

Copernicus, N. (1543). *De revolutionibus orbium coelestium*. Nuremberg: Johannes Petreius.

Copernicus, N. (1992). *On the revolutions* (Translated from the Latin by E. Rosen). Baltimore: Johns Hopkins University Press.

Descartes, R. (1637). *Discours de la Méthode pour bien conduire la raison...plus la Dioptrique, les Météores et la Géométrie....* Leiden: Jan Maire.

Donahue, W. H. (1996). Kepler's approach to the oval hypothesis of 1602. *Journal for the History of Astronomy, 27*, 281–295.

Davis, A. E. L. (1975). Systems of conics in Kepler's work. *Vistas in Astronomy, 18*, 673–685.

Dürer, A. (1525). *Underweysung der Messung*. Nuremberg. Online at https://commons.wikimedia. org/wiki/Category:Underweysung_der_Messung

Galilei, G. (1610). *Sidereus Nuncius*. Venice: Thomas Baglioni.

---

[98] Published in many editions and translations well into the seventeenth century.

Galilei, G. (1960). *The controversy on the comets of 1618: Galileo Galilei, Horatio Grassi, Mario Guiducci, Johann Kepler* (Edited and translated by S. Drake and C. D. O'Malley). Philadelphia: University of Pennsylvania Press.

Galilei, G. (1989). *Sidereus Nuncius, or the Sidereal Messenger* (Translated from the Latin by A. van Helden). Chicago: Chicago University Press.

Gillespie, C. (Ed.). (1970–1980). *Dictionary of scientific biography.* New York: Charles Scribner's Sons.

Huygens, C. (1690). *Traité de la Lumière.* Leiden: Pierre van der Aa.

Huygens, C. (1912). *Treatise on light* (Translated from the French by S P. Thompson). London: Macmillan and Company.

Jardine, N. (2009). 'God's ideal reader: Kepler and his serious jokes. In R. L. Kremer & J. Włodarczyk (Eds.), *Johannes Kepler: From Tübingen to Żagań* (pp. 41–51). Warsaw: Polish Academy of Science (*Studia Copernicana*, 42).

Kepler, J. (1604). *Ad Vitellionem paralipomena quibis astronomiae pars optica traditur.* Frankfurt: Claudius Marnus and the heirs of Johannes Aubrius.

Kepler, J. (1609). *Astronomia* nova aitiologêtos seu physics coelestis tradita commentariis de motibus stellae Martis ....* Heidelberg: Vogelin.

Kepler, J. (1610). *Dissertatio cum Nuncio Sidereo.* Prague: Daniel Sedesanus.

Kepler, J. (1611a). *Dioptrice.* Augsburg: David Frank.

Kepler, J. (1611b). *Narratio de observatis a se quatuor Iovis satellitibus erronibus.* Frankfurt: Zacharias Palthenius.

Kepler, J. (1904). *Johannes Keplers Dioptrik* (Translated into German by F. Plehn). Leipzig: Wilhelm Engelmann. Available online: https://play.google.com/books/reader?id=JXtemtSrI 80C&pg=GBS.PP6&hl=de

Kepler, J. (1937). *Johannes Kepler Gesammelte Werke, Band III: Astronomia Nova.* Munich: C. H. Beck.

Kepler, J. (1939). *Johannes Kepler Gesammelte Werke, Band. II: Astronomiae pars optica* (F. Hammer, Ed.). Munich: C. H. Beck.

Kepler, J. (1941). *Johannes Kepler Gesammelte Werke, Band IV: Kleinere Schriften 1602–1611 / Dioptrice* (F. Hammer, Ed.). Munich: C. H. Beck.

Kepler, J. (1945). *Johannes Kepler Gesammelte Werke, Band XIII: Briefe 1590–1599* (M. Caspar, Ed.). Munich: C. H. Beck.

Kepler, J. (1949). *Johannes Kepler Gesammelte Werke, Band XIV: Briefe 1599–1603* (M. Caspar, Ed.). Munich: C. H. Beck.

Kepler, J. (1951). *Johannes Kepler Gesammelte Werke, Band XV: Briefe 1604–1607* (M. Caspar, Ed.). Munich: C. H. Beck.

Kepler, J. (1955). *Johannes Kepler Gesammelte Werke, Band XVII: Briefe 1612–1620* (M. Caspar, Ed.). Munich: C. H. Beck.

Kepler, J. (1965). *Kepler's Conversation with Galileo's Sidereal Messenger* (Translated from the Latin with and introduction and notes by E. Rosen). New York: Johnson Reprint Corp.

Kepler, J. (2000). *Optics: Paralipomena to Witelo and the optical part of astronomy* (Translated by W. H. Donahue). Santa Fe: Green Lion Press.

Kepler, J. (2015). *Astronomia nova* (Translated by W. H. Donahue and with Foreword by O. Gingerich). Santa Fe: Green Lion Press.

Lindberg, D. C. (1970). *John Peckham and the science of optics.* Wisconsin: University of Wisconsin.

Lindberg, D. C. (1976). *Theories of vision from Al-Kindi to Kepler.* Chicago: University of Chicago.

Pitzer, A. (2021). *Icebound: Shipwrecked at the edge of the world.* New York: Charles Scribner's Sons.

Proclus Diadochus. (1992). *A commentary on the first books of Euclid's elements* (Translated from the Greek by G. R. Morrow). Princeton: Princeton University Press.

Sabra, A. I. (1994). *Optics, astronomy and logic.* Variorum. Farnham: Ashgate.

Scheiner, C. (1626–1630). *Rosa Ursina sive Sol.* Bracciano: Andreas Phaeus.

Straker, S. M. (1970). *Kepler's optics: A study in the development of seventeenth-century natural philosophy*, Ph.D. Dissertation, Indiana University.

Thoren, V. E. (1990). *The Lord of Uraniborg*. Cambridge: Cambridge University Press.

Westfall, R. S. (1980). *Never at rest*. Cambridge: Cambridge University Press.

Chapter 8
Kepler and Galileo

J. V. Field

Galileo Galilei (1564–1642) was born about seven years before Kepler, but he began to publish only in his forties, by which time Kepler, as Imperial Mathematician and the author of several books, already had an international reputation as an astronomer. Despite the age difference, Galileo outlived Kepler, and some of the works for which he is now best remembered were not published until after Kepler's death: for instance the *Dialogue concerning the Two Chief World Systems* (*Dialogo sopra i due massimi sistemi del mondo*, Florence, 1632), which considers geocentric and heliocentric systems largely in their relation to terrestrial physics, and the *Discourses and mathematical demonstrations concerning Two New Sciences* (*Discorsi e dimostrazioni matematiche intorno a due nuove scienze*, Leiden, 1638), which deals with projectile motion and the strength of materials. Galileo had become famous with the publication of his account of the observations he made with the telescope, *Sidereus nuncius* (Venice, 1610)—the title is awkward to translate because '*nuncius*' means both 'message' and 'messenger'. The book seems to have been very widely read and received the commercial accolade of appearing in several pirated editions.

Kepler and Galileo came from different backgrounds: Galileo, whose father was a professional musician, and whose family had been settled in Florence for several generations, was born and lived in Italy; Kepler, the son of a mercenary soldier and the daughter of an innkeeper, travelled a great deal, but always lived north of the Alps and within the Holy Roman Empire. This social and geographical separation was overridden by their professional concerns and did not prevent the exchange of letters. The religious divide was of greater importance. Galileo was a Catholic and Kepler a Protestant. In a period when religion was pervasive and important, and sectarian allegiances could be highly significant, it is probable that Galileo felt he needed to be a little circumspect in his dealings with Kepler. Kepler, whose relations

J. V. Field (✉)
Birkbeck, University of London, London, UK
e-mail: jv.field@hart.bbk.ac.uk

© Springer Nature B.V. 2024
A. E. L. Davis et al. (eds.), *Reading the Mind of God*, Springer Praxis Books,
https://doi.org/10.1007/978-94-024-2250-4_8

with his own church were always rather complicated,[1] seems to have had no qualms about regarding Galileo simply as a fellow Copernican and therefore in principle an ally. For instance, since Copernicus' theory made the Earth a planet, both Galileo and Kepler were highly suspicious of the standard Aristotelian distinction between celestial and terrestrial physics. Their sharing this attitude works against one division that contemporaries might have made between them: Kepler being a mathematician (that is an astronomer) and Galileo a natural philosopher.

Today, Galileo's use of the telescope tends to cause him to be regarded as an astronomer, but in his own time this was not so. Apart from casting some horoscopes, the only standard astronomer's task he ever undertook was teaching mathematics at the University of Padua (the University of the Republic of Venice) from 1592 to 1610. Unlike Kepler, Galileo never made systematic calculations of planetary positions or produced astronomical tables or ephemerides. These were the normal tasks for astronomers of the time.

Apart from their shared Copernicanism, and their both taking an interest in the New Star that appeared in October 1604 (now known as 'Kepler's supernova'), the most obvious overlap in Galileo's and Kepler's research was in optics, specifically in connection with telescopes in the years 1610–11. Later they both became involved in a controversy over the comets observed in 1618–19, which raised questions about physics and about the planetary system. However, their first exchange of letters was in the late 1590s when Kepler sent two copies of his newly-published *Mysterium cosmographicum* (Tübingen, 1596) to Italy.

8.1 Cosmology

In a letter written from Graz in early October 1597, Kepler tells his former astronomy teacher Michael Maestlin (1550–1631) how the correspondence with Galileo began:

> Recently I sent 2 copies of my little work (or rather yours) to Italy, they were accepted in a most gracious and welcoming spirit by a Mathematician at Padua, by name Galilaeus Galilaeus, as he signs himself. For he himself has been of the Copernican sect for many years. He sent one copy to Rome and asked to have more.[2]

In view of what happened later, Kepler's choice of the term 'haeresis'— here translated 'sect' but also used to denote 'heresy'—looks unfortunate.

From how Galileo begins his letter of thanks, it seems that an acquaintance of Kepler's was travelling to Italy, so Kepler gave him two copies of the *Mysterium cosmographicum* with instructions to pass them on to professors of mathematics (which would have included astronomy). The professor in Padua just happened to be Galileo.

---

[1] See Methuen, 'Kepler, religion and natural philosophy: a theological biography' (this volume, Chap. 1).

[2] Kepler to Michael Maestlin, beginning of October 1597, in Kepler (1945), letter 75, ll. 119–23, p. 143.

As regards the book itself, from what we know of his opinions later in life, it seems extremely unlikely that Galileo was sympathetic to Kepler's use of the five regular polyhedra to explain the number and spacing of the planetary orbs, an explanation given prominence by the introduction of an elegant fold-out engraved plate. If he read so far, Galileo might perhaps have been interested by the questions Kepler poses to show the superior explanatory power of the Copernican theory.[3] Galileo's favoured argument for Copernicanism, that is for the motion of the Earth, was that it could explain the tides, namely as being caused by the combined action of the diurnal rotation of the Earth and its annual revolution about the Sun, motions which sometimes reinforce one another and sometimes counteract one another. This theory is described in detail, with experimental support, in Galileo's *Dialogue concerning the Two Chief World Systems* of 1632.[4]

It is, in any case, obvious that in 1597 there was no instant meeting of minds. Galileo's letter thanking Kepler for the book, dated 4 August 1597, is a classic of the genre.[5] For example, after polite preliminaries, in the rather elaborate style that is normal in this period, Galileo says

> of the book I have so far looked at nothing except the preface, from which, however, I have gathered something of your intention … .

Galileo does, nevertheless, go on to say he is 'gratified to have an associate in the investigation of truth'; and adds 'I shall read through your book in a fair spirit, since I am certain that I shall find in it most beautiful things'. So far so conventionally polite.

The letter shows Galileo's literary skill, which was to be evident in his published writings also, but its chief interest is that Galileo says that he too believes the Copernican theory to be true:

> many years ago, I came to be of Copernicus' opinion, and from this position I have also been able to find the causes of many effects in natural things, which beyond doubt are inexplicable by the common hypothesis … .

It is possible that here 'natural things' is being used in the Aristotelian sense of 'things in the sublunary world'. In any case, this passage is our earliest evidence for Galileo's Copernicanism. Unfortunately, he does not tell Kepler which 'effects in natural things' he has in mind. It is tempting to suppose, but impossible to prove, that he meant the tides. Galileo adds that he has not yet written about these opinions, there being general disapproval of Copernicanism. Here it is relevant that at Padua Galileo was being paid to teach standard geocentric astronomy. The university had a large medical faculty, and students of medicine needed to learn some elementary astronomy in order to become competent at applying astrology in their medical practice. At this time, astrology played an important part in medicine, as a guide in diagnosis,

---

[3] *Mysterium cosmographicum* is discussed in more detail in Field, 'Kepler's cosmology' (this volume, Chap. 2). See also Field (1988).

[4] Galilei (1632, 1953).

[5] Galileo to Kepler, 4 August 1597, letter 73, in Kepler (1945), pp. 130–31. An English translation of the whole letter can be found in the appendix to this chapter.

prognosis and treatment. Since it is concerned with the effects of heavenly bodies on terrestrial ones, astrology is effectively geocentric. Accordingly, there would have been little sense in Galileo teaching his students about Copernicus' work, which was at once too difficult and of doubtful relevance.[6]

Kepler replied to Galileo on 13 October 1597. His reply, which is considerably longer than Galileo's letter, tells Galileo that

it is not only your Italians who cannot believe that they move unless they feel it; but here in Germany also we do not enter into this belief with the warmest gratitude.

Kepler goes on to urge Galileo to discuss Copernicus' work with his colleagues and friends and in print, which he (Kepler) has found to have an effect.

Then, noting that some time has elapsed since Galileo wrote to him, Kepler suggests that by now Galileo has read the book, and asks for his comments on it.

After some further urging of Galileo to discuss his opinions in public, Kepler then asks about the reasons for supporting Copernicus' theory that Galileo had hinted at:

You can at least communicate with me in writing privately, if you do not wish to do so publicly, if you have found something in support of Copernicus.

Next, Kepler asks Galileo to make some observations for him, since he has not got any instruments with which to make them himself. It seems most unlikely that Galileo was in fact in a position to undertake this task.

Kepler's letter is now in Florence, so we have good reason to suppose Galileo received it. But it seems he did not reply. When he came to write to Galileo again, after the publication of *Sidereus nuncius* in 1610, Kepler would politely refer to this earlier exchange as 'our interrupted correspondence'.

Despite his apparent reluctance to tell Kepler what he thought of the work, Galileo seems to have kept his copy of the *Mysterium cosmographicum*. At least, he uses numbers from it in his *Dialogue concerning the Two Chief World Systems* (1632), though the numbers were by then somewhat out of date.[7] Had he kept up with Kepler's subsequent publications, by 1632 Galileo should have been using the revised, and much more accurate, sizes of the orbits and extreme speeds given in book 5 of *Harmonice mundi* (1619). Not that it was wise to cite Kepler at all. In 1630 the entry for Kepler in the *Index of Prohibited Books* published in Rome says 'Epitome of Copernican Astronomy and all other works by this author'. Although events turned out otherwise, Galileo had hoped that his Dialogue would not attract opprobrium from the Church. In the end, Galileo's references to Kepler seem to have passed unnoticed: the *Dialogue* provided the religious authorities with far more important things to which they wished to raise objections.

But let us return to the period before Galileo received an official warning from the church authorities that he should not discuss Copernicanism in public. That message was to be delivered in 1616. Meanwhile, the exchange of letters with Kepler does

---

[6] On the relations between Copernicanism and astrology see Westman (2011).

[7] See Drake (1973), pp. 174–191.

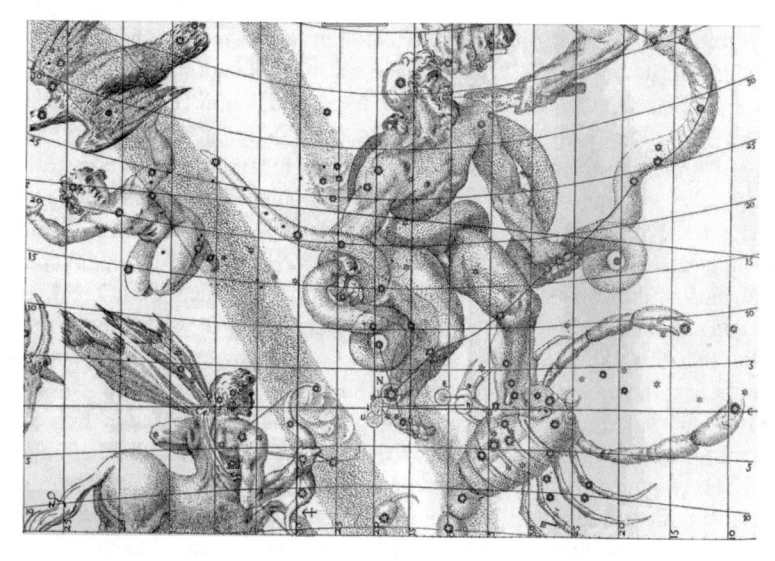

Fig. 8.1 Star map showing the position of the New Star in the right ankle of Serpentarius (now called Ophiuchus). Reproduced from J. Kepler, *De stella nova,* 1606

not seem to have encouraged Galileo to think about astronomy. However, in October 1604 the Universe and the university authorities at Padua took a hand in the matter.

8.2   The New Star of 1604

On 9 October 1604, a new star was observed in the constellation of Serpentarius (see Fig. 8.1). It was very bright, but it was rather low in the sky, which meant that finding its position involved making corrections for atmospheric refraction. Since astronomers were not in agreement about such corrections there was room for dispute about the corrected positions. In contrast, the star that had appeared with similar suddenness in 1572 (now known as 'Tycho's supernova') had been more convenient to observe, since it was in the constellation of Cassiopeia, which at latitudes in Europe meant that the star never set and was always high in the sky, well above the altitude at which corrections for atmospheric refraction were applied at the time. (Kepler discussed atmospheric refraction, and corrections for it, in his first book on optics, *Things not in Witelo that belong to the optical part of astronomy,* 1604.)[8]

As with its predecessor in 1572, measuring the position of the New Star of 1604 in the sky, that is finding its position relative to nearby fixed stars, was crucial as providing the only observational evidence for its position in the Universe. Repeated measurements of the star's position, or simultaneous measurements from different

---

[8] Kepler (1604); reprinted Kepler (1939); see also Donahue, 'Kepler's work on optics' (this volume, Chap. 7).

places on the Earth, allowed one to calculate its diurnal parallax, and hence its distance from the Earth (its 'altitude'). That distance was a key factor in decisions about the nature of the Star. On the positive side, it so happened that in 1604 there were many observers of the New Star. It had appeared close to a Great Conjunction, that is a meeting of Jupiter and Saturn, a rare event that was regarded as of great astrological significance. Accordingly, a large number of astronomers were looking at that region of the sky at the appropriate time. As a result, to put it in today's terms, we have an unusually large amount of data for reconstructing the light curve of the first phase of the explosion of the 1604 supernova, making it of lasting scientific interest.

Today's explanation of the New Stars of 1572 and 1604 is fairly simple and very dramatic: the stars exploded. The explanations offered by astronomers and natural philosophers at the time were much tamer. The idea of a star exploding was perhaps excessively non-Aristotelian. The various explanations are collected by Kepler in his book about the star of 1604, *On the New Star* (*De stella nova*), published in Prague in 1606.[9] Having recorded the observations of the star's brightness and position, and having devoted considerable space to astrological interpretations, Kepler carefully refutes several of the theories put forward to explain the star. The theory that the star is newly created is rejected on methodological grounds. Kepler says 'However, before we come to [special] creation, which puts an end to all discussion, I think we should try everything else.'[10]

Soon after the appearance of the star—to which he was not an early witness, thanks to cloudy weather in Prague—Kepler had written a brief popular account, in German, presumably intended for local circulation.[11] His more substantial work on the star was no doubt delayed by the pressure of other work, specifically his work on the orbit of Mars. Kepler reached the conclusion that the orbit was elliptical in May 1605 and, after he had finished writing *Astronomia nova*, he then expended considerable effort on getting his work into print. (It was eventually published in Heidelberg in 1609.) Hence, probably, the delay in Kepler's turning his attention to writing a book on the New Star. The delay may partly explain why, conveniently for historians, Kepler gives what is effectively a literature survey. The delay perhaps also partly accounts for the reflective nature of the book.

Meanwhile, in Padua, Galileo was teaching astronomy, and the university authorities decided his duties extended to giving some formal public lectures on the New Star that was attracting so much attention. The surviving evidence about the lectures is scrappy and does not suggest that Galileo embraced the task with much enthusiasm.[12] Nor did he have much time to plan. It seems the lectures were delivered in late November or early December 1604. They were in Latin and therefore addressed to a learned audience, including his university colleagues, rather than to the public at large.

---

[9] Kepler (1606); reprinted in Kepler (1938). For essays on topics connected with the New Star, see Boner (2020).

[10] Kepler (1606), Chap. 22; Kepler (1938), p. 257, lines 23–24.

[11] For an English translation see Field and Postl (1977).

[12] On the lectures and their aftermath, see Cosci (2018a).

No complete text of the lectures has been found, but some of their content can be reconstructed from notes taken by a friend of Galileo's, and from later publications. At the time—and predictably, since there was much dispute concerning the New Star itself—the public lectures excited considerable discussion (much of it strongly critical of Galileo). His explanation for the sudden appearance of the star avoided directly contradicting Aristotle's assertion that the heavens were unchanging, instead suggesting that the New Star was a body of vapours that rose from the region below the Moon and moved upward through the planetary spheres to the sphere of Jupiter, where it was ignited by the Great Conjunction. With hindsight, this looks like a salute to Aristotle's ideas about comets, which as transitory phenomena were not considered to belong to the celestial region above the Moon. However, for Galileo's opponents his theory clearly did not settle the matter of the origin of the New Star. Specifically, at Padua there were thoroughgoing and very learned Aristotelian natural philosophers, led by Cesare Cremonini (1550–1631), several of whom were as skilled in arguing for their opinions as Galileo was for his.

It may have been partly a wish to avoid direct conflict with colleagues that made Galileo choose to publish his ideas in collaboration with others and under pseudonyms. There was in fact a tradition at Padua of carrying on disputes in this way, and it is possible that contemporaries were expected to work out the identities of authors with only these polemical texts to their names: Cecco di Ronchiti da Bruzene (Padua, 1605), Astolfo Arnerio Marchiano (Padua, 1605) and Alimberto Mauri (Florence, 1606).[13] We shall come across pseudonymous writings and use of intermediaries again in connection with the controversy over the comets of 1618–19 (see Sect. 8.4 below). For the New Star of 1604, there are surviving manuscript notes that link Galileo's lectures to these pseudonymous publications, but there are no publications under Galileo's own name. Although there is nothing very extraordinary in Galileo's adoption of this system of pseudonymous argumentation, his choice may perhaps also reflect the difficulties in having open discussions with his colleagues that were hinted at in his letter to Kepler in 1597.

There are indications in the third of these pseudonymous publications, the one by Alimberto Mauri, that Galileo had been reading Kepler's book on the New Star. There is, however, no evidence that the correspondence abandoned in 1597 was renewed, or that Kepler realized that Galileo was responsible for any of the very numerous writings on the New Star. However, the New Star had turned Galileo's attention to astronomy, that is to astronomy beyond the standard material he was employed to teach. And it may be that, in connection with its discussion of corrections for atmospheric refraction, Galileo also read Kepler's treatise on optics of 1604. The work of Kepler and Galileo on the New Star of 1604 shows some confluence of interests, but it did not bring Kepler and Galileo together. Galileo's work with the telescope did.

---

[13] See Drake (1976, 1978); Cosci (2018b, 2019); Cosci (forthcoming).

8.3 The Telescope

By the time *Sidereus nuncius*, Galileo's short description of the astronomical observations he had made with his telescopes, was published in Venice in March 1610, Kepler was established as Imperial Mathematician in Prague, having succeeded Tycho Brahe in the post after the latter's death in 1601. The situation between Kepler and Galileo was now more or less reversed from what it had been in 1597: this time it was Galileo who was eager to have Kepler's support. From the first, *Sidereus nuncius* had been enthusiastically received, and the first printing had sold out in days. The work is indeed immensely readable. Galileo writes vivid descriptions and provides clear drawings of the Moon. He also gives a careful step by step account of his discovery of four small 'stars' that move round Jupiter. But Galileo was keen that his work should be taken seriously by professional astronomers. He had made sure Kepler would see the book by having the Florentine ambassador, Giuliano de' Medici (1574–1636), deliver a copy to the Emperor. As it happened, this was just as well, since the Imperial ambassador in Venice, Georg Fugger (1577–1643), a member of the famous banking family of Augsburg, had decided that, although there was a great deal of fuss being made about it, the book was not really worthy of Rudolf's attention.

It is clear from the beginning of Fugger's letter to Kepler about the *Sidereus nuncius*, dated 16 April, that Kepler had written to him to enquire about it, but this letter has not survived. The first paragraph of Fugger's reply speaks of Kepler's works. The second turns to Galileo:

> In relation to Galileo's aetherial messenger [*nuncium aethereum* i.e. *Sidereus nuncius*], I had a copy in my hands some time ago, but because to many experts in the study of mathematics it seemed to be a dry account or lacking a philosophical basis [i.e. lacking a basis in Natural Philosophy], a mere show (*palliata ostentatio*), I did not make so bold as to send it to his Holy Roman Majesty.[14]

The significance of the words 'palliata ostentatio' is not clear. The word 'palla' is the Latin term for a Greek style of cloak, and by extension 'palliata' refers to a form of Latin drama based on a Greek text. It is possible the use of the word here implies Galileo is borrowing from an unacknowledged Greek source. The Greek author Fugger presumably had in mind, but does not name, was Plutarch (b. AD 45, d. 119–23), who had written of the Moon as Earth-like in *The Face in the Moon* (published as one of the *Moralia*). The work was well known in this period. Galileo is most likely to have known either the Latin version translated by Xylander (Venice, 1572) or the Italian one translated by Gandino (Venice, 1598).[15] Fugger immediately goes on to say that Galileo's character is not above suspicion:

> The man [i.e. Galileo] knows the process [of pretence] and is in the habit of decorating himself with others' plumes, collected here and there, as the crow in Aesop, to the point that he also claims to be the inventor of this ingenious spy-glass (*perspicilla*), whereas, however, a Belgian who came to these parts through France (*Galliam*) first brought [it] here, [and] it

---

[14] Georg Fuggert to Kepler, 16 April 1610, letter 566, in Kepler (1949), p. 302.

[15] See Casini (1983).

was shown to me and to others, and as Galileo saw it he made others in imitation of it, and perhaps, since it was easy to do so, added some things he had invented.

In his general irritation with Galileo's book, Fugger goes a little beyond what historians would now accept as facts. First, he suggests that Galileo saw the telescope brought to Venice, which was probably not the case, since Galileo would have been in Padua at the time. Second, Fugger seems to understand Galileo as claiming to have invented the telescope. In fact, such instruments, that is tubes fitted with combinations of lenses that gave magnified views of distant objects, had been known for some time and Galileo does not claim to have invented his spyglass. But he does claim, and *pace* Fugger, apparently with justification, that using his knowledge of optics he carried out experiments and made a considerable improvement to the performance of the instrument, increasing the magnification with each new modification.[16]

If, in his letter to Kepler, Fugger is indeed describing a common reaction in Venice, then it is no wonder Galileo was keen to hear from a fellow Copernican. Kepler was, of course, as accustomed as Galileo to encountering Aristotelian modes of thought. And he was probably irritated by Georg Fugger's emphasis on describing the popular reception of Galileo's work rather than its astronomical content. In the event, Kepler did indeed give Galileo's work an enthusiastic endorsement. His response was not in a private letter, but took the form of a short book, essentially an open letter to Galileo, *Conversation with the sidereal messenger* (*Dissertatio cum nuncio sidereo*, Prague, 1610). A 'final' draft of the text, written out by a copyist, with Kepler's corrections, is dated 19 April.[17] To put it briefly: Kepler greatly admired Galileo's work. Since spyglasses had a history as fairground entertainment, it was by no means given that what was seen through a telescope should be taken seriously as showing bodies truly present in the sky. Kepler, who of course understood the relevant optics, does not discuss the matter. He simply says he believes Galileo's accounts of his observations and takes them to describe reality.[18]

In setting out his work to persuade others, Galileo had chosen wisely in starting with his detailed description of the Earth-like character of the Moon, a matter in which he could be seen as defending an opinion already known from Plutarch as well as adding considerable detail to it. In any case, as a Copernican Kepler had his own reasons for accepting the non-Aristotelian notion that the Moon is Earth-like, not 'perfect' (that is spherical and smooth) but marked by mountains and dark flat areas (Galileo called them 'seas'). Kepler was also happy to accept that the Milky Way was made up of small stars, which again was an opinion expressed by some Ancient natural philosophers, and that there were four small bodies which moved round Jupiter and moved with it. This last is important, to both Galileo and Kepler, as establishing that there are motions among the bodies of the planetary system that are certainly not centred on the Earth. In the following year, 1611, Kepler was

---

[16] See van Helden (1977), reprinted with a new introduction in van Helden (2008).

[17] For details see Field (1988), esp. Chap. 4, and Field, 'Kepler's cosmology' (this volume Chap. 2). Kepler's *Dissertatio* is translated by Edward Rosen in Kepler (1965).

[18] A fuller account of Kepler's *Dissertatio* is given in Donahue, 'Kepler's work on optics' (this volume, Chap. 7).

able to use one of Galileo's telescopes to observe the moons of Jupiter, to which he gave the name 'satellites' (literally 'attendants').[19] The telescope Kepler used was one made by Galileo and delivered to the Elector Archbishop of Cologne, Ernst of Bavaria (1554–1612), by Galileo's brother Michelangelo Galilei (1575–1631), a professional musician who was employed as a lute-player at the court in Munich.[20]

Galileo's description of the telescope itself and its design, turned Kepler's attention back to optics, this time to consider the action of lenses. This led him to suggest a new form of telescope, using two convex lenses. In contrast to the long title of Kepler's first work on optics, the title of this one is a single word *Dioptrice* (*Dioptrics*), a word taken from Greek but not in common usage.[21] Though Kepler had not concerned himself with this side of things, it happened that this new type of telescope gave a considerably larger field of view than the Galilean design, making it easier to line the telescope up on the object one wished to observe. Kepler's telescope also had various other advantages, such as allowing the introduction of crosshairs (which opened up the possibility of measuring positions more exactly). The advantages eventually led to its becoming the preferred design for astronomical work—so that it is now generally known not by the name 'Keplerian', after its inventor, but simply as 'the astronomical telescope'. The disadvantage of the Keplerian design is that it gives an inverted image, which does not matter for astronomy but was impractical for most terrestrial uses, such as watching approaching ships. But the wider field of view and the introduction of crosshairs were very convenient, so for terrestrial work it became usual to use the Keplerian design but to insert an additional biconvex lens so as to obtain an upright image, or to employ some other system of rectification.

Galileo's telescopes, which used a convex lens for the object glass and a concave one for the eyepiece, gave an upright image, and one of the uses Galileo suggested for the instrument was indeed watching the approach of ships—a matter of obvious importance to the Republic of Venice. Galileo himself, however, continued to turn his telescopes to the skies. Having made an interesting discovery about Jupiter, he naturally looked at the other planets, and in late 1610 he saw that Venus, whose brightness varies a great deal, shows phases like those of the Moon. That is, Venus shows a complete set of phases, including a phase in which its whole face is illuminated, as the Moon's face is at Full Moon. Such a 'Full Venus' phase can occur only when, as seen from the Earth, Venus lies beyond the Sun (see Fig. 8.2). This proves that Venus moves round the Sun, because if it moved round the Earth, as in the standard Ptolemaic geocentric model of the planetary system, Venus would always lie between the Earth and the Sun. This simple refutation of the Ptolemaic model was accepted very rapidly by the astronomical community. The fact probably goes some way to explaining the increasingly wide acceptance of the Tychonic planetary system, in which Venus, together with Mercury, Mars, Jupiter and Saturn, moves round the Sun, while the Sun takes them all with it as it moves round the Earth. The

---

[19] Kepler (1611a), reprinted in Kepler (1941), pp. 312–25.

[20] Schmid (2022).

[21] Kepler (1611b). On this work see the essay by Donahue, 'Kepler's work on optics' (this volume, Chap. 7).

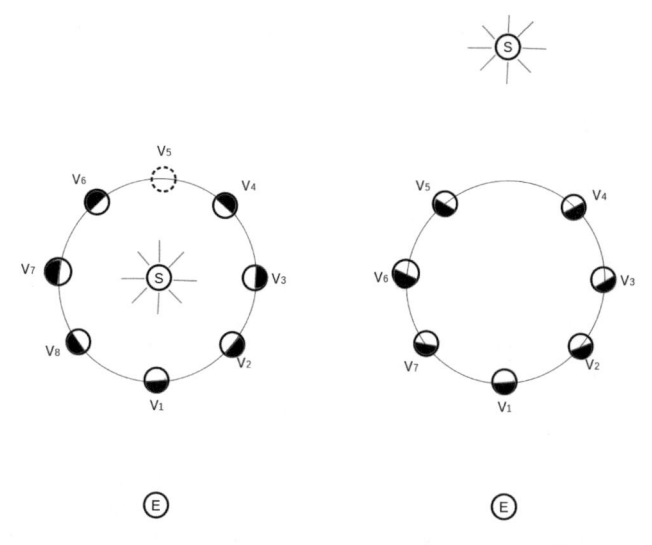

Fig. 8.2 Phases of Venus. *E* represents the Earth, *S* the Sun and V_1–V_8 Venus at eight points in its orbit. *Left:* Heliocentric planetary system. Venus shows a full range of phases resembling those of the Moon. When Venus is in position 1, the dark side of the planet faces Earth and Venus is 'new'; at positions 2 and 3 Venus is in crescent phase; at positions 3, 4, 6 and 7 Venus is gibbous. When Venus is full (dashed circle at position 5) it is unobservable from Earth owing to the glare from the Sun. *Right:* Geocentric planetary system. The phases of Venus range from new to crescent and can never attain full

Jupiter system could also be seen as evidence for this Tychonic model because it showed a moving body as the centre of the motions of other bodies.

Another phenomenon rapidly discovered by the use of the telescope, though this time using a projected image, was that of sunspots. Like the roughness of the surface of the Moon, sunspots also represented a departure from Aristotelian notions about the perfection of heavenly bodies. There was, almost inevitably, a priority dispute about who 'discovered' sunspots, that is who first identified these apparent imperfections in the bright surface of the Sun. When the Jesuit astronomer Christoph Scheiner (1573/5–1650) saw dark circular patches that moved over the Solar disc, he interpreted them as small opaque bodies moving round the Sun close to its surface. Galileo made an extended series of observations, following one dark patch as it moved across the Sun day by day. These observations showed that there were progressive changes in the shape of the dark area that indicated it formed part of the spherical surface of the Sun. That is, the changes in shape were what we should now call 'projection effects'. If, as Galileo claimed, the spots were part of the surface of the Sun than their motion proved the Sun was rotating, and Galileo duly estimated its period of rotation, which he put at 27 days.[22] In his work on the orbit of Mars, Kepler had proposed that the Sun rotated, thus communicating motion to the planets, sweeping them round with its rays. This is far from being a precise or detailed description of how the Sun

---

[22] On the discovery of sunspots and rotation of the Sun, see Galilei and Scheiner (2010).

causes the motion of the planets, either in terms of mathematics or in those of natural philosophy. It is notable that Kepler does not give the description a mathematical form, but Galileo's proof that the Sun rotates must surely have appeared to confirm the idea behind it.[23]

Kepler also provides a footnote to the priority dispute about the observation or discovery of sunspots. He scooped both Scheiner and Galileo. Unfortunately, he did so unknowingly. After proving that the orbit of Mars was elliptical and, on the way, that the same was true for the orbit of the Earth, Kepler set about finding orbits for the other planets, not starting only from observations as he had with Mars, but in each case trying an ellipse to see if that would give sufficiently good agreement with observations, and putting the plane of the proposed orbit through the Sun.[24] Isaac Newton (1642–1727) unkindly referred to this procedure as Kepler guessing the remaining orbits were elliptical. In fact, the available observations were not good enough to settle the matter. The orbit of Mercury was, of course, one of the most uncertain. The planet is awkward to observe because it is never seen far from the Sun, and tables of its motion had always been unreliable. So Kepler was pleased to see that the orbit he had found, which included relatively reliable values of the ecliptic latitude of the planet, predicted that Mercury would be seen to cross the Solar disc in October 1608, that is, he could predict a transit of Mercury. Detailed observations of a transit would enable astronomers to construct a more accurate path for the planet.[25] So Kepler observed the Sun on the appropriate date, using a projected image in a camera obscura. The observation procedure is described in his short book *A strange phenomenon or Mercury in the Sun* (*Phænomenon singulare seu Mercurius in Sole*, Leipzig, 1609).[26]

News that what he had observed was in fact not Mercury but (probably) a sunspot seems to have reached Kepler through friends in early 1612, and in 1613 Galileo mentioned Kepler's mistake in his *History and demonstrations concerning sunspots and their behaviour* (*Istoria e dimostrazioni intorno alle macchie solari e loro accidenti*, Rome, 1613). Kepler told a friend that he had written to Galileo about the observation, but the letter seems not to have survived. Kepler apologized in print for his mistake, in an ephemeris for 1617. Later, when writing further ephemerides, he predicted a transit of Mercury for 1631 and went so far as to write a book to draw astronomers' attention to the event. This transit was successfully observed.[27]

As time wore on, the continuing ability of Kepler's *Rudolphine Tables* (Ulm, 1627) to predict transits, that is to predict ecliptic latitudes as well as longitudes, became one of the elements that led to an increasingly wide acceptance of the orbits Kepler had calculated and of the laws of planetary motion that were used in calculating the

---

[23] See Aiton (1972).

[24] See Bialas (1971).

[25] On transits and the one Kepler predicted for 1631, see Pasachoff, 'Johannes Kepler, the *Kepler* spacecraft and transits' (this volume, Chap. 10).

[26] Reprinted in Kepler (1941), pp. 77–98; see also 'Nachbericht', in Kepler (1941), pp. 429–33.

[27] See Pasachoff, 'Johannes Kepler, the *Kepler* spacecraft and transits' (this volume, Chap. 10).

tables. Kepler's involvement with transits had started rather badly but it ended very well indeed.

8.4 Comets and Kepler's Defence of Tycho Brahe (1625)

It is clear from the toing and froing described in the previous section, that in the years immediately following the publication of his work with the telescope Galileo, who left the University of Padua to become Mathematician and Natural Philosopher to the Grand Duke of Tuscany, established himself as an active member of the astronomical community as well as a participant in debates about terrestrial physics. The two interests came together in the controversy that arose over the bright comets that appeared in the years 1618–19. Kepler too became involved in the controversy, though not until it was well under way, with Galileo as one of the active participants. However, Kepler's interest in comets was of long standing, being bound up with his respect for Tycho Brahe's skill as an observer and his estimates of the accuracy of his observations.

In studying comets, as for the New Stars of 1572 and 1604, measurements of parallax were crucial, since they provided estimates of the distances of the comet from the Earth. In his wide-ranging *On recent phenomena in the aetherial world* (*De mundi aetherii recentioribus phaenomenis*, Uraniborg, 1588), Tycho described investigations of the New Star of 1572 and of the bright comet of 1577; Kepler, who had used Tycho's methods in studying the New Star of 1604, went on to use measurements of parallax to estimate distances of comets and thus to find their paths. Unfortunately, as we have seen in connection with the New Star of 1604, measurements of the parallax of a comet are not always easy to make. For a comet, as for a New Star, it may be necessary to make measurements of its position when the object is inconveniently close to the horizon, or (particularly for a comet) close to the Sun. Further, comets move relatively fast and at variable speed, making it difficult to use new observations to check earlier ones.[28] In an additional twist, one of the comets of 1618–19 appeared to break in two. Kepler had, however, taken an interest in an earlier comet that had first appeared in late September 1607 and ceased to be visible in December of the same year.[29] In December he wrote to his friend Joachim Tanckius (1557–1609) in Leipzig enclosing an account of observations of the comet and a calculation of the path it followed through the planetary system. Kepler's investigation seems to be the first in which anyone had attempted to find the path of a comet in three dimensions rather than simply tracing its motion against the

---

[28] The scientific facts (as we now know them) conspire against accurate measurement of the parallax of comets. The real motion of comets, especially when they are relatively nearby (i.e. inside the orbit of Jupiter) can be very great, thus introducing a possibly huge systematic error in any attempt to use diurnal parallax. Even in less than half a day the displacement of nearby comets on the sky can be quite substantial.

[29] See Hellman (1975). The paper now seems dated but Hellman provides a useful account of earlier literature.

pattern of the fixed stars. Kepler's short non-technical book, written in German and thus presumably intended for a local readership, was published in Leipzig early in 1608 under the straightforward title *Report on the comet that appeared in the year 1607 (Bericht von dem in Jahre 1607 erscheinenen Kometen*, Leipzig, 1608).[30] The comet had been a very prominent object in the sky and Kepler was by no means alone in writing about it. Many other astronomers and astrologers did the same. But it seems that—unless historians have failed to detect his authorship of a pseudonymous text— these authors did not include Galileo. Opinion among philosophers at Padua was presumably strongly in favour of the standard Aristotelian position that comets were a meteorological phenomenon, belonging to the sublunary sphere, so that Galileo, as a professor of mathematics, was not called upon to comment as he had been in the case of the New Star of 1604.

In 1619 a greatly expanded Latin version of Kepler's 'Report' on the comet of 1607, incorporating (among other new material) accounts of the comets of 1618– 19, and with tables of positions and diagrams of paths, was printed as *Three short books on comets (De cometis libelli tres*, Augsburg, 1619).[31] The three short books have titles that describe their subject matter as astronomy, physics and astrology (see Fig. 8.3). The books are of very different lengths: 91, 11 and 20 pages respectively in the modern reprint. Together with 'new' (*novus*) two of the titles use the Greek adjective 'paradoxos' (meaning 'contrary to received opinion') thus making an explicit claim to originality, the first book for its demonstration of the appearances and heights of comets (that is their distance above the Earth) and the second for its new 'physiologia' (that is natural history) of comets. The title page also supplies a quotation from the Roman philosopher Seneca (d. AD 65) to the effect that some day someone will identify the place of comets and why they move as they do.[32]

Kepler's investigation of the distances of these comets is a development of what Tycho had done in regard to the comet of 1577, but using the Copernican planetary system rather than the Tychonic one. Kepler may also have had some knowledge of work by Christoph Rothmann (1550/60–c. 1605), the astronomer employed by Land- graf Wilhelm IV of Hessen (1532–1592), since one of Rothman's fellow employees at the Landgraf's court was the instrument and clock maker Jost Bürgi (1552–1632), with whom Kepler became friendly when they met in Prague at the court of Rudolf II.[33] The nature of Kepler's originality can be seen in the first of his four introductory definitions, where he distinguishes two notions of movement:

Definitions

　　1. We shall use the term motion for what appears to sight, [motion] against the fixed stars. The name trajectory [is used] for the true motion through the spaces of the world.[34]

---

[30] Reprinted in Kepler (1941), pp. 55–76; see also Kepler (1941), p. 426 ff. Tanckius was a professor of medicine at Leipzig University and had been Rector of the University in the 1590s.

[31] Reprinted in Kepler (1963), pp. 128–262; 'Nachbericht' in Kepler (1963), pp. 457–76.

[32] Seneca (1988), lib 2, Chap. 26.

[33] For Rothmann's work, see Rothmann (2014).

[34] Kepler (1619), p. 7; Kepler (1963), p. 142.

DE COMETIS
LIBELLI TRES.

I. *ASTRONOMICVS*, *Theoremata continens de motu Cometarum, vbi Demonstratio Apparentiarum & altitudinis Cometarum qui Annis 1607. & 1618. conspecti sunt, noua & παράδοξ⊙.*

II. *PHYSICVS*, *continens Physiologiam Cometarum nouam & παράδοξον.*

III. *ASTROLOGICVS*, *de significationibus Cometarum Annorum 1607. & 1618.*

AVTORE
IOHANNE KEPLERO,
SAC. CÆS. MAIEST. MA-
thematico,

Seneca Nat. Quæst. lib. 6. cap. 26.

Erit qui demonstret aliquando, in quibus Cometæ partibus errent, cur tam seducti à cæteris eant, quanti qualesq̃, sint. Contenti simus inuentis: aliquid veritati & posteri conferant.

Cum Priuilegio Sac. Cæsareæ Maiest.
ad Annos XV.

AVGVSTÆ VINDELICORVM,
Typis Andreæ Apergeri, Sumptibus Sebastiani Mylii Bibliopolæ Augustani, M. DC. XIX.

Fig. 8.3 Title page of Johannes Kepler, *De cometis libelli tres*, Augsburg, 1619. The book deals with the comet of 1607 and the two comets of 1618–19

As in his work on Mars, Kepler proposes to find the true path in physical space. The definitions are immediately followed by 'Assumptions', the first being that

> The Earth moves through the spaces of the World with that annual motion about the Sun that Aristarchus and Copernicus have attributed to it.[35]

Kepler then turns to the motion of comets. Assumption 2 is that their trajectories are not uniformly along straight lines; the third is that at first comets travel with constant speed, then little by little the daily increments of the trajectory increase, increasing according to a law of equal parts of tangents of an arc of a circle, or in some similar way, that is in an orderly fashion.[36]

There follow theorems, problems and diagrams showing trajectories for the comet of 1607 and those of 1618–19 (see Fig. 8.4). Kepler explicitly rejects the possibility of constructing the paths as combinations of circles.[37]

The second book, on the physics or natural history of comets has a separate title page though the numbering of pages continues without interruption. The text begins

> It seems that the origin and nature of Comets is as follows: as waters, particularly salt water, provide for fish, similarly the aether provides for the origin of Comets: and as fish move around in the waves so comets move in aether[38]

Kepler goes on to describe comets as being formed as dense parts in the aether, which need to be expelled from the system:

> Then there is accordingly a need for defecation and purgation, which is provided for by the faculty that is in the substance of the aetherial flow, [acting] in the same way as animal and vital faculties.[39]

The use of the term *faeces* to denote any form of solid or dense residue, not necessarily organic in origin, derives from alchemy. Kepler's analogies with the living world may have seemed helpful to contemporary readers. In the twenty-first century, when we are accustomed to the idea that the behaviour of inert matter is better understood than the complicated processes that govern living things, Kepler's analogy may prove distracting. However, in context it is clear that he is making two points, important and (in his time) unconventional points: first (against standard Aristotelian opinion) that comets are formed in, and belong to, the aetherial region, that is the region above the Moon, second that what we see is a process of comets being expelled from the vicinity of the Sun. This second idea explains why comets are transient phenomena.

The suggestion that comets are subject to repulsion may have arisen from observation of their tails. Kepler describes the tail as flowing out from the body of the comet, eventually destroying it, in the same way that silk thread flows from the bodies of silkworms.[40] (Kepler apparently shared the common opinion of the time that after

---

[35] Kepler (1619), p. 7; Kepler (1963), p. 142.

[36] Kepler (1619), p. 7; Kepler (1963), p. 142.

[37] Kepler (1619), p. 85; Kepler (1963), p. 210.

[38] Kepler (1619), p. 99; Kepler (1963), p. 225.

[39] Kepler (1619), p. 99; Kepler (1963), p. 225.

[40] Kepler (1619), p. 99; Kepler (1963), p. 225.

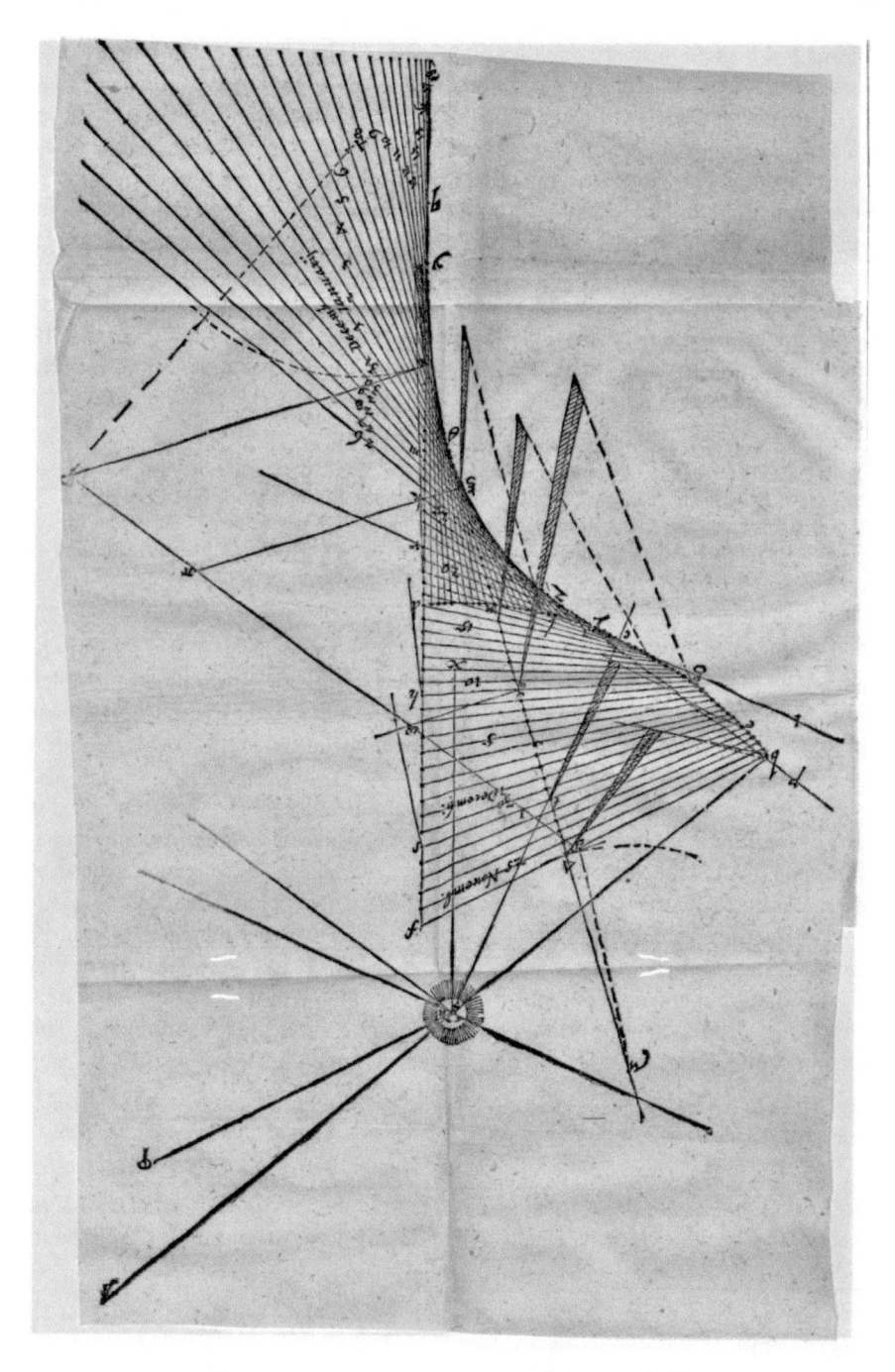

Fig. 8.4 Trajectory of the comet of 1607, from Johannes Kepler, *De cometis libelli tres*, Augsburg, 1619

spinning its cocoon the silkworm dies. Ideas were to change later in the century after experimental investigations by Jan Swammerdam (1637–1680).)

Then, in a paragraph with a marginal note asking 'Why do tails point away from the Sun?', Kepler explains that he thinks 'material is continually expelled from the body of the comet by the rays of the Sun through the force of the rays of the Sun'.[41] He explains that the rays themselves are invisible because the aether is transparent, pointing out that even on Earth we see sunlight only when it falls on something such as a wall, clothing, the surface of water, mountains, or clouds, 'or in thick air'.

The next question, again with a marginal note, is why comet tails are curved. Kepler quotes examples to show the fact is well attested, while pointing out that the Sun's rays are necessarily straight. He explores various explanations and even goes so far as to say

> As, for example, if I were to declare the thing possible, suppose a wind blows in between, although there is no wind in the heavens. This fiction is rendered very plausible (valdè verismilis) by the appearance of the Comet in Libra in December 1618 since there the tail deviated from pointing away from the Sun, and at the same time the shape of the tail ran out to a tip: and from the side of the tail, from which it was wafting away, the tail was denser and more sharply defined, as when a transverse wind, catching heaped up grain, piles it into a line and spreads it out along that line.[42]

There follows a suggestion that one could also suppose the head of the comet to be moving so fast that the newest part of the tail, pointing away from the Sun, is left behind, something that might explain what was seen in the tail of the second comet of 1618. He adds that Tycho had observed something similar in the case of the comet of 1577.

Kepler then turns his attention to the material of comets. The question posed in his marginal note is 'whether comets burn' (Cometae an ardeant). The ambiguous wording is presumably intentional since he first examines the astrological association of comets with heat waves, but he then looks at the Ancient opinion that comets are fire or firebrands, with the flame forming the comet's tail. He argues that this theory arose from the Ancient belief that comets were sublunary, whereas once one realizes they are superlunary it becomes obvious from the unevenness of their light, the brightest parts being those facing the Sun, that comets shine by reflection of sunlight.[43] This rather discursive section is followed by a brisk summing up:

> So, Reader, never in any circumstances doubt that the form of the tail of a Comet is from the Sun, taking material from the head, and the brightness is that of illumination by the Sun, never actual fire.[44]

The remainder of the book is about the effects comets have on the Earth. That is, we are concerned with astrology, which Kepler regards as a part of physics.[45] Marginal

---

[41] '... per vim radiorum Solis.' Kepler (1619), p. 101, Kepler (1963), p. 216.

[42] Kepler (1619), p. 101; Kepler (1963), p. 227.

[43] Kepler (1619), p. 103; Kepler (1963), p. 228.

[44] Kepler (1619), p. 103; Kepler (1963), p. 228.

[45] See Rabin, 'Kepler's reform of astrology' (this volume, Chap. 6).

notes indicate the topics: whether comets cause winds, bad harvests, earthquakes, the way in which comets modify the state of the air, the animal faculty in the sublunary world (with a reference to the astrological fourth book of Kepler's *Harmony of the World*, published in 1619) and so on. Taking the work on comets as a whole, the clear message is that Copernicanism, taken to the extent of relating comets to the Sun, makes sense of their observed behaviour.

Meanwhile in Italy, and apparently unknown to Kepler, the bright comets of 1618 were tempting scholars into print and sometimes to make philosophical points. Contributors to the ensuing exchanges of opinions eventually included Galileo and several members of the Jesuit Order (the Society of Jesus, SJ). The involvement of Jesuits was to be expected since the remit of the Order was partly educational, with emphasis on mathematics and natural philosophy.[46] Early in 1619 the Jesuit college in Rome published an anonymous book about the comets, with the sober title *On the three comets of 1618* (*De Tribus Cometis Anni MDCXVIII*, Rome, 1619). The author, Orazio Grassi SJ (1583–1654), professor of mathematics at the Collegio Romano, argued that since they show no parallax, comets are very distant from the Earth and must be celestial, that is not of meteorological (sublunary) origin as Aristotelian natural philosophers supposed and not merely produced by reflected or refracted light. Grassi also followed Tycho Brahe's example in suggesting that comets followed circular paths. At about the same time, Scipione Chiaramonti (1565–1652)—a scion of a wealthy and well-connected Florentine family, a graduate of the university of Ferrara, who had taught natural philosophy in Perugia and had been employed as a mathematician by the Duke of Modena—wrote a book called *Discourse on the bearded comet of the year 1618* (*Discorso della cometa pogonare dell'anno 1618*, Venice, 1619), which came to conclusions very different from Grassi's. Chiaramonti argued that the comet was made of sublunary material and that it showed parallax, which proved its spatial location was close to the Earth.[47]

Galileo wanted to reply to Grassi, whose book he thought very foolish. But in 1616 he had received a discreet (but official) warning, delivered by Cardinal Roberto Bellarmino (1542–1621), that he should not defend the Copernican hypothesis in public. This warning probably influenced Galileo's decision to write the book jointly with his friend Mario Guiducci (1583–1646) and have it published with Guiducci alone named as author. The book, *Discourse on comets* (*Discorso delle comete*, Florence, 1619), was presented to the Accademia Fiorentina (of which Guiducci had become Consul in 1618) and was published in May 1619. As with Galileo's pseudonymous pamphlets on the New Star of 1604, surviving manuscripts make the authorship clear: most of the text appears in Galileo's hand, and Galileo made corrections to parts written by Guiducci.[48] This may, of course, give a less than fair picture of Guiducci's contribution, since there could have been extensive oral discussion, but it is clear that Galileo played a large part in the writing, and this was rapidly and widely recognized by readers. While Galileo and Guiducci were

---

[46] John Donne (1572–1631) sends up the Jesuit remit in *Ignatius His Conclave* (Donne, 1611).

[47] On Chiaramonti, see Rothmann (2014).

[48] See Shea and Artiga (2003).

engaged in writing their *Discourse*, another short book on the comets appeared, in Milan and again sponsored by the Jesuits. This was *A celestial assembly recently met together in Parnassus on the new comet* (*Assemblea Celeste Radunata Nuovamente in Parnasso Sopra la Nuova Cometa*, Milan, 1619). The unnamed author argued in favour of Tycho's geoheliocentric model of the planetary system and against the standard geocentric one. Like Grassi's book, this new one also relied on arguments about parallax in relation to the comet, so Galileo and Guiducci attacked it in the same way as they attacked Grassi's work. For good measure, they also attacked another Jesuit: their *Discourse* is extremely critical of Christoph Scheiner's work on sunspots. Perhaps Galileo knew that a fuller and fully illustrated account of Scheiner's work was being prepared for publication. Scheiner named his book *Rosa Ursina*, in tribute to his patrons the Orsini family, and the first part eventually appeared in 1628.[49]

Grassi chose to reply to his opponents under a pseudonym, though one that is an anagram of the full version of his name in Latin, 'Horatius Grassius Salonensis' becoming 'Lothario Sarsi Sigensano'. Anagrams were a recognized way of sending less-than-explicit messages to colleagues; for instance, Galileo had first announced his observations of the phases of Venus by circulating an anagram. Although he gave a vernacular name, Sarsi/Grassi's reply to Guiducci/Galileo was in Latin, with the title *Scales for astronomy and philosophy* (*Libra Astronomica ac Philosophica*, Perugia, 1619) (see Fig. 8.5). The implication of the title is presumably that Sarsi is weighing up the issues, though there may also be a punning reference to the fact that the first comet had appeared in the constellation of Libra. Sarsi sees Guiducci as deriving his views from Galileo and addresses his reply to the latter. As the reference to both astronomy and natural philosophy implies, the reply is wide ranging and it is probably not only hindsight that gives a sinister sense to Sarsi's remarking that some of the views expressed by Galileo resemble those of predecessors whose works have been considered suspect by the Church. Whereas Guiducci/Galileo had replied promptly to Grassi's short anonymous publication on the comets (which may have been based on a lecture given at the Collegio Romano), it took Galileo until 1623 to reply to Sarsi.

Meanwhile, Scipione Chiaramonti had not finished thinking about comets and in 1621 he published a substantial volume, in Latin, with the succinct title *Anti Tycho* (*Anti Tycho*, Venice, 1621), in which he attacked both Tycho and Grassi/Sarsi for their belief that comets were celestial.

There was continuing pressure on Galileo to reply to Sarsi. Not least because Galileo had become a leading member of the Accademia dei Lincei—named after the lynx, famous for its acuity of vision—an academy founded by Federico Cesi (1585–1630) in 1603 and based in Rome. As a *Linceo* Galileo was expected to stand up for new ideas, and his affiliation to the Academy appears on the title page of his reply to Sarsi, *The Assayer* (*Il Saggiatore*, Rome, 1623), where the panel of text is flanked by symbolic figures of Natural Philosophy (left) and Mathematics (right; see Fig. 8.6). The book is only 128 pages long and has little obvious structure beyond the numbering of sections, each of which starts with a quotation from Sarsi, sometimes

---

[49] Scheiner (1628–1630). The volumes have excellent illustrations of the apparatus Scheiner used.

Fig. 8.5 Title page of Lotherio Sarsi (Orazio Grassi), *Libra Astronomica ac Philosophica*, Perugia, 1619

after a few words of introduction by Galileo. Despite the lack of formal headings, Galileo refers to the numbered sections as 'chapters' (*capitoli*). This is a common way of structuring works that are contributions to a controversy. Even those who do not wish to see each *capitolo* as a head in an indictment are compelled to read the text as a series of short essays.

Galileo quotes passages from Sarsi in their original Latin, and makes his own comments in Italian vernacular, specifically that of the country people around Padua. This switch of language and style of expression makes it easy to identify whose words one is reading. It may also carry a further message, such as that someone from Padua is not intimidated by a professor, or that sturdy common sense is a match for

Fig. 8.6 Title page of Galileo Galilei, *Il Saggiatore*, Rome, 1623. Lettering on their pedestals identifies the flaking figures as Natural Philosophy (left) and Mathematics (right)

philosophy. We may note also that, thanks to the efforts made by the Medici, Galileo's native dialect, Tuscan, was rapidly becoming the favoured vernacular among the upper classes and would not have jarred so much against Sarsi's Latin. It is of course all but certain that Galileo knew who 'Sarsi' was, but the pseudonym gave him some freedom in his replies. The point made by the title of the book is no doubt that an assayer weighs things with great precision, so Galileo is suggesting the Paduan countryman will outdo the learned academic in his assessment of the matters in hand. Meanwhile, despite the combative tone, Galileo's style is engaging and witty. His opponents needed not to mind being laughed at.

One impetus to Galileo replying to Sarsi at this time was that the warnings about possible difficulties with the religious authorities now seemed less cogent. In 1623 a new pope was elected. He was Maffeo Barberini (1568–1644), a Florentine who was on friendly terms with many associates of Galileo and Guiducci.

So Galileo wrote *The Assayer* and it was published in Rome in autumn 1623. Presumably with permission, the members of the Accademia dei Lincei dedicated the book to the new pope, Urban VIII, who was later reported to have said he greatly enjoyed it.

In the summer of the following year, 1624, Kepler, who had just completed the manuscript of his *Rudolphine Tables*, received a brief visit from a friend who happened to be passing through Linz (where Kepler was then living). This unidentified friend—perhaps the Jesuit mathematician Paul Guldin (1577–1643)—showed Kepler copies of two books, Chiaramonti's *Anti Tycho* and Galileo's *Assayer*. Kepler seems to have read the former, with at least enough care to get an idea of its style and content, but then had no time to do much more than glance at the *Assayer*.

Near the end of a very long letter to the astronomer Peter Crüger (1580–1639), written in September 1624, Kepler says Chiaramonti's *Anti Tycho* is about comets being sublunary and describes it as both bold and worthless (perhaps meaning rash and pointless), but he adds that it needs to be refuted.[50] Kepler's refutation was published under the title *Tychonis Brahei Dani Hyperaspistes* (Frankfurt, 1625) (see Fig. 8.7). The title means 'Tycho's defender', with legal connotations (but a literal translation of the Greek would give something like 'one who holds a shield over Tycho Brahe'). Personal feelings and philosophical or astronomical convictions aside, Kepler probably saw it as his duty to write in defence of his predecessor as Imperial Mathematician. Tycho's name would appear prominently on the title page of the *Rudolphine Tables* (Ulm, 1627). Further, Tycho's heirs were still making difficulties about allowing Kepler to use Tycho's observations, and Kepler may have thought matters might be helped by his not only writing in defence of Tycho but also dedicating the book, in the elaborate style of the time, to 'the renowned and magnanimous (*generosae*) Brahe family old in honours and possessions and famous throughout the kingdoms of Denmark and Sweden'.[51]

---

[50] Kepler to Peter Crüger, 9 Sept 1624, letter 993, in Kepler (1959), pp. 197–213, paragraph numbered 21, see p. 211.

[51] Kepler (1625), p. *2; Kepler (1963), p. 267.

TYCHONIS BRAHEI DANI

HYPERASPISTES,

ADVERSVS SCIPIONIS CLARAMONTII
Cæsennatis Itali, Doctoris & Equitis

ANTI-TYCHONEM,

In aciem productus
à

IOANNE KEPLERO, IMP. CÆS.
FERDINANDI II. MATHEMATICO.

Quo libro doctrina præstantissima de Parallaxibus, deque Nouorum
siderum in sublimi æthere discursionibus, repetitur, con-
firmatur, illustratur.

Cum INDICE *rerum memorabilium.*

FRANCOFVRTI,
Apud Godefridum Tampachium.
M. DC. XXV.

Fig. 8.7 Title page of Johannes Kepler, *Tychonis Brahei Dani Hyperaspistes*, Frankfurt, 1625

After he had written his reply to Chiaramonti Kepler travelled to Vienna (a relatively easy journey by boat down the Danube) where he borrowed a copy of Galileo's *Assayer* and read it through. It raised a number of questions and objections that eventually formed a substantial appendix to his draft defence of Tycho. But before looking at that we need first to go back to Chiaramonti.

As we have seen, Kepler seems to have had a low opinion of Chiaramonti's book. Later he said he had taken it for the work of a rather young man. He certainly appears to assume, throughout his response, that Chiaramonti has much to learn about astronomy—which may indeed have been true. For example, at the end of his Preface Kepler points out that the first book of Chiaramonti's *Anti Tycho*, about a hundred pages, is taken up with a long discussion of parallax, all of which could have been found in standard works on astronomy such as those of Regiomontanus (1436–1476), Tycho and others or, he adds, 'in chapter 40 of book 9 of my own *Optics*'.[52] Kepler accordingly chooses to begin his consideration of Chiaramonti's work with the first three chapters of its second book. We are sometimes supplied with a page reference but Kepler's style is largely expository rather than argumentative. Nevertheless, the organization of Kepler's text situates it as part of an exchange: the headings for Kepler's quasi-chapters tend to refer to Chiaramonti's chapters rather than to the astronomical matters that appear in Kepler's text, and within each chapter Kepler's sections are numbered. On occasions, however, Kepler's tone is indeed polemical. For instance, in discussing Chaps. 19 and 20 of Chiaramonti's second book, Kepler starts by saying these two chapters are worthless (*futiles*). He proceeds to claim that Chiaramonti contradicts himself when developing his account of Tycho's reasoning, and then likens the resulting text to the Augean stables.[53] As Kepler's original readers would have known, cleaning these stables was one of the labours of Hercules. It is only after rather heavy preliminaries that, in Sect. 22, Kepler provides a detailed account of Tycho's 'hypotheses', that is the pattern of spheres Tycho proposes for his geoheliocentric planetary system.[54] As might be expected, the account is admirably clear; but it spares the reader nothing by way of complexity. The next section explains how Tycho fitted the path of the comet of 1577 into this system.

The work ends with a short 'Conclusion and Presentation', whose first paragraph addresses the Brahe family and concludes by saying of Chiaramonti's book that 'Were it dealing with things of mine, it has so much that is worthless that I would have been ready to despise it.'[55] The next paragraph tells Chiaramonti that he should learn some more astronomy before publishing again about the Tychonic or Copernican systems, and points to Ancient precedent for a conflict with religious beliefs. On a much more conciliatory note, the section ends with a prayer, in which Chiaramonti is invited to join, praising God the Creator of all things visible and invisible—though

---

[52] Kepler (1625), Preface, Sect. 7, p. 4; Kepler (1963), p. 272. The *Optics* to which Kepler refers is Kepler (1604); reprinted in Kepler (1939).

[53] Kepler (1625), p. 93; Kepler (1963), p. 342.

[54] Kepler (1625), p. 98; Kepler (1963), p. 346, line 5 seqq.

[55] Kepler (1625), p. 183; Kepler (1963), p. 413.

the full wording suggests Kepler is not directly quoting any version of the Creed and later we move back to pagan times with a reference to Aristotle as the father of the peripatetics. The text nevertheless ends with 'Amen'. As with the endings to many of Kepler's works, this one is a reminder of his religious convictions.

Religious beliefs did not, however, blunt his scientific judgment. Nor, as we shall see, was his judgment swayed by knowing his opponent was also a Copernican. Maybe reading Chiaramonti had made Kepler a little irritable when he came to read Galileo.

Galileo's *Assayer*, to which, as we have seen, Kepler turned next, is recognized by today's historians and philosophers of science as a highly significant text, particularly for its emphasis on the use of experimentation. Moreover, like most of Galileo's works, it is written in an entertaining and persuasive style—that is persuasive unless one was (or supposed one was) among the opponents being held up to ridicule. Kepler appears to have been impervious to the literary qualities of the text.

As we have seen, Galileo's *Assayer* is the subject of an Appendix to Kepler's *Tychonis Hyperaspistes*. This Appendix has no separate title page, and is put together with the main text in Kepler's index, though some Galileo scholars have nevertheless treated it as an independent work.[56] Under the heading 'Appendix to the *Hyperaspistes* or notes from the Assayer (*ex Trutinatore*) of Galilei', the Appendix is prefaced by a brief introduction. Kepler first tells us that at the time of his initial encounter with the work of Chiaramonti he also came across Galileo's 'book written in Italian vernacular (*Italico scriptum idionate*) against Lotharius Sarsius' *Libra Astronomica* in which Galileo had been mentioned repeatedly'.[57] That is to say, Kepler seems to know no more of Sarsius or the earlier part of the controversy than Galileo tells him. On the other hand, the pieces that Galileo quotes from Sarsi do indeed contain many references to Tycho, as do the comments from Galileo (who calls him 'Ticone'), and there are also some references to Kepler, for instance for his discussion of the shapes taken by the tails of comets (*Assayer*, Chap. 34). In the circumstances, it is rather hard to see how Kepler could have kept to the idealistic line he proposes in the third paragraph of his introduction, namely that

> In this controversy in which Sarsi and Galileo are engaged I shall not be a judge, because that would exceed my limits as writing a defence [sc. of Tycho]; but where Galileo touches on Tycho's case I should not pass over those places, nor make a halfhearted defence or appear to be acting in bad faith.[58]

Perhaps Kepler wrote this before he saw some of the errors? He is, however, generally polite and he does keep it short. In the original edition Kepler's text takes up only eighteen pages.[59]

---

[56] Kepler's Appendix has been translated by Stillman Drake and C. D. O'Malley in Galileo et al. (1960), pp. 337–55.

[57] Kepler (1625), Appendix, p. 185; Kepler (1963), p. 413.

[58] Kepler (1625), Appendix, p. 185.

[59] Kepler (1625), pp. 183–201; Kepler (1963), pp. 413–25.

Occasionally, he argues against the opinions of Sarsi, for instance the passage quoted by Galileo referred to in item 11 (see below)—and because in Kepler's discussion the whole text is in Latin, each quotation (the ones from Sarsi are sometimes slightly condensed) is introduced by the author's name[60]—but on the whole Kepler is addressing Galileo, and that is the aspect that is of most interest in the present context.

Kepler starts with some asperity. The first item is not clearly keyed to a particular passage in the *Assayer* and it is aimed at others besides Galileo, namely Chiaramonti and Sarsi, on the subject of an error by Tycho. The passage in *The Assayer* to which Kepler is responding seems to be the first part of Chap. 6. As he points out, he has already put Chiaramonti right about this error by Tycho, since Chiaramonti mentioned it in his *Anti Tycho*.[61] As Kepler told Chiaramonti, Tycho's error—made in dealing with a straightforward measurement of parallax—consisted of his providing an inadequate diagram (one apparently originally designed for a different purpose) and expressing himself in incorrect terms. The error is described as an example of *pseudographum*. This form of error, mentioned in Plato's dialogue *The Sophist*, is listed by Aristotle, so it must have been familiar to all concerned. In replying to Sarsi and Galileo, Kepler adds some detail and supplies all terms in Greek.[62] He also supplies a page reference to the relevant passage in Tycho's work. No doubt Kepler had copies of Tycho's writings on his bookshelf, but his meticulous response seems bad-tempered. Perhaps he suspected all three Italians of a degree of bad faith in making elaborate complaints about such a minor matter.

To historians of science the *Assayer* is of particular significance for Galileo's advocacy of the experimental method in natural philosophy and for its showing Galileo's belief that it was appropriate to use mathematics in all investigations of nature. There is a clear statement of this belief in Chap. 6 of *The Assayer*, about a page after the passage about Tycho to which Kepler responded in the first item of his Appendix. In this passage about mathematics, which is much quoted by historians and philosophers of science, Galileo's spokesman says

> Philosophy is written in that greatest of books which always lies open before our eyes (I mean the Universe), but it cannot be understood unless we first learn the language, and know the characters in which it is written. It is written in mathematical language, and the characters are triangles, circles, and other geometrical figures, without which it is impossible to understand even a single word; without these it is going round and round in a dark labyrinth.[63]

In the 1620s this idea may have been new to natural philosophers concerned with what today would be called physics, but it was of course not news to mathematicians, that is to astronomers, and Kepler makes no comment—though the attitude

---

[60] Kepler (1625), Appendix, item 11, pp. 193–94; Kepler (1963), p. 419. Galilei (1623), Chap. 10, pp. 21–22. Kepler says 'fol. 35', but he seems not to be referring to the first edition of *The Assayer*. The editor of Kepler (1963), Franz Hammer, does not mention this discrepancy.

[61] Chiaramonti (1621), Book II, Chap. XX, p. 219; Kepler's comments at Kepler (1625), section headed *Ad Cap. XIX. fol. 218. et XX. fol. 219*, item 8, p. 95; Kepler (1963), p. 343–44.

[62] Kepler (1625), Appendix, pp. 185–86; Kepler (1963), pp. 413–14.

[63] Galilei (1623), Chap. 6, pp. 16–17.

to mathematics is different from what we find in Kepler's own works. From these it is clear that Kepler sees geometry not as providing characters for a language but as itself determining the structure of the Universe.[64] This difference, which mirrors that between Plato and Aristotle on the status of mathematics (with Galileo as Aristotle) probably goes a very long way to explaining why Galileo never responded explicitly to Kepler's *Mysterium cosmographicum* (1596) or his *Harmony of the World* (1619), despite the fact that the latter contains numerous references to the work of Galileo's father, the musician and music theorist Vincenzo Galilei (*c.* 1520–1591).

In all, Kepler's Appendix makes nineteen comments of *The Assayer* and, apart from the first and last, all are explicitly keyed to passages in Galileo's text. As we have already noted, some address Sarsi, or the passage quoted from him, rather than Galileo's response. For instance, Kepler's second comment begins

> Sarsius assumes, on page 20, that Mars is closer to the Earth than the Sun, as proved by Tycho.[65]

And promptly adds that one of the things the Tychonic system and the Copernican systems have in common is that the eccentric orb of Mars circles the Sun. Kepler then quotes Tycho's measurements of the parallaxes observed for Mars and for the Sun. He adds that he now believes Tycho's value of 3 arcmin for the Solar parallax is too large and that the true figure is about 1 arcmin. After sketching the story of its determination (going back to the geocentric system of Ptolemy), Kepler comes to a conclusion that reads more like a warning. It is that measurements of parallax cannot be used to distinguish the various models of the planetary system: 'The Parallax of the Sun is rather to be found from the Hypotheses [that is from the model of the planetary system] than the Hypothesis from the Parallax', and he adds that Galileo seems to wish to deny this astral determinism in the art of astronomy.[66]The decidedly astrological term I have translated as 'astral determinism', *astrotelesma*, appears in Greek. Kepler's further comments make it clear that he thinks Galileo has not looked beyond the schematic diagrams of planetary orbs to the combinations of spheres within each orb that determine the path of the planet, that is not only its path along the ecliptic but also the path bringing planets closer to the Earth or taking them further from it. The 'astral determinism' is partly a matter of observational facts. Thus Galileo seems to be underestimating the explanatory powers of the models of the planetary system he does not believe in. That hindsight endorses Galileo's unbelief does not make him a competent astronomer by the standards of his own time. As historians have noted, in his use of telescopes, Galileo was pursuing a new kind of astronomy, introducing a visual element.[67] Kepler's comments underline the fact that Galileo seems never to have taken much interest in the older form of astronomy whose rudiments he had been teaching in Padua.

---

[64] See, for instance, Field, 'Kepler's cosmos' (this volume, Chap. 2).

[65] Kepler (1625), Appendix, item 2, p. 186; Kepler (1963), p. 414; Galilei (1623), Chap. 6, p. 14.

[66] Kepler (1625), Appendix, item 2, p. 188; Kepler (1963), p. 415.

[67] Winkler and van Helden (1992); see also Winkler and van Helden (1993).

A group of Kepler's comments, items 9–14, mainly refer to the idea that comets were merely reflections, or that details of their appearance involved reflection.[68] So here we are concerned with optics. This is an area in which Kepler's and Galileo's professional competence is in better balance than in technical mathematical astronomy.

Kepler's items 9, 10, 11 and 12 all address Galileo's Chap. 10. Kepler separates his response on the subject of reflection (and refraction), which mainly relates to the passage quoted from Sarsi (about 300 words), from responses concerning comets' tails (mainly addressed to Sarsi) and other matters such as the supposed rectilinear motion of comets, mainly relating to Galileo's answer to Sarsi, which inclines to the discursive (about 1300 words). Kepler's item 9 begins by asserting that 'Galileo asserts that Sarsi, having dismissed Aristotle's opinion, inclines to the opinion of Kepler that a comet could be a reflection.'[69] Kepler then jumps to the passage, 'in [my] *Optics* of 20 years ago', that he believes Sarsi has in mind. The reference is clearly to the *Optics* of 1604, where Chap. 6 includes both a discussion of the appearance of comets and a diagram of an optical image showing the characteristic curve now known as a 'caustic'.[70] In the text written in 1624 Kepler explains that the shape he discussed was indeed produced by reflection, using a glass ball or a round-bodied flask full of water, lit by sunlight admitted into a camera obscura.[71] Having added that these experimental conditions do not correspond to anything that could apply for comets, he ends 'So reflection alone would not yield the shape of a comet'.

In item 10 Kepler turns to the formation not of comets themselves but of their tails, which he had also mentioned in his earlier work on optics,[72] pointing out that he has already considered this in his reply to chapter XXXI of book II of Chiaramonti's *Anti Tycho*.[73] He immediately makes a distinction between reflection and refraction, pointing out that the passage in his *Optics* cited in his previous item was concerned with the latter,[74] and that refraction could not give rise to a tail, since the tail, as Galileo has correctly noted, always points away from the Sun. In fact, Galileo mentions the tail only once, and very briefly; he is far more interested in the shape of the comet's path, a matter Kepler turns to in his next comment (item 11). So it is presumably for the sake of completeness in his answer to Sarsi that Kepler summarizes his theory that the tail of the comet arises from material driven out from the head by rays from

---

[68] Kepler (1625),*i* Appendix, items 9–14, pp. 192–97; Kepler (1963), pp. 418–22.

[69] Kepler (1625), Appendix, item 9, p. 192; Kepler (1963), p.418. Galilei (1923), Chap. 10, p. 22.

[70] Kepler (1604), Chap. VI, pp. 264–66; Kepler (1939), pp. 231–33 (diagram on p. 231).

[71] Although the processes are in fact distinguished from one another, in this period the words reflection (*reflexio*) and refraction (*refractio*) are sometimes used as if they were interchangeable. Aristotle uses the same term, *anaclasis*, for both processes. Kepler explicitly distinguishes between these two senses in item 10 of his *Appendix*, see below.

[72] Kepler (1604), Chap. VI, pp. 264–66.

[73] Kepler (1625), p. 123; Kepler (1963), p. 365.

[74] In 'A little appendix on the curved tail of comets' (*Appendicula de curua Cometarum cauda*), Kepler (1604), Chap. IX, Sect. 6, pp. 323–24; Kepler (1939), p. 278.

the Sun, ending with the concise summary 'so that the tail is as it were the death of the head'.[75]

In item 11 Kepler looks at the problem of the motion of comets, where uncertainty in estimates of the object's distance from the Earth (its 'altitude'), and hence its position in space, makes it difficult to convert observations into what in his books on comets Kepler had decided to call a 'trajectory'. Again, Kepler starts with Sarsi, and, presumably to avoid confusion, since both texts are in Latin, presents a series of exchanges set out as dialogue. The first subject is a second passage in the *Optics*, an Appendix to Chap. 10, in which Kepler considered the true motion of comets in space. The passage is headed 'Appendix on the motion of comets'.[76] Sarsi remarks that Kepler realized that postulating straight-line paths leads to difficulties. Kepler admits that he did at first use straight-line motion, when he had 'not yet tried numbers' (*nondum tentatis numeris*).[77] 'Trying the numbers' is Kepler's habitual phrase for the process of matching observed positions, or other values, against the predictions of theory. In this case he found it was not possible to use uniform rectilinear motion. As the discussion proceeds it becomes clear that Sarsi has given some professional thought to comets and that Kepler takes his opinions seriously. At one point Sarsi says Kepler wanted the motion of a comet to start slowly and die away at the end, the most rapid motion being in the middle. Kepler replies

> Indeed at that time it was not on account of any geometrical proofs from observations of a comet, but purely from watching fireballs or fireworks (*ignes arificiales*) that we Germans call 'Raketuli'.[78]

(Kepler is using a diminutive, but the word has lasted: 'Raketa' is the modern German for 'rocket'.)

As we have seen, Kepler was later to change his methods and his mind. In the constructed dialogue with Sarsi, matters take what with hindsight is a darker turn in response to Sarsi's saying that there are things 'no reasoning can allow Catholics to accept'.[79] Kepler's response is that this is 'perverse or reprehensible [even] if you are in the right, or enslavement if [you are] in the wrong'.[80] Kepler puts forward some arguments of his own, but closes by quoting a piece of Galileo's text (translated into Latin) to show that Galileo agrees with him. Kepler's translation appears to be correct but the rhetorical effect of his terse and intricate Latin is very different from that of the

---

[75] *Ut sit cauda veluti mors capitis.* Kepler (1625) Appendix, item 10, p. 193; Kepler (1963), p. 419.

[76] Kepler (1604), Chap. X, Appendix, p. 335; Kepler (1939), pp. 287–88, *Appendix de motu Cometarum.*

[77] Kepler (1625), Appendix, item 11, p. 193; Kepler (1963), p. 419. Galilei (1623), Chap. 10, pp. 21–22.

[78] Kepler (1625), Appendix, item 11, p. 195; Kepler (1963), p. 420.

[79] *Quae nobis Catholicis nulla ratione permittuntur.* Kepler (1625), Appendix, item 9, p. 192; Kepler (1963), p.418. Galilei (1623), Chap. 10, p. 22.

[80] *Pravam vel querelam, si recte, vel servitutem, si male.* Kepler (1625), Appendix, item 9, p. 192.

diffuse vernacular that may have been intended to make Galileo's statement appear less direct (viz. less obviously pugnacious).[81]

In his following comment, item 12, Kepler turns from considering Sarsi's opinions to examining those of Galileo, who had written at length on the possibility that the motion of comets was rectilinear. Galileo accused Tycho of being equivocal on the matter, by first suggesting the motion took place along a great circle and then switching to saying it followed a straight line. Kepler points out that this apparent inconsistency can be attributed to the fact that straight-line motion in physical space will appear as motion along a great circle on the celestial sphere. Thus what Galileo seems inclined to make a matter of Aristotelian natural philosophy, in which circular motion is associated with celestial bodies and rectilinear motion with sublunary ones, has instead been dealt with as merely a matter of geometry. (Though Kepler does not say so, it is surely possible Galileo was not reading with enough care and had become confused about whether Tycho was referring to motion in space or on the celestial sphere.) Kepler also points out that the same applies to 'straight-line' motion on the surface of the Earth since the Earth is spherical.

In view of his low opinion of Chiaramonti's work, it is rather damning that Kepler then refers Galileo to the passage in the main text of the *Hyperaspistes* where he explained the matter at some length to Chiaramonti in connection with the fourth chapter of the second book of his *Anti Tycho*.[82] Though Kepler does not comment on it, presumably because he had already discussed the matter with Sarsi, Galileo also remarks that Sarsi has failed to mention that Kepler believed comets moved in straight lines. If Galileo derived this information directly from Kepler, we have an indication that he read Kepler's optical treatise of 1604. The book was indeed widely read, but Galileo (like most of his contemporaries) rarely gives references to 'modern' sources, so this reference, in a non-optical context, is of interest to historians, though the evidence it supplies is unfortunately not conclusive. Galileo's remark also suggests that when he wrote *The Assayer*, in mid-1623, he probably had not read Kepler's book about comets, published in 1619, in which (as we have seen) the paths of comets are shown as far from rectilinear.

With Kepler's item 13, which is very short, we return to Tycho because, accepting the common opinion that comets were self-luminous, he had agreed with Hagecius that the purity of their light implied they were celestial. Kepler regards this as a respectable conjecture (*conjectura laudabilis*) and does not mention that Galileo pokes fun at it.[83] Kepler is more inclined to dwell on the possibility that the brightness of comets is due to their reflecting the radiance of the Sun.

The following section, item 14, is addressed directly to Galileo and concerns material celestial spheres. Kepler says Galileo rejects Tycho's arguments against the

---

[81] Kepler (1625), Appendix, p. 195; Kepler (1963), p. 421. The passage translated is Galilei (1623), Chap. 10, p. 24.

[82] Kepler (1625), Appendix, item 12, p. 196; Kepler (1963), p. 421. Referring to Kepler (1625), pp. 6–19; Kepler (1963), pp. 274–84.

[83] Kepler (1625), Appendix, item 13, p. 196; Kepler (1963), p. 421. And see Galilei (1623), Chap. 19, p. 48.

existence of such spheres—but adds that later on in *The Assayer* Galileo appears nevertheless to deny there are any such spheres. Tycho had argued that celestial spheres would cause refraction, which (like the refraction caused by the Earth's atmosphere) would lead to changes in the observed positions of stars. Kepler quotes Galileo as saying

> [Rays] perpendicular to the spheres reach the Earth as perpendicular rays and in fact are not refracted.[84]

These words—there are only ten in the original Latin—are introduced by 'Galileo says' (*Galilaeus inquit*). However, it turns out we have been given not a Latin translation or a passage in Galileo's vernacular text but a summary of a page of it (about 375 words) at the end of Chap. 22 of *The Assayer*.[85] Perhaps Kepler—who had read the book in Vienna and was now presumably back home in Linz—mistook his own notes on the text for a direct translation? In any case, the drastic abbreviation has not deformed the sense and Kepler's reply is apposite as well as damning:

> But, Galileo, if there are orbs, it must be eccentric ones. Accordingly no rays come onto the Earth perpendicularly to the spheres, except only at apogee and perigee. So the argument [from refraction] is valid, and it has its revenge, since you yourself deny the existence of solid orbs on f. 129.[86]

In Chap. 37 of *The Assayer*, Galileo does indeed make out a case against solid spheres. He starts by directly addressing his opponent: 'Signor Lothario, supposing that the celestial orbs were made of solid material …', and then argues that such spheres would cause refraction. This argument apparently remains unanswered.[87]

Apart from making the remark we quoted above, Kepler seems inclined to let Galileo's self-contradiction stand. The following comment appears to be addressed to Galileo's Chap. 26, in which Galileo first quotes Sarsi on the subject of measurements of the (diurnal) parallax of comets, and specifically the comet of 1577, as a way of finding their distances from the Earth.[88] Galileo, citing Guiducci in support, attacks Sarsi for discussing such measurements, particularly those made by Tycho Brahe, which are useless because comets are optical phenomena, like the haloes we sometimes see round the Moon. In responding to this, Kepler asks whether all or only some of the observations are supposedly worthless, but (apparently) cannot decide what Galileo is getting at. He adds that he cannot see why Tycho is being singled out, but does not argue against Galileo's theory or offer any arguments in favour of comets being real bodies rather than accumulations of reflections. It may be that in this case Galileo's concentrating his criticisms on Tycho was occasioned by Sarsi's original remarks, but Galileo does in fact seem to have had a lower opinion of Tycho

---

[84] Kepler (1625), Appendix, item 14, p. 196; Kepler (1963), p. 421.

[85] Galilei (1623), Chap. 22, pp. 57–58.

[86] Kepler (1625), Appendix, item 14, p. 196; Kepler (1963), pp. 421–22. Galileo's reference to solid orbs is in Galilei (1623) Chap. 37, p. 82.

[87] See Galilei (1623), Chap. 37, p. 82.

[88] Kepler (1625), Appendix, item 15, pp. 196–98; Kepler (1963), p. 422. Galilei (1623), Chap. 26. pp. 67–68.

than many of his contemporaries did. It is not clear how this originated but it certainly extended to Tycho's model of the planetary system—whose details, as we have seen, Galileo had not mastered. And the distrust also extended to Tycho's skill and reliability as an observer. It may be that some of the astronomers Galileo knew were critical of Tycho. It is (after all) hindsight, based on the long-lasting success of the *Rudolphine Tables* (1627), that makes it clear Kepler was right to trust Tycho. On the matter of the Tychonic planetary system, we may note that Galileo's preferred argument in favour of Copernicanism involved the motion of the Earth (as explaining the tides) and he therefore saw the Tychonic system, with its stationary Earth, as equivalent to the traditional geocentric system. As he had been ready to point out, his work with the telescope had made the old geocentric system untenable. Galileo saw this as an argument for the Copernican alternative and presumably looked on the increasing popularity of the Tychonic system with some distaste. In his *Dialogue concerning the Two Chief World Systems* (1632) he was simply to omit all mention of Tycho's system.

The following comment in Kepler's Appendix, item 16, is addressed to Sarsi. Kepler reassures him that if he does not want to accept the motion of the Earth, curved paths (*orbitae*) for comets can be accommodated in the Tychonic system, and he gives a reference to where this matter is dealt with in answering Chiaramonti.[89] Kepler's next comment, again addressed to Sarsi, merely supplies references to earlier writings, going back as far as Regiomontanus, in which comets are observed to have fairly regular motion.[90] The next comment, item 18, is rather more discursive and brings us back to Tycho. Kepler remarks that although they are opponents in other matters, Sarsi and Galileo agree in attacking Tycho on the subject of the curvature of the tails of some comets. Kepler defends Tycho's explanation and then moves on to discuss his own (concerning the comets of 1607 and 1618), pointing out that some of what he is saying here repeats ideas he has already presented in his *Three short books on comets* (1619).[91] The final item of Kepler's Appendix returns to the same subject but in more general terms.[92] Thus the Appendix ends like an appendix, a piece tacked on to a larger work, simply a supplement to the Defence of Tycho with no independent conclusion to draw.

In his response to Galileo's *Sidereus nuncius*, his *Conversation with the Sidereal Messenger* (1610), Kepler was at pains to assess Galileo's description of what he had observed with his telescope and the significance it had for astronomers. In contrast, Kepler's series of comments on *The Assayer* do not amount to an overall judgement on the book or its significance. But an overall judgment of *The Assayer* was not Kepler's aim. As the title *Defence of Tycho* says, Kepler saw his book as parrying attacks on the work of Tycho Brahe, first by replying to the criticisms made in

---

[89] Kepler (1625), Appendix, item 16, p. 198; Kepler (1963), p. 423. Galilei (1623), Chap. 38, p. 84.

[90] Kepler (1625), Appendix, item 17, p. 198; Kepler (1963), p. 423. Galilei (1623), Chap. 33, pp. 77–78.

[91] Kepler (1625), Appendix, item 18, pp. 199–200; Kepler (1963), pp. 423–24. Galilei (1623), Chap. 34, pp. 78–79.

[92] Kepler (1625), Appendix, item 19, pp. 200–202; Kepler (1963), pp. 425.

Chiaramonti's *Anti Tycho* (1621), which he had seen only in mid-1624, at the same time as he first saw Galileo's book. Although the latter is conceived as a contribution to the controversy over the comets of 1618 and 1619, Kepler's *Hyperaspistes* does not appear to have been planned as a contribution to that dispute except insofar as the dispute also concerned Tycho's work on the comet of 1577 and, though only in passing, Kepler's own work on later comets. Further, as we have seen, the only part of Kepler's work on comets Sarsi and Galileo seem to have known about was two short passages in the *Optics* of 1604 that put forward ideas that Kepler, in 1624, had no wish to defend. The nearest Kepler gets to joining in the comet controversy is explaining some of his current thinking to Sarsi and later discussing the shapes of comets' tails (set out a few years previously in the *Three short books on comets* of 1620). On these, his remarks show his usual unwillingness to wave away observations.

The Appendix to the *Hyperaspistes* tells us something about relations between Kepler and Galileo at this time and over the longer term. Obvious differences of temperament apart—Kepler tending to the cooperative, Galileo to the competitive—each recognizes the other as a fellow Copernican but Kepler is uncomfortable with Galileo's not always thinking astronomical matters through, and Galileo seems to be uneasy with Kepler's reliance on Tycho.

In relation to comets, Kepler seems more comfortable when addressing Sarsi than when addressing Galileo. And it is notable that Kepler, who is of course dealing with a live controversy—either about comets or about the reliability of Tycho Brahe—gives Galileo a much rougher ride than historians now tend to do. In arguments, Galileo tends to play it point by point, as if he were engaged in a fencing match. One moral would seem to be that if you do not intend to take great care about consistency then you should write a lot less clearly than Galileo habitually does. Not that anything would save you from a critic like Kepler.

8.5   In Conclusion

There is considerably more written evidence for what Kepler thought of Galileo than vice versa. This is almost certainly at least partly a result of the religious divide, a schism initiated by the actions of Martin Luther (1483–1546) in 1517. In sixteenth- and seventeenth-century Italy, and elsewhere, the division led to the (Catholic) Church authorities taking an increased interest in developments in what would today be called Science, that is in what was then called Natural Philosophy and Mathematics, and particularly astronomy where some of the new ideas seemed to be in direct contradiction with parts of the Bible. In Northern Europe confessional divisions made a considerable contribution to the outbreak of war in 1618. It was to be a widespread and brutal war that lasted until 1648 and affected the civilian population as no previous war had done. In 1616 Galileo, then living in Florence, received a quiet warning that he should not discuss Copernicanism in public; he did not need to be told that it was almost equally inadvisable to be seen to be too close to a Protestant astronomer who apparently had no inhibitions about publicly declaring his belief that the planetary system was heliocentric. Kepler, as Imperial Mathematician, and living within the Holy Roman Empire, was in a more secure position than Galileo, at least while the Emperor's army was led by Albrecht von Wallenstein (Albrecht Václav Eusebius z Valdštejna, 1583–1634), though when Linz was besieged Kepler found his work was interrupted by soldiers passing through his house carrying gunpowder and weaponry to the city walls. In 1612 Kepler was excommunicated by the Lutheran Church, on theological grounds. This hurt him deeply but probably made him feel he had little to lose by expressing his opinions about the work of Galileo.

The first contact between Kepler and Galileo is effectively accidental and a standard form of academic contact. In sending copies of his first book, the *Mysterium cosmographicum* (1596), to Italy with a friend who was presumably simply asked to give them to an astronomer, Kepler found himself in contact with Galileo. But the exchange of letters—one from each party—while in some respects revealing of their two characters, could well have led nowhere. Galileo, having caught Kepler's attention by saying he too was a Copernican, did not explain why.

Their second contact was set up by Galileo. Though he probably need not have worried, it seems very likely that Galileo was simply in search of an endorsement of his work with the telescope from the eager young Copernican who had, rather quickly, risen to the rank of Imperial Mathematician. And it turned out that in relation to optics, and Galileo's discoveries, the interaction between Galileo and Kepler—which took place before Galileo's warning or Kepler's excommunication—was a happy and fruitful one for both parties. But in some ways Galileo was not a natural member of the astronomical community and his apparent distrust of Tycho Brahe guaranteed that there would be clashes of opinion with Kepler. They are indeed on the same side in regard to Copernicanism, but their approaches to proving the theory is true are completely different. Their contributions to the controversy concerning comets are very revealing in this respect. Galileo prefers to look for answers in terrestrial physics, whereas Kepler prefers to trust to astronomy.

Kepler's tendency to speak his mind is helpful in the present context. But it angles our discussion towards astronomy, and (as can be seen in connection with the comets) specifically towards the kind of technical astronomy that seems to have held very little interest for Galileo. In contrast, today's historians and philosophers of science, who, unlike Kepler, are chiefly interested in Galileo's work on terrestrial physics, and in particular its emphasis on experimentation, tend to forget Galileo published *The Assayer* as a contribution to a controversy about the nature of comets that was concerned with a problem in Aristotelian natural philosophy: whether comets should be regarded as belonging to the sublunary realm or to the celestial region above the Moon. Tycho came into the story because of his work on the comet of 1577, which he claimed established that its path lay between Venus and the Sun. And it seems to have been the attack on Tycho that brought in Kepler.

It is in the nature of hindsight, even the variety exercised cautiously by historians, to see things differently from how they appeared to the actors involved. Among historians, the general view is that Galileo's *Assayer* is of significance largely for its advocacy of experimentation (which seems to have been of little interest to Kepler, though he had done some optical and weighing experiments for his own purposes)[93] and that Kepler's contribution to the controversy is of no historical significance at all (which is very probably what Galileo thought of it). The simplification has something to be said for it, but the more complicated reality is much more interesting.

First, if we examine the exchanges in the controversies that Galileo engaged in, we can see that in regard to the New Stars and the comets Galileo appears curiously reluctant to accept that there were transitory phenomena in the celestial region. This does not square very well with his Copernicanism, in which (since the Earth is a planet) there should be no division between terrestrial and celestial physics. Like other Copernicans Galileo does appear to have believed that the traditional Aristotelian distinction between terrestrial and celestial physics was false. Yet when it comes to comets it is not Galileo but Kepler (whose physics is mainly Aristotelian in spirit) who accepts the phenomena as celestial but starts to look to explanations that have counterparts on Earth. In contrast, Galileo's explanations, for both New Stars and comets, show an Aristotelian determination to locate at least the origins of change within the sublunary sphere.

Further, the controversy itself looks rather different if we examine the part played by comets in the astronomy of the remainder of the seventeenth century. Astronomers in general came round to Tycho's position and comets were accepted as celestial objects within the planetary system. Meanwhile, the continuing reliability of the *Rudolphine Tables* strongly suggested the system was indeed heliocentric, with the planets moving according to Kepler's laws. That of course served to vindicate Kepler's high opinion of Tycho's skill as an observer.

Meanwhile, Galileo's determination of the law of free fall and his investigations of the movement of the pendulum helped to establish new ideas about terrestrial physics that, after some further development, allowed Newton to postulate an inverse square law of universal gravitation that explained Kepler's laws of planetary motion.[94] Thus Kepler's and Galileo's projects of proving Copernicanism was true were fulfilled together. And by way of a footnote, the comet Kepler had observed in 1607 was one of the comets (with those of 43 BC, AD 531, 1106 and 1680) that in 1705 Edmond Halley (*c.* 1656–1743) identified as having orbits so similar that he thought they must in fact be returns of the same comet. He successfully predicted its return in 1758, thereby (posthumously) providing confirmation of Newton's inverse square law.

Appendix: Letters Exchanged Between Galileo and Kepler in 1597

Translations J. V. Field

---

[93] On Kepler's experiments to determine specific gravities, see Pastorino (2020).
[94] Newton (1687).

A Note on the Translations

The original letters are in Latin, so it is not generally possible to follow word order exactly. However, as far as possible the translation preserves the order in which ideas are presented.

For Latin of this period, it can be difficult to know how closely the writer is adhering to the sense a particular word would have in ancient writings such as those of Cicero (106–43 BC) or Tacitus (*c.* AD 56–after 113). The present translations have been guided largely by context.

Except where it seemed likely to impede understanding, original punctuation has been retained. For instance, a colon regularly plays the part that in the twenty-first century would be given to a full stop.

Texts have been taken from *Johannes Kepler Gesammelte Werke,* ed. Max Caspar et al., volume XIII, Munich, Beck, 1945.

1.   Galileo Galilei to Johannes Kepler, from Padua 4 August 1597.

Letter 73, Kepler (1945), pp. 130–131. The original is in the National Library, Vienna.

I received your book, most learned sir, sent to me through Paulus Ambergerus, in fact not a few days, but a few hours ago; and since the same Paul told me about his return to Germany I judged that I would truly be ungrateful if I did not thank you with this letter for the gift I have received. Therefore I am doing so, and I do so again, as much as I am able, for the honour you do me by inviting me to [join in] friendship with you in such a manner: of the book I have as yet examined nothing but the preface, from which, however, I have understood at least a little of your intention, and I am exceedingly gratified to have a companion in searching out the truth, and one who is very much a friend of truth itself; for it is wretched how very few there are among students of truth who do not follow a perverse line of reasoning in Philosophizing: but since this is not the place to lament the troubles of our day, but rather [the place] to rejoice with you in very beautiful discoveries in confirmation of the truth: to which I shall add only this, and promise that I shall read through your book in a fair spirit; since I am certain that I shall find very beautiful things in it, moreover I shall do so the more willingly because many years ago I came to be of Copernicus' opinion, and from this position I have, also, found the causes of many things and natural effects, which beyond doubt are inexplicable according to the general hypothesis. I have written down many reasons and refutations of arguments to the contrary, which however I have not so far dared to publish, thoroughly frightened by the reception accorded to our teacher Copernicus himself, who although, for some, he has earned himself immortal fame, yet among an infinity of others (for so great is the number who are stupid), he appears as someone to be laughed at and hissed off the stage. I should certainly dare to disclose my thoughts if more such as you step forward, but while they do not, I shall hold back in this kind of business. I am under pressure from lack of time, and from eagerness to read your book, so making an end to this, therefore in bringing this to an end I sign myself your most affectionate and in all things your most willing servant. Padua, the day before the nones of August 1597.

The greatest friend to your honour and reputation,

Galileo Galilei, Mathematician at the University of Padua.

2.   Johannes Kepler to Galileo Galilei, from Graz, 13 October 1597. Letter 76, Kepler
(1945), pp. 144–146. The original is in the National Library, Florence.

Most courteous Sir, I received your letter written on 4 August on 1 September, and
it indeed gives me a double pleasure: first, on account of the friendship entered
into with you, an Italian; then on account of our agreement concerning Copernican
cosmography. Since therefore in the body of the letter you courteously invite me to
write frequent letters, and even had there been no impulse and spur of my own, I could
not do otherwise than write to you through the present noble young man. For I think
that since that time, if you have had the leisure, you will know my book through and
through, from which arises my strong desire to be made aware of your comments:
for it is my habit, with anyone I write to, to earnestly request honest judgements of
my work; and please believe me that I prefer the comment of one wise man, even if
it is harsh, to the thoughtless applause of the whole crowd. Would that you, endowed
with such intelligence, had put forward a different proposition! For although you
wisely and privately, using the example of your own person, advise yielding to the
universal ignorance, and that one should neither rashly press nor oppose the frenzies
of the crowd of the learned, in which [advice] you follow Plato and Pythagoras, our
true teachers, though in this century first by Copernicus, and then by very many,
and also by the most learned of mathematicians, let a beginning be made on a huge
undertaking, nor do I say this is something new, to move the Earth; it would perhaps
prevail by common acclaim once this waggon is constantly pulled to its goal, so that,
since no crowd measures the weight of reason, we may perhaps begin by the exercise
of intelligence to succeed in leading it to a recognition of the truth: For it is not
only your Italians who cannot believe that they move unless they feel it; but here in
Germany also we do not enter into this belief with the warmest gratitude. Indeed there
are reasons with which we may protect ourselves against these difficulties. First, I
have separated myself from that huge multitude of humanity, and I do not by any act
take cognizance of the noise of so many exclamations. Next, those who are near to me,
that is the common people, while they do not understand these (as they say) abstruse
matters, nevertheless marvel at them, without ever thinking whether they believe
them or no. The moderately educated, thus more prudent, are more cautious about
involving themselves in these mathematical disputes; they can be drawn to what is
said by an expert, on the authority of those versed in mathematics: so that when they
hear what ephemerides we already have, constructed on Copernicus' hypotheses;
[and that] those who write ephemerides today all follow Copernicus; and by them it
is postulated that they concede what can only be demonstrated in mathematical terms,
that the phenomena cannot occur without the motion of the Earth. For even if these
postulates or statements are not credible in themselves[95] they are however matter that
may be conceded by non-mathematicians; and since they are true why should they
not be put forward as irrefutable? So there remain only the mathematicians, dealing
with whom is harder work. They, since they have the same name, do not concede a

---

[95] Kepler uses the Greek word αὐτοπίστα.

postulate without a proof: of these, he who is the more inexperienced will make more or the business. But yet even here a remedy can be applied: solitude. In any place there is a mathematician; where this happens, that is best. Then if he has a colleague elsewhere he will receive letters from him; for this reason, if the letter is shown [to others] (which your letter would give me the means of doing), he can stir up this opinion in the spirits of the learned, as if all professors of mathematics everywhere were in agreement. Indeed what need is there of trickery? Be confident, Galileo, and step forward. If my deductions are correct, few of the leading mathematicians of Europe would wish to distance themselves from us: such is the force of truth. If Italy is less appropriate for publication, and if you will have some difficulties, perhaps Germany would allow us that freedom. But enough of these matters. You can at least communicate with me in writing privately, if you do not wish to do so publicly, if you have found something in support of Copernicus.

Now I should like to ask you for some observations: that is, since I lack an instrument, I need to have recourse to others. Have you a quadrant with which you can see minutes and quarters of minutes? If so then observe, around 19 December coming, the heights of the end of the tail of the Bear,[96] the maximum and minimum on the same night. And in the same way round 26 December, similarly observe both [maximum and minimum] heights of the pole star. Observe the first star also around 19 March of the year '98, at its nocturnal height at the hour of 12; the second [star] around 26 September also at the hour of 12. For if, as I hope, some slight difference between the pair of observations is of one minute or more, particularly if ten or fifteen, it would be a matter for discussion widely spread through the whole [community of] astronomy, however if on the other hand we notice nothing by way of difference, however, having given a proof of a most noble problem, together we shall win the victory that no one so far has attained. A word is enough to the wise.

I am, in any case, sending you two more copies since Hambergerus told me you wanted more. Whoever you send them to, he will discharge his obligation to me by writing me a letter about the book. Farewell, Most Eminent man, and rebalance the account with a very long letter to me. 13 October in the year '97, Graz.

With warmest appreciation for your civility, Magister Johan Kepler.

References

Aiton, E. J. (1972). *The vortex theory of planetary motion.* London and New York: Macdonald and Elsevier.

Bialas, V. (1971). Die quantitative Beschreibung der Planetenbewegung von Johannes Kepler in seinem handschriftlichen Nachlaß. In *Kepler Festschrift 1971.* Regensburg.

Boner, P. J. (Ed.). (2020). *Kepler's new star (1604): Context and controversy.* Leiden: Brill.

---

[96] '*caudae in Ursa. Eductio*': Lewis and Short say post classical 'setting out, departure'. This seems to be Kepler's rendering of Ptolemy's name for the star in *Almagest* VII.4, for either the first star in Ursa Minor or the 27th in Ursa Major, see Ptolemy (1984), pp. 341 and 343. As Caspar notes, the same task is given to Maestlin (letter 75, Kepler to Maestlin, beginning of Oct 1597, Kepler (1945), pp. 140–44, esp. pp. 142–43), and in that letter the explanation is clearer, and has some diagrams. It seems that Kepler wants parallax measurements.

Casini, P. (1983). Il Dialogo di Galileo e la Luna di Plutarco. In *Novità celesti e crisi del sapere* (P. Galluzzi, Ed., pp. 57–62). Florence. (Supplementary volume to *Annali del Istituto e Museo di Storia della Scienza di Firenze*).

Chiaramonti, S. (1621). *Anti Tycho*. Venice.

Cosci, M. (forthcoming). Galileo alias Alimberto Mauri e la disputa fiorentina sulla Stella Nuova.

Cosci, M. (2018a). Le fonti di Galileo Galilei per le Lezioni e studi sulla Stella Nuova del 1604. *Archives Internationales D'histoire Des Sciences, 68*, 6–70.

Cosci, M. (2018b). Astronomia pavana nel "Doalogo de Cecco di Ronchiti da Bruzene in perpuosito de la Stella Nuova" tra commedia, satira, *disputatio* accademica e poesia. In *I generi letterari dell'aristotelismo volgare rinascimentale* (M. Sgarbi, Ed., pp. 125–187). Padua: Cleup.

Cosci, M. (2019). Galileo alias Astolfo Arnerio Marchiano e la disputa padovana sulla Stella Nova. In *Atti del XVI Gionnata Galileiana (Padova 19 Gennalo 2019)*, Parte III: *Memorie delle Classe di Scienze Morali, Lettere ed Arti*, pp. 35–83.

Donne, J. (1611). *Ignatius His Conclave*. London: Thomas Morton.

Drake, S. (1976). *Galileo against the philosophers in his Dialogue of Cecco di Ronchitti and Considerations of Alimberto Mauri* (With English translations, introductions and notes by S. Drake). Los Angeles: Zeitlin and Ver Brugge.

Drake, S. (1973). Galileo's "Platonic" cosmogony and Kepler's prodromos. *Journal for the History of Astronomy, 4*, 174–191.

Drake, S. (1978). *Galileo at work: His scientific biography*. Chicago: University of Chicago Press.

Field, J. V. (1988). *Kepler's geometrical cosmology*. London and Chicago: Athlone Press and University of Chicago Press (reprinted London: Bloomsbury Press, 2014).

Field, J. V., & Postl, A. (1977). A thorough description of an extraordinary New Star which first appeared in October of this year, 1604. *Vistas in Astronomy, 20*, 333–339.

Galilei, G. (1623). *Il Saggitore*. Rome: A. G. Mascardi.

Galilei, G. (1632). *Dialogo sopra i due massimi sistemi del mondo*. Florence: Giovanni Battista Landini.

Galilei, G. (1953). *Dialogue concerning the two chief world systems—Ptolemaic & Coperican* (Translated from the Italian by Stillman Drake and with Foreword by Albert Einstein). Berkeley and Los Angeles: University of California Press.

Galilei, G., & Scheiner, C. (2010). *On sunspots* (Translated and with introductions by A. van Helden & E. Reeves). Chicago: University of Chicago Press.

Galilei, G., Grassi, H., Guiducci, M., & Kepler, J. (1960). *The controversy of the comets of 1618*. Pennsylvania University Press.

Hammer, F. (Ed.). (1941). *Johannes Kepler Gesammelte Werke, Band IV: Kleinere Schriften 1602–1611 / Dioptrice*. Munich: C. H. Beck.

Hellman, C. D. (1975). Kepler and comets. *Vistas in Astronomy, 18*, 789–796.

Kepler, J. (1604). *Ad Vitellionem paralipomena quibis astronomiae pars optica traditur*. Frankfurt: Claudius Marnus and the heirs of Johannes Aubrius. Frankfurt.

Kepler, J. (1606). *De stella nova*. Prague: Paul Sessius.

Kepler, J. (1611a). *Narratio de observatis a se quatuor Iovis satellitibus errantibus*. Frankfurt: Zacharias Palthenius.

Kepler, J. (1611b). *Dioptrice*, Frankfurt.

Kepler, J. (1619). *De cometis libelli tres*. Augsburg: Andreas Aperger (paid for by Sebastian Mylius, bookseller of Augsburg).

Kepler, J. (1625). *Tychonis Brahei Dani Hyperaspides*. Frankfurt: Gottfried Tampach.

Kepler, J. (1938). *Johannes Kepler Gesammelte Werke, Band I: Mysterium cosmographicum/De stella nova* (M. Caspar, Ed.). Munich: C. H. Beck.

Kepler, J. (1939). *Johannes Kepler Gesammelte Werke, Band. II: Astronomiae pars optica* (F. Hammer, Ed.). Munich: C. H. Beck.

Kepler, J. (1941). *Johannes Kepler Gesammelte Werke, Band IV: Kleinere Schriften 1602–1611 / Dioptrice* (F. Hammer, Ed.). Munich: C. H. Beck.

Kepler, J. (1945). *Johannes Kepler Gesammelte Werke, Band XIII: Briefe 1590–1599* (M. Caspar, Ed.). Munich: C. H. Beck.

Kepler, J. (1949). *Johannes Kepler Gesammelte Werke, Band XIV: Briefe 1599–1603* (M. Caspar, Ed.). Munich: C. H. Beck.

Kepler, J. (1959). *Johannes Kepler Gesammelte Werke, Band XVIII: Briefe 1620–1630* (M. Caspar, Ed.). Munich: C. H. Beck.

Kepler, J. (1963). *Johannes Kepler Gesammelte Werke, Band VIII: Mysterium cosmographicum (editio altera cum notis) / De cometis, Hyperaspides* (F. Hammer, Ed.). Munich: C. H. Beck.

Kepler, J. (1965). *Kepler's conversation with Galileo's Sidereal Messenger* (Translated from the Latin by E. Rosen). New York and London: Johnson Reprints.

Newton, I. (1687). *Philosophiae Naturalis Principia Mathematica.* London: Joseph Streete (The Royal Society).

Pastorino, C. (2020). Johannes Kepler and the exploration of the weight of substances in the long sixteenth Century. *Early Science and Medicine, 20,* 328–359.

Ptolemy, C. (1984). *Ptolemy's Almagest* (Translated from the Greek and annotated by G. J. Toomer). London: Duckworth.

Rothmann, C. (2014). *Christoph Rothmann's Discourse on the Comet of 1583* (An edition and translation with accompanying essays by M. A. Granada, A. Mosley & N. Jardine). Leiden: Brill.

Scheiner, C. (1628–30). *Rosa Ursina sive Sol.* Bracciano: Andreas Phaeus.

Schmid, A. (2022). *Die Münchner Galilei: Eine italienische Künstlerfamilie am Wittelsbacherhof im 17. Jahrhundert (Vergessenes Bayern).* Munich: Volk Verlag.

Seneca. (1988). *Naturales quaestiones,* vol. 1: Books 1–3 (Loeb Classical Library 450. Seneca, 7). London: Heinemann.

Shea, W. R., & Artiga, M. (2003). *Galileo in Rome: The rise and fall of a troublesome genius.* Oxford: Oxford University Press.

van Helden, A. (1977). The invention of the telescope. *Transactions of the American Philosophical Society, 67* (part 4). Philadelphia.

van Helden, A. (2008). The invention of the telescope. *Transactions of the American Philosophical Society, 98* (part 4). Philadelphia.

Westman, R. S. (2011). *The Copernican question: Prognostication, skepticism, and celestial order.* Berkeley: University of California Press.

Winkler, M. G., & van Helden, A. (1993). Johannes Hevelius and the visual language of astronomy. In *Renaissance and revolution: Humanists, craftsmen and natural philosophers in Early Modern Europe* (J. V. Field & F. A. J. L. James, Eds. , pp. 97–116). Cambridge: Cambridge University Press (reprinted 1997).

Winkler, M. G., & van Helden, A. (1992). Representing the heavens: Galileo and visual astronomy. *Isis, 83,* 195–217.

Chapter 9
The Long Life of the *Rudolphine Tables*

J. V. Field

In 1624 Kepler wrote a book in defence of Tycho Brahe (1546–1601), whose reputation as an astronomer had been attacked by Scipione Chiaramonti (1565–1652) and Galileo Galilei (1564–1642), in particular for his work on the comet of 1577, which (to Tycho's mind) had shown the comet moved between Venus and the Sun and thus established the impossibility of there being solid spheres in the heavens.[1] The books by Chiaramonti and Galileo had been published in Italy, in Venice in 1621 and Rome in 1623 respectively, but copies of them had only just reached Kepler, who was then living in Linz, in Austria. The work he wrote in reply, *Tychonis Brahei Dani Hyperaspistes* ('Tycho's defender'), was published in Frankfurt in 1625. In the first paragraph of the preface, Kepler tells us that when he read these books, in 1624, the moment seemed opportune because he had just completed the manuscript of the *Rudolphine Tables*.[2]

The tables were eventually published in Ulm in 1627. The difficulty had apparently been to find a printer who could take on a work containing so many numbers. Moreover, the war that had broken out in 1618 (and was to end only in 1648) was disrupting almost all aspects of civilian life. But there had also been difficulties before Kepler began compiling the tables.

These difficulties were with Tycho's heirs. And the problem had arisen before. After Tycho's death in 1601, Kepler had wanted to continue to use Tycho's observations to complete his work on the orbit of Mars, a task he had been given by Tycho himself. However, Tycho's heirs regarded the books of observations as private property that they had inherited. In their defence it should be said that Tycho seems

---

[1] See Donahue (1981) and Field, 'Kepler and Galileo' (this volume, Chap. 8).

[2] Kepler (1625), p. 1; Kepler (1963), p. 271. In the modern reprint of the *Rudolphine Tables*, published in 1969, the editor, Franz Hammer says Kepler's Preface was written in 1625, see Hammer in Kepler (1969), p. 5*.

J. V. Field (✉)
Birkbeck, University of London, London, UK
e-mail: jv.field@hart.bbk.ac.uk

© Springer Nature B.V. 2024
A. E. L. Davis et al. (eds.), *Reading the Mind of God*, Springer Praxis Books,
https://doi.org/10.1007/978-94-024-2250-4_9

to have had a similar attitude, as if the observations were arcane items in a private cabinet of curiosities.[3] The list of possible alternatives to Kepler as the author of the projected tables had included Tycho's son-in-law Tengnagel (Frans Gansneb, called Tengnagel van de Camp, 1576–1622) and Longomontanus (also called Severinus: Christen Sørensen, 1562–1647), both of whom had worked with Tycho. But Kepler was, after all, Tycho's successor as Imperial Mathematician.[4] The hierarchy of ownership leaves its mark on the title page of the printed Tables (Fig. 9.1).

The largest type on the title page is, of course, reserved for the word 'Rudolphinae', referring to the patron who gave his name to the tables, the Holy Roman Emperor Rudolf II (b. 1552, reigned 1576–1612). Tycho Brahe, introduced as 'the Phoenix of astronomers', has his name in slightly smaller type, and his name is followed by an assertion of the nobility of his family. The terms of this are presumably conventional since they echo those Kepler used in dedicating *Tychonis Hyperaspistes* to the Brahe family.[5] In contrast, Kepler's name appears, in smaller type, below a horizontal rule, and after the statement that publication is on the order of and paid for by the Emperor Ferdinand (1578–1637, Holy Roman Emperor 1619–1637). Kepler is described as first employed by Tycho and then successively by the three Emperors Rudolf, Matthias and Ferdinand. Readers are not invited to imagine him as having ideas of his own. In view of this, it comes as no surprise that Kepler occupies a similarly modest position in the elaborate frontispiece that shows a temple to astronomy (see Fig. 9.2).

It is perhaps also an echo of the disputes between Kepler and Tycho's heirs that the *Tables* contain two Dedicatory Letters. Both are addressed to the same Dedicatee, the Holy Roman Emperor Ferdinand II (b. 1578, reigned 1619–1637). The first is signed by 'Tycho Brahe's heirs and children',[6] the second, which is much longer, is signed by Kepler, who signs off with a formula that describes him as a humble servant in matters of mathematics.[7]

9.1 The Frontispiece

As befitted a book bearing the name of a Holy Roman Emperor and dedicated to one of his successors, the *Rudolphine Tables* were printed in folio. The large format afforded an opportunity for a display of humanist classical learning in a decorative frontispiece, see Fig. 9.2. The engraving depicts a building whose twelve columns support a domed roof. This kind of twelve-sided structure, called a 'monopteros' because it has only one line of columns, is described by Vitruvius (*fl. c.* 40 BC)

---

[3] See Mahoney, 'Measuring the heavens: how Tycho Brahe revolutionized observational astronomy' (this volume, Chap. 3).

[4] For a fuller account of the complicated story see Gingerich (1971).

[5] See Field, 'Kepler and Galileo' (this volume, Chap. 8).

[6] Kepler (1627), p.): (2v; Kepler (1969), p. 10.

[7] *Ad excolenda Mathemata conductus servulus.* Kepler (1627), p.): (4r; Kepler (1969), p. 14.

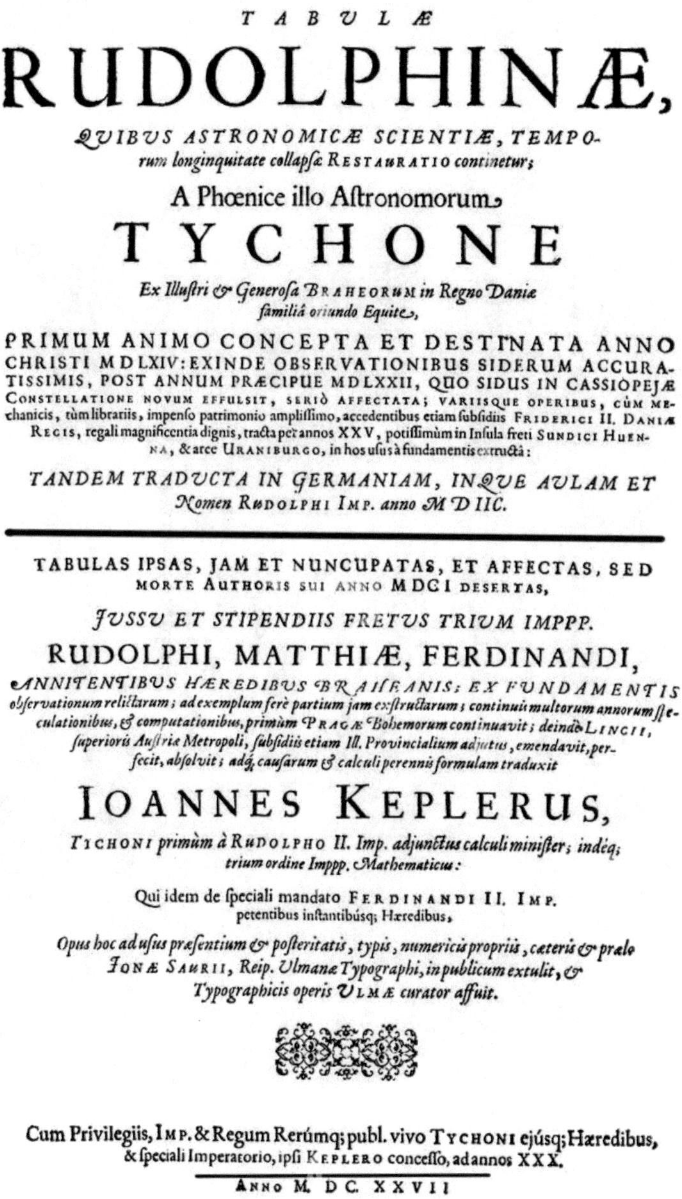

TABULÆ

RUDOLPHINÆ,

QUIBUS ASTRONOMICÆ SCIENTIÆ, TEMPO-
rum longinquitate collapsæ RESTAURATIO *continetur;*

A Phœnice illo Aſtronomorum

TYCHONE

Ex Illuſtri & Generoſa BRAHEORUM *in Regno Daniæ*
familiâ oriundo Equite,

PRIMUM ANIMO CONCEPTA ET DESTINATA ANNO
CHRISTI MDLXIV: EXINDE OBSERVATIONIBUS SIDERUM ACCURA-
TISSIMIS, POST ANNUM PRÆCIPUE MDLXXII, QUO SIDUS IN CASSIOPEJÆ
CONSTELLATIONE NOVUM EFFULSIT, SERIÒ AFFECTATA; VARIISQUE OPERIBUS, CÙM ME-
chanicis, tùm librariis, impenſo patrimonio ampliſſimo, accedentibus etiam ſubſidiis FRIDERICI II. DANIÆ
REGIS, regali magnificentia dignis, tractâ per annos XXV, potiſſimùm in Inſula freti SUNDICI HUEN-
NA, & arce URANIBURGO, in hos uſus à fundamentis extructâ:

TANDEM TRADUCTA IN GERMANIAM, INQUE AULAM ET
Nomen RUDOLPHI *IMP. anno* MDIIC.

TABULAS IPSAS, JAM ET NUNCUPATAS, ET AFFECTAS, SED
MORTE AUTHORIS SUI ANNO MDCI DESERTAS,

JUSSU ET STIPENDIIS FRETUS TRIUM IMPPP.

RUDOLPHI, MATTHIÆ, FERDINANDI,

ANNITENTIBUS HÆREDIBUS BRAHEANIS; EX FUNDAMENTIS
obſervationum relictarum; ad exemplum ferè partium jam exſtructarum; continuũ multorum annorum ſpe-
culationibus, & computationibus, primùm PRAGÆ *Bohemorum continuavit; deindè* LINCII,
ſuperioris Auſtriæ Metropoli, ſubſidiis etiam Ill. Provincialium adjutus, emendavit, per-
fecit, abſolvit; adq́ causarum & calculi perennis formulam traduxit

IOANNES KEPLERUS,

TYCHONI *primùm à* RUDOLPHO II. *Imp. adjunctus calculi miniſter; indéq;*
trium ordine Imppp. Mathematicus:

Qui idem de ſpeciali mandato FERDINANDI II. IMP.
petentibus inſtantibúsq; Hæredibus,

Opus hoc ad uſus præſentium & poſteritatis, typis, numericis propriis, cæteris & prælo
JONÆ SAURII, *Reip. Ulmanæ Typographi, in publicum extulit, &*
Typographicis operis ULMÆ curator affuit.

Cum Privilegiis, IMP. & Regum Rerúmq; publ. vivo TYCHONI ejúsq; Hæredibus,
& ſpeciali Imperatorio, ipſi KEPLERO conceſſo, ad annos XXX.

ANNO M. DC. XXVII.

Fig. 9.1 Title page of Kepler's *Rudolphine Tables*. (Reproduced from Kepler, J., *Tabulae Rudol-phinae*, Ulm, 1627, folio)

Fig. 9.2 Frontispiece of Kepler's *Rudolphine Tables*. (Reproduced from Kepler, J., *Tabulae Rudolphinae*, Ulm, 1627)

in his treatise *On Architecture* (*De architectura*) book 4, Chap. 8.[8] The standard edition of this work in Kepler's time was that by the Venetian humanist scholar Daniele Barbaro (1513–1570), which is beautifully illustrated.[9] Buildings like this

---

[8] Many Latin editions. English translation: Vitruvius (2009).

[9] Vitruvius (1566).

have survived from ancient times, and detailed drawings of them can be found in many sixteenth- and early seventeenth-century books about architecture. The drawings of ancient buildings supplied in such books are often very detailed, because it was the custom to copy such ancient structures not only in the design of new buildings but also in the design of temporary architecture used for festivities such as those welcoming visiting princes. The authors of such books included many famous architects, such as Sebastiano Serlio (1475–1554) and Andrea Palladio (1508–1580), and books of this kind seem to have been very widely distributed.[10] We may note, however, that in Vitruvius, and in the ancient buildings taken to exemplify his theories, the columns are fluted. Un-fluted columns became popular in the fifteenth century, partly through their use by Filippo Brunelleschi (1377–1446), and are (then) considered to be modelled on an ancient Tuscan variant of Vitruvius' Roman style. The designer of the temple in the frontispiece of the *Rudolphine Tables* has adopted this Renaissance revisionist version of architectural history.

The little temple of the frontispiece is thoroughly in accord with what we know of the tastes of Rudolf II, who was a keen collector of works of art—perhaps in competition with his cousin King Philip II of Spain (b. 1527, reigned 1556–1598)—and a patron of all the arts as well as the sciences.[11] That is, the Imperial court, as Kepler knew it, was (with hindsight) a good example of the Italianate humanist culture that prevailed in many courts at the time. The collections of paintings and *objets d'art* held by the Kunsthistorisches Museum in Vienna and museums in Prague show much evidence of Rudolf's sometimes idiosyncratic tastes.

In the best traditions of court-culture humanism, we are given a descriptive poem (an 'idyll') of about 460 lines, complete with marginal notes, that serves as a kind of key to the contents of the frontispiece.[12] The temple is proposed as showing the history of the science of astronomy, whose Muse, Urania, is apostrophized in the first few lines of the poem. More prosaically, in the first pages of the main text of the volume, and in prose, Kepler provides a Preface on the history of astronomical tables.[13] There is an extensive overlap of personnel: Kepler mentions all the people who appear in the frontispiece as astronomers.

For our present purposes, the main interest is the identity of the figures who are presented as pillars of astronomy, in the near-literal sense of each being associated with one of the columns. However, it may be as well to start reading the frontispiece from the top. Above the temple there hovers the Imperial eagle, from which coins are falling to the floor of the temple below, and a few as far as Kepler's writing table in the relief immediately to the left of centre on the twelve-sided platform the temple stands on. Symbolic figures stand at the lower edge of the shallow dome, one above

---

[10] Serlio (1537–51) and many subsequent editions and translations; Palladio (1570) and many subsequent editions and translations. See also Field (1999).

[11] On Rudolf see Evans (1997); on his collection of works of art, see Kaufmann (1993), especially Ch. 6, 'Ancients and Moderns in Prague' (pp. 151–173). Unfortunately, Kaufmann is not reliable on matters of science or technology or their history.

[12] Kepler (1627), pp. (4v -): ():(2v; Kepler (1969), pp. 15–26. For detailed analysis and a translation of the poem see Jardine, Leedham-Green and Lewis (2014a, 2014b); Jardine et al. (2014a, 2014b).

[13] Kepler (1627), *Praefatio*, pp. 1–8; Kepler (1969), pp. 36–44.

each of the visible columns. They include personifications of Astronomy (with a telescope), Geometry and varieties of Arithmetic involving logarithms. The column capitals below them are of various kinds, the one for the column behind the figure of Copernicus (1473–1543) being a simple Doric capital, while the capital to the right, on the column associated with Tycho Brahe, bristles with the acanthus leaves of the Corinthian order. In this mixture of orders, symbolism is taking precedence over classical correctness.

Tycho would have approved of the emphasis on observation. On the columns whose plinths each carry the name of a particular astronomer, there are representations of the instruments they used for making observations. The design and use of these instruments and estimates of the degrees of precision they could attain are discussed in Tycho's book about his own instruments: *Astronomiae instauratae mechanica* (Wandsbeck 1598).[14] Most of Tycho's instruments were far too large to be suspended on a column and there has in fact been no attempt to convey the relative sizes of the instruments that are depicted. But they are presented as significant. Moreover, the front face of the platform supporting the temple displays a map of the island of Hven, in Copenhagen Sound, on which Tycho's observatory had been situated.

Names of astronomers appear on the plinths under the six columns at the front of the temple. The two most prominent, under the middle two columns, are the most recent: Copernicus to the left of centre and Tycho to the right. Copernicus is shown seated and is identified not only by the name on the plinth he is leaning against but also by the book resting on his knees, whose title seems to be 'DE REV LIB V.'. Presumably Kepler was not able to correct the drawing, since *De revolutionibus* (Nuremberg, 1543, several subsequent editions) has six books not five (see Fig. 9.3). (In this period, it was not usual for authors to see and correct proofs.) To the left, that is behind Copernicus, leaning against the plinth supporting a rather time-worn column, is Hipparchus (Hipparchus of Nicaea, *c*. 190–*c*. 120 BC). At the far left there is another dilapidated column (its inner brickwork is exposed by the loss of the stucco covering), standing on a plinth that has the letters 'ARA', identified in the Idyll as referring to Aratus (*c*. 315–240 BC), famous for his poem about the heavens. To the right, behind Tycho, we have 'PTOLEMAEUS' (Claudius Ptolemy, *fl*. AD 129–141) and 'VETER.' (presumably an abbreviation for 'veteres', meaning 'Ancients') which refers to pre-Greek Babylonian (and perhaps also Egyptian) astronomers since the standing man in the background behind Tycho and Copernicus, who is apparently measuring an angle by means of his fingers, is identified in the Idyll as a 'Chaldean'. Both the Idyll and Kepler's Preface cite Hipparchus for his work on the sphere of the fixed stars, which (Kepler remarks) was used by Ptolemy in the *Almagest*.

On the vertical faces of the platform supporting the temple we can see five pictures presented as sculpted in relief. The outer two are too foreshortened to read. The one in the centre is, as already noted, a map of the island of Hven, the site of Tycho's observatory. To the left we see Kepler, identified by the heavily abbreviated book

---

[14] Reprinted Nuremberg 1602. English translation: Brahe (1946). See also Mahoney, 'Measuring the heavens: how Tycho Brahe revolutionized observational astronomy' (this volume, Chap. 3).

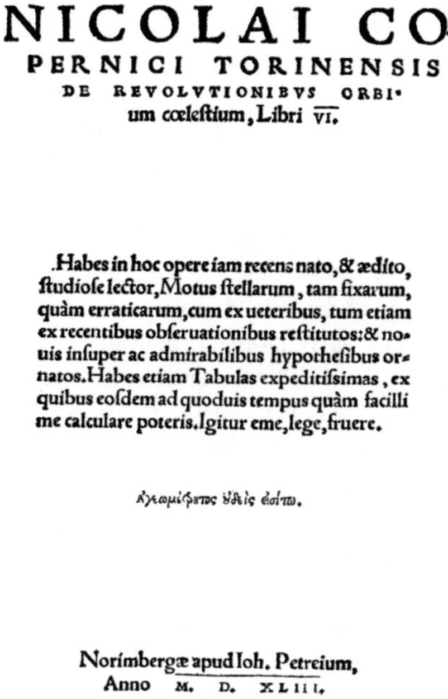

Fig. 9.3   The title page of
Copernicus' *De
revolutionibus*. (Reproduced
from Copernicus, N., *De
revolutionibus orbium
cœlestium libri VI,*
Nuremberg, 1543)

NICOLAI CO⟨
PERNICI TORINENSIS
DE REVOLVTIONIBVS ORBI·
um cœleftium, Libri $\overline{VI}$.

.Habes in hoc opere iam recens nato, & ædito,
ftudiofe lector, Motus ftellarum, tam fixarum,
quàm erraticarum, cum ex ueteribus, tum etiam
ex recentibus obferuationibus reftitutos: & no·
uis infuper ac admirabilibus hypothefibus or·
natos. Habes etiam Tabulas expeditifsimas, ex
quibus eofdem ad quoduis tempus quàm facilli
me calculare poteris. Igitur eme, lege, fruere.

Ἀχαμίϕετος ὕδλις ἐσίτω.

Norimbergæ apud Ioh. Petreium,
Anno   M.   D.   XLIII.

titles on the panel behind him, which appear to refer to *Mysterium cosmographicum,
Optics, Commentaries on Mars* (Kepler's usual way of referring to *Astronomia nova*)
and *Epitome of Copernican Astronomy*. The lighting makes the image to the right of
centre rather hard to read but it seems to show two men at work in a printer's shop. The
Idyll confirms this identification, its final section being concerned with the printing
of the *Table*s, emphasising Kepler's involvement with this process and ending with
further expressions of gratitude to the Emperor for his financial support.[15]

Astronomical tables were directed to the practical side of astronomy, providing a
basis for calculating future positions of planets (see the next section), for astrological
as well as astronomical purposes. It is thus appropriate—as well as being a reflection
of Tycho Brahe's interests—that the frontispiece displays a concern with observation. As already noted, observing instruments hang on the columns of the temple and
there is a map showing the location of Tycho's observatory. Further, the astronomers
associated with the columns are apparently honoured in this way for their contributions to observation as much as for establishing advances in understanding. The
apparent exception is Copernicus but that is partly a matter of hindsight; at the time
his observational work was seen as much more important than it now appears to
historians. In any case, the explanatory poem makes it clear that Copernicus and
Tycho are in dispute about theory in interpretation of the observations, with Tycho
pointing upwards and saying 'What if it is like this?' (*Quid si sic?*). In so saying he is
pointing to the ceiling under the dome of the temple, which carries a diagram of his

---

[15] Kepler (1627), p.): ():(2v; Kepler (1969), pp. 23–26; Jardine et al. (2014a, 2014b), Part 2, pp. 32–
33.

own version of the planetary system, in which the Sun moves round the Earth while Mercury, Venus, Mars, Jupiter and Saturn move round the Sun and are carried round the Earth with it.[16] This system was given additional plausibility by the discovery that Jupiter has four moons which it carries with it round the Sun.[17]

Thus the frontispiece as printed (or at least its central part) is concerned with promoting Tycho's world system as well as his achievement as an observer. This emphasis on cosmology does not reflect Kepler's original intentions. In April 1627, Tycho's heirs had asked Kepler to propose a design for the frontispiece of the *Tables*, and a sketch (probably not by Kepler himself) is preserved in Vienna.[18] Kepler's design looks to astronomical tables. Like the final version, it shows an open 'monopteros', domed and surmounted by the munificent Imperial eagle. However, the structure is octagonal, contains only five astronomers and no observing instruments. From left to right the figures are Copernicus, Regiomontanus (Johannes Müller of Königsberg, 1436–1476), Ptolemy, Tycho and Albategnius (al-Battani, *c.* AD 858–929). This list corresponds with what we find in Kepler's Preface, where the emphasis is on astronomical tables. Al-Battani was the author of a very famous set of tables (a *zij*) which was translated into Latin in the twelfth century.[19] Copernicus' connection with tables is that his work provided the basis for the *Prutenic Tables* (1551) though Kepler (entirely reasonably) describes his models of planetary motion (based on combinations of circles) as derived from the work of Regiomontanus (see below).

Returning to the printed version of the frontispiece, Tycho's question 'What if it is like this?'—in a context presumably intended as a rhetorical question inviting assent—can hardly be described as an argument. A more radical departure from the emphasis on observational work is provided by the figure of Kepler, whom the explanatory poem refers to only in connection with the printing of the tables. However, the books listed behind him in the 'relief' panel tell a different story. One title explicitly mentions Copernicus, and apart from the Optics all the others are directly conceived as arguments for heliocentrism. It is in fact difficult to find any reasonably long book by Kepler that is not largely concerned with promoting Copernicanism. And, as it turned out, the *Rudolphine Tables*, though apparently not so conceived, did in fact serve that end.

---

[16] For details of the Tychonic planetary system see Thoren (1990) and Mahoney, 'Measuring the heavens: how Tycho Brahe revolutionized observational astronomy' (this volume, Chap. 3).

[17] See Field, 'Kepler and Galileo' (this volume, Chap. 8).

[18] Kepler to Tycho and Georg Brahe, letter 1043, 10/20 April 1627, in Kepler (1959), pp. 292–93; Kepler to Wilhem Schickard, letter 1044, 25 April/5 May, in Kepler (1959), p. 294; and Hammer in Kepler (1969), pp. 3*–88*, esp. p. *36. The sketch is reproduced in Kepler (1969), p. [278].

[19] Kepler also refers to another Islamic astronomer, Geber (Abu Musa Jabir ibn Hayyam, d. between *c.* AD 806 and 816).

9.2   Astronomical Tables and Their Use

In twenty-first century usage, the term 'astronomical tables' has a spread of possible meanings. That was not so in the seventeenth century, or indeed in preceding centuries. A table showing the position of a planet day by day was called an 'ephemeris', a term that survives in technical usage. The general term for numbers, such as sines, presented in tabulated form was *canon* (plural *canones*), a Latin version of the Greek term used by Ptolemy in the *Almagest*. A single piece of tabulation might be called a *tabella*. The specialized term 'astronomical tables' referred to works that supplied the information required to calculate positions of the Sun, Moon and planets, such as the positions found in ephemerides. Which is to say that astronomical tables contain a quantity of cosmological ideas, or assumptions about celestial physics, as well as much detailed description of the motion of celestial bodies (that is 'theories', *theoricae*, of their motion). Thus Kepler's heliocentrism and his laws of planetary motion are built into the *Rudolphine Tables*.

The user of astronomical tables needed to start with some (observed) positions of whichever celestial body was in question; the astronomical tables provided the tools for finding further positions. Usually these would be future positions, so the tables were giving instructions for proceeding along the path of extrapolation (in principle risky). Thus, though one would probably prefer to start from relatively recent observations, which had to be reduced into a suitable standard form, there was no need to use particularly recent tables. The *Alfonsine Tables*, dating from the thirteenth century and named after Alfonso X King of Castile and Leon (1221–1284, reigned from 1252), were still in widespread use in the sixteenth century.[20] They were based on models of planetary motion taken from Ptolemy and developed by Islamic astronomers. After the publication of *De revolutionibus* (1543), Erasmus Reinhold (1511–1553) published tables based on Copernicus' models of planetary motion, which were taken from those found in the work of Regiomontanus (that is, not very different from Ptolemy's).[21] Reinhold's tables, called the Prutenic (i.e. Prussian) tables: *Prutenicae tabulae coelestium motuum* (Tübingen, 1551, second edition 1556, reprinted 1571, 1585), were reliable for a few years—which with hindsight we might expect since Copernicus had used recent observations in fixing the numerical details of his models. But by the 1570s the *Prutenic Tables* were proving to be sometimes only slightly more reliable than the Alfonsine ones. Tycho, for instance, complained the time of a conjunction that he observed was incorrect by several days.[22]

After many years of painstaking observation, Tycho had started work on the *Rudolphine Tables* shortly before he died in 1601. The tables as published in 1627 rely very heavily on his observations but not on his interpretations of them, except for adopting his tables of positions of fixed stars and a table showing corrections for atmospheric refraction. Apart from the table of refraction, on which he expresses uneasiness (see below), Kepler has taken care to make everything up to date and

---

[20] See Kremer (2023a).

[21] On Regiomontanus' tables see Kremer (2023b).

[22] See Field on Tycho and Kepler (in Chap. 8 of this volume).

the models of planetary motion used in the new tables are his own. That was, of course, the secret of the tables' success, but before we come to the elliptical orbits let us briefly look at what else Kepler thought his readers would need to know. The chapters of text that precede the pages of tabulations make up a sizeable treatise on astronomy. There are 34 chapters printed in double columns and covering more than 120 pages, folio, in the first edition. The close setting of the type is almost certainly an economy measure, reflecting the difficulty of getting an adequate supply of suitable paper in wartime.[23] Most chapters contain numbered 'precepts' (*praecepta*) which provide instructions on how to use the corresponding tables that appear in the second part of the volume. The large number of marginal notes, which also supply cross-references to precepts, are no doubt intended to make these introductory chapters easier to use.

The subjects of the chapters include methods of calculation, for instanceChaps. 3 and 4 describe how to use logarithms—whose invention is correctly ascribed to John Napier (1550–1617)—and give worked numerical examples.[24] Kepler supplies logarithms and antilogarithms that he calculated himself (see below).

Kepler's model for planetary motion, that is his ellipse and area laws appear, in that order,[25] in Chapter 20, which has a long and perfectly conventional title stating that it is concerned with calculating 'anomalies', that is how planetary positions differ from those one would find by assuming their motion was uniform.[26] Nevertheless, the chapter starts not with astronomy but with the mathematics of the ellipse. We are given a method of constructing points of the curve and a method of finding the area of a sector whose apex lies at one of the foci. The accompanying diagram is shown in Fig. 9.4. It is only several paragraphs later, after a history of the construction of planetary paths using increasingly elaborate combinations of circles, that we come to Kepler's own astronomical work. Then, accompanied by the marginal note 'Why multiplication of circles was abandoned',[27] there is a very brief account of how he arrived at an elliptical orbit, ending with the remark that 'a reader who is curious can find out the rest from my commentaries on Mars [that is *Astronomia nova*, 1609] and from the Epitome of Astronomy [*Epitome of Copernican astronomy*, 1618–21]'.[28]

In the next paragraph there is a shift of gear. The lettering of the points of the ellipse drawn in the opening section is explained: the focus S is identified as the Sun, while the ends of the major axis, A and P, represent Aphelion and Perihelion, and

---

[23] As already mentioned, the Thirty Years War, as it came to be called, started in 1618. It was the first 'modern' war in the sense that it caused huge suffering to civilians. There was also disruption of normal agricultural, industrial and commercial activity.

[24] Kepler (1627), Precepts 2–8, pp. 9–13; Kepler (1969), pp. 47–53.

[25] That is the inverse order of their discovery, see Davis 'Kepler's discovery of the planetary orbit: the Goldilocks solution' (this volume, Chap. 4).

[26] '*De tabulis prosthaphaereseon et de ratione excerpendi ex iis motus anomaliae, vel etiam aequationes eccentrici*', Kepler (1627), pp. 55–61; Kepler (1969), pp. 129–139.

[27] *Cur contempta circularum multiplicatio*, Kepler (1627), p. 57, Kepler (1969), p, 132.

[28] Kepler (1627), p. 57, Kepler (1969), p. 132, line 11. See also Davis, 'Kepler's discovery of the planetary orbit: the Goldilocks solution', and Gregory, 'The title of the *Astronomia nova*' (this volume, Chaps. 4 and 5 respectively).

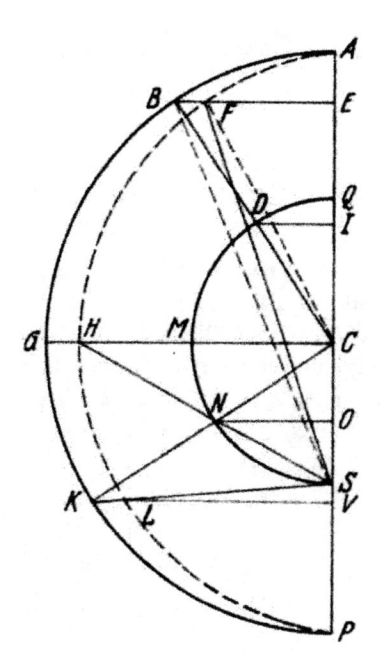

Fig. 9.4 Diagram to show how to construct points of an ellipse, and also used to find areas of sectors. Kepler supplied a separate plate referring to *Tabulae Rudolphinae* chapter 20, pp. 55–56. This plate supplies no lettering. Kepler later mentions lettering of points and readers were obviously expected to add the letters themselves. The lettering shown here was presumably added by the editor of *KGW* X. (Copyright Bavarian Academy of Sciences)

attention is drawn to the importance of the line joining the planet to the Sun and there is a reference to areas. But these things are mentioned only as matters that will appear in the tables proper—which are designed to allow one to find the deviation from perfectly uniform motion. The laws (as we now call them) are not marked as 'precepts' since they are, in effect, merely components in the calculation procedures. There is no explicit statement of the area law, though it could of course be worked out from the instructions given for calculation. Later Kepler does go so far as to supply detailed page references to the *Epitome*.[29] So, while the laws are not given much prominence, it would have been clear to a serious reader that they were embedded in the book, and it would have been possible to follow the clues to finding them there or elsewhere in Kepler's works.

Matters are much simpler in regard to the third law. The third law, which Kepler says is useful because it allows one to use ratios of periods instead of ratios of distances, appears in Chap. 24 as Precept 102, with a marginal reference to the relevant passages in *Harmonice mundi* and *Epitome*.[30]

The final formal chapter of precepts, Chap. 34 'On the variation of the plane of the ecliptic',[31] is followed by a short section called a 'Present sent to authors of birth horoscopes' (*Sportula genethliacis missa*) about how to use the *Rudolphine Tables*

---

[29] E.g. Kepler (1627) p. 57; Kepler (1969), p. 132, marginal note at line 40.

[30] Kepler (1627), p. 70; Kepler (1969), p. 159. See also Field, 'Kepler's cosmology' (this volume, Chap. 2).

[31] Kepler (1627), pp. 116–120; Kepler (1969), pp. 235–243.

in astrological computations; it contains a further twelve precepts (numbers 198 to 209) and some suggestions for improvements in astrologers' practices.[32]

9.3   The Tabulations

The chapter addressed to astrologers is followed by an Appendix written by Kepler's son-in-law Jakob Bartsch (*c.* 1600–1633), serving as an introduction to the tabulations.[33] It includes a series of short poems forming a dialogue between Kepler and Tycho, but otherwise deals with technical matters.

The tabulations themselves are introduced with a contents page set out rather as if it were a supplementary title page, though only for Part 1 of the set of tabulations, in which there are thirteen sets of tables of items that are also applicable in matters other than planetary astronomy (see Fig. 9.5). There are, for instance, as already mentioned, several sets of tables of logarithms and antilogarithms. These are what are now called 'natural logarithms' (base *e*) but they are not taken from Napier; the tables are to seven figures and have been newly calculated by Kepler himself.[34] The chief recommendation for the *Rudolphine Tables*, and (as we shall see) their historical importance, lay in the tables directly relating to the motion of the planets, but, as is made clear in the series of introductory chapters in the first part of the work, Kepler is at pains to be as up to date as possible in all respects.

We see another example of this concern in the list of places, with their geographical coordinates on pages 33 to 36, which is supplemented by a fold-out world map. As Kepler tells us in Chap. 16 the list and the map are derived from the map Guglielmus Janssonius of Alkmaar—that is Willem Janszoon Blaeu (1571–1638)—made for Rudolf II in 1605.[35] Blaeu had been an assistant to Tycho on Hven in 1594–96.[36] Kepler's list of places is printed in three columns.[37] Each name is followed by its 'difference in meridian', in hours and minutes, followed by the letter 'a' (for *additio*) for places that lie to the East of the zero line or 's' (for *subtractio*) for places to the West of it. (Inconveniently, the Latin terms for East and West, *Oriens* and *Occidens* both begin with 'O'.) These differences are equivalent to geographical longitude. They are followed by geographical latitudes, 'the height of the pole', in degrees and minutes. In Chap. 16 Kepler stresses that the differences in meridian are difficult to measure precisely and gives a description of the various methods available for

---

[32] Kepler (1627), pp. 121–125; Kepler (1969), pp. 244–254.

[33] Kepler (1627), Ch. XVI (section 'De Mappa Mundi universali', with a marginal reference 'fol. 36'), p. 41 (near end of second column); Kepler (1969), p. 103.

[34] On Kepler's enthusiasm for logarithms, see Field (1988).

[35] Kepler (1627), Ch. XVI (section 'De Mappa Mundi universali', with a marginal reference 'fol. 36'), p. 41 (near end of second column); Kepler (1969), p. 103.

[36] See Mahoney, 'Measuring the heavens: how Tycho Brahe revolutionized observational astronomy' (this volume, Chap. 3) and Mosley (2007).

[37] In his Afterword, Franz Hammer says the list contains 530 items (Kepler, 1969, p. 71*).

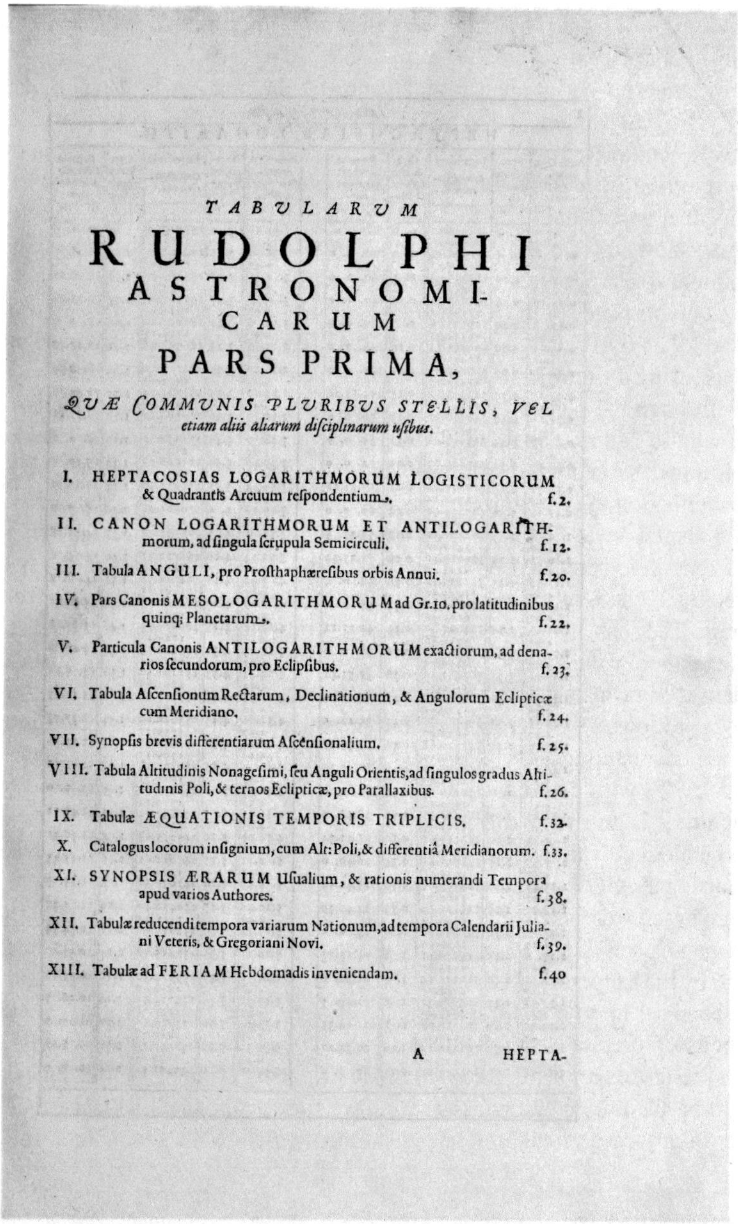

TABVLARVM

RUDOLPHI
ASTRONOMI-
CARUM

PARS PRIMA,

QVÆ COMMVNIS PLVRIBVS STELLIS, VEL etiam aliis aliarum diſciplinarum uſibus.

A HEPTA-

Fig. 9.5 Contents page for the first part of the tabulations. The section of the work containing the tabulations is numbered separately. (Reproduced from Kepler, J., *Tabulae Rudolphinae*, Ulm, 1627, p. [1])

finding them, for instance by comparing the local times of an occultation observed at two different places.

A few names appear in uppercase: Hvenna (p. 34, col. 2, with the comment 'the seat of Tycho's Astronomy') 0.0, 55°55′; Lincium (Linz, p. 34, col. 3, 'where these Tables were completed') 0.10 a, 48°10′; Roma (Rome, p. 35, col. 3) 0.0, 42°2′; Uraniburgium (p. 36, col. 3, 'the seat of Astronomy') 0.0, 55°55′. So Tycho has been given his due, twice, and his observatory defines the zero meridian.[38] The fold-out world map shows the globe in stereographic projection, a projection that would have been familiar to astronomers from its use on astrolabes.

The Contents pages of the following sets of tabulations, Parts 2 to 4, are shown in Figs. 9.6, 9.9 and 9.10. Part 2, which covers 47 pages,[39] gives tables for the motions of planets (using the term in its traditional sense), starting with the Sun, then working inwards through the planetary system: Saturn, Jupiter, Mars, Venus, Mercury, and ending with the Moon. The first few lines of the tables for Mars and for Mercury are shown in Figs. 9.7 and 9.8 respectively.

The inclination to economize on paper reasserts itself in Part 5, starting with the lack of a display-style first page. There are only two tables. The first is Tycho's list of the fixed stars—prefaced by a note to say that not all the stars in the list were observed by Tycho himself, because some of them are too far south to be seen from Uraniborg. The stars are arranged according to constellations and individual stars are identified either by their names or in the discursive style that goes back at least as far as Ptolemy.[40] For instance, the first star in Cassiopeia is described as 'in the head' (*in capite*) followed by its ecliptic coordinates (with longitudes referred to zodiac signs) and its magnitude.[41] The last star in the list for Cassiopeia is 'the new star of the year 1572', with coordinates 6°52′ Taurus, 55.54 B (that is *Borealis* = North). No magnitude is given.

The catalogue of fixed stars is followed by a short final item, Tycho's table of refractions.[42] Kepler's introductory note says that the observations on which the tables are based were made on Hven, in Copenhagen Sound, between the Baltic and the German Ocean (i.e. the North Sea) and in Bohemia at Benátky (the estate given to Tycho by the Emperor). At Benátky there is archaeological evidence of equipment having been set up to make observations. In one of the rooms on the second floor of the house, a precise North–South groove has been cut into the floorboards to line up with the left edge of the floor-length south-facing window. Unfortunately, that seems to be the only solid evidence we have for astronomical work carried out at the site. In any case, though he has given the tabulation the title 'Three-fold table

---

[38] London, effectively on the meridian of Greenwich, appears as *Londinium Angliae* at 0.48 s, 51°32′ (p. 34, col. 3); Prague (*Praga Bohemiae*) is 0.6 a, 50°6′ (p. 35, col. 3); Kassel (*Cassellae Hessiae*) is 0.13 s, 51°19′ (p. 33, col. 3); Rostock (*Rostockium Meckelburgica Duc.*) is 0.0, 54°10′ (p. 34, col. 3); Munich (Monachium Bavariae) is 0.1 s, 48°2′ (p. 35, col. 1).

[39] Kepler (1627) and Kepler (1969), pp. 42–88.

[40] Kepler (1627), pp. 105–119; Kepler (1969), pp. 105–141.

[41] Kepler (1627), p. 107; Kepler (1969), pp. 110–111.

[42] Kepler (1627), p. 119 [printed as 115]; Kepler (1969), p. 142.

Fig. 9.6 Contents page for the second part of the tabulations, giving tables for the motions of the Sun, Moon and planets. (Reproduced from Kepler, *Tabulae Rudolphinae*, Ulm, 1627, p. 41)

of refractions' (*Tabula refractionum triplex*) Kepler's note does not comment on the fact that Tycho has supplied three parallel tables, one relating to the Sun, one to the Moon and one to the fixed stars.

However, there is a longer comment in the earlier part of Kepler's work, in the final section of Chapter 34, under the heading 'On Refractions of the rays of heavenly bodies in the air' (*De Refractionibus radiorum sideralium in acre*).[43] Here Kepler expresses unease with Tycho's results. Indeed, though he does not say so here, in his treatise on optics of 1604, Kepler had not adopted Tycho's results, but had supplied his own tables of atmospheric refraction[44]; so his distrust of Tycho's work on atmospheric refraction is not new. in 1627 his defence of there being three tables rests on the practical issue that the viewing conditions for the Sun, Moon and fixed stars are

---

[43] Kepler (1627), p. 120, col. 2; Kepler (1969), pp. 242–43.

[44] See Donahue, 'Kepler's work on Optics' (this volume, Chap. 7).

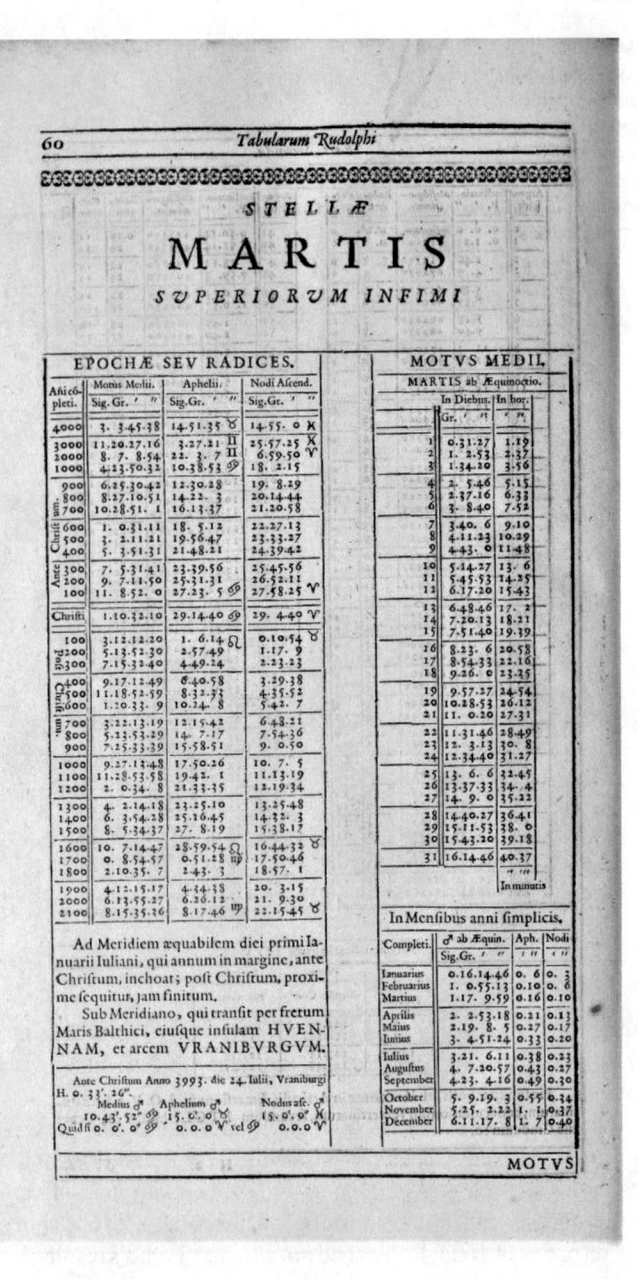

Fig. 9.7 The start of the table for the motion of Mars. (Reproduced from Kepler, J., *Tabulae Rudolphinae*, Ulm, 1627, p. 60)

STELLÆ
MERCURII
INFERIORVM SECVNDI

EPOCHÆ SEV RADICES.

Ani cô-pleti.	Motus Medii. Sig.Gr. ' "	Aphelii. Sig.Gr. ' "	Nodi Ascend. Sig.Gr. ' "
4000	7.10.49.41	19.46.47 ♊	29.49.16 ↗
3000	8. 4.44.58	28.53.45 ♋	23.30. 0 ♈
2000	8.28.40.16	28. 0.43 ♋	17.10.44 ♋
1000	9.22.35.33	27. 7.41 ♍	10.51.27 ♏
Ante Christum			
900	0. 6.59. 5	0. 2.23 ♎	13.12.31 ♓
800	2.21.22.37	2.57. 5	15.35.36
700	5. 5.46. 9	5.51.46	17.57.40
600	7.20. 9.40	8.46.28	20.19.45
500	10. 4.33.12	11.41.10	22.41.49
400	0.18.56.44	14.35.52	25. 3.53
300	3. 3.20.16	17.30.34	27.25.58
200	5.17.43.47	20.25.15	29.48. 2 ♓
100	8. 2. 7.19	23.19.57 ♎	2.10. 7 ♈
Christ.	10.16.30.51	26.14.39 ♎	4.32.11 ♈
Post Christum			
100	1. 0.54.23	29. 9.21 ♎	6.54.15 ♈
200	3.15.17.54	2. 4. 3 ♏	9.16.20
300	5.29.41.26	4.58.44	11.38.24
400	8.14. 4.58	7.53.26	14. 0.29
500	10.28.28.30	10.48. 8	16.22.33
600	1.12.52. 1	13.42.50	18.44.38
700	3.27.15.33	16.37.32	21. 6.42
800	6.11.39. 5	19.32.13	23.28.46
900	8.26. 2.37	22.26.55	25.50.51
1000	11.10.26. 8	25.21.37	28.12.55 ♈
1100	1.24.49.40	28.16.19 ♏	0.35. 0 ♉
1200	4. 9.13.12	1.11. 1 ↗	2.57. 4
1300	6.23.36.44	4. 5.42	5.19. 9
1400	9. 8. 0.15	7. 0.24	7.41.13
1500	11.22.23.47	9.55. 6	10. 3.18
1600	1. 6.47.19	12.49.48	12.25.22
1700	4.21.10.51	15.44.29	14.47.26
1800	7. 5.34.22	18.39.11	17. 9.31
1900	9.19.57.54	21.33.53	19.31.35
2000	0. 4.21.26	24.28.35	21.53.40
2100	2.18.44.58	27.23.17 ↗	24.15.44 ♉

MOTVS MEDII.
MERCVRII ab Æquinoctio.

	In Diebus. Sig.Gr. ' "	In horis. Gr. ' "
1	0. 4. 5.32	0.10.14
2	0. 8.11. 5	0.20.28
3	0.12.16.38	0.30.42
4	0.16.22.10	0.40.56
5	0.20.27.43	0.51. 9
6	0.24.33.16	1. 1.23
7	0.28.38.48	1.11.37
8	1. 2.44.21	1.21.51
9	1. 6.49.53	1.32. 5
10	1.10.55.26	1.42.19
11	1.15. 0.58	1.52.32
12	1.19. 6.31	2. 2.46
13	1.23.12. 4	2.13. 0
14	1.27.17.36	2.23.14
15	2. 1.23. 9	2.33.28
16	2. 5.28.41	2.43.41
17	2. 9.34.14	2.53.55
18	2.13.39.47	3. 4. 9
19	2.17.45.19	3.14.23
20	2.21.50.52	3.24.37
21	2.25.56.24	3.34.51
22	3. 0. 1.57	3.45. 4
23	3. 4. 7.30	3.55.18
24	3. 8.13. 2	4. 5.32
25	3.12.18.35	4.10.46
26	3.16.24. 7	4.16. 0
27	3.20.29.40	4.21.14
28	3.24.35.13	4.26.27
29	3.28.40.45	4.31.41
30	4. 2.46.18	4.36.55
31	4. 6.51.50	4.42. 9
		In minutis

In Mensibus anni simplicis.

Completi.	☿ ab Æquin. Sig.Gr. ' "	Aph. ' "	Nodi ' "
Ianuarius	4. 6.51.50	0. 9	0. 7
Februarius	8. 1.27. 3	0.17	0.13
Martius	0. 8.18.53	0.26	0.10
Aprilis	4.11. 5.11	0.35	0.27
Maius	8.17.57. 1	0.44	0.35
Iunius	0.20.43.19	0.52	0.42
Iulius	4.27.35. 9	1. 1	0.49
Augustus	9. 4.26.59	1.10	0.57
September	1. 7.13.17	1.19	1. 4
October	5.14. 5. 7	1.27	1.11
November	9.16.51.25	1.36	1.18
December	1.23.43.15	1.45	1.25

Ad Meridiem æquabilem diei primi Ianuarii Iuliani, qui annum in margine, ante Christum, inchoat; post Christum, proxime sequitur, jam finitum.

Sub Meridiano, qui transit per fretum Maris Balthici, eiusque insulam HVEN-NAM, et arcem VRANIBVRGVM.

Ante Christum Anno 3993. die 24. Iulii, Vraniburgi H. 0. 33'. 26".

Medius ☿	Aphelium ☿	Nodus asc. ☿
0. 0'. 0" ♈	0. 0. 0" ♋	0. 0'. 0" ♈

MOTVS

Fig. 9.8 The start of the table for the motion of Mercury. (Reproduced from Kepler, J., *Tabulae Rudolphinae*, Ulm, 1627, p. 72)

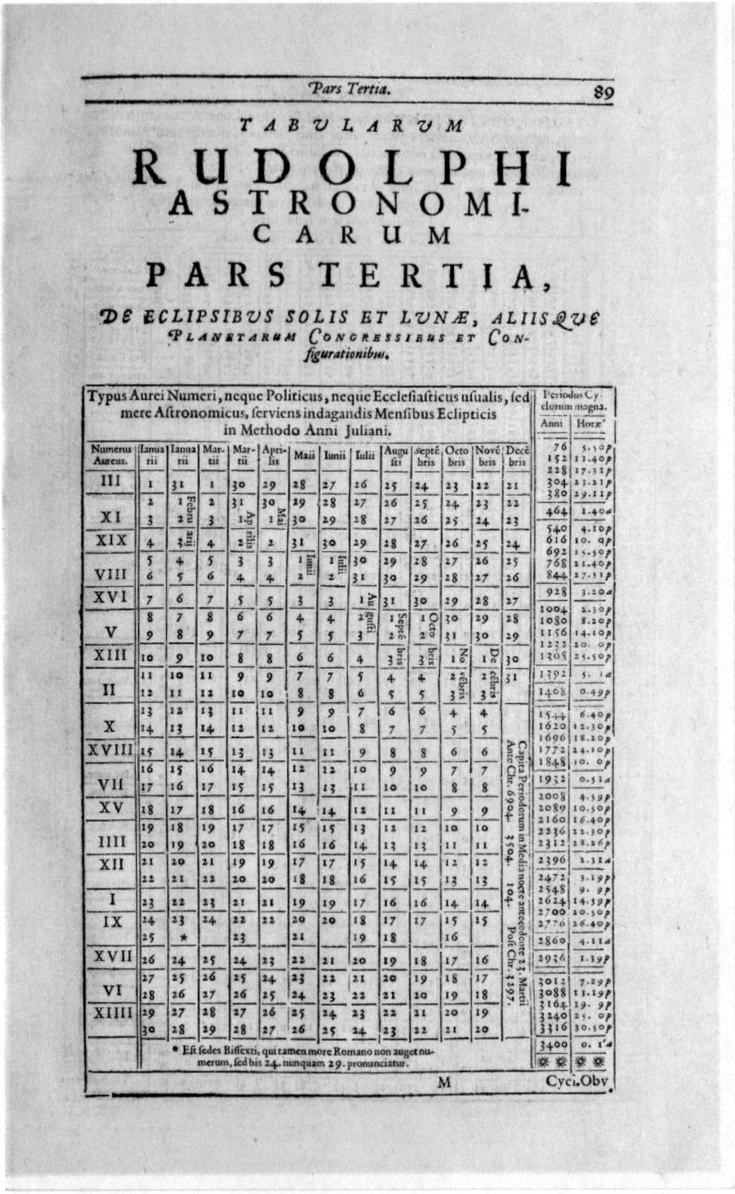

T A B V L A R V M

R U D O L P H I
A S T R O N O M I-
C A R U M

P A R S   T E R T I A,

DE ECLIPSIBVS SOLIS ET LVNÆ, ALIISQVE
PLANETARVM CONGRESSIBVS ET CON-
figurationibus.

Typus Aurei Numeri, neque Politicus, neque Ecclefiasticus ufualis, fed mere Aftronomicus, ferviens indagandis Menfibus Eclipticis in Methodo Anni Juliani.

Numerus Aureus	Ianuarii	Ianuarii	Martii	Martii	Aprilis	Maii	Iunii	Iulii	Augufti	Septebris	Octobris	Novebris	Decebris
III	1	31	1	30	29	28	27	26	25	24	23	22	21
	2	1	2	31	30	29	28	27	26	25	24	23	22
XI	3	2	3	1	1	30	29	28	27	26	25	24	23
XIX	4	3	4	2	2	31	30	29	28	27	26	25	24
	5	4	5	3	3	1	1	30	29	28	27	26	25
VIII	6	5	6	4	4	2	2	31	30	29	28	27	26
XVI	7	6	7	5	5	3	3	1	31	30	29	28	27
	8	7	8	6	6	4	4	2	1	30	29	28	
V	9	8	9	7	7	5	5	3	2	1	31	30	29
XIII	10	9	10	8	8	6	6	4	3	3	1	1	30
II	11	10	11	9	9	7	7	5	4	4	2	2	31
	12	11	12	10	10	8	8	6	5	5	3	3	
X	13	12	13	11	11	9	9	7	6	6	4	4	
	14	13	14	12	12	10	10	8	7	7	5	5	
XVIII	15	14	15	13	13	11	11	9	8	8	6	6	
	16	15	16	14	14	12	12	10	9	9	7	7	
VII	17	16	17	15	15	13	13	11	10	10	8	8	
XV	18	17	18	16	16	14	14	12	11	11	9	9	
	19	18	19	17	17	15	15	13	12	12	10	10	
IIII	20	19	20	18	18	16	16	14	13	13	11	11	
XII	21	20	21	19	19	17	17	15	14	14	12	12	
	22	21	22	20	20	18	18	16	15	15	13	13	
I	23	22	23	21	21	19	19	17	16	16	14	14	
IX	24	23	24	22	22	20	20	18	17	17	15	15	
	25	*		23		21		19	18		16		
XVII	26	24	25	24	23	22	21	20	19	18	17	16	
	27	25	26	25	24	23	22	21	20	19	18	17	
VI	28	26	27	26	25	24	23	22	21	20	19	18	
XIIII	29	27	28	27	26	25	24	23	22	21	20	19	
	30	28	29	28	27	26	25	24	23	22	21	20	

* Eft fedes Biffexti, qui tamen more Romano non auget numerum, fed bis 24. nunquam 29. pronunciatur.

M

Caput Periodorum in Medii nocte antecedente 12. Martii Anni Chr. 6904.   4604.   104.   Pof Chr. 1977.

Periodus Cyclorum magna.

Anni	Horæ'
76	5.10p
152	11.40p
228	17.31p
304	21.31p
380	29.11p
464	1.40a
540	4.10p
616	10. 0p
692	15.50p
768	21.40p
844	27.31p
928	1.20a
1004	2.10p
1080	8.20p
1156	14.10p
1232	20. 0p
1208	25.50p
1392	5. 1a
1408	0.49p
1544	6.40p
1620	12.30p
1696	18.20p
1772	24.10p
1848	10. 0p
1932	0.51a
2008	4.19p
2084	10.50p
2160	16.40p
2236	22.30p
2312	28.26p
2396	1.31a
2472	3.19p
2548	9. 9p
2624	14.59p
2700	20.50p
2776	26.40p
2860	4.11a
2936	1.19p
3012	7.29p
3088	13.19p
3164	19. 9p
3240	25. 0p
3316	30.50p
3400	0. 1a

Cycl. Obv

Fig. 9.9 First page of the third part of the tabulations ('On eclipses of the Sun and Moon …'). (Reproduced from Kepler, J., *Tabulae Rudolphinae*, Ulm, 1627, p. 89)

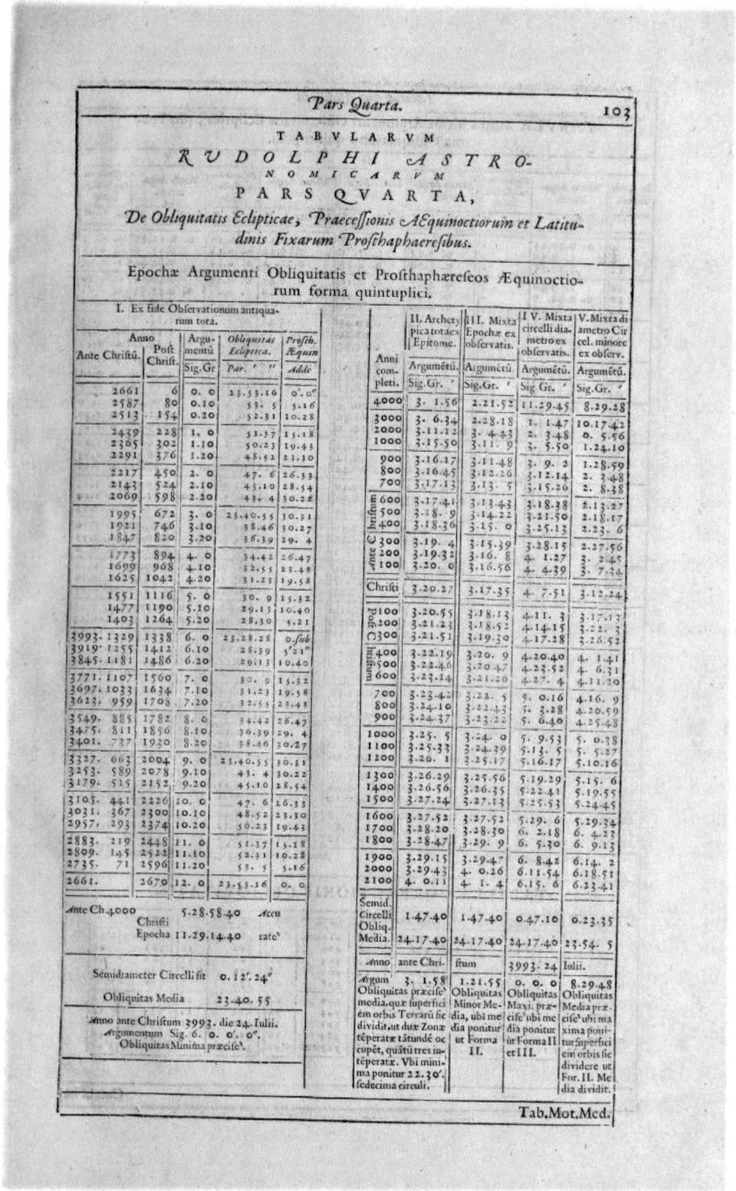

Pars Quarta. 103

TABVLARVM
RVDOLPHI ASTRONOMICARVM
PARS QVARTA,
De Obliquitatis Eclipticae, Praecessionis Æquinoctiorum et Latitudinis Fixarum Prosthaphaeresibus.

Epochæ Argumenti Obliquitatis et Prosthaphæreseos Æquinoctiorum forma quintuplici.

I. Ex fide Observationum antiquarum tota.

Anno Ante Christū.	Post Christ.	Argumentū Sig.Gr	*Obliquitas Ecliptica.* *Par. ' "*	*Prosth. Æquin* *Addē*
2661	6	0. 0	23.53.16	0'.0"
2587	80	0.10	53. 5	5.16
2513	154	0.20	52.31	10.28
2439	228	1. 0	51.37	15.18
2365	302	1.10	50.23	19.43
2291	376	1.20	48.52	23.30
2217	450	2. 0	47. 6	26.33
2143	524	2.10	45.10	28.54
2069	598	2.20	43. 4	30.28
1995	672	3. 0	23.40.55	30.31
1921	746	3.10	38.46	30.27
1847	820	3.20	36.39	29. 4
1773	894	4. 0	34.42	26.47
1699	968	4.10	32.55	23.48
1625	1042	4.20	31.23	19.58
1551	1116	5. 0	30. 9	15.32
1477	1190	5.10	29.13	10.40
1403	1264	5.20	28.30	5.21
3993. 1329	1338	6. 0	23.28.28	0.*sub*
3919. 1255	1412	6.10	28.39	5'21"
3845. 1181	1486	6.20	29.13	10.40
3771. 1107	1560	7. 0	30. 9	15.32
3697. 1033	1634	7.10	31.23	19.58
3623. 959	1708	7.20	32.55	23.43
3549. 885	1782	8. 0	34.42	26.47
3475. 811	1856	8.10	36.39	29. 4
3401. 737	1930	8.20	38.46	30.27
3327. 663	2004	9. 0	23.40.55	30.31
3253. 589	2078	9.10	43. 4	30.22
3179. 515	2152	9.20	45.10	28.54
3105. 441	2226	10. 0	47. 6	26.33
3031. 367	2300	10.10	48.52	23.30
2957. 293	2374	10.20	50.23	19.43
2883. 219	2448	11. 0	51.37	15.18
2809. 145	2522	11.10	52.31	10.28
2735. 71	2596	11.20	53. 5	5.16
2661.	2670	12. 0	23.53.16	0. 0

Ante Ch. 4000 Christi Epocha 5.28.58.40 *Accu*rate
Epocha 11.29.14.40

Semidiameter Circelli sit 0.12'.24"

Obliquitas Media 23.40.55

Anno ante Christum 3993. die 24. Iulii.
Argumentum Sig. 6. 0. 0'. 0".
Obliquitas Minima præcise.

Anni completi.	II. Archetypica tota ex Epitome. Argumētū. Sig.Gr. '	III. Mixta Epochæ ex observatis. Argumētū. Sig.Gr. '	IV. Mixta circelli diametro ex observatis. Argumētū. Sig.Gr. '	V. Mixta diametro Circel. minore ex observ. Argumētū. Sig.Gr. '
4000	3. 1.56	2.21.52	11.29.45	8.29.28
3000	3. 6.34	2.28.18	1. 1.47	10.17.42
2000	3.11.12	3. 4.43	2. 3.48	0. 5.56
1000	3.15.50	3.11. 9	3. 5.50	1.24.10
900	3.16.17	3.11.48	3. 9. 2	1.28.59
800	3.16.45	3.12.26	3.12.14	2. 3.48
700	3.17.13	3.13. 5	3.15.26	2. 8.38
Ante Christum 600	3.17.41	3.13.43	3.18.38	2.13.27
500	3.18. 9	3.14.22	3.21.50	2.18.17
400	3.18.36	3.15. 0	3.25.13	2.23. 6
300	3.19. 4	3.15.39	3.28.15	2.27.56
200	3.19.32	3.16. 8	4. 1.27	3. 2.45
100	3.20. 0	3.16.56	4. 4.39	3. 7.34
Christi	3.20.27	3.17.35	4. 7.51	3.12.24
Post Christum 100	3.20.55	3.18.13	4.11. 3	3.17.13
200	3.21.23	3.18.52	4.14.15	3.22. 3
300	3.21.51	3.19.30	4.17.28	3.26.52
400	3.22.19	3.20. 9	4.20.40	4. 1.41
500	3.22.46	3.20.47	4.23.52	4. 6.31
600	3.23.14	3.21.26	4.27. 4	4.11.20
700	3.23.42	3.22. 5	5. 0.16	4.16. 9
800	3.24.10	3.22.43	5. 3.28	4.20.59
900	3.24.37	3.23.22	5. 6.40	4.25.48
1000	3.25. 5	3.24. 0	5. 9.53	5. 0.38
1100	3.25.33	3.24.39	5.13. 5	5. 5.27
1200	3.26. 1	3.25.17	5.16.17	5.10.16
1300	3.26.29	3.25.56	5.19.29	5.15. 6
1400	3.26.56	3.26.35	5.22.41	5.19.55
1500	3.27.24	3.27.13	5.25.53	5.24.45
1600	3.27.52	3.27.52	5.29. 6	5.29.34
1700	3.28.20	3.28.30	6. 2.18	6. 4.23
1800	3.28.47	3.29. 9	6. 5.30	6. 9.13
1900	3.29.15	3.29.47	6. 8.42	6.14. 2
2000	3.29.43	4. 0.26	6.11.54	6.18.51
2100	4. 0.11	4. 1. 4	6.15. 6	6.23.41
Semid. Circelli	1.47.40	1.47.40	0.47.10	0.23.35
Obliq. Media.	24.17.40	24.17.40	24.17.40	23.54. 5
Anno	ante Chri-	stum	3993. 24	Iulii.
Argum	3. 1.58	1.21.55	0. 0. 0	8.29.48
	Obliquitas præcise media, quæ superficiem orbis Terrarū sic dividit, ut duæ Zonæ tēperatæ tātundē occupēt, quātū tres intēperatæ. Vbi minima ponitur 22.30'. sedecima circuli.	Obliquitas Minor Media, ubi media ponitur ut Forma II.	Obliquitas Maxi. præcise ubi media ponitur ut Forma II et III.	Obliquitas Media præcise ubi maxima ponitur superficiem orbis sic dividere ut For. II. Media dividit.

Tab. Mot. Med.

Fig. 9.10 First page of the fourth part of the tabulations ('On the obliquity of the Ecliptic …'). (Reproduced from Kepler, J., *Tabulae Rudolphinae*, Ulm, 1627, p. 103) different. For instance, the Sun is seen during the day, when the air is relatively warm. It may be that the references to the maritime surroundings of Tycho's observatory in the note immediately above the tables is intended to imply possible influence from water vapour and atmospheric turbulence. This downbeat ending is a final echo of the proclamation on title page of the *Rudolphine Tables* that Tycho was their author.

9.4 Reception and Consequences

The prominence given to Imperial patronage and to Tycho on the title page and frontispiece of the *Rudolphine Tables* may have helped to ensure their initial reception was favourable. At the time the *Tables* were published, Tycho's geoheliocentric planetary system, whose truth he had hoped his observations would establish, was popular among astronomers. It was, after all, reasonable—and acceptable to the Church—to believe the Earth was at rest. Moreover, as Kepler had repeatedly noted, for the purposes of mathematical astronomy, the Tychonic system was equivalent to the Copernican one. For instance, it allowed the relative dimensions of planetary orbits to be determined from observation. Galileo, who saw the motion of the Earth as providing an explanation for the tides, disliked the Tychonic system, regarding it as equivalent to the Ptolemaic one because it showed the Earth as at rest. Galileo seems in fact to have distrusted Tycho even as an observer.[45] But the first readers of his *Dialogue concerning the two chief world systems* (*Dialogo sopra i due massimi sistemi del mondo*, Florence, 1632)—the book that brought him into conflict with the Inquisition—must have included many who thought it strange that he did not discuss the Tychonic system along with the Ptolemaic and Copernican ones.

The *Rudolphine Tables* were, of course, expected to be more accurate in their predictions than their predecessors. As we have already noted in connection with Reinhold's *Prutenic Tables* (1551), this was always true for new tables, since the models they used were based on more recent data. By the time the *Rudolphine Tables* were published, the data on which Kepler had based his models was no longer very new, since Tycho's observations had been made in the 1580s and 1590s. However, the orbits of planets are stable and Kepler had been strong-minded in deducing the rules he proposed as governing them. As he put it in the last two paragraphs of chapter 19 of *Astronomia nova* (Heidelberg, 1609), in response to finding a difference between the prediction from a Ptolemaic model and the position of Mars observed by Tycho:

> And from this difference of eight minutes [of arc], so small as it is, the reason is clear why Ptolemy, when he made use of bisection [of the linear eccentricity], was satisfied with a fixed equalizing point. … Now Ptolemy professed not to go below 10′, or the sixth part of a degree, in his observation. The uncertainty or (as they say) the 'latitude' of the observations therefore exceeds the error in this Ptolemaic computation.

---

[45] This may be because of Tycho's work on the New Star of 1572. Galileo did not believe the New Star of 1604 belonged among the fixed stars. See Cosci (2018); and Field, 'Kepler and Galileo' (this volume, Chap. 8).

Since the divine benevolence has vouchsafed us Tycho Brahe, Tycho (1546–1601) observational accuracy, a most diligent observer, from whose observations the 8′ error of this Ptolemaic computation is shown in Mars, it is fitting that we with thankful mind both acknowledge and honour this favour of God. For it is in this that we shall carry on, to find at length the true form of the celestial motions, supported as we are by these proofs showing our suppositions to be fallacious. In what follows, I shall myself, to the best of my ability, lead the way for others on this road. For if I had thought I could ignore eight minutes of longitude, in bisecting the [linear] eccentricity I would already have made enough of a correction in the hypothesis found in ch. 16. Now, because they could not be ignored, these eight minutes alone will have led the way to the reformation of all of astronomy, and have become the material for a great part of the present work.[46]

As Kepler says, his intransigence pays off. Forty chapters later, having meanwhile shown that the area swept out by a line joining the planet to the Sun can be used as a measure of elapsed time, Kepler has deduced that the path of Mars is an ellipse. These two rules are then, as we have seen, put to use in the calculation instructions supplied in the *Rudolphine Tables*. So, with hindsight it is not surprising that ephemerides based on the new tables proved to be reliable. However, the length of time for which they continued to be reliable surprised contemporaries. And as ephemerides based on the *Table*s gave accurate predictions, year after year, astronomers naturally saw this as an indication that Kepler's rules for calculation, unambiguously Copernican as they were, might indeed be physically correct.

Kepler gave no name to the first two calculation rules (in Chapter 20 of the *Tables*) but he does use the word 'law' in connection with the third (in Chap. 24). The idea of 'natural laws' became established through writings on Natural Philosophy by René Descartes (1597–1650), who was a heliocentrist (apparently because he thought the Copernican system simpler than the Tychonic one), but gave a non-mathematical account of planetary motion.[47] As the *Rudolphine Tables* became standard reading for astronomers, thereby fulfilling Rudolf II's hopes for them, they seem to have eclipsed Kepler's other works, coming to serve as the main source for knowledge of what are now known as Kepler's laws of planetary motion. The third law, which is really a law of the Solar System since it relates the sizes of orbits to the periods of the planets concerned, can be taken directly from Chapter 24 of the introductory section of the *Tables* (though Kepler does provide a reference to the chapter of *Harmonice mundi* (1619) in which the law had first appeared. But the first two laws are not stated explicitly in Chapter 20 of the *Rudolphine Tables*, so anyone in search of them would have had to follow up Kepler's marginal reference to the *Epitome of Copernican Astronomy* book IV (1621).

Some such procedure almost certainly explains the way the three laws were known to the English astronomer Thomas Streete (1621–1689). Like Descartes, Streete is apparently casual in his acceptance of heliocentrism. And he is matter-of-fact about Kepler's three laws, which he states in his textbook, *Astronomia Carolina* (London,

---

[46] Kepler (1609), ch. XIX, pp. 113–14; Kepler (1990), pp. 177–78; Kepler (2015), p. 211. See also Davis, 'Discovery of the planetary orbit: the Goldilocks solution' (this volume, Chap. 4) and Mahoney, 'Measuring the heavens: Tycho Brahe and the reform of observational astronomy' (this volume, Chap. 3).

[47] On Descartes' astronomy see Aiton (1972); more generally, see Gaukroger (2022).

1661, 1664). This work, in which, notwithstanding the Latin title, the main text is in English, became a standard university-level introduction to astronomy, and there is strong evidence it served as such for the young Isaac Newton (1642–1727).[48] Kepler's laws eventually appear in Book 1 of Newton's *Mathematical Principles of Natural Philosophy* (*Philosophiae Naturalis Principia Mathematica*, London, 1687) where—together with new ideas about force partly derived from Descartes and partly original to Newton himself—the two laws are used to prove that the planets are moving under the influence of a force that attracts them to the Sun and obeys an inverse square law of distance (Book 1, proposition 11). Kepler's third law is used to show that the four satellites of Jupiter are subject to a similar force attracting them to Jupiter (Book 1, prop. 15).[49] In this latter case, Newton uses the third law—which he has shown is a mathematical consequence of the first two and equivalent to them—because observations of the orbits of the satellites showed them as very nearly circular, which made it impossible to use the argument he had applied to the motion of the planets in the earlier proposition. Proving that this attractive force is the same as the gravity that pulls the falling apple to the Earth, a proof given in *Principia* Book 3, involved terrestrial physics and up to date estimates of distances in the Solar System, and used experimental results obtained by Christiaan Huygens (1629–1695) that were ultimately based on experiments carried out by Galileo and his pupils.[50]

The connection with the work of Newton might provide a heroic note on which to end an account of the *Rudolphine Tables* and their importance in the history of astronomy. But let us instead turn to the *Rudolphine Tables* themselves and some matters of observational astronomy. That was what the *Tables* were intended to serve.

One of the novelties presented by the *Rudolphine Tables* was that predictions of the (ecliptic) latitudes of planets proved to be as reliable as those for their longitude. Astronomers did not expect this; they were accustomed to predictions of latitudes being rather unreliable. However, Kepler had shown that the paths of planets were plane curves—with all such planes passing though the Sun. The planes of the orbits could be found, and seemed to be fixed, so ecliptic latitude and longitude were found together. In earlier times, latitudes were the subject of separate calculation and had tended to be neglected, except for the Moon, whose latitude is significant because eclipses of the Sun or Moon occur only when the Moon is on the ecliptic (hence its name). It was again in relation to the Sun that latitudes for planets proved to have a use: they allowed astronomers to predict transits, that is occasions when the apparent paths of the inner planets, Mercury and Venus, took them across the disc of the Sun (as special cases of inferior conjunction just as a Solar eclipse is a special case of a New Moon).

The story of transits starts badly. By May 1605, Kepler had completed his calculations of the orbit of Mars and by early 1607 he had an orbit for Mercury.[51] But

---

[48] Whiteside (1970).

[49] Newton (1687), Book 1, Sect. 9.3, Prop. 11, p. 50–51; and, Book 1, Sect. 9.3, Prop. 15, p. 56.

[50] Newton (1687), Book 3, Props 3 and 4, p. 406.

[51] For a chronology of Kepler's determination of orbits see Bialas (1971).

Mercury is awkward to observe. Being physically close to the Sun, it is never seen far from it in the sky, so observations may need to be made when the planet is close to the horizon, requiring the application of corrections for atmospheric refraction. Kepler noticed his orbit for Mercury predicted a transit, which might allow him to check the accuracy of the orbit. So he set up a *camera obscura*—which was the usual way of making observations involving the Sun—and on 29 May 1607 he saw a dark spot approximately where he expected Mercury to be. His account was published in a short book called *A Strange Phenomenon, or Mercury in the Sun (Phœnomenon singulare seu Mercurius in Sole*, Leipzig, 1609).[52] Kepler soon realized that he had made a mistake. The apparent size and motion of the supposed planet were not plausible, and a couple of years later he decided it was probable that what he had seen was a sunspot. The first knowing observations of sunspots were by Christoph Scheiner (1573/5–1650) in 1611 and by Galileo in 1612.[53]

Kepler went on to publish ephemerides, and the one for 1631 predicted transits of both Venus and Mercury in that year. Unfortunately, the calculations were not accurate in predicting from which parts of the Earth it would be possible to see the transit of Venus. Observations of the transit of Mercury were made in Paris by Pierre Gassendi (1592–1655).[54] A transit of Venus, predicted for 1639, was observed by Jeremiah Horrocks (1618–1641) in the North of England.

These observations seem to have been undertaken with the purpose of making corrections to the orbits of the planets concerned. Later it was recognized that observations of a transit, taken from different locations on the Earth, would also allow one to obtain a more accurate estimate of solar parallax, which is a measure of the distance between the Sun and the Earth. Double observation became the rule, so that, for example, while Richard Towneley (1629–1707) was observing a transit of Mercury from North West England (Lancashire) on 7 November 1677 (Julian 28 October),[55] Edmond Halley (*c.* 1656–1743) was making similar observations in mid Atlantic on St Helena.[56]

Kepler's values for the dimensions of the orbits of the planets are exceedingly accurate. Indeed they are so accurate that in assessing them one needs to make allowances for secular changes in the orbits since the time Tycho made his observations,[57] but all Kepler's results are expressed in terms of the Earth–Sun distance. The observations of transits led to increased estimates of this distance, which increased estimates of the size of the Solar System and thus showed the Earth as a smaller component of it. This in turn tended to make the Copernican planetary system look

---

[52] Reprinted in Kepler (1941), pp. 78–98. Kepler provides illustrations of the face of the Sun and of the observational set up. The delay in publication was probably due to Kepler's time being taken up with the publication of *Astronomia nova* (which eventually appeared in 1609).

[53] See Field, 'Kepler and Galileo' (this volume, Chap. 8).

[54] Gassendi (1632).

[55] England did not adopt the Gregorian calendar reform (of 1582) until 1752.

[56] For further details see Pasachoff, 'Johannes Kepler, the *Kepler* spacecraft and transits' (this volume, Chap. 10).

[57] See Bialas (1971).

more plausible. One way and another, the *Rudolphine Tables* made a major contribution to fulfilling Kepler's life ambition as an astronomer, that of proving the planetary system was centred on the Sun. And they helped to do this not only for astronomers and natural philosophers but also for the popular imagination.

References

Aiton, E. J. (1972). *The vortex theory of planetary motion.* London and New York: Macdonald and American Elsevier.

Bialas, V. (1971). Die quantitative Beschreibung der Planetenbewegung von Johannes Kepler in seinem handschriftlichen Nachlaß. In *Kepler Festschrift 1971.* Regensburg.

Brahe, T. (1946). *Tycho Brahe's description of his instruments and scientific work* (H. Raeder, E. Strömgren, & B. Strömgren, Trans.), Copenhagen.

Cosci, M. (2018). 'Astronomia pavana nel "Doalogo de Cecco di Ronchiti da Bruzene in perpuosito de la stella nuova" tra commedia, satira, *disputatio* accademica e poesia'. In: *I generi letterari dell'aristotelismo volgare rinascimentale* (pp. 125–187). Padua: Cleup.

Donahue, W. H. (1981). *The dissolution of the celestial spheres.* New York: Ayer Co.

Evans, R. J. W. (1997). *Rudolf II and his world, a study in intellectual history 1576–1612* (pbk). London: Thames and Hudson.

Field, J. V. (1988). What is scientific about a scientific instrument? *Nuncius, 3*(2), 3–26.

Field, J. V. (1999). Why translate Serlio? In F. Ames-Lewis (Ed.), *Thomas Gresham and Gresham College: studies in the intellectual history of London in the sixteenth and seventeenth centuries* (pp. 198–221). Aldershot: Ashgate.

Gassendi, P. (1632). *Mercvrivs in sole visvs, et Venvs invisa Parisiis, anno 1631: pro voto, & admonitione Keppleri* (*Mercury seen in the Sun and Venus not seen at Paris, in 1631*). Paris.

Gaukroger, S. (2002). *Descartes' system of natural philosophy.* Cambridge: Cambridge University Press.

Gingerich, O. (1971). Johannes Kepler and the *Rudolphine Tables. Sky and Telescope, 12*(1971), 328–333.

Jardine, N., Leedham-Green, E., & Lewis, C. (2014a). Johann Baptist Hebenstreit's Idyll on the Temple of Urania, the Frontispiece Image of Kepler's *Rudolphine Tables*, Part 1: Context and Significance. *Journal for the History of Astronomy, 45*(1), 1–19.

Jardine, N., Leedham-Green, E., Lewis, C., & Fay, I. (2014b). Johann Baptist Hebenstreit's Idyll on the Temple of Urania, the Frontispiece Image of Kepler's *Rudolphine Tables*, Part 2: Annotated translation. *Journal for the History of Astronomy, 45*(1), 21–34.

Kaufmann, T. da C. (1993). *The mastery of nature: Aspects of art and science, and humanism in the Renaissance.* Princeton: Princeton University Press.

Kepler, J. (1609) *Astronomia nova aitiologêtos, seu physica coelestis, tradita commentariis de motibus Stellae Martis.* Heidelberg: E. Vogelin.

Kepler, J. (1625). *Tychonis Brahei Dani Hyperaspides.* Frankfurt: Gottfried Tampach.

Kepler, J. (1627). *Tabulae Rudolphinae.* Ulm: Jonas Saurius.

Kepler, J. (1941). *Johannes Kepler Gesammelte Werke, Band IV: Kleinere Schriften 1602–1611 / Dioptric.* F. Hammer (Ed.). Munich: C. H. Beck.

Kepler, J. (1959). *Johannes Kepler Gesammelte Werke. Band XVIII: Briefe 1620–1630.* M. Caspar (Ed.). Munich: C. H. Beck.

Kepler, J. (1963) *Johannes Kepler Gesammelte Werke, Band VIII: Mysterium cosmographicum (editio altera cum notis) / De cometis, Hyperaspides.* F. Hammer (Ed). Munich: C. H. Beck.

Kepler, J. (1969). *Johannes Kepler Gesammelte Werke. Band X: Tabulae Rudolphinae.* F. Hammer (Ed.). Munich: C. H. Beck.

Kepler, J. (1990). *Johannes Kepler Gesammelte Werke. Band III: Astronomia nova* (2nd edn). M. Caspar and Kepler-Kommission. Munich: C. H. Beck.

Kepler, J. (2015). *Astronomia nova* (W. H. Donahue,& O. Gingerich, Trans.). Santa Fe: Green Lion Press.

Kremer, R. L. (2023a). *Alfonsine astronomers at work: Computing planetary positions in Europe, 1350–1560*. Turnhout: Brepols.

Kremer, R. L. (2023b). Controlling errors in the first printed book of astronomical tables: Regiomontanus' *Ephemerides* (Nuremberg 1474). In G. D. Rocca de Candal, P. Sachet, & A. Grafton (Eds.), *Oxford companion to printing and misprinting*. Oxford: Oxford University Press.

Mosley, A. J. (2007). *Bearing the Heavens: Tycho Brahe and the astronomical community of the late sixteenth century*. Cambridge: Cambridge University Press.

Newton, I. (1687). *Principia*. London: Royal Society.

Palladio, A. (1570). *Quattro libri dell'architettura*. Venice.

Serlio, S. (1537–51). *L'Architettura*, 5 vols published separately Venice, Paris and Lyon.

Thoren, V. E. (1990). *The Lord of Uraniborg: A biography of Tycho Brahe*. Cambridge: Cambridge University Press.

Vitruvius, M. (1566). *De architectura libri decem*. D. Barbaro (Ed.). Venice.

Vitruvius, M. (2009). *On architecture* (R. Schofield, & R. Tavernor, Trans.). Harmondsworth: Penguin Books.

Whiteside, D. T. (1970). Before the *Principia*: The maturing of Newton's thoughts on dynamical astronomy, 1634 to 1684. *Journal for the History of Astronomy, 1*(1), 5–19.

Chapter 10
Johannes Kepler, the *Kepler* Spacecraft and Transits

Jay M. Pasachoff

10.1 Introduction

Johannes Kepler's (1571–1630) work has been basic to astronomy since his *Astronomia nova* of 1609,[1] but nobody could have thought that his third law, published in book V of his *Harmonice mundi* of 1619,[2] would be so fundamental to the astronomy and astrophysics of the twenty-first century. Yet his third law is the key to uncovering the planetary content of the Universe, and the study of exoplanets is perhaps the hottest topic in astronomy today, with NASA's *Kepler* spacecraft at the forefront.[3] Johannes Kepler is a contemporary hero, the subject even of an opera by Philip Glass.[4]

---

[1] That is, since the publication of his first two laws of planetary motion (Kepler, 1609, 1990); see also Davis, 'Kepler's discovery of the planetary orbit: the Goldilocks solution' (this volume, chapter).

[2] See Kepler (1619), book 5; Kepler (1940).

[3] Full information on the Kepler mission can be found at http://kepler.nasa.gov/.

[4] For a review of the opera Kepler, see Pasachoff and Pasachoff (1999).

J. M. Pasachoff (Deceased) (✉)
Williams College, Williamstown, MA, USA

Caltech, Pasadena, CA, USA

© Springer Nature B.V. 2024
A. E. L. Davis et al. (eds.), *Reading the Mind of God*, Springer Praxis Books,
https://doi.org/10.1007/978-94-024-2250-4_10

10.2 Kepler and Transits

In his *Rudolphine Tables* of 1627,[5] Kepler used his laws of planetary orbits, which developed Copernicus' heliocentric astronomy to produce an order of magnitude improvement in predicted planetary positions.[6] But not everyone accepted that his work was an advance, or even that Copernicanism was correct. Tycho's rejection of the Copernican hypothesis, for example, stemmed from both physical reasoning and religious conviction (see Chap. 3 of this volume). Christoph Clavius SJ (1538–1612), a major figure in Rome's adoption of the Gregorian calendar, was influential in the incorporation of the Ptolemaic hypothesis and Aristotelian natural philosophy into Catholic doctrine.[7] Nevertheless, Copernicanism began to acquire adherents in the face of both Catholic and Protestant conservatism.

Kepler's *Rudolphine Tables* were not ephemerides; that is, they did not predict positions of objects in the sky. But the tables could be used to make such predictions, and Kepler did predict (Fig. 10.1) that Mercury would transit across the face of the Sun in 1631.[8] The end of this particular transit was observed by Pierre Gassendi (1592–1655)[9] and provided validation for Kepler's work and for heliocentrism in general.[10]

Transits of inferior planets (i.e. either Mercury or Venus) occur when the planet crosses the solar disc (see Fig. 10.2). First contact occurs when the planet appears to touch the Sun at ingress; second contact is established when the following limb (edge) of the planet just touches the limb of the Sun before it progresses across the solar disc; third contact occurs as the leading limb of the planet touches the boundary of the solar limb, and fourth contact, when the following limb of the planet touches the solar limb at the moment the planet leaves the solar disc.

Transits of Venus occur in pairs separated by 8 years (1631 and 1639), with gaps of 122.5 (1761 and 1769) or 105.5 (1874 and 1882) years. No transits of Venus occurred during the twentieth century, but the early twenty-first century was graced with a pair in 2004 and 2012.[11]

But Kepler did not predict the transit of Venus of 1639. It was the young astronomer Jeremiah Horrocks (1618–1641) in England who, on restudying the *Rudolphine Tables*, realized that there would be a transit of Venus in 1639. Only Horrocks and one correspondent of his viewed that transit.[12] But Kepler's tables were fully vindicated, and the use of transits to study phenomena in the Universe became possible, even though the next transits of Venus were not going to be until 1761 and 1769. Horrocks's

---

[5] Kepler (1627, 1969); see also Field, 'The long life of the *Rudolphine Tables*' (this volume, Chap. 9).

[6] See, for example, Gingerich (2009); Rosa (2010).

[7] See Lattis (1994) for an in-depth study of Clavius.

[8] Kepler (1629).

[9] Gassendi (1632); Grant (1852); Helden (1985), Chap. 9; Westfall and Sheehan (2015); Gingerich (2005).

[10] Gingerich (2013); Pasachoff (2011).

[11] Pasachoff (2012); Aughton (2004), esp. Chap. 7.

[12] Westfall and Sheehan (2015); Olson and Pasachoff (2019); Hevelius (1662).

JOANNIS KEPLERI
Mathematici Cæsarei,&c.

De raris mirisq;

Anni 1631. Phænomenis,

*Veneris putà & Mercurii in Solem
incursu,*
ADMONITIO AD ASTRONO-
MOS, RERUMQVE COELESTIUM
STUDIOSOS:
Excerpta.
Ex EPHEMERIDE Anni 1631. & certo
AUTHORIS confilio huic præmifla, & edita
à
*M. JACOBO BARTSCHIO,
Laubano, Mathem.& Med.C.*

LIPSIÆ,
JOAN-ALBERTUS MINZELIUS excudebat.
ANNO M DC XXIX.

Fig. 10.2   Stages of Venus
transiting the Sun: first
contact (1), second contact
(2), third contact (3) and
fourth contact (4)

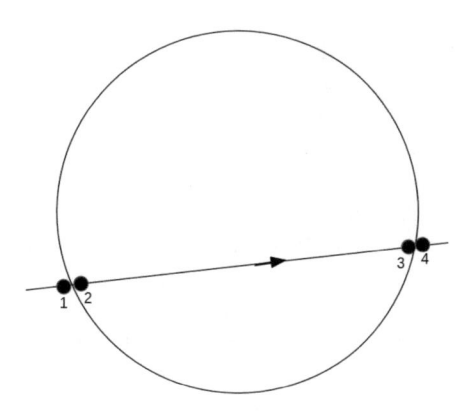

report on the transit of Venus was not published until it was included in a publication
23 years later about a transit of Mercury.[13]

In the eighteenth century, expeditions were sent all over the world to see and time
transits of Venus, because Edmond Halley (1656–1741) had advanced in 1716 a
method of determining the size of the solar system.[14] The importance of this method
traces back to the work of Kepler, since his third law had given proportionalities

---

[13] Halley (1716); see also Halley (1929), pp. 96–100, for an abridged English translation.

[14] Pasachoff (2012); Westfall and Sheehan (2015). Fred Espinak has produced useful lists of past
and future transits of Mercury and Venus. There are two tables for Mercury covering the period AD
1601–2300 and two for Venus covering the period 2000 BC–AD 4000, together with a straightforward formula for calculating visibility. The tables and a full explanation of their use can be found at

Fig. 10.3 The black drop
effect. *Courtesy* Institute for
Astronomy, University of
Vienna

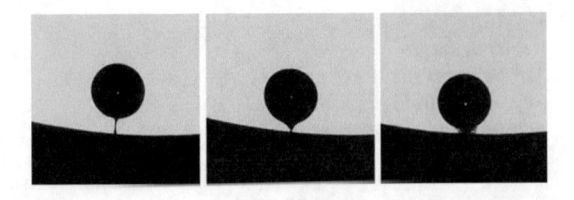

for the sizes of the orbits of the planets. Newton's version of Kepler's third law incorporated mass and the gravitational constant and demonstrated that the law was a natural consequence of Newton's law of gravitation, so that the Kepler's third law in its Newtonian reformulation could be used with confidence. Kepler's third law gave the correct scale of planetary distances from the Sun, yet no absolute distance was yet known; once an absolute distance is known for any pair of Solar System objects, Kepler's third law can then be used to find all the distances in the Solar System.[15]

Yet, when the observations were made in 1761, Kepler having died decades previously, the results were somewhat unsatisfactory. For Halley's method to work well, accurate timing to about 1 s was needed for both ingress and egress, and from sites as far north and as far south on the Earth as possible to provide a long, skinny triangle from the terrestrial baseline to Venus, given its projection on to the Sun that would lead to a slight difference in timing because of the different lengths of the chords traced as a result of the curvature of the solar surface. But to the surprise of the observers, the silhouette of Venus did not separate cleanly when it entered the Sun; instead a dark band linked the dark silhouette with the dark sky outside the Sun. That band, known as the 'black-drop effect' (see Fig. 10.3), lasted almost a minute of time, some 60 or so times longer than desired, and prevented the timing of the transit to the required accuracy.

Using the 1999 transit of Mercury, we explained the black-drop effect as a combination of the effects of the point-spread function of the telescope and the extreme variation of the solar limb darkening right at the apparent solar edge.[16] Indeed, we detected a black drop effect from the 1999 transit of Mercury as observed with a NASA spacecraft; since Mercury has no atmosphere and the spacecraft was above the Earth's atmosphere, we showed that no atmosphere was necessary to provide the black-drop effect. Most recently, we saw a minor black drop effect even with a 1.6 m solar telescope at the 2016 and 2019 transits of Mercury, observed from California with the Goode Solar Telescope of the Big Bear Solar Observatory of the New Jersey Institute of Technology.

When the Russian astronomer Mikhail Lomonosov (1711–1765) saw a distortion near the solar limb (edge) as Venus entered the Sun in 1761, he claimed the discovery of an atmosphere around Venus. (However, he 'knew' in advance that there would be an atmosphere both there and on the Moon to support the inhabitants living there,

---

https://eclipse.gsfc.nasa.gov/transit/catalog/Visible.html. Halley's method of finding the solar parallax is described in detail in Young (1888), pp. 382–384.

[15] Schneider et al. (2004).

[16] Pasachoff et al. (2005). http://transitofvenus.info. See also Pasachoff et al. (2013, 2017a, 2017b).

in line with the beliefs of other Enlightenment natural philosophers.) Pasachoff and Sheehan[17] have shown, rather, that what he saw was an optical effect, akin to the black-drop effect, and undoubtedly not the Cytherean atmosphere.

10.3 The Transit Phenomenon

10.3.1 Transits of Mercury and Venus

The story of measuring the 'solar constant'—the total solar radiation per square metre at the top of Earth's atmosphere—is an interesting one, dating back to the nineteenth century. The ground-based measurements made over many decades by a Smithsonian Institute 'Secretary' (head), Charles Greeley Abbot (1872–1973), eventually proved useless, with our atmosphere varying the measurements. Ultimately, spacecraft discovered a variation of the total solar irradiance over the solar activity (i.e. sunspot) cycle, so the term 'solar constant' fell out of use and has been replaced by Total Irradiance Measurement (TIM).

By the 1980s, spacecraft were launched by the European Space Agency (Virgo on the *Solar and Heliospheric Observatory*, launched in 1995) and NASA (*ACRIMsat*) following an original Active Cavity Radiometer for Irradiance Measurement (ACRIM) by Richard Willson on NASA's *Solar Maximum Mission* (1980). Later, Greg Kopp[18] of the Laboratory for Atmospheric and Space Physics (LASP) of the University of Colorado had a Total Irradiance Measurement (TIM) instrument on *Solar Radiation and Climate Experiment* (*SORCE*), launched in 2003, and most recently another TIM on the U.S. Air Force's *Total Solar Irradiance Calibration Transfer Experiment* (*TSCE*), launched in 2013.

For the transit of Venus of 2004, we described the 0.1% drop in the TSI resulting from the 0.1% coverage of the solar disc by the silhouette of Venus as seen from Earth.[19] For the 2012 transit of Venus, both Willson's[20] and Kopp's instruments (Fig. 10.4) gave excellent observations of the drop in the TSI.

Schneider and I also attempted to observe transits of Venus as seen from Jupiter in reflection off that giant planet with the *Hubble Space Telescope* in September 2012,[21] and as seen from Saturn by direct viewing in the visible and near infrared from NASA's *Cassini* mission in December 2012. But from those greater distances from Venus and the Sun, the expected obscuration was closer to 0.01%, and we did

---

[17] Pasachoff and Sheenan (2012, 2013).

[18] Kopp, G. (2016) 'Greg Kopp's TSI Page', http://spot.colorado.edu/~koppg/TSI/.

[19] Schneider et al. (2006).

[20] http://acrim.com/Venus%20Transit.htm.

[21] Karalidi et al. (2015).

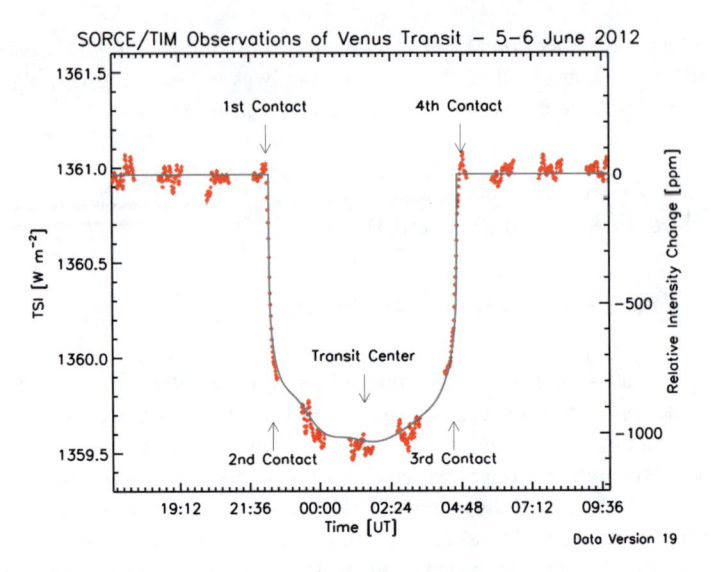

Fig. 10.4 The 0.1% drop in the total solar irradiance (TIM) measured with NASA's *SORCE* spacecraft. *Credit* Data and graph from Greg Kopp, LASP/U, Colorado

not succeed, except probably for the strongest infrared channel from *Cassini* around Saturn.[22]

Transits of Mercury as seen from Earth occur closer to a dozen times a century, and we observed the transits in white light of 2003, 2006, 2009, 2016, which we observed from Maui, and, most recently, 11 November 2019, which we observed from Big Bear. The transit obscures only about 0.1% of the incoming sunlight, and we were unable to definitively see the effect in the earliest sets of data, though it seems marginally apparent in the 2016 event. (The silhouette of Mercury takes up much of the field of view in the observations with the giant telescope at the Big Bear Solar Observatory.) This detection and the non-detections set limits on the search for transiting planets around other stars, so-called exoplanets.

10.3.2 Transits and the Discovery of Exoplanets

The study of transits of Venus and Mercury in our Solar System is analogous to the studies made of planets around other stars by the transit method.[23]

A NASA spacecraft to detect planets around other stars from their transits was developed over many years under the guidance of W. J. Borucki of NASA's Ames

---

[22] Pasachoff et al. (2013).

[23] Schneider and Pasachoff (2006); Pasachoff (2009); Pasachoff et al. (2012).

Research Center, starting as far back as 1984.[24] The discovery of extrasolar planets—starting with observations by Michel Mayor and Didier Queloz in 1995[25] by their spectral variations caused by the Doppler effect, and especially the observations of transiting exoplanets by Charbonneau et al.[26]—helped the proposal succeed.

The spacecraft was launched by NASA in 2009 and named *Kepler*. Its telescope pointed at one region of the sky that contained about 150,000 stars, and imaged it over and over, searching for transits (see Fig. 10.5). By 2021, its data had led to the discovery of over 4000 exoplanets,[27] with thousands of other candidates that had to be vetted, though with the expectation that about 90% were really exoplanets as opposed to spurious detections.

The mission's procedure for discovering and characterizing exoplanets is as follows[28]:

1. Estimate the size of the exoplanet from the *transit depth* (the magnitude of the drop in brightness of the host star during transit).
2. Determine the orbital period from the time between successive transits by the exoplanet.
3. Use Newton's reformulation of Kepler's third law, $4\pi^2 a^2/T^2 = G(m_s + m_p)$, where a is the semi-major axis of the exoplanet's orbit, T its period, m_p its mass and m_s the mass of the host star, to calculate the semi-major axis of the exoplanet's orbit.
4. The distance of the exoplanet from the host star, and the host star's temperature and radius being known, Planck's blackbody formula is then used to derive the temperature of the exoplanet.

The *Kepler* spacecraft lost too many of its reaction wheels, which kept it accurately pointed, in 2013. After a call for proposals, it was repurposed in 2016 as the K2 mission, which used the same telescope and electronics—they were after all already in space—to look at a varying set of stellar regions, still for transits. Hundreds of new exoplanets were discovered by the K2 extended mission until the spacecraft was entirely retired in 2018.[29] The Kepler filter at https://exoplanets.nasa.gov/discovery/exoplanet-catalog/, as of 2021, lists 2414 objects.

In 2018, NASA launched its *TESS (Transiting Exoplanet Survey Satellite)*, especially to use transits to discover exoplanets around nearer stars than those observed by *Kepler*, making it easier to follow up the discoveries with observations from a variety of ground-based telescopes, including the largest in the world, to gather the most starlight. The European Space Agency's *CHEOPS (CHaracterising ExOPlanet Satellite)*, launched in 2019, is designed to determine the size of known extrasolar

---

[24] Borucki and Summers (1984).

[25] Mayor and Queloz (1995).

[26] Charbonneau et al. (2000).

[27] https://www.nasa.gov/image-article/kepler-by-numbers-mission-statistics/.

[28] Gould et al. (2015). See https://www.nasa.gov/kepler/presskit and https://exoplanets.nasa.gov.

[29] https://science.nasa.gov/mission/kepler/.

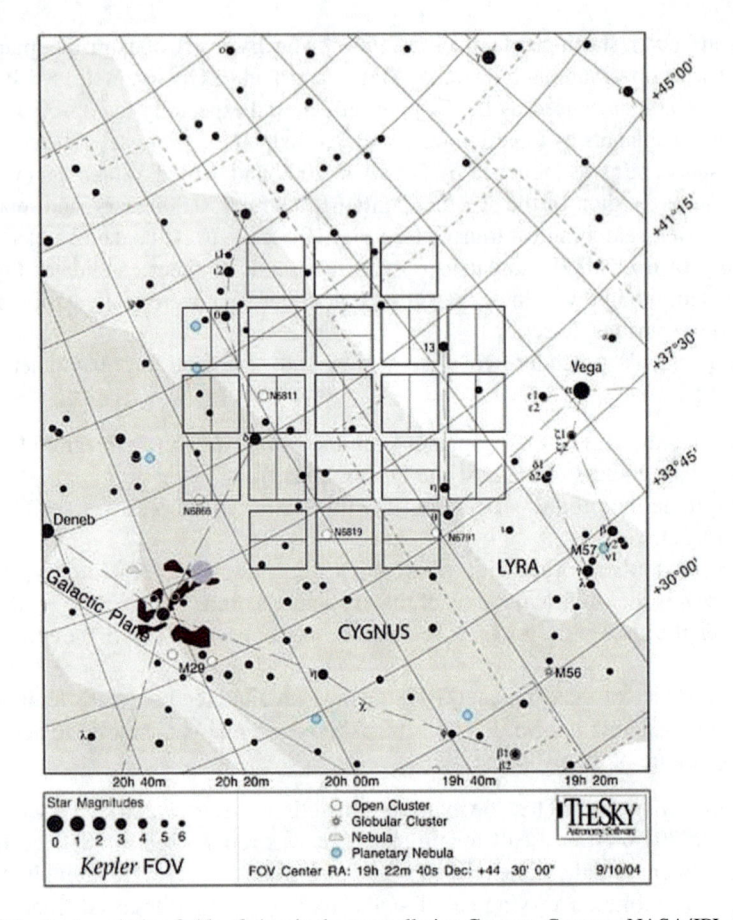

Fig. 10.5 *Kepler* mission fields of view in the constellation Cygnus. *Courtesy* NASA/JPL-Caltech

planets, which will thus allow their mass, density, composition and formation to be determined.

It is fair to say that the study of exoplanets is one of the most exciting areas of contemporary astronomy, with young astronomers and graduate students flocking to the field.

10.4   Johannes Kepler in Our Times

Kepler's life and work are a standard topic in all books about seventeenth-century astronomy or about the mainline development of western astronomical knowledge.[30] Historians of astronomy have indeed not been idle in reminding the astronomical community (which occasionally needs reminding) of important anniversaries related to Kepler's achievements. Special mention must be made of the 400th anniversary of Kepler's birth, which saw the publication by an entire volume of *Vistas in Astronomy*.[31] There were further 2009 celebrations by historians of astronomy of the 400th anniversary of the publication of *Astronomia nova* at meetings held in Prague,[32] Zielona Góra[33] and Rio de Janeiro.[34] An important outcome of Special Session 9 of the General Assembly of the International Union in Rio de Janeiro was the setting up of the IAU's Johannes Kepler Working Group (later renamed the Johannes Kepler Task Group). It was through the actions of this group that the present volume has been produced. A major milestone in Kepler studies has been the digitization by the Kepler-Kommission in Munich of the voluminous *Gesammelte Werke*,[35] which enables quick and cost-free access to what is now the standard edition of all of Kepler's writings.

The year 2009, however, also coincided with the 400th anniversary of Galileo's first telescopic observations, a coincidence that rather sidelined Kepler during the 2009 International Year of Astronomy (IYA2009).[36] There were indeed Special Task Groups dedicated to both Kepler[37] and Galileo,[38] but the emphasis of the organizers was overwhelmingly on Galileo's achievements.

At the Smithsonian Institution's National Air and Space Museum, the astronomy exhibit has a wall plaque describing Galileo Galilei's (1564–1642) contribution and a life-size diorama showing Tycho Brahe (1546–1601) observing with one of his pre-telescopic instruments. Astonishingly, however, Kepler was not mentioned.

In protest, I arranged with senior curator, David DeVorkin, to record two pieces of approximately two minutes each, one about Johannes Kepler and one about how and why he and his work led to a NASA spacecraft being named for him. They were for about two years near the end of the set of rooms in the astronomy exhibition, near the current events that discussed the *Kepler* spacecraft. Finally, in late 2015, official plaques, discs about 20 cm in diameter, stating *Ask an Astronomer*, were placed near the Galileo wall plaque and the Tycho diorama, on two topics (written

---

[30] See, for example, Voelkel (2000); Pasachoff (2000); Pasachoff and Pasachoff (2012); Pasachoff and Filippenko (2019).

[31] Beer and Beer (1975).

[32] Hadravová et al. (2010).

[33] Kremer and Włodarczyk (2009).

[34] Mahoney (2010).

[35] Available online and in downloadable PDF format at https://kepler.badw.de/kepler-digital.html.

[36] https://astronomy2009.org/.

[37] https://astronomy2009.org/organisation/structure/taskgroups/kepler/index.html.

[38] https://astronomy2009.org/organisation/structure/taskgroups/galileo/index.html.

by me): 'What did Johannes Kepler find out about planetary orbits?' and 'How do astronomers today use Johannes Kepler's findings about planetary orbits?'[39] Similar *Ask an Astronomer* plaques for answers by Owen Gingerich about Galileo and about early telescopes are nearby.

Johannes Kepler, with his discovery of three laws of planetary motion and then with his *Rudolphine Tables* (Gattei[40] has viewed over 200 copies of the interesting frontispiece of the *Rudolphine Tables* and discusses his findings there) that allowed the prediction and observation of transits of Venus and of Mercury, started a branch of astronomy that is flourishing to this day. Its appeal is even increasing, with the observations of transits of Mercury and of Venus from Earth and, further, with the discovery of thousands of planets around other stars, so-called exoplanets. Johannes Kepler was honoured by NASA's naming its first exoplanet-hunting satellite *Kepler*. So the name 'Kepler' resonates not only for today's astronomers but also increasingly for the general public.

References

Aughton, P. (2004). *The transit of Venus: The brief life of Jeremiah Horrocks, father of British astronomy*. London: Weidenfeld and Nicolson.

Beer, A., & Beer, P. (Eds.). (1975). Kepler: Four hundred years. *Vistas in Astronomy, 18*.

Borucki, W. J., & Summers, A. L. (1984). The photometric method of detecting other planetary systems. *Icarus, 58*, 121ff.

Charbonneau, D., Brown, T. M., Latham, D. W., & Mayor, M. (2000). *Astrophysical Journal, 529*, L45ff.

Gassendi, P. (1632). *Mercurius in Sole visus et Venus invisa Parisiis anno 1631. Pro voto et admonitione Keppleri, per Petrum Gassendum, cuius heic sunt ea de re epistolae duae cum observatis quibusdam aliis*. Paris: Cramoisy.

Gattei, S. (2014). On Tycho's shoulders, with Vesalius' eyes: Speaking images in the engraved frontispiece of Kepler's *Tabulae Rudolphinae*. In A. Albrecht, G. Cordibella, & V. R. Remmert (Eds.), *Tintenfass und Teleskop* (pp. 327–367). Berlin: De Gruyter.

Gattei, S. (2020). The finger and the tongue of god: Johannes Kepler, reformation theology, and the new astronomy (Chap. 12). In E. Ardissino & É. Boullet (Eds.), *Lay readings of the Bible in early modern Europe* (pp. 260–275). Leiden: Brill.

Gingerich, O. (2005). Credentialing Kepler: Transits in the 17th century. *Bulletin of the American Astronomical Society, 37*, 62.

Gingerich, O. (2009). Kepler versus Lansbergen: On computing ephemerides, 1632–1662. In R. L. Kremer & J. Włodarczyk (Eds.), *Johannes Kepler from Tübingen to Żagań, Studia Copernicana, XLII* (pp. 113–117).

Gingerich, O. (2013). Transits in the seventeenth century and the credentialling of Keplerian Astronomy. *Journal for the History of Astronomy, 44*(3), 303–312.

Gould, A., Komatsu, T., DeVore, E., Harman, P., & Koch, D. (2015). Kepler's third law and NASA's Kepler mission. *The Physics Teacher, 53*, 201–204. https://doi.org/10.1119/1.4914556

Grant, R. (1852). *History of physical astronomy from the earliest ages to the middle of the nineteenth century* (pp. 415–417). London: Robert Baldwin.

---

[39] Pasachoff (2015).

[40] Gattei (2014, 2020).

Hadravová, A., Mahoney, T. J., & Hadrava, P. (Eds.). (2010). *Kepler's heritage in the space age: 400th anniversary of Astronomia nova*. National Technical Museum in Prague.

Halley, E. (1716). Methodus singularis qua Solis Parallaxis sive Distantia a Terra, ope Veneris intra Solem conspicienda, tuto determinari poterit. *Philosophical Transactions, 29*(348), 454–464.

Halley, E. (1929). The parallax of the Sun by the transit of Venus. In H. Shapley & H. E. Howarth (Eds.), *A source book in astronomy* (pp. 96–100). New York: McGraw-Hill.

Hevelius, J. (1662). *Mercurius in Sole Visus Gedani, Anno Christiano MDCLXI, d. III Maji ... cui Annexa Est, Venus in Sole Pariter Visa, Anno 1639, d. 24 Nov. St. V. Liverpoliae, a Jeremia Horroxio, Nunc Primum Edita* .... Gdansk: Published by the author, printed by Simon Reiniger.

Karalidi, T., Apai, D., Schneider, G., Hanson, J. R., & Pasachoff, J. M. (2015). Aeolus: An MCMC code for mapping Brown Dwarf and other ultra cool atmospheres. *Astrophysical Journal, 814*, 65ff.

Kepler, J. (1609). *Astronomia nova AITIOLOGETOS, seu physica coelestis*. Heidelberg: Vogelin.

Kepler, J. (1619). *Harmonice mundi*. Linz: Johannes Plank.

Kepler, J. (1627). *Tabulae Rudolphinae*. Ulm: Jonas Saurius.

Kepler, J. (1629). *De raris mirisque anni 1631 phaenomenis, Veneris puta Mercurii in Solem incursu, admonitio ad astrónomos, rerumque coelestiium studiosos*. Leipzig: Albertus Minzelius.

Kepler, J. (1940). In M. Caspar (Ed.), *Johannes Kepler Gesammelte Werke. Band VI: Harmonice mundi*. Munich: C. H. Beck.

Kepler, J. (1969). In F. Hammer (Ed.), *Johannes Kepler Gesammelte Werke. Band X: Tabulae Rudolphinae*. Munich: C. H. Beck.

Kepler, J. (1990). In M. Caspar & Kepler-Kommission (Eds.), *Johannes Kepler Gesammelte Werke. Band III: Astronomia nova* (2nd ed.). Munich: C. H. Beck.

Kremer, R. L., & Włodarczyk, J. (Eds.). (2009). *Johannes Kepler from Tübingen to Żagań, Studia Copernicana, XLII*.

Lattis, J. M. (1994). *Between Copernicus and Galileo: Christoph Clavius and the collapse of Ptolemaic cosmology*. Chicago: Chicago University Press.

Mahoney, T. J. (2010). Marking the 400th anniversary of Kepler's *Astronomia nova*. *Highlights of Astronomy, 15*, 821–827.

Mayor, M., & Queloz, D. (1995). A Jupiter-mass companion to a solar-type star. *Nature, 378*(6555), 355–359.

Olson, R. J. M., & Pasachoff, J. M. (2019). *Cosmos: The art and science of the universe*. London and Chicago: Reaktion Books and Chicago University Press.

Pasachoff, J. M. (2000). Brief lives of some star performers. In J. Voelkel (Ed.), *Book review of Johannes Kepler and the new astronomy*. Times Higher Education Supplement.

Pasachoff, J. M. (2009). Transits of Mercury and Venus and their implication for exoplanet transits and the Kepler mission. In R. M. Ros (Ed.), *Adventures in teaching astronomy* (p. 23). European Association for Astronomy Education–International Astronomical Union Course on Astronomy Education, EAAE-IAU.

Pasachoff, J. M. (2011, May/June). Catch a Pass! (of Venus with the Sun). *Odyssey*, pp. 40–42.

Pasachoff, J. M. (2012). Transit of Venus: Last chance from Earth until 2117. *Physics World, 25*(5), 36–41.

Pasachoff, J. M. (2015). Audio tours at Smithsonian's National Air and Space Museum: "What did Johannes Kepler find out about planetary orbits?" (#3); "How do astronomers today use Johannes Kepler's findings about planetary orbits?" (#48). http://airandspace.si.edu/exhibitions/explore-the-universe/audio-tour/index.cfm, or s.si.edu/ETUaudio; transcript: http://airandspace.si.edu/files/pdf/exhibitions/etu-audio-transcript

Pasachoff, J. M., Backhaus, U., Gährken, B., & Schneider, G. (2017a, October 17). The 2016 transit of Mercury and the solar parallax. DPS, Provo, session: historical astronomy: Rosetta, Cassini, Transit of Mercury, 200.04.

Pasachoff, J. M., & Filippenko, A. (2019). *The cosmos: Astronomy in the new millennium* (5th ed.). Cambridge: Cambridge University Press.

Pasachoff, J. M., Gährken, B., & Schneider, G. (2017b). Using the 2016 transit of Mercury to find the distance to the Sun. *The Physics Teacher, 55*, 3, cover illustration plus article (pp. 137–141). https://doi.org/10.1119/1.4976653

Pasachoff, J. M., & Pasachoff, N. (1999). Third physics opera for Philip Glass. *Nature, 462*, 724. http://www.nature.com/nature/journal/v462/n7274/full/462724a.html

Pasachoff, J. M., Schneider, G., Babcock, B. A., Lu, M., Edelman, E., Reardon, K. P., Widemann, T., Tanga, P., Dantowitz, R., Silverstone, M. D., Ehrenreich, D., Vidal-Madjar, A., Nicholson, P. D., Willson, R. C., Kopp, G. A., Yurchyhyn, V. B., Sterling, A. C., Scherrer, P. H., Schou, J., Golub, L., McCauley, P., & Reeves, K. (2013). Three 2012 transits of Venus: From Earth, Jupiter, and Saturn. In *221st American Astronomical Society Meeting*, Long Beach, CA, 315.06. 2013AAS...22131506P.

Pasachoff, J. M., Schneider, G., & Golub, L. (2005). The black-drop effect explained. In D. W. Kurtz (Ed.), *Transits of Venus: New Views of the Solar System and Galaxy. Proceedings IAU Colloquium, No. 196*. Cambridge: Cambridge University Press.

Pasachoff, J. M., & Sheehan, W. (2012). Lomonosov, the discovery of Venus's atmosphere, and eighteenth-century transits of Venus. *Journal for the History and Heritage of Astronomy, 15*(1), RP1, 1–12.

Pasachoff, J. M., & Sheehan, W. (2013, January). A major discovery in doubt. *Sky & Telescope, Focal Point*, p. 86.

Pasachoff, N., & Pasachoff, J. M. (2012). Kepler. In A. Robinson (Ed.), *The scientists: An epic of discovery* (pp. 26–31). London: Thames and Hudson.

Pasachoff, J. M., Schneider, G., Babcock, B. A., Lu, M., Reardon, K. P., Widemann, T., Tanga, P., Dantowitz, R., Willson, R., Kopp, G., Yurchyshyn, V., Sterling, A., Scherrer, P., Schou, J., Golub, L. & Reeves, K. (2012). The 2012 transit of venus for cytherean atmospheric studies and as an exoplanet analogue. In *AAS/Division for Planetary Sciences Meeting Abstracts# 44* (Vol. 44, pp. 508–506)

Rosa, M. R. (2010). How really precise and accurate are Tycho Brahe's data? In A. Hadravová, T. J. Mahoney, & P. Hadrava (Eds.), *Kepler's heritage in the space age: 400th anniversary of Astronomia nova* (pp. 102–113). National Technical Museum in Prague.

Schneider, G., & Pasachoff, J. M. (2006). Kepler-mission analog study using ACRIMSAT. In *Working Group on the Transits of Venus, General Assembly of the International Astronomical Union, Prague (Commission 41 on History of Astronomy), XXVIth International Astronomical Union General Assembly, Meeting on Transit 2004 and the Future* (pp. 17–18). IAU.

Schneider, G., Pasachoff, J. M., & Golub, L. (2004). TRACE observations of the 15 November 1999 transit of Mercury and the Black Drop effect: Considerations for the 2004 transit of Venus. *Icarus, 168*, 249–256.

Schneider, G., Pasachoff, J. M., & Willson, R. C. (2006). The effect of the transit of Venus on ACRIM's total solar irradiance measurements: Implications for transit studies of extrasolar planets. *Astrophysical Journal, 641*, 565–571.

van Helden, A. (1985). *Measuring the universe: Cosmic dimensions from Aristarchus to Halley*. Chicago: Chicago University Press.

Voelkel, J. (2000). *Johannes Kepler and the new astronomy*. Oxford: Oxford University Press.

Westfall, J., & Sheehan, W. (2015). *Celestial shadows: Eclipses, transits, and occultations*, with a foreword by J. M. Pasachoff. New York: Springer.

Young, C. A. (1888). *A text-book of general astronomy for colleges*. Boston and London: Ginn.

Chapter 11
Kepler's Contributions to Mathematics

Eberhard Knobloch

11.1 Introduction

When in 1594 the twenty-two-year-old Kepler took on his first paid position in Graz, it was the position of a mathematics teacher and of a provincial mathematician. After that Kepler called himself mathematician on nearly all of the title pages of his published works; at first 'mathematician of the illustrious provincials of Styria' (*Illustrium Styriae provincialium mathematicus*), later on 'Mathematician of the Holy Imperial Majesty' (*S(acrae) C(aesareae) M(ajestatis)*), or only—very seldom— 'Mathematician'. Only on the title pages of his *World harmony*[1] and of the second (book 4) and third volume (books 5–7) of his *Short explanation of the Copernican astronomy*[2] did he omit this expression.

Yet, during his lifetime the notion of mathematics included disciplines such as chronology, music theory, astrology, and astronomy that are no longer included therein today. Kepler contributed important results to all of these disciplines, which will not be dealt with in this chapter. He himself emphatically pleaded for another classification of astronomy. He called his famous book *New astronomy looking for causes or celestial physics.*[3] For him astronomy was a part of physics, no longer of mathematics[4] though arithmetic and geometry were and remained indispensable mathematical tools of astronomy.

---

[1] Kepler (1619, 1940).

[2] Kepler (1618–1621, 1991).

[3] Kepler (1609, 1990).

[4] Kepler (1991), p. 23: *Est pars Physices.*

E. Knobloch (✉)
Berlin Academy of Sciences and Humanities, Berlin, Germany
e-mail: eknobloch@bbaw.de

© Springer Nature B.V. 2024
A. E. L. Davis et al. (eds.), *Reading the Mind of God*, Springer Praxis Books,
https://doi.org/10.1007/978-94-024-2250-4_11

Kepler's mathematical achievements and studies must be seen against the background of four characteristics of theoretical mathematics at the transition from the sixteenth to the seventeenth century: (1) the availability of the works of the Greek authorities such as Apollonius, Archimedes, Euclid, and Pappus; (2) the textbooks on plane and spherical trigonometry by Johannes Regiomontanus, Nicolaus Copernicus, and Bartholomaeus Pitiscus (1561–1613), (3) the development of algebra especially thanks to François Viète (1540–1603); and (4) the invention of logarithms by Jost Bürgi, John Napier and Henry Briggs (1561–1630).

11.2   Philosophy of Mathematics, General Aspects

11.2.1   Foundations

Kepler was a philosophically minded thinker and mathematician. Presumably in Graz (1594–1600), yet certainly shortly after his studies in Tübingen (1589–1594) he elaborated his three books *On quantities*.[5] The first two books are still extant. This fragmentary work reveals the influence of the scholastic professor of physics, medicine and logic Jacob Schegk (1511–1587) in Tübingen. It is based on Aristotle's *Categories*, *Metaphysics* and *Physics*.

Kepler adhered to Aristotle's doctrine that the subject of mathematics is deduced by abstraction from sensible things, thus explicitly rejecting Plato's theory of a priori mathematical perception. The mathematical sciences are divided into arithmetic (the origin of perception), geometry, and the sciences occupying themselves with secondary quantities such as poetics, harmonics, optics, mechanics, astronomy, etc. There is no actually infinite quantity. There is potential infinity by adding in the case of numbers (discrete quantities) or dividing in the case of continuous quantities. In other words Kepler's infinite is the syncategorematic infinite of the Scholastic that can never be completely realized.[6] The second book explains the origin and generation of figures and solids by the motion of points, lines, planes or sections, and compositions respectively.

In geometry existence was equivalent for him with constructibility. Only constructible objects (points, lines) can be known. Non-constructible objects cannot be known either by the human mind or by God; that is, by the creator and first geometer. Therefore, God did not use such objects when he created the world. Kepler explained his standpoint in his correspondence and in his works as well, especially in his *World harmony*.[7] A heptagon cannot be constructed. Kepler's 'proof' of this

---

[5] Kepler (1594–1600); Kepler (2002), writings M1, M2.

[6] Aiton (1975), pp. 671–672.

[7] Kepler (1619); Kepler (1940), p. 55.

affirmation remained insufficient.[8] Hence the side of a heptagon is a 'non-being' (*non ens*). He criticized Albrecht Dürer's 'construction' of the regular heptagon because it was necessarily false. Only an approximate solution could be realized.

Thus the cossists, as the algebraists of his time were called, used a circular argument in Kepler's eyes. They presupposed the existence of the side of the regular heptagon when they established an equation for this side though there was no truly geometrical—that is, exact—construction of such a side. Therefore, Kepler rejected algebra to his own disadvantage, even though in 1591 François Viète's *In artem analyticen isagoge* had led algebra to new heights by his *Introduction into the analytical art*[9] (*In artem analyticen isagoge*).

His colleague in Prague, Jost Bürgi, was well acquainted with algebra. Kepler referred to him in his *World harmony* when he wrote down Bürgi's equation of sixth degree for the side of the heptagon:

$$7-14x^2 + 7x^4-x^6 = 0.$$

Kepler knew the cossic symbols but preferred to write this equation in the following way[10]:

$$7-14\text{ij} + 7\text{iiij}-1\text{vj};$$

that is, he denoted the powers of the unknown x by their exponents. In his posthumous writings Kepler wrote Roman numerals (the exponents) over the numerical coefficients.[11] His algebraic studies regarding quadratic and cubic roots are mistaken.

11.2.2 Methodology

Kepler was creatively inspired by ancient mathematicians such as Euclid and Archimedes. Yet he surpassed their results because he admitted new methods by amply employing analogical thinking in order to make evident the hidden strength of his methods as he formulated Archimedean theorems and demonstrations. He introduced the infinitely small into mathematics though this notion was not yet well defined. He replaced the Archimedean indirect proofs by direct proofs. He used the

---

[8] Kepler (1619); Kepler (1940), p. 50.

[9] Viète (1591).

[10] Kepler (1619); Kepler (1940), 52.

[11] Kepler (2002), writings M4, M6.

general principle of continuity that was closely related to Nicolaus Cusanus' principle of the coincidence of opposites in the infinite. Kepler was strongly influenced by this German cardinal. While Cusanus identified a circle of infinite radius with a straight line, Kepler identified the infinitesimal chord and arc, thus regarding the circle as the limiting case of a polygon.

Kepler could not know that he used the same or similar heuristic means as Archimedes had used in his *Approach related to mechanical problems* (*Peri ton mechanikon theorematon ephodos*) usually falsely called *Method* that which had to be rediscovered several centuries after his death; that is, analogies, indivisibles, and mechanical ingredients.

11.2.3 Mathematical Terminology

Kepler plays a certain role in the history of mathematical terminology. He introduced the Latin expression 'focus' in the case of conic sections. He applied or created many German technical terms which he gathered in a special list at the end of his German written *Excerpts from the ancient art of measuring of Archimedes*.[12] Some of them later became established, such as *Würffel* instead of *Cubus* ('cube'), many of them were never really used like *Spießeckich* instead of *Trapezium* ('trapezium'), *Schnitz* instead of *Segmentum solidum* ('solid segment') or *Durchschneider* instead of *Secans* ('secant').

11.3 Kepler's Contributions to Geometry

11.3.1 Geometry of Regular Polygons and Polyhedra

Already Kepler's cosmological speculations in his first book *Cosmographical mystery*[13] were based on Euclid's theory of the five convex regular polyhedra, usually called the Platonic solids, that was to be found in the thirteenth book of his *Elements*: cube and octahedron, and dodecahedron and icosahedron were reciprocal solids in the sense that the one has as many corners as the other has faces and vice versa, as he said. The tetrahedron is its own reciprocal solid (see Fig. 11.1).

In his *World harmony* Kepler generalized these Platonic solids in three different directions. Following Archimedes, the first generalization admitted solids having regular faces of more than one kind with the same arrangement round each vertex.

---

[12] Kepler (1616); Kepler (1960), pp. 135–274.
[13] Kepler (1596); Kepler (1938), pp. 1–80.

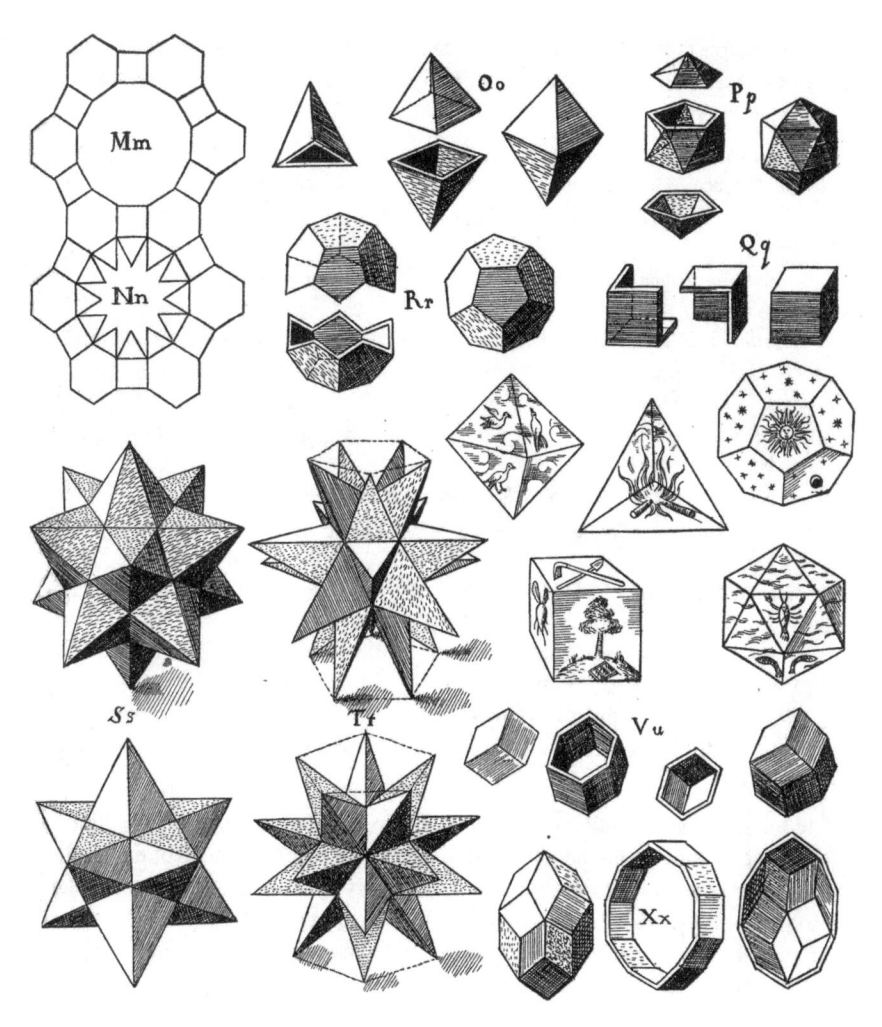

Fig. 11.1 The Platonic solids (Oo, Pp, Qq, Rr)[14]

There are thirteen such Archimedean solids. Kepler knew Pappus' *Collection Synagoge* thanks to the edition that had appeared in 1588 in Venice (Pappus of Alexandria, 1588). Hence he knew that Pappus had attributed the invention of the thirteen semiregular solids to Archimedes.[15] In his booklet *A New Year's gift or on the six-cornered snowflake*[16] he erroneously spoke about 'fourteen' Archimedean solids.

---

[14] Kepler (1619); Kepler (1940), p. 79.

[15] Pappus of Alexandria (1588), Book V, p. 34.

[16] Kepler (1611); Kepler (1941), pp. 259–280, here p. 266.

In the second book of his *World harmony* Kepler elaborated the first systematic account of these solids, constructed them and gave them the names by which they are still known (Figs. 11.2 and 11.3)[17]:

cubus truncus ('truncated cube': '1') having eight triangles and six octagons; *tetraedron-truncum* ('truncated tetrahedron': '2') having four triangles and four hexagons;

dodecaedron truncum ('truncated dodecahedron': '3') having twenty triangles and twelve decagons;

icosiedron truncum ('truncated icosahedron': '4') having twelve pentagons and twenty hexagons;

octaedron truncum ('truncated octahedron': '5') having six squares and eight hexagons;

cuboctaedron truncum ('truncated cuboctahedron': '6') having twelve squares, eight hexagons, and six octagons;

icosidodecaedron truncum ('truncated icosidodecahedron': '7') having thirty squares, twenty hexagons, and twelve decagons;

cuboctaedron ('cuboctahedron': '8') having eight triangles and six squares;

icosidodecaedron ('icosidodecahedron': '9') having twenty triangles and twelve pentagons;

rhombicuboctaedron ('rhombicuboctahedron': '10') having eight triangles and eighteen squares;

rhombicosidodecaedron ('rhombicosidodecahedron': '11') having twenty triangles, thirty squares, and twelve pentagons;

cubus simus ('snub cube': '12') having thirty-two triangles and six squares; and

dodecaedron simum ('snub dodecahedron': '13') having eighty triangles and twelve pentagons.

The second generalization regarded the Platonic solids as regular tessellations of the surface of a sphere and then, by analogy, the regular tessellations of the Euclidean space. Kepler was indeed the first writer to deal systematically with the problem of constructing all the tessellations found by regular polygons (Fig. 11.4).[18]

He had already mentioned the three possible cases in his *New Years's gift*[19] and repeated them in his *World harmony*[20]:

triangles, six round each vertex,

squares, four round each vertex,

hexagons, three round each vertex.

In his *New Year's Gift* he added an important remark taken again from Pappus' *Collectio*: The hexagon includes a larger area than either a triangle or a square with the same perimeter.[21] This true affirmation became an optimization problem of the calculus of variations. The general mathematical honeycomb conjecture asserting

---

[17] Coxeter (1975), esp. p. 667.

[18] Field (1988), p. 105.

[19] Kepler (1611); Kepler (1941), p. 266.

[20] Kepler (1619); Kepler (1940), p. 71, Proposition 18.

[21] Kepler (1611); Kepler (1940), p. 269.

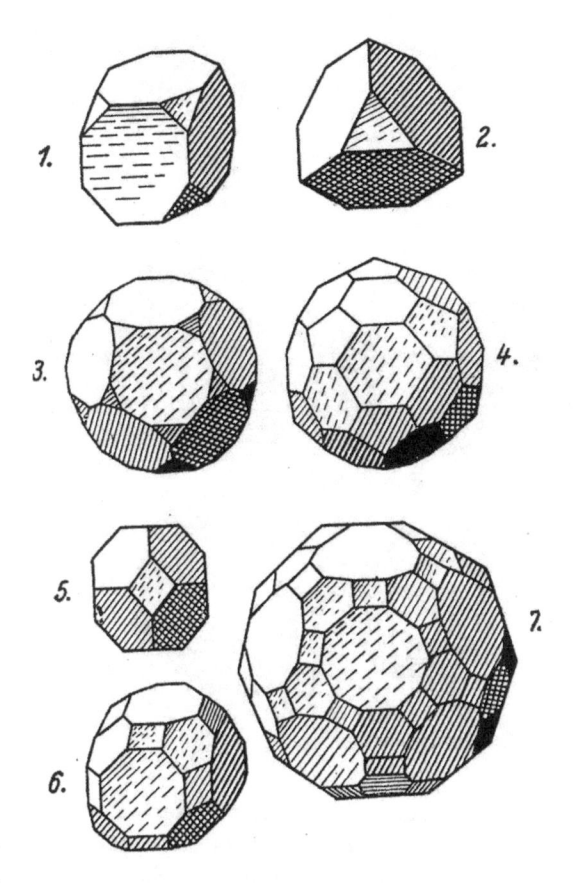

Fig. 11.2 The first seven Archimedean solids[22]

that the most efficient partition of the plane into equal areas is the regular hexagonal tiling was proved only in 1999 by Thomas Hales.[23] The problem can be expressed as a minimization of perimeters for fixed areas or as a maximization of areas for fixed perimeters.[24]

In his *World harmony* Kepler enumerated the six uniform tessellations (analogous to the Archimedean solids, see Fig. 11.5). Now the tiles are regular polygons of several kinds, but with the same arrangement at each vertex (Fig. 11.4 'L', 'N', 'S', 'V'; Fig. 11.1 'Mm', and Fig. 11.5 'Ii'.

Moreover, Kepler considered three-dimensional packing problems. He claimed that the face-centred cubic packing of congruent balls forms the densest possible

[22] Kepler (1619); Kepler (1940), p. 87.

[23] Hales (2000).

[24] Knobloch (2005), here p. 19.

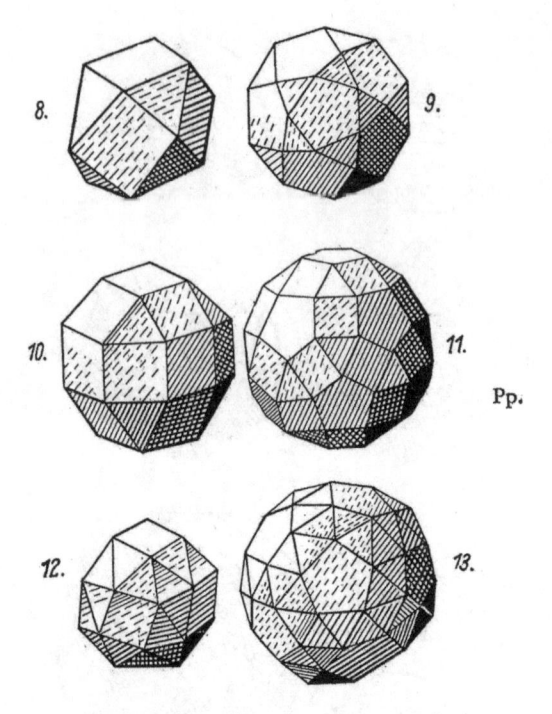

Fig. 11.3 The last six Archimedean solids[25]

configuration.[26] Such a packing is constructed by setting one layer of balls upon another (Fig. 11.6). Again, it was only in 2000 that Thomas Hales proved this so-called Kepler conjecture. Figure 11.6 is taken from Hales's publication[27] because Kepler did not include such an illustration in his booklet.

Another packing problem is connected with rhombic regular polyhedra. Kepler discovered two of the four possibilities[28] and dealt with them in his *New Year's gift*, in his *World harmony*, and again in his *Short explanation of the Copernican astronomy*[29]: the rhombic dodecahedron of the first kind can be repeated to fill the whole Euclidean space without gaps and the triacontahedron (see Fig. 11.7).

Kepler's discovery dates from November 1599, as is documented by a letter to Michael Maestlin.[30] The two missing rhombic polyhedra were discovered only much later: the rhombic icosahedron by E. S. Fedorov in 1885 and the rhombic dodecahedron of the second kind by Stanko Bilinski in 1960.[31]

[25] Kepler (1619); Kepler (1940), p. 85.

[26] Kepler (1619); Kepler (1941), p. 268.

[27] Hales (2000), p. 440.

[28] Coxeter (1975), p. 667.

[29] Field (1988), Appendix 4.

[30] Kepler (1949), 87, letter 142.

[31] Coxeter (1975), p. 667.

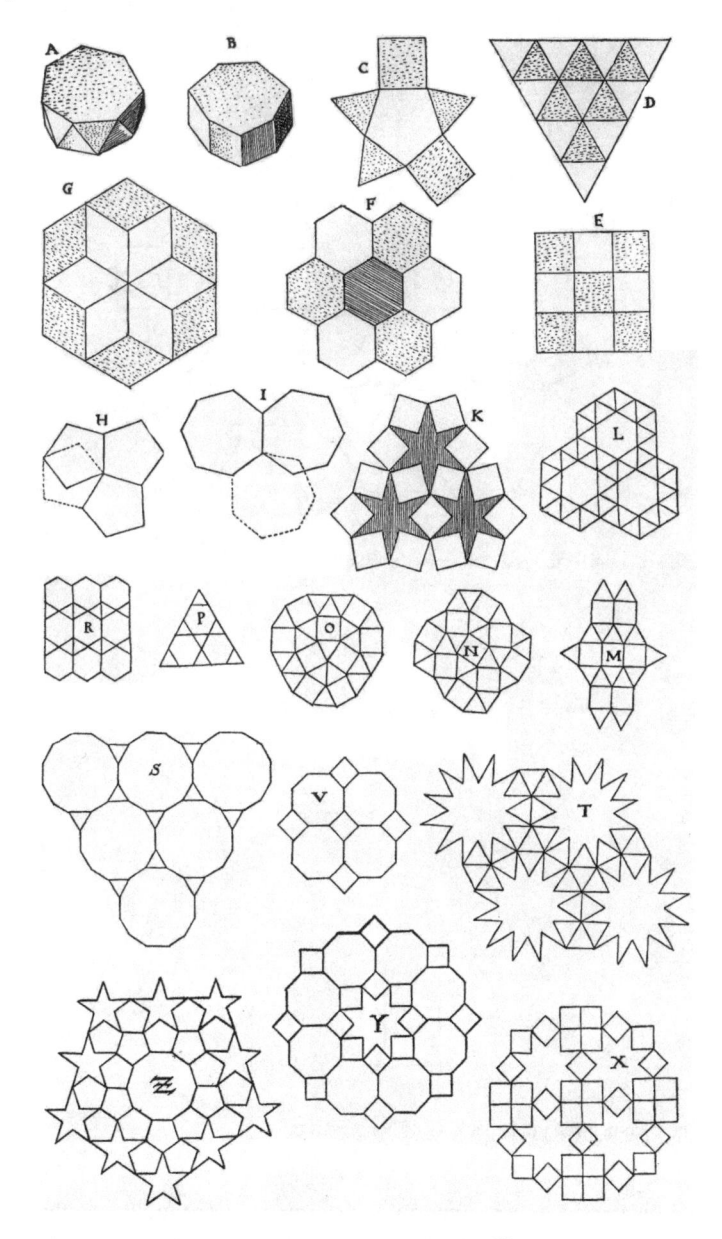

Fig. 11.4 The three regular tessellations of the Euclidean plane[32]

Kepler's third generalization of the Platonic solids admitted the faces to be star polygons instead of ordinary convex polygons. It led him to the most important

---

[32] Kepler (1619); Kepler (1940), p. 73.

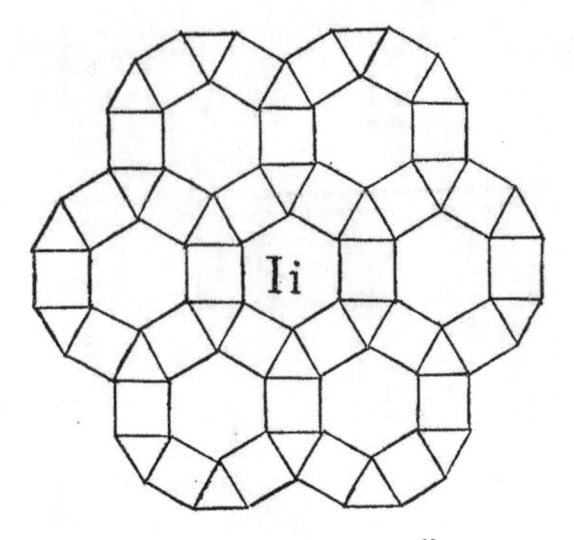

Fig. 11.5 The sixth uniform tessellation of the Euclidean plane[33]

Fig. 11.6 The face-centred cubis packing of congruent balls[34]

discovery of his inquiry into regular polygons; that is, to the regular stellated dodec-ahedra. When he fitted twelve pentagrams, three or five round each vertex, he got the great stellated dodecahedron and the small stellated dodecahedron (Fig. 11.1 'Ss', 'Tt'). If he had considered their reciprocal, as he did in the case of the Platonic regular solids, he would have found also the two so-called Kepler–Poinsot polyhedra.

---

[33] Kepler (1619); Kepler (1940), p. 74.

[34] Hales (2000), p. 440.

Fig. 11.7 Kepler's two rhombic regular polyhedra[35]

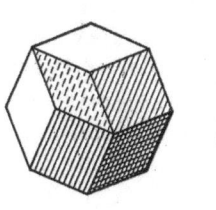
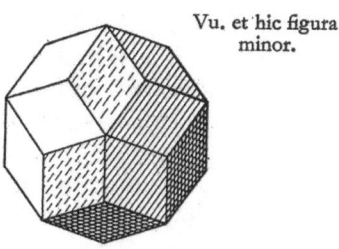

Vu. et hic figura minor.

Kepler's achievement remained unknown for several centuries. When in 1810 Louis Poinsot published his results about polygons and polyhedra he did not know Kepler's inquiries in this regard.

11.3.2 Theory of Conic Sections

In 1603 Kepler intensively studied Apollonius' *Conic sections* (*Conica*) in the Latin translation made by Federico Commandino in 1566.[36,37] He made ample use of conic sections in his *Optical part of astronomy* usually called *Optics*,[38] but also in his *New stereometry of wine casks, especially of the Austrian which has the most suitable figure of all*,[39] and in his *Short explanation of the Copernican astronomy*.[40] A. E. L. Davis has shown this and how Kepler invented the first 'non-cone-based' system of conics that remained unconnected with his contemporaneous work in astronomy; that, with his *New astronomy*.[41] Kepler made an original attempt to correlate and generalize the concept of foci for all conic sections.

As far as we know, Kepler himself invented the name 'focus' in his *Optics*.[42] In his *New astronomy* he called the same point '*punctum eccentricum*'. He used the new terminology again in his *New Stereometry*[43] and in his *Short explanation* as well.[44]

Though Kepler acknowledged—as usual—the works of his predecessors Apollonius and Vitelo, he consciously deviated from their approach, saying that he preferred to study conic sections from 'a mechanical, analogical, and popular point of view'.[45] The geometer should be lenient with him as he explains: 'For the geometrical voices

---

[35] Kepler (1619); Kepler (1940), p. 83.

[36] Apollonius of Perga (1566).

[37] Kepler (2002), p. 669.

[38] Kepler (1604, 1939).

[39] Kepler (1615); Kepler (1960), pp. 5–133.

[40] Kepler (1618–1621); Kepler (1991).

[41] Davis (1975).

[42] Kepler (1604); Kepler (1939), p. 91.

[43] Kepler (1615); Kepler 1960), p. 38.

[44] Kepler (1618–1621); Kepler (1991), p. 372.

[45] Kepler (1604); Kepler (1939), p. 90.

of analogy must serve us. I love indeed analogies most of all, my most reliable masters, they know all secrets of nature. They have to be considered especially in geometry, when they—though by most absurd designations—comprehend infinitely many cases which are inserted between the extreme cases and the middle, and when they make clearly evident the whole nature of every thing'.[46]

Kepler thus proceeded from the straight line to the infinitely many hyperbolas, parabolas, ellipses, and finally to the circle (see Fig. 11.8).

The same illustration was repeated in the *New stereometry*.[48] As far as is known, Kepler thus explicitly stated the principle of analogy that in the case at issue was based on the principle of geometrical continuity. His comments on his classification of conic sections can be read as concepts which later became underlying principles of projective geometry:

(1) Coincident points at infinity occur at both ends of a straight line.
(2) Parallel lines meet in a point at infinity.

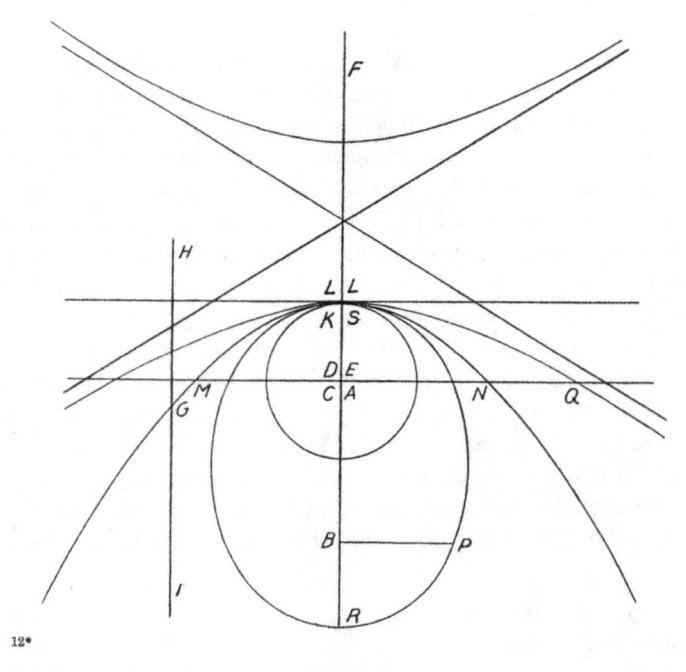

Fig. 11.8 Kepler's analogically generated conic sections[47]

[46] Kepler (1604); Kepler (1939), p. 92.

[47] Kepler (1604); Kepler (1939), p. 91.

[48] Kepler (1615); Kepler 1960), p. 37.

Girard Desargues (1591–1661) might have been acquainted with Kepler's work; at least he adopted the term *'foyer'* as French equivalent of the Latin word *'focus'* ('hearth', 'burning-point').

11.3.3 The Planned Geometrical Textbook and Related Posthumous Writings

Even before he published his *World Harmony* Kepler began to write a *Geometrical textbook*,[49] partly in the scholastic style of the *Short explanation*. The fragments include the second and third book dealing especially with the five Platonic and thirteen Archimedean semiregular solids, with the procedures of finding mean proportionals, and with an analysis of the first four books of Euclid's *Elements*.

Five fragments[50] consider the theory of the circle and sphere, the hexahedron, and deal with spherical geometry. Kepler relied on Johannes Regiomontanus' (1533) and Nicolaus Copernicus' excerpts on spherical trigonometry from *De revolutionibus* edited by Georg Joachim Rheticus (1514–1576) under the title *De lateribus et angulis triangulorum* (Wittenberg, 1542).[51] Fragment M18 enquires into various problems regarding the Platonic solids. He inscribed for example different solids (tetrahedron, cube) in the same sphere or decomposed the icosahedron into twenty triangular pyramids, the dodecahedron into twelve pentangular pyramids, and the octahedron into eight tetrahedra.

11.4 Kepler as a Precursor of Infinitesimal Mathematics and of the Calculus

Already in his *New Stereometry* Kepler came across the calculation of the area of the segment of an eccentric circle:

> And since I knew that there are infinitely many points of the eccentric circle and infinitely many distances [from the Sun] the idea occurred to me that all these distances are contained in the area of the eccentric circle. For I remembered that Archimedes, too, had subdivided the circle into infinitely many triangles long ago when he sought the ratio of the circumference to the diameter. For that was the hidden strength of his indirect proof.[52]

Kepler said, 'That all these distances are contained in the area' (*in plano…has distantias omnes inesse*), he did not speak of a sum as sometimes is falsely translated[53]

---

[49] Kepler, *Institutiones geometricae*, in Kepler (2002), writings M9–M12.

[50] Kepler (2002), writings M13–M17.

[51] Regiomontanus (1533); Rheticus (1542), p. 155.

[52] Kepler (1609); Kepler (1990), p. 264.

[53] Edwards (1979), p. 101.

nor did he equate all these distances with the area concerned. Distances—that is, lines—could not be added up. He was well aware of this matter of fact though he did not avoid such a misleading manner of speaking as: 'The area of the circle measures the sum of the distances'.[54] By distances he then meant infinitely small triangles.

He did indeed deal with the summation of an unbounded number of such infinitely small triangles without having at his disposal or inventing the integral calculus. Yet his ideas paved the way for its coming into being.

Kepler's main mathematical work was his *New stereometry*.[55] One year after the Latin version he published a German version under the title *Excerpt from the ancient art of measuring of Archimedes*.[56] This version coincided to a large extent with the Latin version. Yet it contained some further interesting results and was considerably rearranged.

First of all Kepler wanted to illustrate Archimedean theorems, to reveal the hidden strength of the Archimedean demonstrations, as he said in his *New astronomy*.[57] To that end he freely used infinitesimals, analogies admitting that absolutely and in every respect perfect demonstrations were to be found in the Archimedean writings.[58] Yet he had great confidence in his proofs which rested on analogies. Once he even said: 'This is the invincible demonstration by means of analogy. However, because the geometers are less accustomed to analogies, we would like to give a more difficult and completely geometrical proof.'[59]

The idea of infinitesimals enabled Kepler to compare lines, areas, and solids by means of smallest lines, smallest areas, and smallest solids. It goes without saying that his notion of 'smallest' was not well defined. It was undefinable because in reality there are no such smallest lines, smallest areas, or smallest solids though Kepler spoke of 'smallest rectangular bases (that is lines)', of 'smallest prisms into which a cylinder is cut', of 'smallest, quasi-linear segments of circular segments'.[60]

He even said: 'When infinitely many regular polygons are inscribed in or circumscribed about a circle (Fig. 11.9), we might argue with regard to *EB* (the arc) as with regard to a straight line, because the strength of the demonstration cuts the circle into smallest arcs which are equated with straight lines' (*quia vis demonstrationis secat circulum in minimos arcus qui aequiparantur rectis*).[61]

Paul Guldin contradicted him emphatically, claiming that no strength of any geometrical proof could effect things of that kind. Such assertions were against the principles of geometry. He underlined the aspect that there are no smallest segments,

---

[54] Kepler (1609); Kepler (1990), p. 367.

[55] Kepler (1615, 1960).

[56] Kepler (1616, 1960).

[57] Kepler (1609); Kepler (1990), p. 264.

[58] Kepler (1615); Kepler (1960), p. 13.

[59] Kepler (1615); Kepler (1960), p. 107.

[60] Kepler (1615); Kepler (1960), pp. 25 and 49.

[61] Kepler (1615); Kepler (1960), p. 14.

Fig. 11.9 The circle as the
limiting case of a polygon[62]

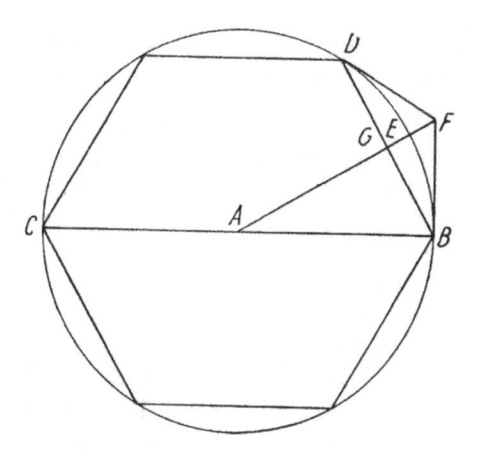

because there will be always a smaller segment than any given ostensibly smallest segment.[63]

For Kepler circle and ellipse perimeters were infinitely divisible but not composed of Euclidean points[64] though he seemed to say that the circumference of a circle has as many parts as it has points, that is, infinitely many. Each part forms the base of an isosceles triangle with vertex at the centre of the circle.[65] The circle is made up of infinitely many small triangles, each with its base on the circumference and with altitude equal to the radius of the circle. The bases are not points but infinitely small straight lines. These triangles can be replaced by a single triangle with the circumference as base. The area of the circle is thus given in terms of the circumference and radius (Fig. 11.10).[66]

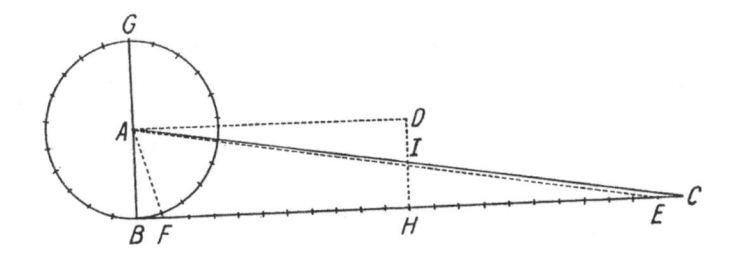

Fig. 11.10 The area of a circle is equal to a triangle[67]

[62] Kepler (1615); Kepler (1960), p. 13.

[63] Guldin (1635–1641), p. 323.

[64] Aiton (1973), p. 303.

[65] Kepler (1615); Kepler (1960), p. 15.

[66] Baron (1969), p. 110.

[67] Kepler (1615); Kepler (1960), p. 15.

The properties of circular cylinder, cones and prisms can be derived by similar considerations. This might be illustrated by four examples:

1. Let a straight circular cylinder be given and a circumscribed quadratic parallelepiped with height equal to the height of the cylinder and faces touching it. Their volumes have the same ratio as their bases. Kepler argued[68]: 'The cylinder and the parallelepiped are so to speak planes which have become corporeal' (cylinder enim et columna aequealta sunt hic veluti quaedam plana corporata).
2. Cones of equal height are to one another as their bases. Kepler argued[69]; 'A cone is here so to speak a circle which has become corporeal' (conus est hic veluti circulus corporatus).
3. Right segments of a cylinder which have been cut off by planes parallel to the axis are to one another as the segments of the base. Kepler argued[70]: 'A cylinder is here so to speak a circle or an ellipse which has become corporeal' (cylinder est veluti circulus aut ellipsis corporata). Hence the same happens to it as to the figures of the base.
4. If the plane cuts the axis without cutting any of the two bases, the cylindrical segments are to each other as the segments of the axis. Kepler argued[71]: 'A right circular cylinder is so to speak a line which has become corporeal, and that by supplying it with a cylindrical body' (cylinder rectus, sectus in plano ad axem recto, est veluti linea corporata, et quidem cylindrico corpore praedita). Hence the same happens to it as to the lines.

Kepler did not use this conception of smallest quantities only in order to make Archimedean theorems and proofs better understandable. He also used it for the demonstration of new theorems neither found nor investigated by the ancients; that is, he was well aware of the heuristic strength of his method. For that reason he gave the second part of his *New stereometry* the title *Supplement to Archimedes* (*Supplementum ad Archimedem*).[72]

His aim was to approximate the form of wine casks by solids of revolution generated by rotating conic sections. The four conic sections circle, ellipse, parabola, hyperbola generated 92 solids of revolution by rotating about lines in the plane other than the principle axes. Let us consider for example rotating circular segments (see Fig. 11.11).

The rotation of the circular segment *MIDKN* about the chord *MN* produces an 'apple'. If the minor segment *IDK* rotates about the chord *MN*, we get an 'apple ring' or 'apple belt' (*zona mali*). If this segment rotates about the chord *IK*, we get a 'lemon'. If this segment rotates about the diameter of the circle which is parallel to *IK*, we get a 'spherical ring', etc.

[68] Kepler (1615); Kepler (1960), p. 17.

[69] Kepler (1615); Kepler (1960), p. 30.

[70] Kepler (1615); Kepler (1960), p. 33.

[71] Kepler (1615, 1960).

[72] Kepler (1615); Kepler (1960), pp. 72–133.

Fig. 11.11 The production
of an apple by a rotating
circular segment[73]

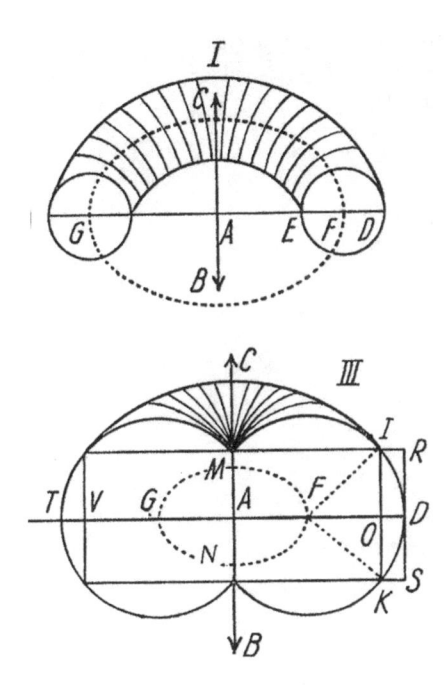

Kepler's method of integration might be explained by his theorem 20. This
theorem describes the volume of such an apple ring in terms of a cylindrical segment
and a spherical ring[74]:

> An apple ring is composed of the belt of the spherical ring (*VTSL*) and the straight segment
> (*ODTV*) of the cylinder whose base is the (circular) segment (*TKD*) lacking that figure which
> produces the apple, while its height is equal to (the circumference of) the circle which is
> described by the centre (*F*) of the greater (circular) segment (*MIKN*).

There are four basic ideas or conceptions, which lead to his really ingenious
solution[75]:

1. *Composition*: The apple and the sphere are solids of rotation which are consisting
   of hollow, coaxial cylindrical sheets (*tunicae*) around the axis of rotation. A
   straight cylindrical ungula is composed of rectangles perpendicular to its base.
2. *Expansion*: Kepler maintained to use the 'same laws' as Archimedes, who, as
   Kepler said, expanded the area of a circle into a rectangular triangle. In the same
   way, the sheets are expanded into rectangles.
3. *Transformation*: Apple and sphere are transformed into cylindrical ungulae
   (Fig. 11.12).
4. *Decomposition*: The cylindrical ungulae are decomposed into three and two
   partial solids, respectively, whose volumes can be geometrically determined.

---

[73] Kepler (1615); Kepler (1960), p. 39.

[74] Kepler (1615); Kepler (1960), p. 49.

[75] Knobloch (2000), pp. 94f.

Fig. 11.12 Transformation
of an apple into a cylindrical
ungula[76]

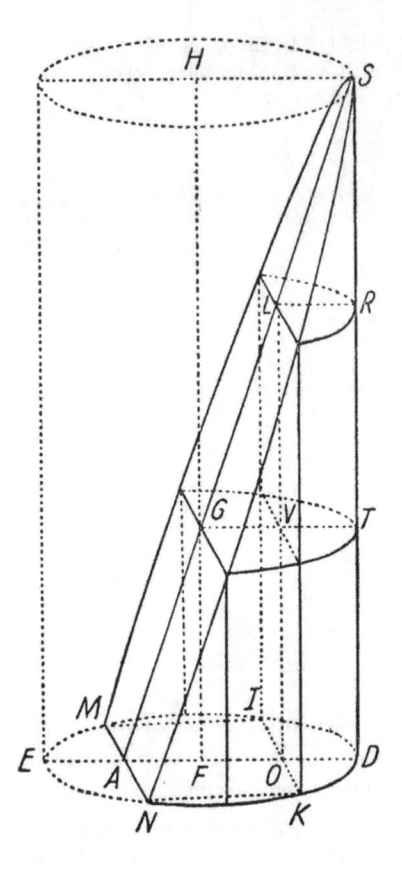

Alexander Anderson and Paul Guldin contested Kepler's first three assumptions
or claims. Let us now consider the structure of Kepler's proof, which consisted of
four steps.

First Step: Generation and Composition of the Apple

Let us assume that the apple is produced by the rotation of the circular segment
MIDKN about the axis *MN*. This segment *MIDKN* is divided by parallels to *MN*
into smallest segments of constant breadth, that is into quasi-linear segments, as
Kepler said. There are as many parallels as points of the straight line *AD*. Every
rotating quasi-linear segment produces a cylindrical sheet. Kepler said: The small
area (*areola*) *MN* creates nearly nothing because it is least moved. The apple consists
of these sheets (*tunicae*) (sheet model of the apple).

---

[76] Kepler (1615); Kepler (1960), p. 50.

Second Step: Construction of the Cylindrical Ungula of the Same Volume as the Apple

Every cylindrical sheet is cut open alongside a perpendicular line and expanded into a rectangle which is erected on the base; that is, on the circular segment *MDN*. The height of every such rectangle or expanded sheet is equal to the circumference of the circle described by the point of the line *AD*, which belongs to the erected rectangle. The construction is based on a transformation of the apple. The strength of this transformation (*vis transformationis*) implies a decomposition of the ungula into (calculable) parts of known volumes.

Third Step: Decomposition of the Ungula

Let *F* be the centre of *AO* or *ED*. Then *FD* = *GT*, *GF* = *TD*. The cylindrical ungula consists of three parts:

1. the cylindrical prism *MNLIK*,
2. the cylindrical segment *ODTV*,
3. the cylindrical part *VTSL*.

The two last parts taken together have the same volume as the ring of the apple produced by the circular segment *IDK*. This is the volume sought.

Fourth Step: Determination of the Volumes of These Three Parts

1. The volume of the cylindrical prism corresponds to the part of the apple which is produced by the rotation of the area *MIKN*. Such a part of the solid is a torus (*tore*). Kepler determined its volume in theorem 19. It is a cylinder, whose base is the area *MIKN* and whose height is equal to the circumference of the circle described by the centre *F*. This result is a special case of the so-called Pappus–Guldin theorem. It is true Kepler knew Pappus' *Collectio*. Yet, he could not find the theorem in Pappus because Commandino's Latin translation used by Kepler was corrupt at the passage in question.[77]
2. The cylindrical segment does not imply any problem.
3. The cylindrical part *VTSL* is at the same time a part of the smaller cylindrical ungula *GST*. Its volume is interpreted as the volume of a sphere produced by the rotation of the semicircle *GT* (a sphere is the limit case of an apple). *ST* is the length of the circumference of the circle described by *T* rotating about *G*:

   (a) *AD*:*DS* = *GT*:*TS* (the triangles *ADS*, *GTS* are similar).
   (b) That is, the radius *AD* is to the circumference of the circle produced by the rotating *D* as the radius *GT* to the magnitude *ST*. Therefore, *ST* is indeed the length of the circumference of the second produced circle.

The sphere is transformed into the smaller cylindrical ungula by the same procedure by which the apple was transformed into the greater cylindrical ungula. Hence the cylindrical part *VTSL* corresponds to the spherical ring which is produced by

---

[77] Hofmann (1973, 1990).

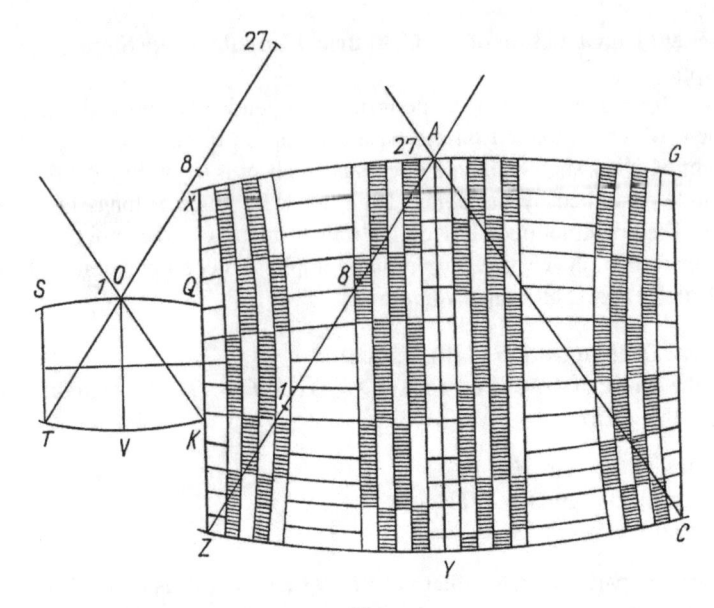

Fig. 11.13 Gauging the volume of a wine cask[78]

the rotating circular segment *VT* being equal to *IKD*. Such a spherical ring does not provide any problem.

To sum up, an ingenious idea led Kepler to his result. His method depended on such an idea, it was not a general integration method that could have enabled him to calculate the volumes of other solids of rotation. He used infinitesimals (that is, infinitely small quantities), not indivisibles like Archimedes, which are—by definition according to Aristotle—not quantities. This is the crucial difference between Kepler and Cavalieri. To what extent Galileo and Cavalieri were directly influenced by Kepler is an unsettled question.

In investigating the best dimensions for wine casks Kepler considered problems concerning maxima and minima. He inscribed circular cylinders with equal diagonals in a sphere. The largest cylinder is that in which the ratio of the diameter to the height = 1. If the Austrian cask had the form of a cylinder, it would have the largest volume given the length of the gauge (see Fig. 11.13).

---

[78] Kepler (1615); Kepler (1960), p. 115.

11.5 Kepler's Contributions to Practical Arithmetic: His Writings on Logarithms

At the beginning of the seventeenth century three mathematicians elaborated different foundations of calculations with logarithms. The Swiss Jost Bürgi occupied himself with them in Prague during the years 1603–1611 (that is, at the same time that Kepler worked there). Yet, Bürgi's *Arithmetic and Geometric Progression Tables together with detailed instructions on how to use these in all sorts of useful calculations and how they should be understood* (*Arithmetische und geometrische Progress-Tabulen, sambt gründlichen unterricht, wie solche nützlich in allerley Rechnungen zu gebrauchen und verstanden werden sol*) appeared only in 1620. Apart from some vague hints Bürgi unfortunately did not reveal details so that Kepler could not profit by it and complained of this secretiveness in his *Rudolphine Tables*.[79]

In the meantime the Scot John Napier had published his *Description of the Wounderful Canon of Logarithms*[80] in Edinburgh and the Englishman Henry Briggs, his *First chiliad of logarithms*.[81] Kepler became acquainted with Napier's logarithms from 1617 on. He decided to replace Napier's mechanical, geometrical foundation by a merely arithmetical foundation that in the Euclidean manner was based on postulates, axioms and propositions. Yet, their common basic idea was the correspondence between a geometrical and an arithmetical sequence: The term of the arithmetical sequence that was assigned to a given term of the geometrical sequence was by definition its logarithm.

Kepler generated such a system of a geometrical and an arithmetical sequence by means of mean proportionals of a ratio $a{:}b$ and the so-called 'measure' of a ratio. For him the logarithm is—according to the naming of this Greek notion—'the number of a ratio'. If a is chosen in a certain way, this 'measure' becomes Kepler's logarithm of b.

In 1624 he published his *Chiliad of logarithms*[82] (Fig. 11.14); that is, thousand logarithms. The book began with a 'legitimate demonstration' (*demonstratio legitima*) of the origin of the logarithms and their use. One year later, *The supplement of the chiliad of logarithms*[83] appeared containing the rules of their use.

Let us denote the measure of $a{:}b$ by $M(a{:}b)$. Kepler did not use such algebraic symbolism but wrote 'the measure of the proportion a to b' (*mensura proportionis a ad b*). Kepler postulated:

1. If $a{:}b = c{:}d$, then $M(a{:}b) = M(c{:}d)$
2. $M(a{:}b) + M(b{:}c) = M((a{:}b).(b{:}c)) = M(a{:}c)$

Hence if b is the mean proportional between a and c or if $a{:}b = b{:}c$,

---

[79] Kepler (1627); Kepler (1969), p. 48.

[80] Napier (1614).

[81] Briggs (1617).

[82] Kepler (1624); Kepler (1960), pp. 77–352.

[83] Kepler (1624); Kepler (1960), p. 319.

$$M(a : c) = 2M(a : b)$$

In order to calculate $M(a:b)$ Kepler calculated mean proportionals $x_1, \ldots, x_n$ between a and b. The terms

$$1, a : x_n, (a : x_n)^2, \ldots, (a : x_n)^{2n} = a : b$$

CHILIAS LOGARITHMORVM

ARCUS Circuli cum differentiis	SINUS seu Numeri absoluti	Partes vicesimae quartae	LOGARITHMI cum differentiis	Partes sexagenariae
— 3. 26			10536. 05	
0. 3. 26	100.00	0. 1. 26	690775. 54	0. 4
3. 27			69314. 72	
0. 6. 53	200.00	0. 2. 53	621460. 82 —+	0. 7
— 3. 26			40546. 51	
0. 10. 19	300.00	0. 4. 19	580914. 31	0. 11
3. 26			28768. 21	
0. 13. 45	400.00	0. 5. 46	552146. 10 —+	0. 14
— 3. 27			22314. 35	
0. 17. 12	500.00	0. 7. 12	529831. 75 —	0. 18
3. 26			18232. 16	
0. 20. 38	600.00	0. 8. 38	511599. 59	0. 22
— 3. 26			15415. 07	
0. 24. 4	700.00	0. 10. 5	496184. 52 —+	0. 25
3. 26			13353. 14	
0. 27. 30	800.00	0. 11. 31	482831. 38 —+	0. 29
— 3. 26			11778. 30	
0. 30. 56	900.00	0. 12. 58	471053. 08	0. 32
3. 27			10536. 05	
0. 34. 23	1000.00	0. 14. 24	460517. 03	0. 36
— 3. 26			9531. 02	
0. 37. 49	1100.00	0. 15. 50	450986. 01	0. 40
3. 26			8701. 14	
0. 41. 15	1200.00	0. 17. 17	442284. 87	0. 43
— 3. 27			8004. 27	
0. 44. 42	1300.00	0. 18. 43	434280. 60	0. 47
3. 26			7410. 80	
0. 48. 8	1400.00	0. 20. 10	426869. 80 —+	0. 50
— 3. 26			6899. 28	
0. 51. 34	1500.00	0. 21. 36	419970. 52 —	0. 54
3. 26			6453. 85	
0. 55. 0	1600.00	0. 23. 2	413516. 67 —	0. 58
— 3. 27			6062. 47	
0. 58. 27	1700.00	0. 24. 29	407454. 20	1. 1
3. 26			5715. 84	
1. 1. 53	1800.00	0. 25. 55	401738. 36	1. 5
			5406. 72	

Fig. 11.14 Kepler's table of logarithms reaching from log sin 0°3′26″ to log sin 1°43′9″[84]

[84] Kepler (1624); Kepler (1960), p. 319.

40	— 3. 27 1. 5. 20	1900.00	0. 27. 22	396331. 64 5129. 33	1. 8
	3. 26 1. 8. 46	2000.00	0. 28. 48	391202. 31 4879. 02	1. 12
	— 3. 26 0. 12. 12	2100.00	0. 30. 14	386323. 29 4652. 00	1. 16
	3. 26 1. 15. 38	2200.00	0. 31. 41	381671. 29 4445. 17	1. 19
50	— 3. 27 1. 19. 5	2300.00	0. 33. 7	377226. 12— 4255. 97	1. 23
	3. 26 1. 22. 31	2400.00	0. 34. 34	372970. 15 —+ 4082. 20	1. 26
H2v	— 3. 26 1. 25. 57	2500.00	0. 36. 0	368887. 95 —+ 3922. 07	1. 30
	3. 27 1. 29. 24	2600.00	0. 37. 26	364965. 88 3774. 03	1. 34
	— 3. 26 1. 32. 50	2700.00	0. 38. 53	361191. 85— 3636. 77	1. 37
	3. 26 1. 36. 16	2800.00	0. 40. 19	357555. 08 —+ 3509. 13	1. 41
60	— 3. 27 1. 39. 43	2900.00	0. 41. 46	354045. 95 3390. 15	1. 44
	3. 26 1. 43. 9	3000.00	0. 43. 12	350655. 80— 3278. 99	1. 48
	— 3. 26				

Fig. 11.14 (continued)

form a geometrical sequence. The measures

$$M(a : x_1) = \frac{1}{2}M(a : b), M(a : x_2) = \frac{1}{4}M(a : b), \ldots M(a : x_n) = \frac{1}{2^n}M(a : b)$$

approach zero for increasing n. Therefore, Kepler defined

$$M(a : x_n) = a - x_n$$

or

$$M(a : b) = 2^n(a - x_n) \text{ for sufficiently large } n.$$

The sequence

$$0, M(a : x_n), 2M(a : x_n), \ldots, (a : b).$$

is the arithmetical sequence corresponding to the geometrical sequence mentioned above.

Now Kepler defined:

$$M(z : x) = \text{Log}\, x.$$

He thought of logarithms of sine values: z is the *sinus totus* whose logarithm should be equal to zero. He chose $z = 10^7$. For that reason his table of thousand logarithms consisted of five columns intitled arcs (*arcus*), sines or absolute numbers (*sinus seu numeri absoluti*), 24th parts (*partes vicesimae quartae*), logarithms (*logarithmi*), sixtieth parts (*partes sexagenariae*).

What is more, whenever Kepler used the notion of logarithm (*logarithmus*) that was denoted by 'Log', he meant what we now call log sin α. Correspondingly his *antilogarithmus* meant log cos α, his *mesologarithmus*, log tan α[85] These notions were used extensively in his *Short explanation of the Copernican astronomy*, and his logarithms above all in his *Rudolphine Tables*. Though his logarithms were outdated by Briggs's decadic logarithms since their publication it was too late for him to replace his own logarithms by the logarithms of the English scholar before the publication of his tables. Thus one has to distinguish between Kepler's original, new foundation of Napier's logarithms and their usefulness. The difference between Kepler's and Napier's logarithms is 3.9 ln.[86] Kepler's German written introduction in his theory of logarithms remained unpublished during his life-time. It was entitled *The inventory of numbers or Scottish practice.*[87]

References

Aiton, E. J. (1973). Infinitesimals and the area law. In F. Krafft, K. Meyer, & B. Sticker (Eds.), *Internationales Kepler-Symposium Weil der Stadt 1971, Referate und Diskussionen*, Hildesheim (pp. 285–305).

Aiton, E. J. (1975). Kepler's ideas on infinitesimals, limits and continuity. *Vistas in Astronomy, 18*, 671–672.

Apollonius of Perga. (1566). *Apollonii Pergaei Conicorum libri quattuor* (Translated by Federicus Commandinus). Bologna: Alexander Banatius.

Baron, M. E. (1969). *The origins of the infinitesimal calculus*. Oxford: Oxford University Press.

Briggs, H. (1617). *Logarithmorum chiliadis prima*. London.

Coxeter, H. S. M. (1975). Kepler and mathematics. *Vistas in Astronomy, 18*, 661–670.

Davis, A. E. L. (1975). Systems of conics in Kepler's work. *Vistas in Astronomy, 18*, 673–685.

Edwards, C. H. (1979). *The historical development of the calculus*. New York: Dover.

Field, J. V. (1988). *Kepler's geometrical cosmology*. London and Chicago: Athlone Press and University of Chicago Press.

Guldin, P. (1635–1641) *De centro gravitatis ... libri IV*. Vienna: Gregorius Gelbhaar.

Hales, T. (2000). Cannonballs and honeycombs. *Notices of the American Mathematical Society, 47*, 440–449.

Hofmann, J. E. (1973). Über einige fachliche Beiträge Keplers zur Mathematik. In F. Krafft, K. Meyer & B. Sticker. Hildesheim (Eds.), *Internationales Kepler-Symposium Weil der Stadt 1971, Referate und Diskussionen* (pp. 261–284)

---

[85] Kepler (1625); Kepler (1960), pp. 353–426.

[86] Hammer in Kepler (1960), *Nachbericht*, p. 472.

[87] Kepler, *Der Zahlen inventarium oder Schottische Practica*, in Kepler (2002), pp. 513–519.

Hofmann, J. E. (1990). Über einige fachliche Beiträge Keplers zur Mathematik. [Reprinted in *Joseph Ehrenfried Hofmann, Ausgewählte Schriften* (Vol. II, C. J. Scriba, Ed.). Hildesheim, et al. (pp. 327–350)].

Kepler. (1594–1600). *De quantitatibus.* In manuscript.

Kepler, J. (1596). *Mysterium cosmographicum.* Tübingen: G. Gruppenbach.

Kepler, J. (1604). *Ad Vitellionem paralipomena quibus astronomiae pars optica traditur.* Frankfurt: Claudius Marnus and the heirs of Johannes Aubrius.

Kepler, J. (1609). *Astronomia nova ....* Heidelberg: E. Vogelin.

Kepler, J. (1611). *Strena seu de nive sexangula.* Frankfurt am Main: Godefried Tampach.

Kepler, J. (1615). *Nova stereometria doliorum vinariorum, in primis Austriaci, figurae omnium aptissimae.* Linz: Joannes Plank.

Kepler, J. (1616). *Außzug auß der Uralten Messekunst Archimedis.* Linz: Hansen Blanden.

Kepler, J. (1618–1621). *Epitome astronomiae Copernicanae.* Books I, II, III (1618). Linz: Johann Plank. Book IV. Linz: Johann Plank. Books IV (1620) Frankfurt: Godefried Tampach.

Kepler, J. (1619). *Harmonices mundi libri V.* Linz: Johannes Plank.

Kepler, J. (1624). *Chilias logarithmorum.* Marburg: Caspar chemin.

Kepler, J. (1625). *Supplementum chiliadis lograrithmorum.*

Kepler, J. (1627). *Tabulae Rudolphinae.* Ulm: Jonas Saurius.

Kepler, J. (1938). *Johannes Kepler Gesammelte Werke. Band I: Mysterium cosmographicum / De stella nova* (M. Caspar, Ed.). Munich: C. H. Beck.

Kepler, J. (1939). *Johannes Kepler Gesammelte Werke, Band. II: Astronomiae pars optica* (F. Hammer, Ed.). Munich: C. H. Beck.

Kepler, J. (1940). *Johannes Kepler Gesammelte Werke. Band VI: Harmonice mundi* (M. Caspar, Ed.). Munich: C. H. Beck.

Kepler, J. (1941). *Johannes Kepler Gesammelte Werke, Band IV: Kleinere Schriften 1602–1611 / Dioptrice* (F. Hammer, Ed.). Munich: C. H. Beck.

Kepler, J. (1949). *Johannes Kepler Gesammelte Werke. Band XIV: Briefe 1599–1603* (M. Caspar, Ed.). Munich: C. H. Beck.

Kepler, J. (1960). *Johannes Kepler Gesammelte Werke. Band IX: Mathematische Schriften* (F. Hammer, Ed.). Munich: C. H. Beck.

Kepler, J. (1969). *Johannes Kepler Gesammelte Werke. Band X: Tabulae Rudolphinae* (F. Hammer, Ed.). Munich: C. H. Beck.

Kepler, J. (1990). *Johannes Kepler Gesammelte Werke. Band III: Astronomia nova* (M. Caspar & Kepler-Kommission, Eds.). Munich: C. H. Beck.

Kepler, J. (1991). *Johannes Kepler Gesammelte Werke. Band VII: Epitome astronomiae Copernicanae* (M. Caspar & Kepler-Kommission, Eds.). Munich: C. H. Beck.

Kepler. (2002). *Johannes Gesammelte werke. Band XXI, 1: Manuscripta astronomica III – De calendario Gregoriano – Manuscripta mathematica* (V. Bialas et al., Eds.). Munich: C. H. Beck.

Knobloch, E. (2000). Archimedes, Kepler, and Guldin: The role of proof and analogy. In T. Rüdiger (Ed.), *Mathesis, Festschrift zum siebzigsten Geburtstag von Matthias Schramm* (pp. 82–100). Berlin-Diepholz.

Knobloch, E. (2005). Mathesis perennis: Mathematics in Ancient, Renaissance and Modern Times. *Studies in the History of Natural Sciences, 24 Suppl.*, 10–22.

Napier, J. (1614). *Mirifici Logarithmorum Canonis Descriptio.* Edinburgh: Andrew Hart.

Pappus of Alexandria. (1588). *Mathematicae collectiones* (Translated from the Greek into Latin by F. Commandino). Pesaro: Hieronymus Concordia.

Regiomontanus, J. (1533). *De triangulis omnimodis libri quinque.* Nuremberg: Johann Petri.

Rheticus, G. J. (Ed.). (1542). *De lateribus et angulis triangulorum.* Leipzig: Wolfgang Gunter.

Viète, F. (1591). *In artem analyticen isagoge.* Tours: Mettayer.

Chapter 12
Kepler's *Dream* and Lunar Astronomical Phenomena

Jarosław Włodarczyk

In this chapter I give an outline of the history of Kepler's writing of *Somnium*, which spanned virtually the entirety of his life. Kepler's *Dream*, published posthumously in 1634, is a short book about a journey to the Moon and what the Universe looked like when seen from the Moon. It seems to have been written, and repeatedly extended or rewritten, over a long period, having apparently been started in the 1580s as a student project set by his astronomy teacher at the University of Tübingen Michael Maestlin (1550–1631). As we shall see later, there were Hellenistic models for describing such a journey and for describing what the traveller found on the Moon. Kepler, however, employed this concept in an innovative way because he was interested primarily in the astronomical phenomena seen from the Moon. His lunar or seleno-centric astronomy, like almost everything else Kepler wrote, was conceived as an argument for Copernicanism (since no one doubted the Moon was moving). Significantly enough, at the turn of the seventeenth century, Kepler was not the only writer who used this argument in the discussion about the Sun-centred cosmology. And yet Kepler's *Dream* was the most comprehensive representation of lunar astronomy, additionally enhanced with some literary merit. As it turned out, Kepler provided an early example of a genre that with various modifications of style and emphasis was to prove popular in the decades after his death.[1]

---

[1] The work is reprinted in Kepler (1993) and is available in English translations (Kepler, 1965a, 1967).

J. Włodarczyk (✉)
Institute for the History of Science, Polish Academy of Sciences, Warsaw, Poland
e-mail: jaroslawwlodarczyk@wp.pl

© Springer Nature B.V. 2024
A. E. L. Davis et al. (eds.), *Reading the Mind of God*, Springer Praxis Books,
https://doi.org/10.1007/978-94-024-2250-4_12

12.1 The Book

The full title of Kepler's *Dream* reads *Somnium, seu opus posthumum de astronomia lunari* (*The Dream, or Posthumous Work on Lunar Astronomy*), which confirms that the book was printed after Kepler's death. However, this title hardly does justice to the content and scope of Kepler's book.

The first part of the book consists of approximately 8000 words. It opens with a brief preface by the narrator, who recalls his observations of the stars and the Moon, and then introduces his dream. The dream consists of reading a book about the young Duracotus who in turn speaks about his life and adventures. Duracotus was an Icelander. His mother, Fiolxhilde, was a herb gatherer and his father had perished before his son got to know him. At the age of 14, Duracotus found himself in the astronomical observatory that Tycho Brahe (1546–1601) had set up on the island of Hven. A few years later Duracotus returns to his mother. Fiolxhilde tells him that she knows wise spirits who can carry people to distant lands or can tell stories about such lands in such a way that people think they have been there themselves. Fiolxhilde is particularly friendly with the Daemon of Levania (as the Daemon calls the Moon)[2] and urges Duracotus to use the Daemon's agency to allow him to explore the lunar world. Summoned by Fiolxhilde, the Daemon appears and takes over the narrative. First the Daemon explains how one can travel to the Moon with the help of daemons. He then describes various celestial phenomena from the point of view of an observer on the Moon. Two separate sections are devoted to the Daemon's descriptions of the differences between the two hemispheres of the Moon. This is a significant matter since from one hemisphere the Earth is always visible, whereas it is not visible at all from the other. The account ends with the description of vegetation and inhabitants of the Moon. At this point the main narrator, Kepler's *alter ego*, wakes up and the first part of *Somnium* comes to a close.

What follows is the central part of *Somnium*, which, as the author himself states in a letter to Matthias Bernegger (1582–1640) in December 1623, contains '[…] as many problems as it does lines. Some of the problems should be solved with the help of astronomy, whereas others will require the help of physics, and still others—by recourse to history.' But readers 'have no wish to fill up their minds with riddles.'[3] For this reason Kepler supplements the text with endnotes to make the story easier to understand. It is up to the reader to decide if the strategy proves successful. And yet it is in these notes that Kepler defines his primary aim:

> The purpose of my *Dream* is to use the example of the moon to build up an argument in favour of the motion of the earth, or rather to overcome objections taken from the universal opposition of mankind.[4]

---

[2] Kepler (1993), p. 337; Kepler (1967), p. 53. Levana' (לבנה) is the name for the Moon in Biblical Hebrew. Kepler learned Hebrew as part of his theological studies at the University of Tübingen, see Methuen, 'Kepler, religion and natural philosophy: a theological biography' (this volume, Chap. 1).

[3] Kepler to Bernegger, 4 December 1623; Kepler (1959), p. 143.

[4] Kepler (1634), Note 4, p. 31; Kepler (1993); Kepler (1967), p. 36.

Kepler then uses the notes to make lunar astronomy part of his earlier works such as *Ad Vitellionem paralipomena* (1604), *Astronomia nova* (1609), *Dissertatio cum nuncio sidereo* (1610), *Epitome astronomiae Copernicanae* (1618–21) and an unfinished treatise *Hipparchus seu de magnitudinibus et intervalis trium corporum Solis, Lunae et Telluris*. Additionally, the notes offer insight into Kepler's biography and his polemical temper. The latter finds full expression in Kepler's epigram, which he wrote in response to the Decree of the Holy Office as of 1620. The Decree specified over a dozen fragments in Copernicus' *De revolutionibus* that were to be deleted or altered. Kepler sneered:

> They wished to keep the poet away from whores
>
> So they castrated him, the awful bores.
>
> Thus of his testicles bereft of force
>
> The poet could live, tormented by remorse.
>
> Poor you, Pythagoras, to feel worse pain,
>
> You who they say gave yourself mental strain.
>
> They took your brain out with their surgeon's knife
>
> And left you what it's wrong to call a life.[5]

Kepler's last note is numbered 223. Taken together, 'Johannes Kepler's Notes on His *Astronomical Dream*' are roughly three times longer than the whole account of the journey to the Moon and the conditions prevailing there.

The third part of *Somnium* is the 'Geographical or, if You Prefer, Selenographic Appendix'. This is a copy of the letter sent to the Jesuit mathematician Paul Guldin (1577–1643) after 17 July 1623. The letter is short (approximately 500 words) and deals with lunar craters observed through a telescope.

Kepler tells a story of how the craters were built by the Moon-dwellers. Following this letter, there is the fourth part, 'Notes on this Appendix' (approximately 5000 words). This part mainly deals with the morphology of the lunar surface and the description of certain phenomena specific to this surface and known from telescopic observations. The book ends with a fifth part: Kepler's new translation of Plutarch's *On the Face which Appears in the Orb of the Moon*.

The great diversity of the texts which constitute *Somnium* reflects the history of its composition. The idea of developing an astronomy of the Moon undoubtedly originated in Tübingen, as Kepler himself informs us. He reveals in Notes 2 and 43 of the *Dream* that in 1593 Christoph Besold (1577–1638) proposed some twenty theses on lunar phenomena—derived from Kepler's essays—as the subject of a public debate to be held by the students in Tübingen:

> For with my quite reliable memory I recall the origins of the individual parts of my tale … I have a very old document which you, most illustrious Christopher Besold, wrote with your hand, when, in the year 1593, on the basis of my essays, you formulated about twenty theses concerning the celestial phenomena on the moon and showed them to Veit Müller,

---

[5] *Ne lasciviret, poterant castrare Poetam,/Testiculis demptis vita superstes erat./Vae tibi Pythagora, Cerebro qui ferris abusus;/Vitam concedunt, ante sed excerebrant.* Kepler (1634), p. 31; Kepler (1993), pp. 333–334; Grafton (1991), p. 22.

who then regularly presided over the philosophical disputations, with the thought that you would engage in a debate over them if he approved.[6]

The proposal was rejected, presumably on the account of the heterodoxy of the subject matter since Aristotelian geocentric cosmology was obligatory in academic teaching, whereas Kepler wished to use the Moon as an argument in support of heliocentrism.

Another date mentioned by Kepler is the year 1608, which appears in the first line of the *Dream*. Does this mean that Kepler returned to his essays on lunar astronomy after fifteen years and reworked them into *Somnium* while he was working in Prague? The trail of evidence for this is not so clear. In his *Conversation with the Sidereal Messenger* (*Dissertatio cum nuncio sidereo*), published in April 1610, Kepler wrote about the nature of the Moon's surface: 'We were deeply engaged in these discussions last summer ... To please Wacker, I even founded a new astronomy for the inhabitants of the Moon, as it were; in plain language, a sort of lunar geography.'[7] Thus 1609 rather than 1608 is the most likely year. While in Prague, the Johannes Matthäus Wacker von Wackenfels (1550–1619) was one of Kepler's closest friends and patrons at the court of Emperor Rudolf II. Their close relationship is attested to by Kepler's letters, in which they discussed many astronomical matters including the lunar world. In 1611 Wackher, a 'patron of writers and philosophers', was the dedicatee of Kepler's essay, *Strena* (*A New Year's Gift or, The Six-Cornered Snowflake*) which, among other things, was a pioneering work on the close packing of equal spheres, a mathematical problem that later proved fundamental to crystallography, and raised questions on the geometry of polyhedra.

In any event 'a new astronomy for the inhabitants of the Moon' must have taken on the form of a dream journey involving Fiolxhilde, Duracotus, and the Daemon of Levania at the end of the first decade of the seventeenth century. In Note 8 Kepler makes it clear that the manuscript of *Somnium* was taken out of Prague in 1611, and that because of its content, the text contributed towards his mother being accused of witchcraft.

Between the years 1593 and 1608 or 1609 Kepler led a very active intellectual life. He wrote and published *Mysterium cosmographicum*—the first open defence of the heliocentric system after *De revolutionibus*. He met Tycho Brahe, with whom he subsequently collaborated closely in Prague. He published *Ad Vitellionem paralipomena*, in which he emphasized that predictions concerning the appearance of the sky in some exotic regions of the Earth made on the basis of the theoretical work of astronomers had eventually been confirmed by eye witnesses, i.e. travellers who had reached previously unexplored places. He discovered the first two laws of planetary motion and announced them in his book of 1609, *Astronomia nova*. As he informs us in Note 2 of the *Dream*, he also read two works from antiquity about the Moon and voyages to the Moon: Plutarch's *On the Face in the Moon* and *A True Story* by Lucian of Samosata (b. *c.* AD 125, d. after 180). All these texts, written

---

[6] Kepler (1634), note 2, pp. 29–30; Kepler (1993), p. 332; Kepler (1967), p. 32. Rosen's comments on his translation of *Somnium* give many details of how the book came to be written.

[7] Kepler (1941), pp. 297–298; Kepler (1965b), pp. 25–26.

and read by Kepler, left their mark on his *Somnium*, though it is impossible to say which parts of his early student notes on lunar astronomy were used in the story and in what manner.

The next phase of the composition of the book concerns the supplementary notes. We are informed by Kepler about the time when this addition to the *Somnium* was written: the ten-year period between 1620 and 1630. The year 1620, being the time of the trial of Kepler's mother, is the most enigmatic of these years. Kepler attended the trial in person during its most intense phase and was away from Linz from September 1620 to November 1621.[8] Is it possible that he took the text with him and continued to work on *Somnium* which was, as he suspected, at least an indirect cause of the trial? In a letter to Matthias Bernegger (1582–1640) dated 4 December 1623 Kepler wrote: 'Two years ago, soon after my return to Linz I started to rework my *Astronomy of the Moon* or rather explain it with the aid of notes.'[9] This would mean that the work on the notes commenced at the end of 1621, which is further confirmed by Note 8 where he mentioned 'last year's journey'.[10]

Some sporadic references to Kepler's further work on the *Somnium* and on the notes have survived in various sources. In a letter to the Emperor's librarian on 21 April 1622, Kepler requests a copy of the Greek manuscript of Plutarch's *On the Face*: 'I am working on a little book that requires the assistance of Plutarch's *On the Face in the Moon*. The Latin translation by Xylander (1532–1576) is full of gaps and the rest is to a large extent unclear.'[11] The request was not met, and more than a year-and-a-half later, in the letter to Bernegger of 4 December 1623 to which we referred earlier,[12] Kepler complained to Bernegger about the grievous lack of a copy of Plutarch in Greek. In February 1624 Bernegger offered his own Greek copy of *On the Face in the Moon* published by Stephanus in 1572, and it was this copy that Kepler most probably used to work on the *Somnium* and to translate Plutarch's work into Latin. This new translation was appended to the *Somnium* and completed presumably at the end of 1628 or the beginning of 1629 (letter to Bernegger dated 2 March 1629)[13] (that is, when Kepler was already in Żagań).[14] Kepler's comments on the 'Geographical or, if You Prefer, Selenographic Appendix' were also written in Żagań. Furthermore, in Note 179 to the *Dream*, Kepler refers to his reply to Jacob Bartsch (1600–c. 1632), the husband of his daughter Susanna (b. 1602, married 1630, d. 1638), regarding the publication and the calculations in his *Ephemerides* which

---

[8] Rublack (2015).

[9] Kepler to Matthias Bernegger, in 4 December 1623; Kepler (1959), p. 143.

[10] Kepler (1634), p. 32; Kepler (1993), p. 334; Kepler (1967), p. 40.

[11] Kepler to Sebastian Tengnagel, 21 April 1622, in Kepler (1959), pp. 87–88.

[12] See footnote 8 above.

[13] Kepler to Bernegger, 2 March 1629, in Kepler (1959), p. 386.

[14] Known as Sagan in Kepler's day. Żagań was founded by King Bolesław IV in the twelfth century and named after a nearby fort. The Żagań area came under successive Saxon, Hapsburg, and Prussian rule. The town once again became part of Poland at the end of World War II and adopted its original Polish name.

went to press at the beginning of 1629, thereby providing evidence that the 223 notes were put together in Lower Silesia.

Having completed his manuscript in Żagań, Kepler began to print the book using his own printing press located in the cellar of the house rented by the Kepler family. In a letter dated 22 April 1630 Kepler wrote to Philip Müller (1585–1659) of Leipzig:

> My workers were handicapped by my absence. Instead of the *Ephemerides*, they therefore printed the *Astronomy of the Moon* with annotations; six sheets are ready now.[15]

Eventually the *Dream* went to press in Żagań, together with the *Ephemerides* that gave the positions of the planets for the years 1621–1628 and 1629–1636. Bartsch actively participated in the enterprise. It is from him that we learn that Kepler never saw the complete printed version of the *Somnium*. On 8 October 1630 Kepler set out from Żagań on what was to be his final journey. In the words of Bartsch:

> [H]e left me as supervisor and director of the printing operation in Żagań ... Before [the printing of the *Somnium*] was finished down to the very last word, that greatest of men, who devoted himself to the study of the heavens with a zeal that was virtually miraculous, fell into his very last sleep ... In the belief that it would be criminal if the work which had been begun was destroyed, I continued to put ... the *Dream* through the press.[16]

Ultimately it was Bartsch who brought the printing of the *Astronomy of the Moon* to completion along with a new expanded edition of Kepler's logarithmic tables, *Logarithmorum logisticorum heptacosias* (a short work that was published in 1631). The *Dream* remained incomplete, lacking a title page and a letter of dedication. It nevertheless began to circulate in this form across Europe. On 5 May 1631, from Lubań, Bartsch informed Philip Müller that he would send him a copy that had 'lost its freshness after having been read here by so many people,' promising another copy in better condition.[17]

The brief account of the publication of the completed edition of the *Somnium* features in the *Letter of Dedication* written by Kepler's son Ludwig (1607–1663) and placed at the beginning his father's book:

> [M]y stepmother, an impoverished widow with four orphans, comes to me at a turbulent time and in a place [Frankfurt am Main] most inconvenient on account of the high cost of living. She brings with her the unfinished copies of this Dream, and asks for my help. But I myself need the aid and support of others. In particular she wants me to complete the copies of this *Dream* ... I could not refuse this request, but instead I have made it my goal.[18]

As a result, the title page of *Somnium* carried the following information: 'printed in part in Żagań, and completed in Frankfurt ... In the Year of Our Lord 1634' (Fig. 12.1).

---

[15] Kepler to Philipp Müller, 22 April 1630, in Kepler (1959), p. 429. See Baumgardt (1952), p. 186, for an English translation.

[16] Kepler (1967), p. xxi.

[17] Jacob Bartsch to Philip Müller, 5 May 1631, in Kepler (1975), p. 238.

[18] Kepler (1634), Dedicatory letter; Kepler (1993), p. 319; Kepler (1967), pp. 6–7.

Fig. 12.1 The title page of the *Somnium* (1634)

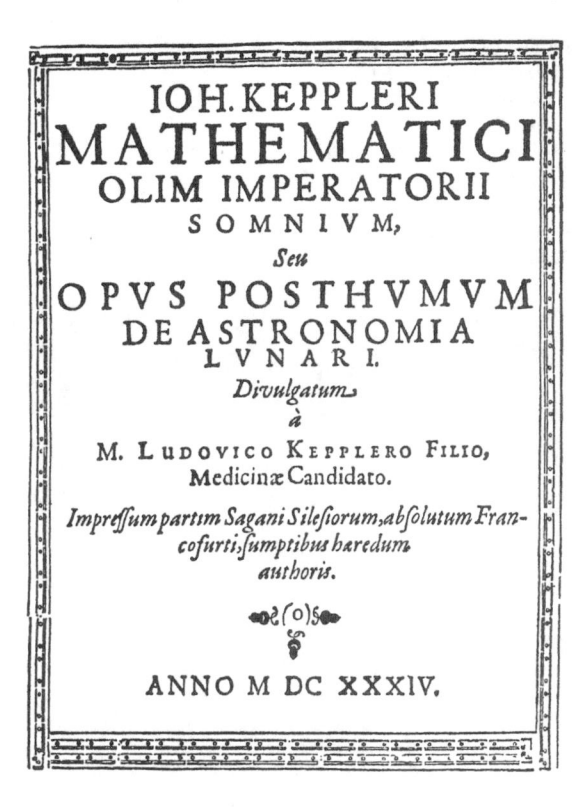

IOH. KEPPLERI
MATHEMATICI
OLIM IMPERATORII

S O M N I V M,

Seu

O P V S   P O S T H V M V M
DE ASTRONOMIA
L V N A R I.

Divulgatum
à

M. Ludovico Kepplero Filio,
Medicinæ Candidato.

Impreſſum partim Sagani Sileſiorum, abſolutum Fran-
cofurti, ſumptibus hæredum
authoris.

ANNO M DC XXXIV.

12.2 Novel Propositions in Regard to the Moon

It seems probable that a substantial, or perhaps even complete, version of the 'Dream, or Lunar Astronomy' was ready in 1609. The same year Kepler published *Astronomia nova*, his work on the orbit of Mars, whose long full title contains a reference to 'celestial physics' (that is, to Kepler's concern with the actual path of the planet in space).[19] Perhaps this concurrence explains why Kepler decided to describe the flight from the Earth to the Moon in practical rather than fantastic terms. Kepler did not propose any specific lunar vehicle—we may suppose he was aware how impossible such enterprise was with the contemporary level of science and technology. He described the flight of the human body in accordance with the physics he knew.[20] According to Kepler, the travellers have to cover the distance separating the Moon from the Earth within four hours, because they are safe as long as they remain in the cone of the Earth's shadow during a lunar eclipse (see Fig. 12.2). The daemons have to push the travellers throughout the duration of the journey since in Kepler's physics

[19] For a discussion of the meaning of the full title of *Astronomia nova* see Gregory, 'The translation of the title of Kepler's *Astronomia nova*' (this volume, Chap. 5).

[20] Kepler (1634), pp. 6–7; Kepler (1993), pp. 323–324; Kepler (1967), pp. 15–17.

Fig. 12.2 A solar eclipse (the shadow of the Moon N is cast on the surface of the Earth) and a lunar eclipse (the Moon P immersed in the shadow of the Earth C). The Sun is located at S. Reproduced from Kepler, J., *Somnium* 1634, p. 51

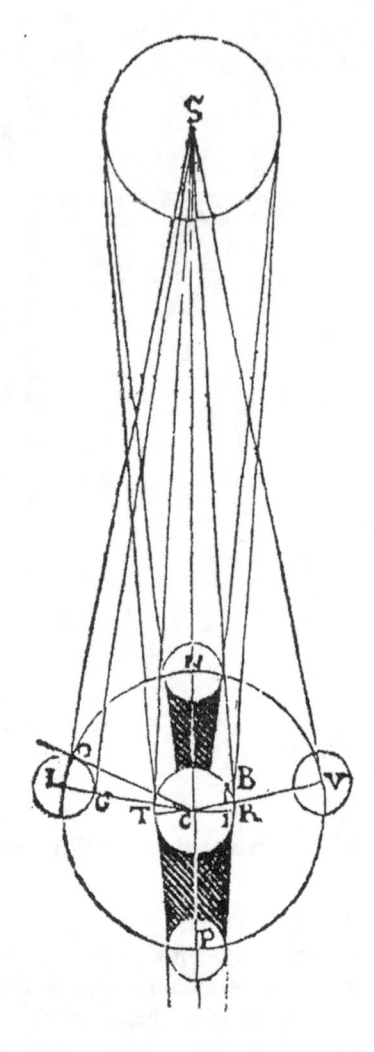

'force' caused motion ('velocity' and 'acceleration' first assume their present-day importance only in the work of Isaac Newton).

To neutralize the effects of a severe shock caused by the flight, daemons served drugs to the travellers. Kepler assumed that the space separating heavenly bodies is devoid of air and terribly cold. 'The cold is relieved by a power which we are born with; the breathing, by applying damp sponges to the nostrils',[21] explained the lunar daemon. Kepler also described the travellers' passage from the gravitational sphere of the Earth, through the liminal state of weightlessness to the gravitation sphere of the Moon—again in accordance with physics as then understood. Such an approach to the subject of cosmic travels made historians of literature hail the *Dream* as the

---

[21] Kepler (1634), p. 16; Kepler (1993), p. 323; Kepler (1967), p. 16.

first science fiction novel, and one that set an example for the whole genre,[22] whereas the visions of voyages to the Moon by earlier authors were relegated to the realm of fantasy.

After safely landing the travellers on the surface of the Moon, Kepler strives to describe the physical conditions prevailing there. As a lunar day lasts half an Earth month and a lunar night is equally long, it must get significantly hotter on the Moon than we experience at the Earth's equator and much colder than at the Earth's poles. Kepler also describes various types of plants and creatures that might live on the Moon, bearing in mind the extreme conditions described above. Both the plants and animal inhabitants of the Moon are first and foremost of gigantic size. The enormous plants are for the most part covered in bark whereas the animals have skin, or something serving the same purpose, that is spongy and porous. In most of the cases the plants start to grow and die on the same day. If anything remains on the surface during the day without taking shelter in the caves its outer layer becomes a hardened and charred crust which falls off when evening comes. The dominant life forms among the Moon's inhabitants are snake-like: they expose themselves to the Sun but only in the vicinity of the mouths of caves so that they can quickly take shelter. An essential feature of the lunar environment ensuring the survival of life are the waters of the Moon's oceans which, however, have to be sought out because they are subject to the powerful effects of the Earth's influence and change location rapidly. Consequently, another general characteristic of the Moon's inhabitants is their ability to survive underwater. In fact, one may risk the statement that in his *Dream* Kepler was a forerunner of the investigation into the adaptation of living organisms to their environment.[23]

The most important part of the *Dream* however, was Kepler's lunar astronomy. As he put it:

> But here take a look at the thesis of this book, and learn that what are for us among the main features of the entire universe: the twelve celestial signs, solstices, equinoxes, tropical years, sidereal years, equator, colures, tropics, arctic circles, and celestial poles, are all restricted to the very tiny terrestrial globe, and exist only in the imagination of the earth-dwellers. Hence, if we transfer the imagination to another sphere, everything must be understood in an altered form.[24]

In the *Dream* Kepler provided a thorough description of the cosmos seen from the Moon. He supplied new definitions of all celestial circles and cycles, taking into account many subtle effects arising from the motion of the Moon. For instance, the line of nodes of the lunar orbit has a retrograde motion with a period of almost 19 years: the point at which the Moon's orbit crosses the ecliptic moves steadily clockwise (viewed from celestial north), closely aligned with Earth's orbital path but in the opposite direction. Kepler saw this phenomenon as an equivalent of the precessional motion of the Earth which results in the motion of equinoxes with a

---

[22] Menzel (1975); Hallyn (1993), pp. 253–280; Poole (2010), pp. 57–69.

[23] The problem has been discussed in the context of other early modern texts in Christie (2019).

[24] Kepler (1634), p. 48; Kepler (1993), p. 345; Kepler (1967), p. 85.

period of about 26,000 years. It allowed him to build an analogy between terrestrial sidereal and tropical years and their lunar equivalents. He wrote:

> For us in one year there are 365 revolutions of the sun, and 366 of the sphere of the fixed stars … Similarly, for them the sun revolves 12 times in one year and the sphere of the fixed stars 13 times … But they are more familiar with the nineteen-year cycle, for in that interval the sun rises 235 times, but the fixed stars 254 times.[25]

These conclusions are further elucidated in note 98:

> We on earth—not people in general, but astronomers—reckon 99 months in 8 years, or 235 months in 19 years. Yet the natural lunations are not as intimately connected with our affairs as are the days and nights. Then what else can we think about the moon-dwellers whom we imagine, if there are any creatures up there capable of counting, than that they adopt those same numbers, since they have no other day? But for them the indication that the period on nineteen years has ended is the rising of the same stars in precisely the previous arrangement.[26]

At this point Kepler proceeds to constructs the following analogy: if on the Earth the relation between the sidereal year and the tropical years is as follows:

$$\text{sidereal year} = (1 + 1/26{,}000) \times \text{tropical year},$$

the Moon-dwellers could observe that:

$$13 \approx (1 + 1/19) \times 12.$$

Describing the apparent motions of the Sun and Earth as seen from the surface of the Moon, Kepler depicted eclipses of the Earth and Sun in astonishing detail which can best exemplified by juxtaposing the picture taken by the NOAA/NASA GOES West satellite (Fig. 12.3) and a corresponding excerpt from Kepler's *Dream*. At this point the narrator is an inhabitant of the Moon and Kepler has him use 'Volva' as the Moon-dwellers' name for the body we call Earth:

> They never see a total eclipse of Volva. However, for them the body of Volva is traversed by a certain small spot which is reddish around the rim and black in the middle. Entering from the eastern side of Volva, it leaves by the western edge, following the same course as the natural spots of Volva, while surpassing them in speed.[27]

In a similarly precise way Kepler reconstructed the course of a solar eclipse as viewed from the Moon (Fig. 12.4):

> Yet among them solar eclipses have the following peculiar feature. Hardly has the sun disappeared behind the body of Volva than, as happens quite frequently, bright light arises on the opposite side. It is as though the sun expanded and embraced the entire body of Volva, whereas at other times the sun appears just as many degrees smaller than Volva.[28]

---

[25] Kepler (1634), pp. 9–10; Kepler (1993), p. 324; Kepler (1967), pp. 17–18.

[26] Kepler (1634), p. 47; Kepler (1993), p. 344; Kepler (1967), pp. 82–83.

[27] Kepler (1634), p. 24; Kepler (1993), p. 329; Kepler (1967), p. 26.

[28] Kepler (1634), p. 24; Kepler (1993), p. 329; Kepler (1967), p. 26.

Fig. 12.3   The moon's shadow gliding across the earth during the total solar eclipse of July 2, 2019, in this photo captured by the NOAA/NASA GOES West satellite. Courtesy of CIRA/NOAA

Fig. 12.4   The Sun eclipsed by the Earth as it would appear from the Moon. Reproduced from J. Nasmyth, J. Carpenter, *The Moon Considered as a Planet, a World, and a Satellite*, 4th edn, London, 1903

This phenomenon was detected for the first time on 24 April 1967 when the unmanned lunar lander *Surveyor 3* recorded, from a crater in Mare Cognitum, the Earth eclipsing the Sun. (Figure 12.4 shows a nineteenth century artist's impression of such an event.)

Kepler not only created general descriptions of lunar phenomena, he went as far as giving numerical values, especially in his Notes. Thus in Note 186 he wrote:

> For the disk of Volva has a radius varying between $63'41''$ and $58'22''$. But the moon's shadow, which causes the eclipse of the Volva for the moon-dwellers, on account of the sun's size narrows down from the moon to the disk of Volva, so that its radius never exceeds $1'22''$.[29]

Naturally, these calculations were in agreement with the dimensions of Kepler's cosmos, in which the distance from the Earth to the Moon was known with reasonable accuracy (54 Earth radii), but the solar distance (3469 Earth radii) was seven times smaller than in reality, and the radius of the Sun was thought to be 15 Earth radii.

The imaginary voyage to the Moon proved instructive for Kepler himself: he realized it would be necessary to accept the Moon's rotation on its own axis while remaining at absolute rest with respect to the fixed stars. In the fourth book of his *Epitome of Copernican Astronomy* (*Epitome astronomiae Copernicanae*), published in 1620, Kepler clearly rules out the possibility that the Moon is rotating with respect to the fixed stars. His argument is teleological and based on the Aristotelian assumptions in his physics whereby the rotation of the central body is necessary to sustain the motion of the bodies revolving around it:

> [T]hat the moon in turn does not wheel around the axis of its own body is argued by the spots. But why is this so? If not because no further planet is seen to go around the moon. Accordingly, the moon has no planet to which it gives movement by the rotation of its body. Accordingly, in the moon, the rotation was left out, as being superfluous.[30]

However, in the notes to the first part of the *Dream*, which, as we have seen were probably written between 1620 and 1630, Kepler makes it absolutely clear that he is describing the cosmos as seen from the rotating Moon. For instance, in Note 128 he states:

> A revolution of the moon's body occurs in the period of a month. For throughout its entire course it turns the same face toward the earth, as we know from the unchanging permanence of its spots. But the earth, that is Volva, seems to traverse the entire zodiac in the period of a month. The face of the moon also travels along with the earth, turning at one time toward the Crab, and at another time toward the Goat, which is the opposite sign. In other words, the moon revolves.[31]

In the seventeenth century this was hardly a trivial answer[32]: the rotation of the Moon would be rediscovered by Isaac Newton in the 1670s to explain the lunar libration in longitude.[33]

---

[29] Kepler (1634), p. 66; Kepler (1993), p. 358; Kepler (1967), p. 118.

[30] Kepler (1991), p. 319; Kepler (1952), p. 920.

[31] Kepler (1634), pp. 54–55; Kepler (1993), p. 350; Kepler (1967), p. 96.

[32] For more on this subject, see Włodarczyk (2009).

[33] Gabbey (1991); Włodarczyk (2011).

12.3 Lunar Astronomy in Context

To what extent is Kepler's *Somnium* a unique work? This seems to be yet another riddle left for us to resolve. To answer this question, scholars habitually focus on the plot and compare it with other literary accounts of fictional travels to the Moon and some fantastic imagery of lunar vegetation, wildlife and Selenites which preceded *Somnium*.[34] We have previously note that Kepler readily was familiar with the two famous Hellenistic texts: Plutarch's *On the Face in the Moon* and the *True Histories* by Lucian of Samosata. The work by Plutarch is not only a rich source of information on ancient ideas about the Moon, it also features a cosmic voyage: an essentially Platonic lunar metempsychosis. The other text is a fable. Descriptions of imaginary journeys beyond the Earth occur in works written before Kepler's day. For instance, scenes set on the Moon play a part in the epic poem *Orlando Furioso* (1516) by the Ferrarese court poet Ludovico Ariosto (1474–1533).

However, Kepler's voyage to the moon, even though described in a novel manner, is a mere pretext to look at the Universe from somewhere that is not on the surface of the Earth. Here some literary predecessors can be named too. One of the best known is *Scipio's Dream* (*Somnium Scipionis*) by Marcus Tullius Cicero (100–43 BC). In this story Scipio Africanus the Younger (185/4–129 BC), in his dream, marvels at the geocentric cosmos. It is possible that Cicero drew his inspiration from the myth of Er, recounted by Plato (428/427 or 424/423–348/347) at the end of Book 10 of his dialogue *The Republic*. There is a Christian example in *The Divine Comedy* of Dante Alighieri (1265–1321), which presents a journey through the cosmos, ending (in *Paradise*) with an ascent through the heavenly spheres, defined in moral—rather than astronomical—terms, upwards towards the light of God (that is, towards Salvation). It is difficult to work out even the order of the planets from the description given after his guide (Beatrice) tells Dante to look back as they move beyond the planets in the Empyrean.[35] A common feature of these Pagan and Christian visions is a divine perspective which makes it possible to contemplate the entirety of the cosmos portrayed as somewhat resembling a geocentric armillary sphere.

What Kepler proposed in *Somnium* is entirely different. He endeavours to make his readers aware that whatever heavenly body they find themselves on, this body will for them become the centre of the Universe, and that they will see a complete range of celestial phenomena corresponding to those we observe in the sky as seen from the Earth. If Kepler's readers grasp the fact that the inhabitants of the Moon could deduce from their observations that the Moon is the centre of the world, and that the Sun and the planets go round it, they would also be ready to go beyond the traditional mode of thought and embrace Copernicus' theory. A similar argument was used by Copernicus in Book I of *De revolutionibus*:

---

[34] Cf. Nicolson (1948); Montgomery (1999); Parrett (2004); Lambert (2002); Cressy (2006), reprinted in Cressy (2016); Aït-Touati (2011).

[35] Dante, *Divina Commedia, Paradiso*, canto 22, Beatrice II. 124–132; planetary spheres II. 133–150.

This situation closely resembles what Vergil's Aeneas says: 'Forth from the harbor we sail, and the land and the cities slip backward' [*Aeneid*, III, 72]. For when a ship is floating calmly along, the sailors see its motion mirrored in everything outside, while on the other hand they suppose that they are stationary, together with everything on board. In the same way, the motion of the earth can unquestionably produce the impression that the entire universe is rotating.[36]

If this were indeed the true purpose of Kepler's lunar astronomy, could it not be possible that others, following Copernicus' insight, might arrive at a similar or the very same idea? Kepler certainly does not seem to be alone in asking these questions.

Let us recall that the concept of writing about astronomy as seen from the Moon originated in Tübingen, where Copernicus' heliocentric astronomy was first introduced to Kepler by Michael Maestlin. Maestlin was an early Copernican and his observations of the new star of 1572 and of the comets contributed to the demise of Aristotelian cosmology.[37] Further, Maestlin in his treatise *A Disputation about Eclipses of the Sun and Moon* (*Disputatio de eclipsibus solis et lunae*, 1596) included a highly suggestive passage, cited *in extenso* by Kepler in his *Ad Vitellionem paralipomena* (1604):

> We therefore say that the earth, by its gleaming light, sent to it from the sun, casts its rays on the opacity or night in the lunar body no less than, in turn (in exactly the same way) the full moon illuminates our nights on earth with its rays received from the sun, and turns them almost to day in proportion to their brightness. It does this with all the greater clarity in proportion to the earth's circle's being greater than the moon's circle. And the ratio of the one to the other is greater than twelve times.[38]

The remarkable theatricality of this description—the observer is placed on the surface of the Moon illuminated by the light of the Sun refracted by the Earth's atmosphere—is enhanced by pointing out that the lunar disc of the Earth seen in the lunar sky is twelve times bigger than the lunar disc observed in our sky (this refers to the area of the disc; the diameter of the Earth seen from the Moon is almost four times bigger than the angular diameter of the Moon seen in our sky). In fact, these descriptions can be interpreted as elements of lunar astronomy based on astronomical calculations. A similar line of reasoning was pursued by Kepler in the *Somnium*.

Giordano Bruno (1548–1600) wrote seven dialogues in Italian that were published in London in 1584. Two of these dialogues, *De l'infinito universo e mondi* (*On the Infinite, Universe and Worlds*) and *La cena de le Ceneri* (*The Ash Wednesday Supper*), featured the motif of the observation of the sky from the Moon, with more extensive discussion in the first dialogue. Bruno began with a general argument for the translation of an observer from one heavenly body to another when he introduced his idea of a homogeneous universe. He next described the Earth as seen from the other vantage points in the cosmos, as well as from the Moon:

> Now just as from our earth (itself a moon) the diverse parts of the moon appear some more and some less bright – so from the moon (itself another earth) can the diverse parts of this

[36] Copernicus (1978), p. 16.

[37] Westman (1972).

[38] Kepler (1604), Chap. 6, p. 255; Kepler (1939), p. 223; Kepler (2000), p. 266.

earth be distinguished by the variety and difference of the portions of her surface. Moreover just as, if the moon were at a greater distance from us, then the diameter of the opaque parts would fail, while the bright parts would tend to unite for us and shrink in our view, giving us the impression of a smaller body of uniform brightness, similar also would be the appearance of our earth as seen from the moon if the distance between them were greater.[39]

Some new elements of lunar astronomy appeared in Bruno's poem *Of Innumerable Things, Vastness and the Unrepresentable* (*De innumerabilibus, immenso et infigurabili*) published in Frankfurt in 1591.[40] Bruno repeated his remarks about the changing appearance of the Earth as seen from the Moon. These ideas were supplemented by an insightful remark concerning the symmetry of the phenomena of eclipses, reflection of light from the Sun and phases within the Earth–Moon arrangement:

> … and when there is an eclipse of the sun on the earth those who live on the moon see an eclipse of the earth, and the earth shines like the face of the moon, returning the rays received from the sun, by the glassy body of the sea into the opposite direction, as the moon does willingly to the earth at night … And from this place you would see how the earth circles and changes [its position] towards the sun, just as from our place you see how the moon changes.[41]

Bruno's account of astronomy as viewed from the Moon limited to fairly basic phenomena, introduced by rule of analogy with the phenomena seen from the Earth, and it is devoid of technical insight. Furthermore, in Bruno's philosophy, these images are merely part of a more general idea of a homogeneous universe, in which the observer placed on any heavenly body will experience the same phenomena as an observer on the Earth.

An even more indefinite reference to lunar astronomy can be found in the treatise *Nova de universis philosophia* by Francesco Patrizi of Cherso (1529–1597), a leading critic of Aristotelianism. His *The New Universal Philosophy* was published in 1591, the same year as *De immenso* by Bruno, and soon afterwards with a spurious retrodating of 1593. In Chap. 20, devoted to the Moon, Patrizi recalls ancient writers and philosophers who referred to the Moon as a second Earth and postulated the existence of cities and other artificial structures on its surface. Patrizi himself indeed thinks that the Moon is a second Earth but is composed of ether; that both bodies have light and dark spots, but one is suspended in the air and the other in ether. He believes that the view of the Moon seen from the Earth would have to be similar to the view of the Earth from the Moon.[42] However, he does not mention observing other heavenly bodies from the Moon, which is the key element of lunar astronomy.

A highly original lunar astronomy was invented by an English astrologer and physician, Edward Gresham (1565–1613), an ardent follower of heliocentric astronomy.[43] Gresham introduced his lunar astronomy in an unpublished treatise

---

[39] Singer (1950), pp. 312–313. Cf. also Tessicini (2002); Fabbri (2016).

[40] For summary cf. Michel (1973), pp. 185–186.

[41] Bruno (1879), pp. 342–343.

[42] Patrizi (1593), ff. 112r–114r. Cf. also Fabbri (2016), pp. 138–139.

[43] Recently there have been several studies of this figure, hitherto absent from the history of astronomy: Włodarczyk et al. (2018); Włodarczyk (2020, 2022).

Fig. 12.5 The terrestrial
globe in *Astronomiae* by
Maestlin (Tübingen, 1588).
Courtesy of A. K.
Wróblewski

entitled *Astrostereon or the Discourse of the Falling of the Planet* completed in
London on 1 September 1603. In the section with the description of what can be
observed from the lunar surface he first discusses the most obvious phenomena which
are easiest to imagine. The hypothetical observer on the Moon would see some spots
on the Earth indicating the presence of continents and oceans (see Fig. 12.5). The
Earth would go through a cycle of phases like those of the Moon as seen in our sky,
maintaining appropriate symmetry: the new Earth would be visible from the lunar
hemisphere facing the Earth which in turn can be seen in our sky at full Moon, and
so on. Naturally, such symmetry would apply to eclipses: '[T]he Sunne eclipsed to
us is the Moone eclipsed to them, and the Moone eclipsed to us is the Sunne eclipsed
to them'[44].

Gresham next deals with the movements of the Earth and the planets in the lunar
sky. He states that the values of the mean motion of the planets in longitude (when
observed from the Moon) will be the same as those observed from the Earth. (He also
relocates the observers from the Earth to Mars, Jupiter and Saturn, all to convince
them that the mean motion of each of these planets in the sky as seen from the
Earth will be equal to the mean motion of the Earth in their skies.) On the whole,
his descriptions are more general than those given by Kepler in the *Somnium*, but
he proves to be more precise in his understanding of one important detail. Gresham
is aware that even though the Earth seen from the Moon will perform the same
daily sidereal motion in longitude, just as the Moon in our sky (in Gresham's view
approximately 13°10'), it will remain motionless in the lunar sky. He also offers the
correct explanation of the phenomenon: the Moon turns around its own axis at the
same speed of 13°10' per day. As noted previously, Kepler was aware of the fact
that the Earth appears motionless in the lunar sky, but he did not at first see it as a
consequence of the rotation of the Moon. As demonstrated earlier, the rotation of

---

[44] Gresham (1603).

the Moon around its own axis was introduced to Kepler's astronomy when he was adding notes to the *Somnium* in the years 1620–1630.

The *Astrostereon* was not printed, and it is extant in five manuscript versions. We still know very little about the early reception of this treatise or its circulation in manuscript. Gresham himself mentions the *Astrostereon* in his almanac for 1607. In fact, Gresham was an acknowledged author of almanacs which he published regularly in the years 1603–1607. And it is precisely in this genre of popular literature which—at a time when Kepler composed his first version of the *Dream*—that lunar astronomy began to be disseminated in English. The word 'dissemination' appears truly appropriate as almanacs were probably the most popular printed material at the time.[45]

Thomas Bretnor (1570/71–1618), another well-known London almanac maker, like Gresham, was in sympathy with heliocentric astronomy, which was becoming popular in England at the time,[46] since he refers to the movement of the Earth around the Sun. In the 1610 almanac, Bretnor went a step further, informing his readers that the eclipses observed from the Earth have their counterparts in the lunar sky. He played with similar switching of astronomical sites for astronomical observation with regard to eclipses in his almanacs for 1613, 1614, 1616 and 1617. This is how he described the 1616 eclipses on the Moon:

> Twise this yeare will most of the Selenian inhabitants, and twise also will some part of our earthy-Orbists, bee depriued of all, or most part of the Sun his sole-heating light; which howsoeuer it may seeme strange and incredible to my friend M. *Dancy*, and others of his ranke and skill (who with S. *Thomas* will beleue nothing but what they see and feele) …[47]

We cannot identify a direct source for Bretnor's discussion of eclipses seen from the Moon. Gresham's *Astrostereon* appears to be a likely candidate though not a unique one. There was yet another text written in London at that time which discussed the idea of the symmetry of eclipses. William Gilbert (1544–1603), the Queen's physician and author of the renowned treatise *On the Loadstone* (*De magnete*, London 1603) and of the first sketch of the moon drawn in about 1600, died in November 1603, leaving in manuscript his treatise *A New Philosophy of Our Sublunary World* (*De mundo nostro sublunari philosophia nova*).[48] The manuscript contained the first map of the Moon to which we referred earlier. However, in Gilbert's *De mundo* one can also find a passage describing the observations made from the lunar surface:

> … the eclipse of the moon which we see is the eclipse of the sun on the moon, and the eclipse of the sun which we see is the eclipse of the earth for the inhabitants of the moon; hence the earth is deprived of light likewise the moon is deprived.[49]

A great majority, if not all of the above-mentioned authors (the only exception is probably Patrizi) who employed lunar astronomy in their texts, both knew and

---

[45] Capp (2008); Cassel (2011).

[46] Johnson (1937).

[47] Bretnor (1616), f. C2r. There is a recent reprint (Bretnor, 2010).

[48] Kelly (1960, 1965).

[49] Gilbert (1651), p. 174.

embraced heliocentric astronomy. Kepler was not the only author who at the turn of seventeenth century would move his observer to the lunar surface to expose the absurdity of the arguments made by the adherents of the geocentric cosmos. Little can be said about possible interactions between the authors writing about lunar astronomy. Kepler knew the works by Bruno and Patrizi,[50] but we do not know if some brief references in these works referring to the universe seen from the moon influenced his lunar astronomy. The English variant of lunar astronomy appears to be an independent phenomenon spread via an astonishing transmission channel of astrological almanacs published not in Latin but in the vernacular. We are not currently in a position to assess the impact of lunar astronomy, featured in both scholarly texts and popular literature, on widening the circle of the adherents to the new model of the universe. In any case, Kepler's *Dream* appears to be the most sophisticated version of lunar astronomy. Given the extraordinary rigour of Kepler's analysis, the *Dream* is far from being a simple experiment in thought. His description of the cosmos as seen from the Moon is still illuminating today. I strongly believe that Kepler's *Dream* should be accepted as a recommended reading for future travellers to the Moon.

References

Aït-Touati, F. (2011). *Fiction of the cosmos: Science and literature in the seventeenth century.* Chicago and London: University of Chicago Press.

Baumgardt, C. (1952). *Johannes Kepler: Life and letters.* London: Victor Gollancz.

Bretnor, T. (1616). *A new almanacke and prognostication, made for the yeare of our lord God, 1616.* London.

Bretnor, T. (2010). *A new almanacke and prognostication, made for the yeare of our lord God, 1616 (Early history of astronomy and space).* Eebo Editions.

Bruno, G. (1879). *Jordani Bruni Nolani opera latine conscripta* (Vol. I.1). Naples: Domenico Morano.

Capp, B. (2008). *Astrology and popular press: English almanacs 1500–1800.* London: Faber and Faber.

Cassel, L. (2011). Almanacs and prognostication. In J. Raymond (Ed.), *The Oxford history of popular print culture: Volume one: Cheap print in Britain and Ireland to 1660* (pp. 431–442). Oxford: Oxford University Press.

Christie, J. E. (2019). *From influence to inhabitation: The transformation of astrobiology in early modern period.* London: Cham.

Copernicus, N. (1978). *On the revolution* (J. Dobrzycki, ed. & E. Rosen, trans.). Warsaw and Cracow: Johns Hopkins University Press.

Cressy, D. (2006). Early modern space travel and the English man in the moon. *American Historical Review, 111,* 961–982.

Cressy, D. (2016). Early modern space travel and the English man in the moon (reprint). In J. A. Hayden (Ed.), *Literature in the age of celestial discovery: From Copernicus to Flamsteed.* London: Palgrave Macmillan.

Fabbri, N. (2016). Looking at an earth-like moon and living on a moon-like earth in renaissance and early-modern thought. In C. Muratori & G. Paganini (Eds.), *Early modern philosophers and the renaissance legacy* (pp. 135–151). London: Cham.

---

[50] Włodarczyk (2022).

Gabbey, A. (1991). Innovation and continuity in the history of astronomy: The case of the rotating moon. In P. Barker & R. Ariew (Eds.), *Revolution and continuity: Essays in the history and philosophy of early modern science* (pp. 95–129).

Gilbert, W. (1651). *De mundo nostro sublunari philosophia nova.* Amsterdam: Ludovicus Elzevier.

Grafton, A. (1991). Humanism and science in Rudolphine Prague: Kepler in context. In J. A. Parente Jr., et al. (Eds.), *Literary culture in the Holy Roman Empire, 1555–1720.* Chapel Hill: The University of North Carolina Press.

Gresham, E. (1603). *Astrostereon,* MS Sloane 3936, f. 13v. British Library. London.

Hallyn, F. (1993). *The poetic structure of the world: Copernicus and Kepler.* New York: Zone Books.

Johnson, F. R. (1937). *Astronomical thought in Renaissance England: A study of the English scientific writings from 1500 to 1645.* Baltimore: Johns Hopkins University Press.

Kelly, O. S. B. (1960). The De mundo nostro sublunari of William Gilbert. *Proceedings of the Oklahoma Academy of Science, 40,* 88–91.

Kelly, O. S. B. (1965). *The 'De mundo' of William Gilbert.* Amsterdam.

Kepler, J. (1604). *Ad Vitellionem paralipomena quibis astronomiae pars optica traditur.* Frankfurt: Claudius Marnus and the heirs of Johannes Aubrius.

Kepler, J. (1634). In L. Kepler (Ed.), *Somnium seu opus posthumum de astronomia lunari.* Sagan in Silesia and Frankfurt: The heirs of the author.

Kepler, J. (1939). In F. Hammer (Ed.), *Johannes Kepler Gesammelte Werke, Band. II: Astronomiae pars optica.* Munich: C. H. Beck.

Kepler, J. (1941). In F. Hammer (Ed.), *Johannes Kepler Gesammelte Werke, Band IV: Kleinere Schriften 1602–1611/Dioptrice.* Munich: C. H. Beck.

Kepler, J. (1952). Epitome of Copernican astronomy. In *Great books of the western world* (C. G. Wallis, trans., Vol. 15). Chicago: University of Chicago.

Kepler, J. (1959). In W. von Dyck & M. Caspar (Eds.), *Johannes Kepler Gesammelte Werke. Band XVIII: Briefe 1620–1630.* Munich: C. H. Beck.

Kepler, J. (1965a). *Kepler's dream.* Translated by P. F. Kirkwood, with commentary by J. Lear. Berkeley and Los Angeles: University of California Press.

Kepler, J. (1965b). *Kepler's conversation with Galileo's sidereal messenger.* Translated from the Latin with and introduction and notes by E. Rosen. New York: Johnson Reprint Corp.

Kepler, J. (1967). *Kepler's Somnium.* Translated with a commentary by E. Rosen. Madison and London: University of Wisconsin Press.

Kepler, J. (1975). In M. List (Ed.), *Johannes Kepler Gesammelte Werke, Band XIX: Documente zu Leben und Werk.* Munich: C. H. Beck.

Kepler, J. (1991). In M. Caspar & Kepler-Kommission (Eds.), *Johannes Kepler Gesammelte Werke. Band VII: Epitome astronomiae Copernicanae.* Munich: C. H. Beck

Kepler, J. (1993). In V. Bialas & H. Grössing (Eds.), *Johannes Kepler Gesammelte Werke, Band XI, 2: Calendaria et prognostica/Astronomica minora/Somnium.* Munich: C. H. Beck.

Kepler, J. (2000). *Optics: Paralipomena to Witelo & optical part of astronomy.* Translated from the Latin by W. H. Donahue. Santa Fe: Green Lion Press.

Lambert, L. B. (2002). *Imagining the unimaginable: The poetics of early modern astronomy.* Amsterdam and New York: Editions Rodopi B.V.

Menzel, D. H. (1975). Kepler's place in science fiction. *Vistas in Astronomy, 18,* 895–904.

Michel, P.-H. (1973). *The cosmology of Giordano Bruno.* Paris: Hermann, Methuen and Cornell University Press.

Montgomery, S. L. (1999). *The moon and the western imagination.* Tucson: University of Arizona Press.

Nicolson, M. H. (1948). *Voyages to the moon.* New York: Macmillan.

Parrett, A. (2004). *The translunar narrative in the western tradition.* Aldershot: Ashgate Publishing Limited.

Patrizi, F. (1593). *Nova de universi philosophia.* Venice: Robertus Merettus.

Poole, W. (2010). Kepler's Somnium and Francis Godwin's the man in the moone: Births of science-fiction 1593–1638. In C. Houston (Ed.), *New worlds reflected: Travel and utopia in the early modern period*. Farnham and Burlington: Routledge.

Rublack, U. (2015). *The astronomer and the witch: Johannes Kepler's fight for his mother*. Oxford: Oxford University Press.

Singer, D. W. (1950). *Giordano Bruno: His life and thought. With annotated translation of his work "On the infinite universe and worlds"*. New York: Schuman.

Tessicini, D. (2002). "Pianeti consorti": La Terra e la Luna nel diagramma eliocentrico di Giordano Bruno. In M. Á. Granada (Ed.), *Cosmología, teología y religion en la obra y en el proceso de Giordano Bruno* (pp. 159–188). Barcelona: Universidad de Barcelona Publicaciones.

Westman, R. S. (1972). The comet and the cosmos: Kepler, Mästlin and the Copernican hypothesis. *Studia Copernicana, 5*, 7–30.

Włodarczyk, J. (2009). Kepler's moon. In R. L. Kremer & J. Włodarczyk (Eds.), *Johannes Kepler: From Tübingen to Żagań. Studia Copernicana, 42*, 119–129.

Włodarczyk, J. (2011). Libration of the moon, Hevelius's theory, and its early reception in England. *Journal for the History of Astronomy, 42*, 495–519.

Włodarczyk, J. (2020). The pre-telescopic observations of the moon in early seventeenth-century London: The case of Edward Gresham (1565–1613). *Notes and Records: The Royal Society Journal of the History of Science, 74*, 35–53.

Włodarczyk, J. (2022). "Out of a greate laborinth of errors": Lunar astronomy in London before Kepler. *Notes and Records: The Royal Society Journal of the History of Science, 76*, 371–386.

Włodarczyk, J., Kremer, R. L., & Hughes, H. C. (2018). Edward Gresham, Copernican cosmology, and planetary occultations in pre-telescopic astronomy. *Journal for the History of Astronomy, 49*, 269–305.

Chapter 13
On Translating Kepler

W. H. Donahue

13.1 Issues of Translation

Scientific writing in general relies upon precision and accuracy of expression. There is little or no room for ambiguity, suggestion, irony, or other literary devices with which other forms of prose abound. It might therefore be expected that scientific translation would require no more than a corresponding precision and accuracy in the translated text. Although this may be true of present-day scientific works, the issues involved in translating early science are much less clear. Precision may be lacking in the original, and accuracy is difficult or impossible where the matter being described has no counterpart in today's world. Rhetorical effects are likely to play an important role in the author's intention. A translator of Kepler must pay close attention to these and other matters; specific instances will be brought out in what follows.

There is, however, another dimension to the translation process, namely, the treatment of elements that are not strictly textual. In translating Kepler, two such elements are the mathematics and the diagrams or illustrations. It might appear that these are not really elements of translation, but of conceptual or visual content that needs only be re-presented correctly. Yet it is a very common practice to 'translate', that is, to 'carry over', these elements into a different idiom when presenting them to a modern readership. The geometrical methods and styles of the sixteenth century are seen as archaic and difficult for modern readers, and algebraic expressions and symbolism may be substituted. Mathematical language has changed considerably over the past four centuries, and typographical standards have evolved along with the language. Therefore, the translator, who today is often also the typesetter, must decide how much modification is appropriate. Further, in regard to visual elements, certain features of diagrams or illustrations may seem to a modern editor to be mere

W. H. Donahue (✉)
St. John's College, Santa Fe, NM, USA
e-mail: william.donahue@sjc.edu

A. E. L. Davis et al. (eds.), *Reading the Mind of God*, Springer Praxis Books,
https://doi.org/10.1007/978-94-024-2250-4_13

matters of style and ornament, or confusing, or even erroneous, and are thus often altered or eliminated in a new edition or translation. These issues are discussed at greater length below.[1,2,3,4]

13.2   Kepler Translations and Their Intended Audiences

Until relatively recently, little of Kepler's writing had been translated from the original languages (usually Latin). The earliest English translation of any of his work appears to be a selection from an astrological prediction, appended to a book addressing certain recently published prophecies.[5] In 1661, Thomas Salusbury (1564–1586) included in his *Mathematical Collections* a short extract from Kepler's preface to *Astronomia Nova*, responding to scriptural objections to the earth's motion.[6] The earliest complete work to appear in English translation, according to the *Bibliographia Kepleriana*, is an astrological work: *Concerning the More Certain Fundamentals of Astrology*.[7] It is not difficult to see why earlier translators passed over astronomical and mathematical works in favour of astrological and theological writings: Kepler suffered the historical misfortune of having his more influential scientific achievements mistakenly subsumed under, and therefore replaced by, the epoch-making *Principia* of Isaac Newton and the *Dioptrique* of Descartes, while astrology was increasingly marginalized and ossified. Thus, ironically, it was Kepler's astrological work that was seen as meriting serious study, but mainly by astrologers. Moreover, those capable of reading Kepler's technical works in the seventeenth, eighteenth, and even the early nineteenth centuries would have learned Latin in school as a matter of course. Indeed, a book on the Gregorian calendar that Kepler originally wrote in German was published in Latin translation in 1726, presumably to make it more accessible.[8]

The turn of the nineteenth century saw a trend towards translating Kepler's letters and nontechnical papers into German, apparently due to their philosophical and historical interest for a more general readership.[9] This unsystematic mining of the Kepler manuscripts continued throughout the century. During this same period, a group of German philosophers and literati (which included the philosopher F. W. J. Schelling, 1775–1854) became interested in Kepler's ideas as a romantic and

---

[1] Small (1804, 1963).

[2] Kepler (1609, 1937).

[3] Davis (2003), p. 357.

[4] Kepler (1609); Kepler (1937), pp. 36–55; Kepler (2015), p. 41.

[5] Lilly (1644).

[6] Salusbury (1661), vol. 1, pp. 461–467.

[7] Kepler (1942).

[8] Kepler (1726); Caspar (1968), no. 105.

[9] The *Bibliographia Kepleriana* lists at least thirty such translations, many in journal articles or as parts of books. See Caspar (1968).

idealist alternative to the mechanistic physics of the day.[10] Although this movement eventually led to Christian Frisch's edition of Kepler's works,[11] the writings that emerged from the circle itself were academic in nature, and no need was felt to present Kepler in the vernacular. In fact, Frisch's introductions and notes, as well as his biography of Kepler, were in Latin, which was evidently thought to be well enough known among likely readers to obviate the need for translation. Ironically, by the time this edition was completed in 1871, three hundred years after Kepler's birth, the view of Kepler as a serious natural philosopher whose ideas could still form the basis for an alternative physics appears to have waned.

English translations of more substantial selections from Kepler's writings began to appear in the late nineteenth century. Part of the preface to the *Dioptrice* (1611) was included in a translation of Galileo's *Sidereus nuncius,* published in 1880.[12] Some of Kepler's early letters were published in English in the American journal *Sidereal Messenger* under the editorship of W. W. Payne in 1887.[13] An astronomical source book edited by Harlow Shapley and Helen E. Howarth in 1929 contained an article by John H. Walden on 'The Discovery of the Laws of Planetary Motion', which included translated selections from Kepler's *Harmonice mundi* (1619).[14] And the first English translation of a complete work (*De Fundamentis astrologiae certioribus,* 1601) came out in 1942,[15] although by this time large portions of the *Epitome* and the *Harmonice mundi* had already appeared in a small mimeographed edition at St. John's College, Annapolis, for use in their Great Books curriculum (see below).

From the mid-nineteenth century through the middle of the twentieth century, appreciation of Kepler's achievement by English speaking readers was skewed by the manner in which he was portrayed by biographers such as Arthur Koestler, Agnes M. Clerke, and others.[16] They tended to view him as a speculative mystic with a fondness for tedious computation who (thanks to Tycho Brahe's superb observations) 'sleepwalked' his way to discovering the laws of planetary motion.[17] Especially striking is Berry's (1898) remark: 'As one reads chapter after chapter without a lucid still less a correct idea, it is impossible to refrain from regrets that the intelligence of Kepler should have been so wasted, and it is difficult not to suspect at times that some of the valuable results which lie imbedded in this great mass of tedious speculation were arrived at by a mere accident.'[18] Such portrayals suggested that his works might make interesting reading, but would hold forth little promise of gain in understanding the process of scientific discovery. Serious scholars might well have

---

[10] This movement is described in Ziche and Rezvykh (2013).

[11] Kepler (1858–1871).

[12] Galilei and Kepler (1880); Caspar (1942, 1968), no. 136.

[13] Galilei and Kepler (1887), pp. 109–112, 133–138, and 212–217; Caspar (1942, 1968), no. 140.

[14] Shapley and Howarth (1929), pp. 30–40; Caspar (1942, 1968), no. 159A.

[15] Kepler (1942).

[16] See Clerke (1910–1911), article on Kepler; Koestler (1959) and other authors; Berry (1898), p. 197.

[17] Clerke (1910–1911), article on Kepler; Koestler (1959) and other authors; Berry (1898), p. 197.

[18] Berry (1898), p. 197.

expected to learn more from reading Robert Small's retelling of *Astronomia nova*, which 'translated' (that is, reinterpreted) Kepler's great work by clearing away most of the erroneous physics and embarrassing Neoplatonic rhapsodizing in an attempt to make it comprehensible and acceptable to contemporary readers.[19] As late as the 1960s Kepler was still viewed as something of a curiosity rather than a serious scientist, as illustrated by the appearance of translations of two minor works (one of them, the *Somnium*, by two different translators).[20] These works, though by no means unworthy of English versions, represent a more whimsical and imaginative side of Kepler, with little hint of his true intellectual power. It was not until important reassessments by such scholars as Caspar, Koyré, Gingerich and Curtis Wilson in the middle decades of the last century that Kepler began to be appreciated as a powerful and subtle natural philosopher whose achievements could not be appreciated without reading them in context. The quadricentennial of Kepler's birth, in 1971, helped spark an increase in serious studies of Kepler,[21] and since then Kepler has come to be more widely recognized as an author whose significant achievements in astronomy, optics, mathematics and other areas make interesting and instructive reading in their own right, and not merely as footnotes to later works. The idea that Kepler's works might actually be worth reading naturally led to an interest in translating them. Caspar led the way with his German translations of *Mysterium cosmographicum, Astronomia nova,* and *Harmonice mundi*[22] followed by Jean Peyroux, whose French translation of *Astronomia nova* appeared in 1979.[23] In the interim, translations of selections from the *Epitome of Copernican Astronomy* and *The Harmonies of the World*, done by C. G. Wallis, were included in Volume 16 of Encyclopaedia Britannica's *Great Books of the Western World* (for more on these selections, see below).[24] A somewhat uneven but still useful English translation of the *Mysterium cosmographicum* by A. M. Duncan, with the Latin text on facing pages and notes by E. J. Aiton, came out in 1981,[25] followed in 1984 by a reliable French translation, with very extensive and valuable notes, by Alain Segonds.[26] The English translation of *Astronomia nova* (my own) was finally published in 1992.[27] *The Harmony of the World,* which was the culmination of Kepler's cosmology, came out in English in a fine edition in 1997, through the collaboration of E. J. Aiton, A. M. Duncan and J. V. Field.[28]

Meanwhile, interest in Kepler's innovative approach in both astrology and optics was met with English translations of two important works: *On the More Certain*

---

[19] Small (1804, 1963).

[20] Kepler (1965, 1966, 1967).

[21] See, for example, Krafft et al., (1973).

[22] Kepler (1923, 1929, 1939a).

[23] Kepler (1979a).

[24] Hutchins (1952), pp. 839–1085.

[25] Kepler (1981).

[26] Kepler (1984).

[27] Kepler (1992) This translation has recently been completely revised and reformatted, and provided with an index and appendices (Kepler, 2015).

[28] Kepler (1997).

Fundamentals of Astrology (translated by Mary Ann Rossi in 1979, and later by J. V. Field in 1984), and *Optics* (my translation, in 2000).[29] While the appearance of a multitude of translations over the last few decades bespeaks an increasing interest in reading Kepler's own writings, there are many important works that are still available only in Latin (or in Kepler's German). Among these are such astronomical works as *De stella nova* (1606), *Epitome of Copernican Astronomy* (1618–1621) as a whole, and *Tabulae Rudolphinae* (1627)[30]; and *Dioptrice* (1611), extending his earlier treatment of optics.

The translations of parts of the *Epitome* and *Harmonies of the World* that were included in the Britannica *Great Books of the Western World* (picked up from earlier mimeograph editions) made a good deal more of Kepler's work widely available. However, the nature of the selections unfortunately tended to reinforce the impression that Kepler was too fond of baseless speculation to be considered a truly scientific thinker.[31]

As one might expect, the history of translations of Kepler reflects the interests of the times. Well into the eighteenth century, those curious individuals who might want to have a look at Kepler would have found his Latin no obstacle. But at the same time, Kepler's achievement had been largely condensed into three laws of planetary motion, which were buried in a mass of difficult computations and inadequate physics. As mathematical physics became more lifeless and mechanical, philosophers in the Romantic tradition made an effort to revive interest in Keplerian views. This was not especially successful (although it may have helped initiate a series of publications of selections from Kepler's correspondence in German translation): in any case, the needs of serious readers, who would know Latin, were met by Christian Frisch's edition of Kepler's works (*Ioannis Kepleri astronomi opera omnia*, 8 vols., 1858–1871).

Frisch's edition was nevertheless soon considered lacking in important respects, and largely through the efforts of Walther von Dyck (1856–1934), beginning around 1910, work began on a new and more complete edition.[32] An important feature of this edition was that all notes and commentary were to be in German. This reflects a change in the anticipated audience, and Max Caspar, chief editor of the initial volumes, followed through by translating several of Kepler's major works into German. The publication of these translations, together with the gradual appearance of volumes of the excellent *Johannes Kepler Gesammelte Werke* (*KGW*) and Caspar's authoritative biography (soon translated into English by C. Doris Hellman (1910–1973)[33] heralded a new era of Kepler studies and translations in modern languages. To read Kepler, one no longer needed to master Latin or early modern German, a change that opened understanding of the development of modern science to a much

---

[29] Kepler (1979b, 2000a, 2000b); Field (1984).

[30] See, however, Gingerich and Walderman (1972).

[31] See Hutchins (1952), pp. 492–3 for the translator's reasons for his selections.

[32] Caspar (1938), 'Einleitung', p. viii.

[33] Caspar (1948, 1959). Caspar did not provide references for the many sources he cited or quoted; however, a new edition of the English translation (Caspar, 1993) has remedied this defect.

wider public. The appearance of English translations was fostered by the publication of the *KGW* as well as the German and French translations. The chronology of publication suggests that neither public demand for translations nor scholarly work on producing them played the leading role in the appearance of modern-language versions. The two seem to have developed in parallel.

13.3   Requirements: Kepler's Latin

Translating Kepler requires a wide range of skills, which will tax the ability of even the most erudite scholar. Most obviously, one must understand Kepler's Latin, as well as his occasional forays into Greek. Neither of these presents any unusual difficulties. Kepler received a good classical education with a Lutheran slant, studying Latin, Greek, and Hebrew. He does not appear to have emulated any particular author. However, when teaching his son Ludwig (1607–1663) the fine points of Latin composition, Kepler translated a selection from Tacitus (*c.* AD 56–after 113) into German for Ludwig to retranslate into Latin.[34] Could this indicate a particular fondness for Tacitus' prose style? I think rather that it is a reflection of the notoriously inimitable nature of Tacitus' writing[35]: the exercise would be a formidable challenge for the young Ludwig. Kepler's prose does not exhibit the extreme terseness of Tacitus, but he also tends to avoid the orotund rhetoric of Cicero (106–43 BC). I have sometimes found, in searching for the best reading of a certain word, that Plautus' (*c.* 254–184 BC) use of it best matches Kepler's. I have no other information about any fondness for Plautus,[36] but it seems to me appropriate that Kepler would find the language of the earthiest of Roman playwrights congenial.

Occasionally one encounters a word or phrase that is puzzling, either because there is really no English counterpart, or because there are several distinct senses in which the Latin can be understood. Kepler's use of *species* to designate the motive power that comes forth from the Sun and moves the planets, is a good example of the former. It can mean 'look' or 'form' or 'shape' or 'kind' or 'image', none of which corresponds to Kepler's meaning. Since Kepler discusses the nature of this entity at some length, there is no need to come up with a translation at all, and some advantage to leaving it in Latin as a Keplerian technical term; hence, in *Astronomia*

---

[34] The German translation was later published by Ludwig Kepler (Kepler, 1625), who related the origin of the book in his letter of dedication; included in *Cornelii Taciti Historischer Beschreibung* (1625); See Kepler (1990), pp. 105–106.

[35] 'Tacitus' literary style is unique. His history can hardly be thought of apart from the style in which it is written, and that style, unlike Cicero's, is so difficult to imitate that it stands alone in Latin literature.' See Mendell (1957), p. 71.

[36] I have found only one clear reference or allusion to Plautus in all of Kepler's works, and this is in a letter to one of Brahe's relatives. Kepler is asking for assistance in reading an Arabic book in his possession, and humorously describes himself as 'Euclio'. Euclio is a character in Plautus's *Aulularia* who was maniacally protective of a pot of gold. The reference is in Kepler (1955), letter 873, line 14.

nova, I have usually left it as is, always italicized to alert the reader. In the *Optics*, in contrast, *species* denotes an image, and to leave it unchanged would obfuscate Kepler's clear meaning. As always, the translator must attend to the context.

Another problematic word appears in the *Optics*. In his account of the nature of colours, in Chap. 1 proposition 15, he ascribes colours partly to 'different degrees of the *lucula*, which is condensed into matter'.[37] I have not found this word elsewhere, and it appears to have been coined by Kepler, on the analogy of *molecula*. It would accordingly mean something like a small particle of light, except that for Kepler the *lucula* is not a particle but a two-dimensional entity. I was tempted to use 'lucule' for the English translation, but this word has its own technical meaning, so I glossed it as a 'spark of light'.

As for the second kind of terminological problem, there is in *Astronomia nova* a pair of terms, *medius* and *longitudo*, that, both separately and together, have created difficulties. *Medius* can simply mean 'middle', describing a place that is intermediate between two other places, or a magnitude that is intermediate in size. Or it can have the more technical sense of a mean or average, and this can be either over time or over space. It is in this sense that Kepler, in common with all other astronomers, most often used the word. *Anomalia media*, as Kepler points out in Chap. 4,[38] is in effect a measure of time, 'expressed according to an arbitrary rule', in which the full time of the planet's period is divided into 360 time degrees. It is 'mean' in the sense that it represents the planet's average motion or position (these two terms often being used interchangeably by Kepler). *Longitudo media*, in its normal astronomical usage, is closely related to *anomalia media*, the only difference being that the latter is the number of time-degrees from aphelion while the former starts from the spring equinox point. In this astronomical sense, *longitudo* is the angular measure of position on the zodiac. There are different measures of *longitudo* depending on the centre about which the angle is measured.

However, there is another, more everyday sense of *longitudo*, namely, 'length'. Kepler uses the word in this way in *Astronomia nova*, ch. 44, although, because he is describing the *longitudo* of a number of lines drawn from the centre of Mars' orbit, one can only discern his meaning by the context. In this instance, he refers to a *longitudo* that was found in a previous chapter, and looking at that chapter one sees that it was a linear, and not an angular, magnitude that was given.[39]

Evidently, then, Kepler did not go to much trouble to clarify ambiguities in his use of terms. The problem is compounded when he puts the words *longitudo* and *medius* together. As was remarked above, *longitudo media* in an astronomical context normally refers to a planet's mean position on the zodiac. However, quite early in *Astronomia nova*, Kepler uses the term in the plural (*longitudines medias*) to denote the region on the planet's eccentric circle about halfway between the two apsides.[40]

---

[37] Kepler (1939b), p. 23; Kepler (2000a), p. 24.

[38] Kepler (1937), p. 75, marginal note.

[39] Kepler (1937), p. 286.

[40] Kepler (1937), p. 177.

What does he mean here? In the first edition of my translation, I rendered the phrase with 'middle longitudes'. Despite the fact that longitude is a position on the zodiac and not on the orbit, I thought that Kepler wanted to describe an angular position halfway between aphelion and perihelion.

In his review of the translation, Stephenson criticized my reading, arguing that careful study of the Latin text shows that Kepler must be referring to the distances of the various places on the orbit from the Sun.[41] In my subsequent reading, and particularly in translating Kepler's Optics, I came to agree with Stephenson. This raised the problem of how to translate this awkward term, which really is the question of why Kepler used it. If he had simply meant 'at the mean distances' he could have written *in distantiis mediis*. What does his avoidance of this expression tell us about how he viewed the orbital dimensions? I confess that I have no answer to this question, but that made it all the more important to provide a translation that reflects Kepler's peculiar term. And even if there is no evident answer, should the translation not reflect Kepler's consistent choice, used throughout *Astronomia nova*, by finding a different English word with the right shade of meaning?

In the end, after trying the literal 'lengths', the inaccurate 'distances', and the perhaps misleading 'intervals' (which has other senses relating to the orbit), I decided in most cases to translate the phrase as 'middle elongations' for the 2015 edition of the translation. Although the word 'elongation' has another angular application, its etymological relation to *longitudo* and its similar ambiguity recommended it. Although this translation may not be entirely satisfactory, its slight oddity may have the benefit of suggesting an odd Latin usage.

In developing his physical account of planetary motion, Kepler uses a pair of terms, *vis* and *virtus*, without clearly distinguishing their meanings. Although *vis* in classical Latin denotes physical force, while *virtus* (derived from *vir*, 'man') primarily refers to manly excellence, which can include physical strength, Kepler has been inconsistent in his use of these terms.[42] Since they are fundamental to Kepler's physics, it is very important to translate these terms correctly.[43] On the other hand, if Kepler's meaning is unclear, a reliable translation is not possible. My solution has been to follow the classical senses consistently, translating *vis* as 'force' and *virtus* as 'power'. This involves the danger of applying a layer of more modern concepts upon Kepler's more fluid ideas. I have considered this risk worth the corresponding advantage of allowing readers to know immediately, without having to refer to the Latin, which word Kepler was using: 'force' always translates *vis* and 'power' always translates *virtus*, and these terms are not otherwise used.

A somewhat less problematic, but still ambiguous, term is *actus*. This is used in chapter 34, where Kepler describes the Sun as the body '*in quo primus actus omnis motus inest*'. I originally translated it as 'impulse', drawing upon the derivation of *actus* from the verb *ago*, to do or drive. However, classicist Perry, who has been

---

[41] Stephenson (1994).

[42] Davis (1981), p. 216.

[43] It may also be worth remarking that since *vis* possesses only three endings to express the five Latin cases, its syntactic function is not always readily evident.

very helpful in revising my translations, questioned this, writing, 'I am tempted to say "actuality"', and referred me to the Aquinas Lexicon,[44] where one definition is 'reality, real being, the opposite of *potentia* and *potestas*'. His criticism was that my use of 'impulse' suggested that the Sun itself is physically pushing things around, while Kepler seems to be describing the Sun as a cause, source, or principle of all motion in the universe. Although I did not adopt his reading, I changed mine to 'activation', which I hope is closer to Kepler's sense.

For any translator, there are places where the text superficially invites a reading that on reflection is unsupportable. An example will suffice. In Chapter 36 of *Astronomia nova* Kepler is considering the effect of changes in distance upon the intensity of illumination from a luminous body. One point he makes is that the change in apparent size of a body when its distance changes has no effect upon the intensity of the rays. It is the same body, whether seen under a large angle or a small one. He writes, '*cum sit tantum deceptio visoriae facultatis, et ex genere rationalium entium; quibus nulla est efficientia*'. I originally translated this as: 'although this is but a deception of the visual faculty, and belongs to the genus of rational thought, of which there is no efficient cause'. Although this may be a defensible translation on purely grammatical grounds, it does not make much sense. A moment's rational thought would have established that the efficient cause of thought is the thinker, who surely cannot be non-existent. Perry flagged this gaffe and suggested the following: '...belongs to the class of theoretical entities that lack any efficacy'. This is much closer to the mark, and the error points to the importance of coming to some degree of understanding of the author's meaning and intent, and not simply a grammatically acceptable English equivalent.

The translator must also consider to what extent the form of the original should be reflected in the translation. Because the Latin syntax is largely determined by the inflections of the words, the arrangement of words in sentences, and the length and complexity of sentences, is not usually the same in good English prose as it is in Latin. It is nonetheless possible to translate each Latin sentence into a single English sentence.[45] However, the result, while perhaps giving the reader some sense of what reading Kepler's Latin is like, is not conducive to ready comprehension, and is in general to be avoided. There are some exceptions, such as where Kepler makes use of a classical rhetorical trope, such as aposiopesis,[46] which can depend upon a long sentence for its effect.

Paragraph breaks, on the other hand, are less problematic. They indicate Kepler's notion of how the text should be segmented, and also make it easier to refer to the Latin text. Therefore, even when adding or changing a paragraph break might be helpful in understanding the text, it has seemed preferable to keep the paragraphing of the original.

---

[44] Deferrari et al., (1985), p. 16.

[45] I provided an example of this sentence-by-sentence translation in the English version of Kepler's dedicatory letter to the Emperor Rudolf II, in Kepler (1992), pp. 7–10.

[46] Aposiopesis is a breaking off of the flow of a sentence for rhetorical effect.

13.4 Requirements: Understanding the Mathematics

A sound knowledge of Latin and proper attention to how the printed text was formatted is only a beginning. As is shown in the examples above, the correct interpretation often depends on the context, and the context is often mathematical. Therefore, a reliable translation depends upon the translator's following the mathematical demonstrations. Kepler's mathematics is mostly classical Greek geometry, though he twice makes use of algebra. Trigonometry of course plays a prominent role, but the techniques used are similar enough to modern methods not to pose a problem. So the main requirement is a sound knowledge of Euclid (*fl. c.* 300 BC)—not what is nowadays called 'Euclidean geometry'. Perhaps surprisingly, Apollonius' (*fl. c.* 230 BC) *Conics* is not helpful in reading *Astronomia nova*: specifically Apollonian properties, such as the 'points of application' (renamed 'foci' by Kepler in his *Optics*) are absent. Instead, Kepler bases his treatment of the elliptical orbit upon Archimedes (*c.* 287–212 BC).[47] Also Archimedean is Kepler's application of very small quantities in formulating rules relating angular intervals to increments of time and similar matters. Here great delicacy is required, as it would be easy to read these passages anachronistically as integrals and other calculus operations. Indeed, it is important *not* to translate Kepler's mathematical expressions, but to keep them exactly as they are, even at the expense of some clarity. We must understand that he was inventing these things as he went along, and we should not demand the vividness and coherence of a well-formulated mathematical argument.

This brings up another very important consideration: the realm of mathematical expression. That mathematics is not just a distinct language, but a panoply of different and contrasting languages, may not be immediately evident. Nonetheless, there is a clear difference between, for example, the ancient and modern definitions and treatment of such curves as conic sections, and the change from the one mode to the other has been described as pivotal in the development of modern European views.[48] Moreover, different mathematical expressions that nonetheless are computationally identical often have very different physical and even metaphysical implications.[49] For example, Kepler's consistent expression of time as a variable dependent upon angular position was important in shaping his investigation of planetary motion.[50] This rhetorical aspect of mathematics places an additional demand on the translator of a mathematical work, particularly one written using the terminology and methods of a past era. A translation should reflect the aims and concerns of the author, while avoiding unnecessary obscurity. To succeed in this requires delicate judgment.

---

[47] *Of Conoids and Spheroids,* Prop. 4, in Archimedes (1897); cf. Kepler (1937) Chap. 57 p. 367. Kepler's use of Archimedean rather than Apollonian properties of the ellipse was first noted by Davis (1992a), p. 162.

[48] Klein (1968), especially Ch. 9 pp. 117–125.

[49] See for example, Simpson (2005), esp. pp. 7–11.

[50] This inversion of the usual order has been noted by Davis (1992b) and Donahue (1994).

In translating Kepler, one needs above all to be aware that Kepler was attempting something nearly unprecedented: the use of mathematics (geometry in particular) as a metaphoric representation of dynamic actions and entities that can at the same time match the best astronomical observations within the limits of observational precision. There was a stage in his enquiry when he doubted that this was even possible,[51] and one must be sensitive to the leaps of imagination required to attempt it. In doing so, Kepler made use of a variety of mathematical models, each of which served as a metaphor for a certain aspect of planetary motion. For example, the eccentric model vividly expresses the time, or mean anomaly, by means of the sum of planetary distances on the circle, or the area swept out by the radius. On the other hand, the epicyclic model, though ambiguously related to the simple anomaly, was useful for generating the planet's distances from the Sun. Neither of these could by itself generate the planet's orbit, but each allowed one constituent of the orbit to be examined by itself. The final solution was the discovery of the correct relation between the two constituents. In contrast, we nowadays tend to think of the discovery of the *form* of the orbit as primary (hence, its designation as the 'First Law'), and the determination of the planetary *position* on this orbit as secondary. This tendency must at all costs be avoided by a translator.

In earlier times, it was the fashion to express Kepler's geometry in analytical terms, using algebra. This is especially prominent in Max Caspar's annotations to his edition of *Astronomia nova* in the *Gesammelte Werke*. Translating the mathematics in this way has the unfortunate effect of obscuring the way Kepler thought about the composition of the orbit. What is originally a geometrical operation, an action, gets turned into a symbol in an algebraic equation, a stasis. Thus, in general, I believe a translator should err on the side of faithfulness to the original mode of expression. Subduplicate ratios, for example, should not be changed into square roots, nor should complements of sines become cosines, nor should a host of other modernizations be introduced. Where necessary, footnotes and glossary entries can be used to explain obscure terms or operations.[52]

There are nevertheless a few places where Kepler's mathematics verges towards the algebraic. I say 'verges' because although the reasoning is clearly algebraic, the mode of presentation is visually the same as his presentation of geometry. A fine example of this appears in Chap. 6 of *Astronomia nova*.[53] The unknown, called the 'figured unit' (*figurata unitas*), is given its own peculiar symbol, which is not an ordinary letter. I have chosen to translate this symbol with the letter 'x', because its role in Kepler's proof is the same as the role conventionally played by x in later algebra. In the translation, I chose to insert many extra line breaks for clarity,

---

[51] For example, in a document titled *Preparatio ad commentaria in theoriam Martis,* written before he had determined the elliptical form of the orbit, he wrote, 'Here a reason is to be given why physics may not agree with experience, and the extent to which it does agree.' Kepler (1937), p. 459.

[52] Here I differ somewhat from Aiton, Duncan, and Field in their translation of *Harmonice Mundi* (Kepler, 1997). In the Translators' Notes, p. xxxix, they regarded Kepler's treatment of ratios and proportions as 'unacceptably confusing to the modern reader', and therefore translate the mathematics as well as the Latin.

[53] Kepler (2015), pp. 115–116.

while leaving Kepler's textual argument unchanged. But I also added a footnote containing the modern algebraic restatement of the proof, which is of course much more compact. In presenting the proof in this way, I am trying to walk a fine line between obscurity and misrepresentation. Subsequent scholars will no doubt pass judgment on the probity of this choice. But the important point here is that this proof is a significant step in the use of some kind of algebraic reasoning in a scientific context, and it is essential that a translation capture both its algebraic essence and its archaic mode of expression. This is a clear example of what is meant by 'mathematical translation'.

13.5   Treatment of Diagrams

One might not think that mathematical diagrams would require translation, or that they could be in any way problematic. Nevertheless, it has in fact been the practice of editors and translators to re-work the diagrams in a way that is at least akin to translation. One reason for redrawing the diagrams is that Kepler's original wood engravings, as printed in the 1609 edition of *Astronomia nova*, are often unclear and their letters are difficult to read. However, the re-working of the diagrams always goes beyond merely cleaning up the lines and lettering. Configurations can be changed, letters repositioned or possibly replaced (e.g. roman instead of Greek), ornaments and other presumably unimportant elements removed. Such liberties, which would not usually be taken with the text, are common in treating diagrams, in both translations and new editions of the texts. The thinking seems to be that the diagrams are subsidiary to the text and should be cleaned up and altered as necessary to express the meaning of the text in the clearest way. This appears to have been the prevailing attitude in, for example, standard editions of Greek geometrical texts.[54] More recent scholars have begun to give diagrams and illustrations the attention they deserve, as meaningful objects that are to some extent independent of the text that they accompany.[55]

Geometrical diagrams in particular are arguably prior to the text, in that they represent the objects (points, lines, planes) whose relations the text is elaborating. In this view, the diagram holds the primary place in the demonstration, while the textual argument is about the diagram. But—however it may be characterized—this relationship requires the translator to attend closely to the diagrams, and to their connection with the text. Frequently a correct understanding of the mathematics, as well as correction of occasional errors, requires reference to the figures. But further, the present availability of computer graphical software provides opportunities

---

[54] In his edition of Apollonius (Apollonius, 1891–1893, II, p. lxv), for example, Heiberg comments, 'it can be demonstrated that as regards the diagrams the manuscripts are not much to be trusted.' It may not be entirely coincidental that, in the former passage, he translates geometrical relations into algebraic expressions.

[55] See for example Crowther and Peter Barker (2013).

that were not open to previous editors and translators, nor to Kepler himself. So the translator at least needs to sketch out the diagrams while working through the mathematics, and would be well advised to draw all the diagrams independently using a computer graphics program. This will help avoid errors that have occurred in past editions, for which the engraving of the diagrams was done by a professional with no knowledge of the underlying mathematics. An example of this is the diagram in chapter 16 of *Astronomia nova*: in every edition and translation before the complete English version, two letters were reversed, making nonsense of the text.

A translator who follows this advice and constructs the diagrams in the course of following the demonstration will sooner or later encounter a figure that is used for two or more different demonstrations that require somewhat different configurations. This was often done to save the considerable expense of having a new wood engraving cut. The most striking example in *Astronomia nova* appears first in Chap. 6 and again in Chapter 67.[56] In the latter case, Kepler needs to refer to Mars' line of nodes, which does not appear in Chap. 6. He therefore refers to two lines that do not appear in the diagram, leaving the reader to imagine them or draw them in.

The translator must therefore decide whether to leave the diagram exactly as Kepler had it engraved (on the grounds of preserving the historical document) or to modify it to show the new lines. In this instance, the new lines are not obtrusive, so they have been inserted as dashed lines, with an explanatory note and interpolated letters in the text identifying them. However, in other cases, a more drastic alteration of the original may be required, and the decision will then depend on the intended audience for the translation. If it is primarily aimed at scholars, faithfulness to the original should be paramount, while if it is for students and non-specialist readers, editing (with suitable advisory notes) will often be appropriate.

13.6   Treatment of Observations

Kepler's treatment of Tycho Brahe's observations calls for separate notice. Kepler presents longitudes, latitudes, right ascensions, declinations, stellar and planetary alignments, and angles, without reference to any diagrams or charts. A correct translation depends upon an accurate understanding of the particular configurations that Kepler describes, and this in turn usually requires reference to a star chart and an approximate sketching out of the positions of the celestial bodies mentioned, along with the locations of the ecliptic and the equator and their poles. Sketching the configuration in this way always clarifies the meaning of the text, and often indicates the reasons for certain statements that might otherwise be obscure. The resulting clarification can in turn improve the aptness of the translation.

---

[56] Kepler (1937), pp. 89 and 401, respectively.

13.7 Style and Ornamentation

There is another aspect of the diagrams, and indeed of the printed text, that, while perhaps not strictly a matter of translation, is an important supplement to the purely verbal content. This is the style and ornamentation of both the text and the diagrams. Throughout his life, Kepler was intimately involved in the production of his works.[57] *Astronomia nova*, in particular, was lavishly produced, in a large format, and shows signs of close cooperation between Kepler and his printer. Most remarkable is having somehow contrived to make the opening of Chap. 8, which covers an entire spread, fall upon the centre sheet of a signature, so that the lines of the table contained therein align perfectly. Also of note is Kepler's having had the woodcuts for the diagrams engraved himself, before arranging for the printing of the book.[58]

The formatting of the text reveals much about Kepler's intentions. Ubiquitous is his use of italics in presenting mathematics, to distinguish it from discussion of other subjects. It was evidently important to him to maintain the integrity of distinct disciplines, or at least to avoid being seen as mingling them inappropriately. We might find such punctiliousness odd nowadays, but this is all the more reason to retain it in the translation. Similarly, in Chapter 59, when Kepler is at last setting forth the mathematical argument for the elliptical form of the orbit, he very strikingly shifts the typography to a much more formal style. The chapter is divided into 'Protheorems', each introduced by an enunciation set in type that is even larger than that used for the chapter title. This tells us that we have reached the climactic moment in the book, when finally all the puzzles and mis-steps of previous chapters are resolved. This style was not emulated in the 1992 edition of the English translation, ironically because I was relying upon Caspar's edition, which in effect had translated the original into a modern book by removing all of Kepler's very deliberate formatting and ornamentation. I had at that time never seen the original. I now view the style as an important and substantive element of the text, and have typeset the 2015 revision accordingly.[59] It is unfortunate that both of the modern editions of the Latin text give no indication of the exuberant typography of the original. The editors clearly believed that the formatting was merely ornamental and had nothing meaningful to convey to the reader. Kepler would surely have disagreed.

The occasional ornamentation of the diagrams has likewise been ignored by editors and translators. Two examples are especially striking.

The ellipse diagram in chapter 59 is adorned with a depiction of Urania, muse of astronomy, arriving in a chariot and holding in her left hand a laurel wreath with

---

[57] One exception was the *Optics,* which he sent to a printer in Frankfurt in order to have it published as quickly and efficiently as possible. The difficulties, delays, and poorly inserted corrections in that edition surely convinced Kepler thenceforth to avoid relinquishing control over the production process.

[58] Caspar (1968), p. 47.

[59] Kepler (2015), pp. 431–443. I initially tried following Kepler by enlarging the type size, but the effect was jarring in the context of a modern printed book. It then occurred to me that the boldface style, although not available to Kepler, would have had the effect he wanted and would not be out of place nowadays.

which to crown Kepler in his triumph. The chariot is riding over rocky ground, presumably to indicate the difficulty of the path that led to victory over Mars.[60] This is clearly an important element of the book, yet a reader would never know of its existence unless consulting the first edition or a facsimile, or (now that it is out) the revised English translation.

A more puzzling emblematic diagram first appears in Chapter 39, and reappears several times again, including a couple of places in Chapter 59. It depicts an epicyclic mechanism whose function is to represent Mars' changing distances from the sun. This epicycle is accompanied by a pair of angels, with wings erect. The angel on our right is standing upon what appears to be flowing water, and is holding a three-columned tablet in her right hand. The one on our left is apparently standing on a stone, and has what appears to be a compass in her right hand and a carpenter's square in her left. The meaning of these presumably emblematic figures is obscure, but their presence is remarkable and should certainly not be omitted from a translation, just as one would not omit a sentence merely because its meaning is not clear.

13.8   Summary and Conclusion

What I hope has emerged from the welter of detail in the present chapter is that neither the history nor the salient issues of Kepler translations is simple and straightforward. I believe enough has already been said regarding the history of translations and editions, but would like to add a few words in conclusion on desirable characteristics of translations emerging from the above discussions, and to suggest directions that future work might take.

It should be clear from what has been said that the production of a good translation of a Kepler work requires much more than an accurate rendering of the Latin text. Kepler took considerable care in the style of expression as well as in the format of the books and the layout of the text and visual elements. He preferred to work closely with the printer to assure the excellence of the result. His mathematical arguments, too, followed a rhetorical plan, and a translator must somehow balance the requirements of clarity and comprehensibility against Kepler's language and intentions. And the diagrams and their ornamental features must also be given full consideration: they too contribute to the meaning of the work as a whole. There are no easy prescriptions here, and no translation is going to be entirely successful in finding the right balance.

In addition, a translator must adapt the tone of the translation to the intended audience. Despite the increasing scholarly interest in Kepler's work, it would be a mistake to suppose that scholars constitute the main readership for translations. My experience has been that a surprising range of readers are interested in Kepler, many of them for reasons having nothing to do with scholarship. I believe that a translator should not attempt to address a particular audience, but should try to

---

[60] This diagram appears on p. 290 of the first edition (Kepler, 1609), and is reproduced on p. 437 of Kepler (2015).

make the translation clear, literate, and lively, while preserving the sense of Kepler's language. A translation done in this way will not be a window on the Latin text: it will not, for example, keep long sentences intact, nor will it strive to preserve the syntax at the expense of clarity and literacy. On the other hand, I believe one should avoid excessive restatement or interpretation of the text. There is a temptation, to which many translators seem to have yielded, to brush aside minutiae of syntax when one feels that the author's meaning is clear. The translator's only excuse for this is that a translation is *always* an interpretation, and cannot hope to be more than an honest one. Serious scholarship, in any case, will always need to refer to the original and treat any translation, however good, with some scepticism.

As for the future, it seems to me that there are two important needs to be met. First, there are significant works that remain untranslated, among them *De stella nova, Dioptrice,* and the complete *Epitome.* Future scholars of Kepler's work are going to want to have these in English, even if they have some knowledge of Latin. Second, Kepler is too good to be constrained within the province of the experts. There is a demand for readable and well annotated selections in translation, for use in university courses as well as for general readership—with the proviso, of course, that the translation should above all be accurate. Production of books for a more general audience may not enhance the prestige of an editor or translator, but will, I believe, be of enduring value.[61]

References

Apollonius of Perga. (1891–1893). In I. L. Heubner (Ed.), *Apollonii pergaei quae graece exstant, cum commentaris antiquis,* (2 vols). Stuttgart: Teubner.

Archimedes. (1897). *Archimedes: Works.* Edited by T. L. Heath (Transl. from the Greek by J. L. Heiberg). Cambridge: Cambridge University Press.

Berry, A. (1898). *A short history of astronomy.* London: John Murray.

Caspar, M., ed. (1938). *Johannes Kepler Gesammelte Werke, band I: Mysterium cosmographicum/ De nova stella.* Munich: C. H. Beck.

Caspar, M. (1948). *Johannes Kepler.* Stuttgart: W. Kohlhammer Verlag.

Caspar, M. (1959). *Kepler.* (Transl. edited by C. D. Hellman). London: Abelard-Schuman.

Caspar, M. (1968). *Bibliographia Kepleriana.* Zweite Auflage, Zweiter Teil, no. 168. Munich: C. H. Beck.

Caspar, M., & Kepler-Kommission, eds. (1937). *Johannes Kepler Gesammelte Werke. Band III: Astronomia nova.* Munich: C. H. Beck.

Caspar, M. (1993) *Kepler.* (Transl. and edited by C. D. Hellman) with a new introduction and references by O. Gingerich; bibliographical citations by O. Gingerich and A. Segonds. New York: Dover Publications.

Clerke, A. M. (1910–1911). In H. Chisholm, art (Ed.), *The encyclopaedia Britannica,* (11th Edn, vol. 15, pp. 749–751). Kepler, Johann.

Crowther, K. M., & Peter Barker, P. (2013). Training the intelligent eye: Understanding illustrations in early modern astronomy texts. *Isis, 104,* 429–470.

---

[61] My little book Kepler (2005) has enjoyed a modest success in bringing some of Kepler's writing to a wider audience, and a recent new translation of *The Six-Cornered Snowflake* (2010) has proved popular. No doubt there is room for other similar publications.

Davis, A. E. L. (1981). *A mathematical elucidation of the bases of Kepler's* Laws. Ph.D. dissertation, imperial college of science and technology. Available from ProQuest (ID: 8822353) or Ethos (E-Theses Online Service).

Davis, A. E. L. (1992a). Kepler's road to damascus. *Centaurus, 35,* 162.

Davis, A. E. L. (1992b). Kepler's distance law: Myth not reality. *Centaurus, 35,* 112.

Davis, A. E. L. (2003). The mathematics of the area Law: Kepler's successful proof in *Epitome Astronomiae Copernicanae* (1621). *Archive for History of Exact Sciences, 57,* 355–393.

Deferrari, R. J., Barry, S. M. I., & McGuiness, I. (Eds.). (1985). *A lexicon of St. Thomas aquinas.* Rinsen: Catholic University of America.

Donahue, W. H. (1994) Kepler's invention of the second planetary law. *British Journal for the History of Science, 27,* 101–102.

Field, J. V. (1984). A lutheran astrologer: Johannes Kepler. *Archive for History of Exact Sciences, 31,* 189–271.

Galilei, G., Kepler, J. (1880). *The sidereal messenger of galileo galilei and a part of the preface to Kepler's Dioptrics.* Translated from the Latin by E. S. Carlos. London: Rivingtons.

Galilei, G., Kepler, J. (1887). *Sidereal messenger,* (vol. 6, pp. 109–112, 133–138, and 212–217).

Gingerich, O., & Walderman, W. (1972). Preface to the *Rudolphine Tables. Quarterly Journal of the Royal Astronomical Society, 13,* 360–373.

Hutchins, R. M. (Ed.). (1952). *Great books of the western world* (Vol. 16). Chicago: Encyclopaedia Britannica.

Kepler, J. (1602). *De Fundamentis astrologiae certioribus.* Prague: Schuman Press.

Kepler, J. (1604). *Ad Vitellionem paralipomena quibus astronomiae astronomiae pars optica traditur.* Frankfurt: Claudius Marnus and the heirs of Johannes Aubrius.

Kepler, J. (1609). *Astronomia nova aitiologêtos seu physica coelestis....* Heidelberg: E. Vogelin.

Kepler, J. (1726). *Joannis Keppleri De calendario Gregoriano liber singularis....* Frankfurt and Leipzig.

Kepler, J. (1858–1871). *Ioannis Kepleri Astronomi Opera Omnia,* (vols. 8, ed. Ch. Frisch). Frankfurt and Erlangen: Heyder & Zimmer.

Kepler, J. (1923). *Das weltgeheimnis.* Augsburg: B. Filser.

Kepler, J. (1929). *Neue astronomie.* Berlin: R. Oldenbourg.

Kepler, J. (1937). *Johannes Kepler Gesammelte Werke, Band III: Astronomia Nova.* Munich: C. H. Beck.

Kepler, J. (1938). In M. Caspar (Ed.), *Johannes Kepler Gesammelte Werke, Band I: Mysterium cosmographicum/De nova stella.* Munich: C. H. Beck.

Kepler, J. (1939a). *Weltharmonik [Harmonice mundi,* Transl. by M. Caspar]. Munich and Berlin: Oldenbourg.

Kepler, J. (1939b). In: F. Hammer (Ed.), *Johannes Kepler Gesammelte Werke. Band II: Astronomiae pars optica.* Munich: C. H. Beck.

Kepler, J. (1942). *Concerning the more certain fundamentals of astrology,* (Transl. by E. Meywald). Clancy Publications.

Kepler, J. (1955). In M. Caspar (Ed.), *Johannes Kepler Gesammelte Werke. Band XVII: Briefe 1612–1620.* Munich: C. H. Beck.

Kepler, J. (1965). *Kepler's dream.* With an introduction by J. Lear and translation by P. F. Kirkwood. Berkeley: University of California Press.

Kepler, J. (1966). *The six-cornered snowflake.* (Transl. by C. Hardie). Oxford: Oxford University Press.

Kepler, J. (1967). *Kepler's Somnium.* (Transl. by E. Rosen). Madison: University of Wisconsin Press.

Kepler, J. (1979a) *Astronomie nouvelle [Astronomia Nova,* Transl. by J. Peyroux]. Paris: Blanchard.

Kepler, J. (1979b). *On the more certain fundamentals of astrology.* Foreword, notes and analytical outline by J. Bruce Brackenridge, (Transl. by M. A. Rossi). Proceedings of the American philosophical society.

Kepler, J. (1981). *Mysterium cosmographicum: The secret of the world*. (Transl. from the Latin by A. M. Duncan). New York: Abaris Books.

Kepler, J. (1984). *Le Secret du Monde*. Introduction, translation and notes by A. Segonds. Paris: Les Belles Lettres.

Kepler, J. (1990). In J. Hübner et al. (Ed.), *Johannes Kepler Gesammelte Werke. Band XII: Theologica/Hexenprozess/Tacitus-Übersetzung/Gedichte*. Munich: C. H. Beck.

Kepler, J. (1992). *New astronomy*. (Transl. from the Latin by W. II. Donahue). Cambridge: Cambridge University Press.

Kepler, J. (1997) *The harmony of the world*, (Transl. with introduction and notes by E. J. Aiton, A. M. Duncan, & J. V. Field). Philadelphia: American Philosophical Society.

Kepler, J. (2000a). *Optics: Paralipomena to witelo, and the astronomical part of astronomy*. (Transl. by W. H. Donahue). Santa Fe: Green Lion Press.

Kepler, J. (2000b). *The six-cornered snowflake*. Philadelphia: Paul Dry Books.

Kepler, J. (2005). *Selections from Kepler's astronomia Nova*. Science classics module for humanities studies. (Transl. by W. H. Donahue). Santa Fe: Green Lion Press.

Kepler, J. (2010). *The six-cornered snowflake*. Latin text with English (Transl. by J. Bromberg). Philadelphia: Paul Dry Books.

Kepler, J. (2015). *Astronomia nova*. Revised English (Transl. by W. H. Donahue). Santa Fe: Green Lion Press.

Kepler, L. (1625). *Cornelii Taciti Historischer Beschreibung*. Johann Blankhen.

Klein, J. (1968). *Greek mathematical thought and the origin of algebra*. MIT Press.

Koestler, A. (1959). *The sleepwalkers: A history of man's changing vision of the universe*. London: Hutchinson.

Krafft, F., Meyer, K., & Sticker, B., (Eds). (1973). *Proceedings of the internationales Kepler-Symposium, Weil der Stadt, 1971*. Hildesheim: Gerstenberg.

Lilly, W. (1644). *A prophecy of the white king, and dreadfull dead-man explained: to which is added the prophecie of sibylla tiburtina, and prediction of John Kepler: All of especial concernment for these times*. London.

Mendell, C. W. (1957). *Tacitus: The man and his work*. London: Oxford University Press.

Salusbury, T. (1661). *Mathematical collections and translations*. London: William Leybourn.

Shapley, H., & Howarth, E. H. (1929). *A source book in astronomy*. New York and London: McGraw-Hill.

Simpson, T. K. (2005). *Figures of thought*. Santa Fe: Green Lion Press.

Small, R. (1804). *An account of the astronomical discoveries of Kepler*. London: Mawman.

Small, R. (1963). *An account of the astronomical discoveries of Kepler* (reprint). With an introduction by W. D. Stahlman. Madison: University of Wisconsin Press.

Stephenson, B. (1994). New astronomy. Johannes Kepler, William H. Donahue. *Isis, 85*, 326–327.

Ziche, P., & Rezvykh, P. (2013). *Sygkepleriazein: Schelling und die Kepler-Rezeption im 19. Jahrhundert* (Schellingiana, 21). Stuttgart: Frommann-Holzboog.

A Kepler Chronology

Entries directly relating to Kepler are in larger type.

1483	Birth of Martin Luther
1533	Grynaeus' Greek edition of Euclid's *Elements*
1543	Death of Nicolaus Copernicus (born 1473)
	Publication of Copernicus' *De revolutionibus orbium coelestium*
1545–1563	Council of Trent
1546	Death of Martin Luther
	Birth of Tycho Brahe
1558	Commandino's Latin edition of *Conics* of Archimedes
1560	Barocius' Latin translation of Proclus' *Commentary on the First Book of Euclid's Elements*
1564	Birth of Galileo Galilei
	Birth of William Shakespeare
	Death of Michelangelo Buonarroti (born 1475)
1566	Commandino's Latin edition of works of Apollonius of Perga
1571 Dec. 27	Kepler Born in Weil der Stadt, Württemberg
1572 Aug. 23–24	St. Bartholomew's Day massacre
1572	Tycho observes New Star of 1572 on 11 November
1575	Kepler contracts smallpox
1577	Kepler's mother takes him to see the Great Comet (C/1577 V1)
1580	Kepler's father takes him to see a lunar eclipse
1582	Gregorian Calendar Reform—adopted in Italy, France, Spain, Portugal
1583	Gregorian Calendar adopted in German Catholic States and others
1584 Oct. 16	Kepler enters Adelberg convent school (instruction in Latin)

© Springer Nature B.V. 2024
A. E. L. Davis et al. (eds.), *Reading the Mind of God*, Springer Praxis Books,
https://doi.org/10.1007/978-94-024-2250-4

1586 Nov. 26   Kepler enters senior seminary, Maulbronn (instruction in Latin)

1587 Oct. 5   Kepler enters Arts Faculty of University of Tübingen to study mathematics (including astronomy) with Michael Mästlin (1550–1631)

1588   Kepler receives baccalaureat

The Armada sent against England by Philip II of Spain

Commandino's Latin edition of the *Collection* of Pappus

1589 Sep.   Kepler enters the Theological Faculty of the University of Tübingen

1591 Aug. 11   Kepler awarded master's degree with distinction

1594   Kepler leaves the University of Tübingen

Mar.   Kepler moves to Graz, as mathematics teacher at the Protestant seminary

1596   *Mysterium cosmographicum* (T¨ubingen)

Birth of René Descartes (died 1650)

1597 Apr. 27   Kepler marries Barbara Müller (aged 23, twice widowed)

1600 Feb. 17   Giordano Bruno burned at the stake in Rome

Feb.—Apr.   Kepler visits Tycho Brahe in Prague

Jul. 10   Kepler observes solar eclipse in marketplace at Graz

31   Non-Catholics banished from Graz

Aug. 3   Kepler expelled from Graz for his espousal of the Augsburg Confession

Oct. 19   Kepler arrives in Prague with family

William Gilbert (1544–1603) publishes *De Magnete*

1601 Oct. 24   Death of Tycho Brahe

26   Kepler appointed Imperial Mathematician to Rudolf II

1603   King James VI of Scotland becomes King James I of England

1604   *Astronomiae pars optica* (Frankfurt)

Oct. 17   Kepler observes New Star in Serpentarius (now Ophiuchus)

1606   *De stella nova* (Prague)

1608 May 14   Protestant Union formed by Frederick IV, Elector Palatinate

1609   *Astronomia nova* (Heidelberg)

Thomas Harriot (1560–1621) makes telescopic observations of Moon (but he did not publish them)

1609 Jul. 10   Catholic League formed to negotiate with the Protestant Union

1610   Publication of Galileo's telescopic discoveries in his *Siderius nuncius*

Dissertatio cum nuncio sidereo (Prague): Kepler's open letter to Galileo

1611 Jul. 3	Death of Barbara Kepler (born *c.* 1574)
	Dioptrice (Augsburg)
	De nive sexangula (Frankfurt)
	Emperor Rudolf abdicates in favour of his brother Matthias
1612 May	Kepler moves to Linz as District Mathematician
	Kepler excommunicated by Lutheran church in Linz
	Death of Rudolf II
1613 Oct. 30	Kepler marries Susanna Reuttinger in Linz
	Kepler summoned to Regensburg to speak on calendar reform
1614	Napier (1550–1617) publishes *Mirifici logarithmorum canonis description*
1615	*Nova Stereometria doliorum* (Linz)
Aug.	Katharina Kepler (Kepler's mother) brings libel suit against Ursula Reinbold for accusing her of witchcraft
1616 Jan. 2	Kepler complains to Leonburg town senate over their treatment of his mother. *Messekunst Archimedis* (Linz)
	Copernicus' *De revolutionibus* put on *Index Librorum Prohibitorum*
	Death of William Shakespeare
1617 Oct.–Dec.	Kepler in Württemberg to defend his mother Katharina against charge of witchcraft
1617	Archduke Ferdinand II named King Designate by Protestant Bohemians
1618	*Epitome astronomiae copernicanae* Lib. I, II, III (Linz)
1618 May 23	Protestant delegates defenestrate Catholic regents in Hradschin Castle (Prague), sparking off the Thirty Years War
1619	*Harmonices mundi libri V* (Linz)
Aug.	The Bohemians offer the crown to the Calvinist Frederick V of the Protestant Palatinate
28	Ferdinand II elected Holy Roman Emperor in Frankfurt
1620	*Epitome astronomiae copernicanae* Lib. IV (Linz)
	Epitome put on *Index Librorum Prohibitorum* (Rome)
1620	Kepler makes numerous visits to Württemberg
	Kepler reads Napier's book (1614) on logarithms
	Kepler reads Vincenzo Galilei's book (1581) on music
Jul. 17	Catholic League troops invade Upper Austria
	Catholic League troops take Linz
Aug. 7	Katharina Kepler arrested for witchcraft and imprisoned
	Kepler leaves Linz for Württemberg for his mother's trial, moves family to Regensburg

Sep. 26	Kepler visits his mother in prison
1621	*Epitome astronomiae copernicanae* Lib. V, VI, VII (Frankfurt)
Sep. 28	Kepler's mother shown the instruments of torture
Oct. 3	Duke of Württemberg orders the release of Katharina Kepler
Dec. 30	Ferdinand II confirms Kepler's appointment as Imperial Mathematician
1622 Apr. 13	Death of Katharina Kepler (born *c.* 1546)
1624	*Chilias logarithmorum* (Marburg)
1627	*Tabulae Rudolphinae* (Ulm)
1628 Jul.	Kepler moved to Sagan, under patronage of Wallenstein
1630 Oct. 8	Kepler leaves Sagan
1630 Nov. 15	Kepler's Death in Regensburg
1631 Nov. 7	Pierre Gassendi observes Transit of Mercury predicted by Kepler
1632	Sack of Regensburg
	Kepler's grave in Regensburg destroyed
1634	*Somnium* (ed. Jakob Bartsch)
	Murder of Wallenstein (born 1583)
1636 Aug.	Death of Kepler's widow, Susanna Kepler (born 1589) in Regensburg
1642	Death of Galileo Galilei
	Birth of Isaac Newton
1648	Peace of Westphalia ends the Thirty Years War

Glossary

A dagger symbol (†) before an entry indicates that the sense is now obsolete in scientific writing. Where a term has more than one sense, the senses are numbered .

altitude^1 In astronomy, the vertical angular distance between the horizon (0°) and a celestial body. It is complementary to the **zenith distance**.

†altitude^2 Height above the Earth.

anomaly An angle measuring the position of an orbiting body from a specific point (usually **periapsis**).

aphelion For a body in orbit around the Sun, the point at which that body is farthest from the centre of the Sun. *See also* **perihelion**.

apoapsis The point at which an orbiting body reaches its greatest distance from its central body.

apogee For a body in orbit round the Earth, the point at which that body is farthest from the centre of the Earth. *See also* **perigee**.

ascending node The point in a planetary orbit where the ecliptic latitude changes from negative to positive.

aspect Two astronomical objects such as planets, stars, zodiac constellations etc., are said to be 'at aspect' to one another if they are separated by some special angle, such as 180° (opposition) or 90° (quadrature).

atmospheric refraction The bending of light entering the Earth's atmosphere which causes the altitude of a celestial body to appear to have an **altitude** higher than its real one.

azimuth An angle measured in the horizontal plane from either the north or the south point of the horizon.

circumpolar star A star that never sets from a location at a given geographical latitude.

circumsphere The sphere that passes through every vertex of a solid (usually a polyhedron).

conjunction In astronomy (and astrology), the alignment, as seen from Earth, of two bodies such that they have the same ecliptic longitude.

© Springer Nature B.V. 2024
A. E. L. Davis et al. (eds.), *Reading the Mind of God*, Springer Praxis Books,
https://doi.org/10.1007/978-94-024-2250-4

culmination The passage of a celestial body across the observer's meridian. **Circumpolar stars** have an upper and a lower culmination.

declination In astronomy, the angular distance of a point on the celestial sphere from the celestial equator.

deferent In a planetary system, a circle whose centre is at, or near the centre of, the system.

descending node The point in a planetary orbit where the ecliptic latitude changes from positive to negative.

diurnal parallax The parallax shown by a celestial body resulting from the rotation of the Earth (or, in geocentric astronomy, of the heavens) between the rising and setting of the body. Also known as 'horizontal parallax'.

eccentric A circle that forms part of the model of the motion of a celestial body, in principle as a **deferent** circle, whose centre is not that of the centre of the planetary system as a whole.

eccentric anomaly If a circle of radius equal to the length of the semimajor axis of an ellipse is concentric with that ellipse, the eccentric anomaly is the angle subtended at the centre of the circle by periapsis and an imaginary point that marks the vertical distance of a planet extended to intersect the circle.

†**eccentricity**[1] In an **eccentric** the distance of the centre of motion from the centre of the circle.

eccentricity[2] Of an orbit, the amount by which it diverges from a perfect circle. More generally, a parameter that determines the shape of a conic section. If e represents eccentricity, then, for a circle, $e = 0$, for an ellipse $0 < e < 1$, for a parabola $e = 1$, and for a hyperbola $e > 1$.

ecliptic The apparent eastward circular path traced out during one year by the centre of the Solar disc against the background stars. More formally, the intersection of the ecliptic plane and the celestial sphere.

ecliptic coordinates A set of coordinates on the celestial sphere comprising latitude (the angular distance of a point on the celestial sphere from the ecliptic) and longitude (the angular distance of a point on the ecliptic from the First Point of Aries).

egress During a transit the moment when the planet parts from the limb of the Sun (or when an exoplanet parts from the edge of the host star). Also known as the **fourth contact**.

elongation The angular distance of an inferior planet from the centre of the Sun.

ephemeris (*pl.* **ephemerides)** A set of tables showing the predicted daily positions of the Sun, Moon and planets.

epicycle In any planetary system based on uniform circular motion, a circle which carries a planet and is in turn carried by a **deferent**.

epoch Generally, the time at which a celestial event occurs. All astronomical measurements must be accompanied by the epoch at which they were made. Star catalogues are referred to universally agreed epoch: B1950.0 before 1984 and J2000.0 after that date.

equant An eccentric point about which a point on the circumference of a circle moves at constant speed.

equated anomaly = **true anomaly**.

equation of time The difference between mean solar and apparent solar time.

equator A great circle on the Earth that is equidistant from both poles. The equator defines the zero of the scale of latitude.

equatorial coordinates A set of coordinates on the celestial sphere comprising declination (measured from the celestial equator) and right ascension (measured along the celestial equator from the First point of Aries).

equinox[1] One of two moments in the year when the Sun crosses the equator. The vernal equinox occurs approximately on 21 March and the autumnal equinox approximately on 21 September. So called because on those dates the length of day equals the length of night all over the Earth.

equinox[2] A celestial coordinate system at a specified epoch. Until 1984 the epoch was B1950.0 (Besselian epoch 1950.0). After 1984, the Julian epoch J2000.0 was applied. All star charts must state the epoch for which their tables are calculated.

exoplanet A planet in orbit about a star other than the Sun.

first contact During a transit of an inferior planet across the face of the Sun, or an exoplanet across the face of its central star, the point at which the limb of the planet (or exoplanet) first makes contact with the limb of the Sun (or central star); the point a which the Moon first touches the limb of the Sun during a solar eclipse.

First Point of Aries The point of intersection of the celestial equator and the ecliptic at which declination changes from negative to positive.

fourth contact During the transit of an inferior planet across the face of the Sun, or of an exoplanet across the face of its central star, the point at which the limb of the planet (or exoplanet) departs from the limb of the Sun (or central star); during a solar eclipse, the point at which the following limb of the Moon separates from the Sun.

geocentrism A world system centred on the Earth around which all celestial bodies revolve.

geographical coordinates A set of coordinates on the surface of the Earth comprising latitude (measured from the equator) and longitude (measured east or west of the IERS (International Earth Rotation System Service) Reference Meridian, which passes 102 m east of the formerly (1884–1984) accepted prime meridian at Greenwich.

geoheliocentric Of a world system centred on the Earth in which the Sun (which revolves around the Earth) orbits the Earth carrying with it all the planets (which revolve around the Sun). Tycho's model of the planetary system is of this type.

gibbous In astronomy a lunar or planetary phase between half and full.

great circle A circle inscribed on the surface of a sphere and concentric with the sphere on which it is inscribed.

heliocentrism A world system centred on the Sun.

horizon A great circle on the celestial sphere representing the plane of the observer.

horizontal coordinates A coordinate system centred at a point on the Earth's surface and comprising altitude (measured from the horizon and azimuth (measured either from the northern or the southern point of the horizon). Telescope mounts based on this system of coordinates are known as 'alt-az(imuth)'.

horoscope In astrology a chart showing the positions of the Sun, Moon and planets, and their mutual **aspects** at the time of a specific event.

Inclination The angle between two planes (measured perpendicularly from the line of intersection of the planes).

inferior conjunction The conjunction of an inferior planet when it is located directly between the Earth and the Sun.

inferior planet A planet whose distance from the Sun is less than that of the Earth from the Sun. Mercury and Venus are inferior planets. In a geocentric system the inferior planets are those that are taken to lie beneath the Sun (hence their name); that is, between the Earth and the Sun.

ingress In a transit the point at which a planet touches the limb of the Sun (or an exoplanet that of its host star).

insphere The greatest sphere that can be drawn inside a solid (usually a polyhedron) touching all of its faces.

latitude (ecliptic) The angular distance of a point on the celestial sphere from the plane of the ecliptic.

line of apsides The straight line joining the periapsis and apoapsis of an orbit.

line of nodes The line of intersection between the planes of two orbits. Its endpoints are the **ascending node** and the **descending node**.

longitude (ecliptic) The angular distance of a point on the ecliptic from the **First Point of Aries**.

major axis The longest diameter of an ellipse.

maximum elongation The greatest angular distance of an inferior planet from the Sun. Eastern elongation occurs after sunset, and western elongation before sunset.

mean anomaly The angle that an imaginary body moving at constant speed and having the same period as a real body makes with respect to perihelion as measured from the focus of the orbit.

mean sun An artificial sun that moves at constant speed along the ecliptic, completing a full revolution in one year. Until the advent of atomic clocks mean solar time (measure with respect to the mean sun) was the basis of civil time.

meridian A great semicircle passing through the north and south points of the horizon and through the zenith.

minor axis The shortest diameter of an ellipse.

nadir The point diametrically opposite to the **zenith** on the celestial sphere.

node Either of two points marking the intersection of an orbit with a reference plane. The node through which the orbit passes from the south to the north of the reference plane is called the **ascending node** and that for which the orbit passes from the north to the south of the reference plane is the **descending node**.

obliquity of the ecliptic The inclination of the Earth's equatorial plane to the plane of the ecliptic.

occultation The partial or complete covering of a celestial body by a nearer one of larger apparent size.

opposition An arrangement of the Earth and a superior planet such that the planet crosses the observer's meridian at midnight, when the planet's ecliptic longitude is 180°. More generally, two celestial bodies are said to be at opposition if they are 180 degrees apart in ecliptic longitude. *See also* **aspect**.

orb In Medieval and Renaissance astronomy, when motions in the heavens were generally taken to be circular or compounded from circular motions, the terms orb and sphere were often used almost interchangeably to designate the part of the heavens taken up by the set of spheers (that is mainly spherical shells) used to construct the path of a planet. Technically, the shape of the planetary sphere or orb is a shell, whose inner and outer surfaces are concentric spheres. Historians tend to prefer the usage in which the term orb designates the spherical shell that contains the apparatus of circles or spheres used to construct the motion of the planet. This is the type of orb shown in the most popular accounts of planetary motion printed in the Renaissance. Kepler, however, from the first, is concerned with the actual path of the planet, not with the circles used to construct it. His definition of a planet's orb is the spherical shell whose inner surface is a sphere that just fits inside the path of the planet, and whose outer surface (concentric with the inner one) is the sphere that just fits outside the path of the planet.

orbit The path of a celestial body around a central body.

parallax The apparent change in position of a distant object as viewed from two separate points. *See also* **diurnal parallax**.

periapsis The point of closest proximity of an orbiting body from its central body.

perigee Of the Moon or a satellite orbiting the Earth, the point in the orbit that is closest to the centre of the Earth.

perihelion The point of closest approach of a body to the Sun.

pole Of the celestial sphere, one of two points of intersection with the sphere of a line perpendicular to the plane of either the equator or the plane of the ecliptic.

precession of the equinoxes The westward drift, at a rate of $50''.3$ per year, of the equinoxes along the ecliptic.

prime vertical A great circle perpendicular to the meridian of an observer passing through the zenith and the points due east and west of the horizon.

prograde motion The motion of a celestial body from west to east (as seen from Earth) or the motion in an anti-clockwise direction as seen from a point north of the ecliptic plane.

quadrant An instrument for measuring angles within the range $0° \geq \theta \geq 90°$.

quadrature The **aspect**, as seen from Earth, of the Moon or a **superior planet** when its angular distance from the Sun is 90°. Houses of the zodiac are also be in quadrature if they are separated by 90°.

refraction Deviation in the path of light when it passes from one medium into another of different density.

retrograde motion Motion of a celestial body from east to west, as seen from Earth, or clockwise motion as seen from a point north of the ecliptic plane.

right ascension Of a celestial body, the angle, measured along the celestial equator, between the **First point of Aries** and that body. It is measured in hours, minutes and seconds of time.

second contact During a transit of a planet across the face of the Sun (or of an exoplanet across the face of its central star), the point at which the following limb of the planet (or exoplanet) is about to separate from the limb of the Sun (or central star); in a solar eclipse the point at which the following limb of the Moon separates from the limb of the Sun.

sesquiquadrate The standard arithmetical term used to describe a ratio of one and a half (that is, three to two or three halves). This is the proportion (in repeated multiplications) that appears in Kepler's third law.

sidereal time Time measured with respect to the stars. The **right ascension** of a star on the **meridian** marks the sidereal time at that moment.

sign (of zodiac) The zodiac is a 15°-wide strip of twelve constellations lying along the **ecliptic**. Each sign occupies 30° and carries the name of the constellation whose area it covers. In the past astronomers regularly used the signs of the zodiac to describe the ecliptic longitudes of celestial objects.

solar time Time as measured from the Sun.

solstices The two points on the ecliptic where the Sun reaches its greatest northern or southern declination; the dates on which these points are reached.

standard deviation (σ) In statistics, standard deviation is a measure of the dispersion of measured values from their mean value. For a set of n measurements x of mean value X, $\sigma = \sqrt{\left[\sum (x - X)^2 / n\right]}$.

superior conjunction A conjunction of an inferior planet when it is directly behind the Sun as seen from Earth.

superior planet In current astronomical parlance, a planet that is more distant from the Sun than the Earth. In geocentric systems, a planet that was thought to be above the Sun.

supernova An extremely violent explosion in which a supergiant star loses its outer layers, reaches a luminosity exceeding that of its entire host galaxy and leave behind an expanding remnant of gas and dust.

syzygy An alignment of three celestial bodies. For the Earth, Sun and Moon syzygy occurs at full and new Moon; for a planet syzygy occurs at conjunction and opposition.

third contact During a transit of a planet across the face of the Sun (or of an exoplanet across the face of its central star), the point at which the leading limb of the planet (or exoplanet) is about to is about to touch the inner limb of the Sun (or central star); during a solar eclipse the point at which the leading limb of the Moon touches the inner limb of the Sun.

transit[1] The passage of an inferior planet in front of the Sun.

transit[2] The passage of an exoplanet in front of its host star.

transversal sight A parallax-free sight designed by Tycho Brahe.

true anomaly In an elliptical planetary orbit, the angle between the planet and perihelion. In models based on uniform circular motion it is the angle subtended at the centre of an **eccentric circle** by the planet and the **line of apsides**.

zenith The point on the celestial sphere that is directly above the observer.

zenith distance The angular distance of a celestial object from the **zenith**.

zodiac A 15°-wide band centred on the ecliptic and spanning the entire ecliptic.

Index

© Springer Nature B.V. 2024
A. E. L. Davis et al. (eds.), *Reading the Mind of God*, Springer Praxis Books,
https://doi.org/10.1007/978-94-024-2250-4

Printed in the United States
by Baker & Taylor Publisher Services